NEW TRENDS IN CANCER FOR THE 21st CENTURY
2nd Edition

ADVANCES IN EXPERIMENTAL MEDICINE AND BIOLOGY

Editorial Board:

NATHAN BACK, *State University of New York at Buffalo*
IRUN R. COHEN, *The Weizmann Institute of Science*
DAVID KRITCHEVSKY, *Wistar Institute*
ABEL LAJTHA, *N.S. Kline Institute for Psychiatric Research*
RODOLFO PAOLETTI, *University of Milan*

Recent Volumes in this Series

Volume 579
IMMUNE MECHANISMS IN INFLAMMATORY BOWEL DISEASE
Edited by Richard S. Blumberg

Volume 580
THE ARTERIAL CHEMORECEPTORS
Edited by Yoshiaki Hayashida, Constancio Gonzalez, and Hisatake Condo

Volume 581
THE NIDOVIRUSES: THE CONTROL OF SARS AND OTHER
NIDOVIRUS DISEASES
Edited by Stanley Perlman and Kathryn Holmes

Volume 582
HOT TOPICS IN INFECTION AND IMMUNITY IN CHILDREN III
Edited by Andrew J. Pollard and Adam Finn

Volume 583
TAURINE 6
Edited by Simo S. Oja and Pirjo Saransaari

Volume 584
LYMPHOCYTE SIGNAL TRANSDUCTION
Edited by Constantine Tsoukas

Volume 585
TISSUE ENGINEERING
Edited by John P. Fisher

Volume 586
CURRENT TOPICS IN COMPLEMENT
Edited by John D. Lambris

Volume 587
NEW TRENDS IN CANCER FOR THE 21ST CENTURY
Edited by J.A. López-Guerrero, A. Llombart-Bosch and V. Felipo

---

A Continuation Order Plan is available for this series. A continuation order will bring delivery of each new volume immediately upon publication. Volumes are billed only upon actual shipment. For further information please contact the publisher.

International Symposium on Cancer

NEW TRENDS IN CANCER FOR THE 21st CENTURY

2nd Edition

Edited by

Antonio Llombart-Bosch
*University of Valencia,
Valencia, Spain*

Vicente Felipo
*Centro de Investigación Prínicpe Felipe,
Valencia, Spain*

and

José Antonio López-Guerrero
*Fundación Instituto Valenciano de Oncología,
Valencia, Spain*

 Springer

*RC270.8
I5286
2005*

ISBN-10 1-4020-4966-8 (HB)
ISBN-13 978-1-4020-4966-8 (HB)
ISBN-10 1-4020-5133-6 (e-book)
ISBN-13 978-1-4020-5133-3 (e-book)

© 2006 Springer

All rights reserved. This work may not be translated or copied in whole or in part without the written permission of the publisher (Springer Science+Business Media, LLC, 233 Spring Street, New York, NY 10013, USA), except for brief excerpts in connection with reviews or scholarly analysis. Use in connection with any form of information storage and retrieval, electronic adaptation, computer software, or by similar or dissimilar methodology now known or hereafter developed is forbidden.
The use in this publication of trade names, trademarks, service marks, and similar terms, even if they are not identified as such, is not to be taken as an expression of opinion as to whether or not they are subject to proprietary rights.

9 8 7 6 5 4 3 2 1

springer.com

Honoring the memory of Prof. Antonio Llombart Rodríguez.
Commemorating the centenary of his birth and the 50th anniversary
of the founding of the Spanish Association
Against Cancer (AECC) in Valencia

Contents

Preface xi
List of Contributors xvii

1. Prognostic and Therapeutic Targets in the Ewing's Family
   of Tumors (PROTHETS) 1
   P. Picci, K. Scotlandi, M. Serra, and A. Rizzi

2. Targeted Therapies in Ewing's Sarcoma 13
   K. Scotlandi

3. The CCN3 Protein and Cancer 23
   B. Perbal

4. Ews-Fli1 in Ewing's Sarcoma: Real Targets
   and Collateral Damage 41
   J. Ban, C. Siligan, M. Kreppel, D. Aryee, and H. Kovar

5. Molecular Karyotyping in Sarcoma Diagnostics and Research 53
   H. Vauhkonen, S. Savola, S. Kaur, M.L. Larramendy,
   and S. Knuutila

6. TuBaFrost: European Virtual Tumor Tissue Banking 65
   P.H.J. Riegman, M.H.A. Oomen, W.N.M. Dinjens, J.W. Oosterhuis,
   K.H. Lam, A. Spatz, C. Ratcliffe, K. Knox, R. Mager,

D. Kerr, F. Pezzella, B. van Damme, M. van de Vijver,
H. van Boven, M.M. Morente, S. Alonso, D. Kerjaschki,
J. Pammer, J.A. López-Guerrero, A. Llombart-Bosch,
A. Carbone, A. Gloghini, I. Teodorovic, M. Isabelle,
A. Passioukov, S. Lejeune, P. Therasse, and E.-B. van Veen

7. Virtual Microscopy in Virtual Tumor Banking..75
   M. Isabelle, I. Teodorovic, J.W. Oosterhuis, P.H.J. Riegman,
   and the TuBaFrost Consortium

8. Harmonizing Cancer Control in Europe 87
   U. Ringborg

9. The Diagnosis of Cancer: "From H&E to Molecular
   Diagnosis and Back" 95
   J. Costa

10. From Morphological to Molecular Diagnosis of Soft Tissue Tumors.......99
    M. Miettinen

11. Prediction of Response to Neoadjuvant Chemotherapy
    in Carcinomas of the Upper Gastrointestinal Tract 115
    H. Höfler, R. Langer, K. Ott, and G. Keller

12. Integrating the Diagnosis of Childhood Malignancies 121
    D. López-Terrada

13. Preclinical Models for Cell Cycle-Targeted Therapies............................139
    M. Malumbres

14. WWOX, a Chromosomal Fragile Site Gene and its Role in Cancer 149
    D. Ramos, and C.M. Aldaz

15. From Genome to Proteome in Tumor Profiling: Molecular
    Events in Colorectal Cancer Genesis 161
    J.K. Habermann, U.J. Roblick, M. Upender, T. Ried,
    and G. Auer

16. Effect of Hypoxia on the Tumor Phenotype: The Neuroblastoma
    and Breast Cancer Models 179
    L. Holmquist, T. Löfstedt, and S. Påhlman

17. Methylation Patterns and Chemosensitivity in NSCLC 195
    J.L. Ramírez, M.F. Salazar, J. Gupta, J.M. Sánchez, M. Taron,
    M. Sánchez-Ronco, Vicente Alberola, and R. de las Peñas

18. Pharmacogenomics and Colorectal Cancer 211
    H.J. Lenz

19. Development of Pharmacogenomic Predictors for Preoperative
    Chemotherapy of Breast Cancer 233
    L. Pusztai

20. Vascular Endothelial Growth Factor Inhibitors
    in Colon Cancer 251
    E. Díaz-Rubio

21. Molecular Imaging of Cancer Using PET and SPECT 277
    A. Kjær

22. MRS as Endogenous Molecular Imaging for Brain
    and Prostate Tumors: FP6 Project "eTUMOR" 285
    B. Celda, D. Monleón, M.C. Martínez-Bisbal, V. Esteve,
    B. Martínez-Granados, E. Piñero, R. Ferrer, J. Piquer,
    L. Martí-Bonmatí, and J. Cervera

23. From Linac to Tomotherapy: New Possibilities for Cure? 303
    G. Storme, D. Verellen, G. Soete, N. Linthout, J. Van de Steene,
    M. Voordeckers, V. Vinh-Hung, K. Tournel, and D. Van den Berge

24. Targeting mTOR for Cancer Treatment..309
    B. Rubio-Viqueira, and M. Hidalgo

25. Dual/Pan-Her Tyrosine Kinase Inhibitors: Focus
    in Breast Cancer 329
    J. Albanell

26. AntiTumor-Associated Antigens IGGS: Dual Positive
    and Negative Potential Effects for Cancer Therapy 341
    E. Barberá Guillem, and J.W. Sampsel

27. Synergistic Molecular Mechanisms in Hormone-Sensitive
    Breast Cancer 375
    A. Llombart Cussac

Index 387

Preface

This volume contains the text of the presentations delivered at the Second International Symposium on Cancer "New trends in Cancer for the 21st Century" held in Valencia, Spain, from November 12 to 15, 2005. This second symposium represents continuity in the scientific activities of the Cátedra Santiago Grisolía and of the Fundación de la Comunidad Valenciana Ciudad de las Artes y las Ciencias.

More than 200 scientists attended this Symposium, not only from Spain but also from different countries of Europe, the USA, and Latin America. As on the last occasion, the venue was the Auditorium of the Museum of Sciences Príncipe Felipe in Valencia. Valencia is an ancient Spanish city with 2000 years of history, the population of which represents almost the entire cultural history of Spain. Its historical background originates from the Phoenician world, with its roots deepened, first by the Roman Empire, and later by a very fertile and influential Arab presence. All these cultural roots are still present and alive in our community, with their enriching influence in buildings, monuments, villages, and street names, and especially in the customs and habits of its citizens.

Moreover, the community of Valencia, with its special dynamism, has provided a solid backing in the building of Spain to its present position. It is a driving force in culture, science, arts, and commerce, within a coherent civil society, that harmonizes in all its manifestations both the responsibilities of work and the joy of life. At present, the Valencian community is experiencing a golden era within the modern history of Spain, producing an example of democratic, common-sense integration, as well as being one of the most attractive regions in this area of the Mediterranean.

In fact, our lecture hall in the Museu de les Ciències Principe Felipe, an interactive science museum, offers the modern view and counterpart to the old city center. The Ciudad de las Artes y las Ciencias, city of arts and sciences complex, forms part of what is becoming, in the fields of education, culture, arts, and sciences, the modern Valencia of the 21st century. This project includes L'Hemisfèric (Imax Dome, Planetarium, and Laserium cinema), L'Umbracle (garden and parking), L'Oceanogràfic (the largest marine park in Europe), the

Museu de les Ciències Principe Felipe, and the Palau de les Arts, an opera house recently inaugurated by the Queen of Spain. This enterprise was designed and directed by the renowned Valencia-born architect Santiago Calatrava, and is but one more example of the commitment of this city and its citizens to the future of Valencia and its community.

During this scientific event, homage was paid to those men who provided the driving force and impulse in creating the Cancer League in Valencia 50 years ago. Among them, present in our memory was Professor Antonio Llombart-Rodríguez, distinguished Valencian pathologist and professor at the Medical School until 1975. This year we commemorated the Centennial of his birth. He belonged to the rich scientific background that formed the Medical School of Valencia in both former and present times. Names such as the anatomists Pedro Gimeno and Jeronimo Valverde pertained to the initial period of the anatomo-clinical correlations in medicine, in the 16th and 17th centuries, even before the famous Italian and central European anatomists. Other prestigious scientists such as Santiago Ramón y Cajal were teachers at this Medical School at the end of the 19th century, founding a renowned school of histologists and pathologists, among whom were Juan Bartual, Luis Urtubey, and Antonio Llombart-Rodríguez, former Director at the Department of Pathology. A short account of his life is included at the end of the introduction.

In the organization of the event, both the University of Valencia and its Medical School through the Department of Pathology, and the Instituto Valenciano de Oncologia (IVO) Foundation, a charitable, nonprofit foundation, provided the scientific support to the Symposium. These institutions represent in some way the ancient and modern aspects of this community.

On the one hand, the University of Valencia is a historical institution, in existence for more than 500 years, while the IVO Cancer Center Foundation was created just 30 years ago with the support of the Spanish Cancer League (Asociacion Española Contra el Cancer). The IVO Cancer Hospital is a modern and prestigious organization, oriented as a Comprehensive Cancer Center, covering prevention, early diagnosis, treatment, and research, and represents the largest cancer institution in our community. Since its creation, it has diagnosed and treated more than 40,000 patients with an average of 3,000 new cancer patients each year. It houses all kinds of modern technology used in diagnosis and treatment, as well as having numerous and well-qualified medical staff. Both these outstanding institutions, illustrations of dynamism and scientific promotion, provide financial support to this activity.

What motivated this second edition of *New Trends in Cancer for the 21st Century*? It is difficult to imagine anyone who has not heard of cancer. This disease can affect our family, friends, or even us at some time in our lives. Every year nearly 3 million Europeans are diagnosed with cancer, leading to around 800,000 deaths per year. This represents a tremendous problem that cannot be ignored by either politicians or citizens, especially considering that these deaths occur not only in an aging population but also in children and in adults during the most active period of their lives. Fortunately, there is a constant growing awareness that cancer is not a problem exclusively for the clinicians and politicians, but for everybody. In this sense, the experience of the Symposium was very enriching, where fundamental researchers, politicians, and representatives from patients' coalitions had the opportunity of sharing their concerns, interests, and knowledge, as well as their own points of view and experiences. We are sure that this meeting served not only to provide state-of-the-art information on cancer but also to look at the problem with other eyes, with greater commitment and with the sensation that we are not alone on this ship.

In the particular case of Europe, the European Parliament, under the auspices of the European Commission Research Directorate-General, has expressed the need for a more active policy. New approaches are required for the development of strategies in prevention, cancer care, and research, by designing evidence-based guidelines for early detection, diagnosis, treatment, and aftercare for the most common types of cancer. These actions will be oriented toward a more interdisciplinary approach in cancer research, with more financial support, more research on courses of treatment, and introducing independent clinical assays on a higher number of patients. Comparative research should be conducted into the cost-effectiveness of various detection and treatment methods in order to increase efficiency and reduce costs in the health care system. European models should be developed for the continued training in research and clinical specialization involved in the various fields involved in the fight against cancer. Finally, the continuance of existing networks with their successful activity should be ensured for the future, in addition to the promotion and creation of new ones.

This European policy has resulted in three projects, each described in the present volume:

(a) The PROTHETS project (of the 6th framework program) has been specifically dedicated to the study of Ewing's sarcoma biology and to design new strategies for treatment of this bone tumor.
(b) The TuBaFrost project (of the 5th framework program) has been useful to constitute the first European Network of fresh frozen tumor banks.
(c) The eTUMOR project (also from the 6th framework program) on the use of Magnetic Resonance Spectroscopy is a tool in the diagnosis and classification of brain and prostate cancers.

The causes of cancer and the many factors that influence its emergence are complex and heterogeneous. From the pathological point of view, the diagnosis and classification of tumors is based mainly upon the recognition of the microscopical tumor patterns, with the support of immunophenotyping and other emerging molecular techniques. The development of novel and sophisticated technologies for the study of tumor profiling, based as much on genetics as on protein expression profiles, is opening doors to what is considered to be the *new molecular pathology*. These novel approaches represent an exceptional opportunity for histopathology, since with the expansion of a large number of assayable markers it is now possible to gather a better molecular portrait of tumors and deliver better oriented therapies.

The development of new therapeutic drugs, directed against specific targets such as tyrosin kinase receptor inhibitors, or monoclonal antibodies that downregulate the hyperactive function of some oncogenes, has considerably improved the outcome for many patients (Phase II and III studies).

Special mention in the fight against cancer must be made of organizations such as the European Organization for Research and Treatment of Cancer (EORTC), the Organization of European Cancer Institutes (OECI) and the independent comprehensive cancer centers. These institutions promote and coordinate high-quality laboratory research and clinical trials and also provide a central facility with scientific expertise and administrative support for this network of scientists and clinical investigators.

The organizers wish to thank all the speakers, chairpersons, and participants for coming again to Valencia for the second Symposium. We express our gratitude for their written contributions and for their enlightened and fruitful discussion. We gratefully acknowledge the financial support provided by the Fundación Instituto Valenciano de Oncología, the Spanish Association Against Cancer (AECC) in Valencia, and the Department of Pathology of the University of Valencia. We would also like to thank the Catedra Santiago Grisolía and the Fundación de la Comunidad Valenciana Ciudad de las Artes y las Ciencias for providing the facilities and organizing both the Symposium and the sessions.

Antonio Llombart-Bosch
Vicente Felipo
José Antonio López-Guerrero

Professor Llombart-Rodríguez – Memorial

Text of a speech delivered by Professor Antonio Llombart-Bosch in memory of Professor Llombart-Rodríguez

It is my pleasure and privilege to deliver a brief portrayal of one of the most outstanding and relevant pathologists of the last century in Spain. As you know, I am speaking of the late Professor Antonio Llombart-Rodríguez (1905–1997). In honor of his memory, and on the centenary of his birth, we are gathered together at this Symposium.

In the second half of the last century, when speaking about cancer in Spain, the name of Professor Llombart-Rodríguez was a point of reference concerning the morphological features of this complex disease. Who was Professor Llombart-Rodríguez?

Born in Valencia, where he obtained his MD degree with the highest awards and developing, still as an MD student, a particular interest in histology he spent several periods in the laboratory of Professor Pio del Rio Hortega, at the Junta de Ampliacion de Estudios in Madrid. Hortega was one of the most distinguished pathologists of the school of Santiago Ramon y Cajal, and described the oligodendroglia and microglia as well as several tumor entities of the CNS.

An enthusiasm for research into the morphology of cancer led Antonio Llombart to extend his training outside Spain. Between 1929 and 1932 he spent three scholarship years in France with Professor P. Champy and Gustave Roussy (founder of the Institute Gustave Roussy), working on technical procedures for silver stains and inflammatory diseases. He also spent time with Professor Rossle at the Charité Krankenhaus in Berlin, working on the Rous sarcoma virus and tissue cultures at the laboratory of Professor Erdmann (Mohawitz Krankenhaus).

Upon his return to Spain his activities were directed towards the Basque Country, organizing the laboratory of pathology at the General Hospital in San Sebastian. The idea of creating a Cancer Institute was formed between a group of scientists and doctors of this city, and in 1934, the Instituto Radio Quirurgico was born, being at that time the second most important Cancer Center in existence in Spain, together with the National Cancer Institute in Madrid. This institution in the Basque Country remains in a prominent position today for cancer diagnosis and therapy, in Spain. He continued as scientific advisor at this Institution until 1980.

The Spanish Civil War and the Second World War affected many of his projects and (as many others) a part of his life. He became full professor of pathology, after a short return to the University of Madrid (1940), at the medical school of Valladolid, where he continued, not without great difficulty, his work on neoplastic diseases.

Called as Professor of Pathology to the University of Valencia in 1945, he restructured his scientific activities, first in the old Medical School building in Guillem de Castro, and later inaugurating, at the new campus, a department of Experimental Cancer research and co-founding the Spanish Cancer League (AECC) in 1955. At the symposium, we also celebrated the 50th anniversary of this event.

His life in the University of Valencia was always oriented towards the same end: investigating cancer by morphological means, fighting cancer through all available systems and preventing cancer by education and communication. I am not going to enumerate the substantial number of books and papers published in International journals (more than 200). In addition, a large number of students grew up during the 30 years of splendid devotion to his activities in the University of Valencia as director of the Department of Pathology and cancer Research Department. Fifteen full professors of pathology present in Spain and Latin America witness the strong position attained by his group in the second half of the last century and which continues today in several universities and research institutions.

For many years he was actively involved in the prevention of cancer through the Spanish Cancer Association and the co-founding of the Research Foundation of the AECC. This was in the early seventies, when in Spain the crystallization of oncology as a medical specialty took place and the Sociedad Española de Oncologia (SEO) was created, unifying diverse clinical cancer societies in Spain. He was elected as its first president, actively promoting this organization at official level.

Nevertheless, one of his major achievements was reached at the end of his academic life, when he was already 70 years old. It was at this point that he founded and guided the construction of the Instituto Valenciano de Oncologia (IVO Foundation), of which he remained president until his death. During this period, he also continued his research activities and actively participated in fighting cancer through the local Cancer league. Furthermore, he published three books: the first was an exhaustive overview of the problems in cancer at that time. The second and third contain his experiences, narrating the history of the Spanish Cancer Association until 1993, while the last was completed one year before his death (1996) describing the origins and evolution of the IVO Cancer Center.

He spent the lasts days of his life surrounded by family, former scholars and friends in a peaceful atmosphere with the satisfaction of having completed a long and fruitful existence. Today his example remains alive in the memory of us all who had the privilege to work under his leadership.

Contributors

ALBANELL, J.
Medical Oncology Department
Hospital del Mar de Barcelona
Barcelona
Spain

ALBEROLA, V.
Hospital Arnau de Vilanova
Valencia
Spain

ALDAZ, C. M.
Department of Carcinogenesis
The University of Texas
M.D. Anderson Cancer Center
Texas
USA

ALONSO, S.
The TuBaFrost Consortium
Centro Nacional de Investigaciones
Oncológicas
Madrid
Spain

ARYEE, D.
Children's Cancer Research Institute
Vienna
Austria

AUER, G.
Unit of Cancer Proteomics
Cancer Center Karolinska
Karolinska University Hospital
Solna
Stockholm
Sweden

BAN, J.
Children's Cancer Research
Institute
Vienna
Austria

BARBERÁ GUILLEM, E.
Celartia Research Laboratories
Powell, Ohio
USA

CARBONE, A.
The TuBaFrost Consortium
Centro di Riferimento Oncologico
Aviano (PN)
Italy

CELDA MUÑOZ, B.
Aplicaciones Biofísicas y
Biomédicas de la RMN
Departamento Química Física
Universidad de Valencia
Valencia
Spain

CERVERA, J.
Instituto Valenciano de Oncología,
(IVO)
Valencia
Spain

COSTA, J.
Department of Pathology and Yale
Cancer Center
Yale University School of Medicine
New Haven, CT
USA

DE LAS PEÑAS, R.
Hospital de Castellon
Castellon
Spain

DÍAZ RUBIO, E.
Hospital Clínico San Carlos
Madrid
Spain

DINJENS, W. N. M.
The TuBaFrost Consortium
Erasmus MC
Rotterdam
The Netherlands

ESTEVE, V.
Aplicaciones Biofísicas y
Biomédicas de la RMN
Departamento Química Física
Universidad de Valencia
Valencia
Spain

FERRER, R.
Aplicaciones Biofísicas y
Biomédicas de la RMN
Departamento Química Física
Universidad de Valencia
Valencia
Spain

GLOGHINI, A.
The TuBaFrost Consortium
Centro di Riferimento Oncologico
Aviano (PN)
Italy

GUPTA, J.
Catalan Institute of Oncology
Hospital Germans Trias i Pujol
Badalona, Barcelona
Spain

HABERMANN, J. K.
Laboratory for Surgical Research
Department of Surgery
University Hospital Schelswig-
Hostein
Lübeck
Germany
and
Genetics Department
National Cancer Institute
NIH
Bethesda, MD
USA

HIDALGO, M.
The Sidney Kimmel Comprehensive
Cancer Center at Johns Hopkins
The Johns Hopkins University
School of Medicine
Baltimore, MD
USA

HÖFLER, H.
Institute of Pathology
Technical University München
München
Germany

HOLMQUIST, L.
Department of Laboratory
Medicine
Molecular Medicine

Lund University
University Hospital MAS
Malmö
Sweden

ISABELLE, M.
The TuBaFrost Consortium
EORTC Data Center
Brussels
Belgium

KAUR, S.
Department of Pathology
Haartman Institute and HUSLAB
University of Helsinki and Helsinki
University Hospital
Helsinki
Finland

KELLER, G.
Institute of Pathology
Technical University München
München
Germany

KERJASCHKI, D.
The TuBaFrost Consortium
Alllgemeines Krankenhaus
University of Vienna
Vienna
Austria

KERR, D.
The TuBaFrost Consortium
National Translational Cancer
Research Network
University of Oxford
Radcliffe Infirmary
Oxford
United Kingdom

KJÆR, A.
Department of Clinical Physioligy.
Nuclear Medicine & PET
Rigshospitalet & Cluster for
Molecular Imaging

Faculty of Health Sciences
University of Copenhagen
Copenhagen
Denmark

KNOX, K.
The TuBaFrost Consortium
National Translational Cancer
Research Network
University of Oxford
Radcliffe Infirmary
Oxford
United Kingdom

KNUUTILA, S.
Department of Pathology
Haartman Institute and HUSLAB
University of Helsinki and Helsinki
University Hospital
Helsinki
Finland

KOVAR, H.
Children's Cancer Research
Institute
Vienna
Austria

KREPPEL, D.
Children's Cancer Research Institute
Vienna
Austria

LAM, K. H.
The TuBaFrost Consortium
Erasmus MC
Rotterdam
The Netherlands

LANGER, R.
Institute of Pathology
Technical University München
München
Germany

LARRAMENDY, M.
Department of Pathology
Haartman Institute and HUSLAB
University of Helsinki and Helsinki
University Hospital
Helsinki
Finland

LEJEUNE, S.
The TuBaFrost Consortium
EORTC Data Center
Brussels
Belgium

LENZ, H. J.
Division of Medical Oncology
University of Southern
California/Norris Comprehensive
Cancer Center
Keck School of Medicine
Los Angeles, California
USA

LINTHOUT, N.
Department of Radiation
Oncology
Oncologic Center AZ-VUB
Brussels
Belgium

LLOMBART BOSCH, A.
The TuBaFrost Consortium
Fundación Instituto Valenciano de
Oncología
Valencia
Spain

LLOMBART CUSSAC, A.
Head Medical Ocnology Service
Hospital Universitario Arnau de
Vilanova
Lleida
Spain

LÖFSTEDT, T.
Department of Laboratory
Medicine
Molecular Medicine
Lund University
University Hospital MAS
Malmö
Sweden

LÓPEZ-GUERRERO, J. A.
The TuBaFrost Consortium
Fundación Instituto Valenciano de
Oncología
Valencia
Spain

LÓPEZ-TERRADA, D.
Texas Children's Hospital and
Baylor Collage of Medicine
Houston, Texas
USA

MAGER, R.
The TuBaFrost Consortium
National Translational Cancer
Research Network
University of Oxford
Radcliffe Infirmary
Oxford
United Kingdom
and
Nuffield Department of Clinical
Laboratory Sciences
University of Oxford
John Radcliffe Hospital
Headington, Oxford
United Kingdom

MALUMBRES, M.
Cell Division and Cancer Group
Centro Nacional de Investigaciones
Oncológicas (CNIO)
Madrid
Spain

MARTÍ-BONMATÍ, L.
Clínica Quirón
Valencia
Spain

MARTÍNEZ-BISBAL, M. C.
Aplicaciones Biofísicas y
Biomédicas de la RMN
Departamento Química Física
Universidad de Valencia
Valencia
Spain

MARTÍNEZ-GRANADOS, B.
Aplicaciones Biofísicas y
Biomédicas de la RMN
Departamento Química Física
Universidad de Valencia
Valencia
Spain

MIETTINEN, M.
Department of Soft Tissue
Pathology
Armed Forces Institute of
Pathology
Washington, DC
USA

MONLEÓN, D.
Aplicaciones Biofísicas y
Biomédicas de la RMN
Departamento Química Física
Universidad de Valencia
Valencia
Spain

MORENTE, M. M.
The TuBaFrost Consortium
Centro Nacional de Investigaciones
Oncológicas
Madrid
Spain

OOMEN, M. H. A.
The TuBaFrost Consortium
Erasmus MC
Rotterdam
The Netherlands

OOSTERHUIS, J.W.
The TuBaFrost Consortium
Erasmus MC
Rotterdam
The Netherlands

OTT, K.
Department of Surgery
Technical University München
München
Germany

PÅHLMAN, S.
Department of Laboratory
Medicine
Molecular Medicine
Lund University
University Hospital MAS
Malmö
Sweden

PAMMER, J.
The TuBaFrost Consortium
Alllgemeines Krankenhaus
University of Vienna
Vienna
Austria

PASSIOUKOV, A.
The TuBaFrost Consortium
EORTC Data Center
Brussels
Belgium

PERBAL, B.
Laboratoire d'Oncologie Virale et
Moléculaire
UFR de Biochimie
Université Paris 7 D
Paris
France

PEZZELLA, F.
The TuBaFrost Consortium
Nufield Department of Clinical
Laboratory Science
University of Oxford
John Radcliffe Hospital
Oxford
United Kingdom

PICCI, P.
Istituti Ortopedici Rizzoli
Bologna
Italy

PIÑERO, E.
Aplicaciones Biofísicas y
Biomédicas de la RMN
Departamento Química Física
Universidad de Valencia
Valencia
Spain

PIQUER, J.
Hospital La Ribera
Alzira, Valencia
Spain

PUSZTAI, L.
Department of Breast Medical
Oncology
The University of Texas
M.D. Anderson Cancer Center
Houston, Texas
USA

RAMIREZ, J. L.
Catalan Institute of Oncology
Hospital Germans Trias i Pujol
Badalona, Barcelona
Spain

RAMOS, D.
Department of Pathology
Medical School
University of Valencia
Valencia
Spain

RATCLIFFE, C.
The TuBaFrost Consortium
National Translational Cancer
Research Network
University of Oxford
Radcliffe Infirmary
Oxford
United Kingdom

RIED, T.
Genetics Department
National Cancer Institute
NIH
Bethesda, MD
USA

RIEGMAN, P. H. J.
The TuBaFrost Consortium
Erasmus MC
Rotterdam
The Netherlands

RINGBORG, U.
Karolinska Institute
Cancer Center Karolinska
Stockholm
Sweden

RIZZI, A.
Istituti Ortopedici Rizzoli
Bologna
Italy

ROBLICK, U. J.
Laboratory for Surgical Research
Department of Surgery
University Hospital Schelswig-
Hostein
Lübeck
Germany
and
Unit of Cancer Proteomics
Cancer Center Karolinska
Karolinska University Hospital
Solna
Stockholm
Sweden

RUBIO-VIQUEIRA, B.
The Sidney Kimmel Comprehensive
Cancer Center at Johns Hopkins
The Johns Hopkins University
School of Medicine
Baltimore, MD
USA

SALAZAR, M. F.
Catalan Institute of Oncology
Hospital Germans Trias i Pujol
Badalona, Barcelona
Spain

SAMPSEL, J. W.
Celartia Research Laboratories
Powell, Ohio
USA

SANCHEZ, J. M.
Catalan Institute of Oncology
Hospital Germans Trias i Pujol
Badalona, Barcelona
Spain

SANCHEZ-RONCO, M.
Autonomous University of Madrid
Madrid
Spain

SAVOLA, S.
Department of Pathology
Haartman Institute and HUSLAB
University of Helsinki and Helsinki
University Hospital
Helsinki
Finland

SCOTLANDI, K.
Growth Factors and Receptors
Oncologic Research Laboratory
Istituti Ortopedici Rizzoli
Bologna
Italy

SERRA, M.
Istituti Ortopedici Rizzoli
Bologna
Italy

SILIGAN, C.
Children's Cancer Research Institute
Vienna
Austria

SOETE, G.
Department of Radiation Oncology
Oncologic Center AZ-VUB
Brussels
Belgium

SPATZ, A.
The TuBaFrost Consortium
Institut Gustave Roussy
Villejuif
France

STORME, G.
Department of Radiation Oncology
Oncologic Center AZ-VUB
Brussels
Belgium

TARON, M.
Catalan Institute of Oncology
Hospital Germans Trias i Pujol
Badalona, Barcelona
Spain

TEODOROVIC, I.
The TuBaFrost Consortium
EORTC Data Center
Brussels
Belgium

THERASSE, P.
The TuBaFrost Consortium
EORTC Data Center
Brussels
Belgium

TOURNEL, K.
Department of Radiation Oncology
Oncologic Center AZ-VUB
Brussels
Belgium

UPENDER, M.
Genetics Department
National Cancer Center
NIH
Bethesda, MD
USA

VAN BOVEN, H.
The TuBaFrost Consortium
Netherlands Cancer Institute
Amsterdam
The Netherlands

VAN DAMME, B.
The TuBaFrost Consortium
U.Z.-K.U.-Leuven
Leuven
Belgium

VAN DE STEENE, J.
Department of Radiation Oncology
Oncologic Center AZ-VUB
Brussels
Belgium

VAN DE VIJVER, M.
The TuBaFrost Consortium
Netherlands Cancer Institute
Amsterdam
The Netherlands

VAN DEN BERGE, D.
Department of Radiation Oncology
Oncologic Center AZ-VUB
Brussels
Belgium

VAN VEEN, E.-B.
The TuBaFrost Consortium
Medlaw Consult
AM Den Haag
The Netherlands

VAUHKONEN, H.
Department of Pathology
Haartman Institute and HUSLAB
University of Helsinki and Helsinki
University Hospital
Helsinki
Finland

VERELLEN, G.
Department of Radiation Oncology
Oncologic Center AZ-VUB
Brussels
Belgium

VINH-HUNG, V.
Department of Radiation Oncology
Oncologic Center AZ-VUB
Brussels
Belgium

VOORDECKERS, M.
Department of Radiation Oncology
Oncologic Center AZ-VUB
Brussels
Belgium

CHAPTER 1

PROGNOSTIC AND THERAPEUTIC TARGETS IN THE EWING'S FAMILY OF TUMORS (PROTHETS)

P. PICCI, K. SCOTLANDI, M. SERRA, AND A. RIZZI

Istituti Ortopedici Rizzoli, Bologna, Italy

Abstract: ProTheTs is a specific targeted research project (STREP) funded by the European Union in the 6th Framework Program (FP6), area: "LSH-2002-2.2.0-8 Translational research on promising predictive and prognostic markers", contract n. LSHC-CT-2004-503036. The research will concentrate on translating the knowledge being created by genomics and other fields of basic research into applications that improve clinical practice and public health

1. PROJECT PRESENTATION

The project, supported by a grant from the European Commission (contract LSHC-CT-2004-503,036), through collaborative studies among eleven European partners:
1. Istituti Ortopedici Rizzoli, Bologna, Italy (Coordinator)
2. Unité INSERM UMR 576, Institut National de la Santé et de la Recherche Médicale, Nice, France
3. Laboratory for Experimental Orthopaedic Research, University Hospital of Münster, Münster, Germany
4. Haartman Institute, Department of Medical Genetics, University of Helsinki, Helsinki, Finland
5. Universitat de València. Estudi General, through the Department of Pathology (Medical School, Hospital Clínico Universitario), Valencia, Spain
6. Children's Cancer Research Institute, St. Anna Children's Hospital, Vienna, Austria
7. Universitè Paris 7 Denis-Diderot, Paris, France
8. CNRS, Délégation Ile de France Est, established in Tour Europa 126, 94,532 Thiais Cedex, France

9. Belozersky Institute of Physico-Chemical Biology, Lomonosov Moscow State University, Moscow, Russia
10. Genx Laboratories s.r.l., Vignate (MI), Italy
11. Mabgène S.A., Ales, France

will define prognostic markers and new therapeutic targets in the Ewing's sarcoma family of tumors (ESFT) to provide rigorous scientific justifications for the development of clinical trials for this rare disease, which is manifested for the most part in children.

The main objective of this project is to evaluate the prognostic relevance of selected markers (EWS/FLI-1, secondary genetic alterations, CD99, IGF-IR, NOVH, erbB-2, and TTF1) and the effectiveness of therapeutic approaches targeting some of these molecules. The prognostic value of these markers will be evaluated in retrospective and prospective series of ESFT patients treated across the participating centres. Through statistical analysis, we will verify which factors have the highest prognostic impact in ESFT patients, in terms of overall survival, disease progression, and chemosensitivity. In order to provide the necessary rationale for the forthcoming application of new therapies, the preclinical effectiveness of new drugs (Herceptin and TRAIL) and strategies targeting molecules (CD99, IGF-IR and EWS/FLI-1) found to be critical for ESFT will be evaluated.

Another major goal of the project is the construction of ESFT c-DNA microarrays and tissue arrays, which will be used for the analysis of different histological subtypes of ESFT, primary and metastatic tumors and poor and good responders to chemotherapy.

Therefore, the expected results are:
1. Identification of prognostic factors in ESFT
2. Definition of patient selection criteria
3. Creation of new therapeutic bullets against ESFT
4. Identification of new therapies
5. Creation of new tools for the diagnosis and screening of high-risk groups

This will lead to:
1. The definition of forthcoming risk-adapted strategies and targeted molecular treatments to be advantageously combined with established therapies
2. Improved quality of life and survival for ESFT patients
3. Prevention of risk in groups at risk

2. AIM

The project will define prognostic markers and new therapeutic targets in the ESFT through collaborative studies to provide rigorous scientific justification for the development of new therapeutic strategies for this rare disease, which is manifested for the most part in children.

ESFT can develop in male and female although the male gender is preferred by 1.5 to 1. Around 85% of the cases occur between 5 and 25 years of age with

a peak of incidence between 5–10 and 10–15 years old. Therefore the project, aiming to ameliorate treatment of ESFT, will have an impact on child health. In particular, the main objective of this project is to develop patient-oriented strategies for Ewing's sarcoma patients by:
1. Integrating different disciplines and advanced technologies to develop effective approaches or new tools for diagnosis, prognosis and treatment
2. Elucidating the contribution of specific molecular and genetic factors to the histogenesis of the disease

3. BACKGROUND

The ESFT includes: (1) Ewing's sarcoma; (2) primitive neuroectodermal tumor; (3) Askin's tumor; (4) paravertrebral small-cell tumor; and (5) atypical Ewing's sarcoma. ESFT represents a peculiar entity in oncology. In spite of its absolute rarity (about 300–400 cases per year in Europe), ESFT is one of the most frequent solid neoplasms in paediatric age. Due to this fact, its impact on the health system is particularly important.

ESFT primarily affects white and Hispanic young people and is extremely rare in individuals of African or Asian origin. The reason for this striking ethnic distribution is not known, although interethnic differences exist for certain alleles of one of the genes consistently disrupted in these neoplasms. Tumors can develop in almost any bone and soft tissue. Approximately 25% of patients have detectable metastatic disease to lung, bone and bone marrow at diagnosis, but nearly all patients have micrometastases, as evidenced by a less than 10% cure rate with local therapy alone.

4. STATE OF THE ART

Past progress: The adoption of multimodal treatments with very aggressive chemotherapeutic regimens have significantly improved the chance of survival of ESFT nonmetastatic patients, shifting the 5-year survival rates to around 60% [1–3]. This improvement was obtained by the standardization of treatment modalities through the design of multicentric trials, which have been adopted by several National and International groups in Europe including most of the project partners.

Present situation and problems: Despite these important clinical results, which are usually difficult to obtain in rare diseases, several problems related to histogenesis, prognosis and treatment response are still open.

5. HISTOGENESIS

The histogenesis of ESFT is still uncertain and the normal counterpart of ESFT cells is still unknown. These neoplasms share the presence of specific chromosomal translocations, which produce an EWS/ets gene rearrangement (in more than 95% of cases, the gene fusion is EWS/FLI-1 due to the t(11;22) (q24;q12) or EWS/ERG due to the t(21;22) (q22;q12)), as well as the expression, at extremely high levels, of an antigen determined by the MIC2 gene (also known as CD99) [4–6], and are currently defined along a limited gradient of neural differentiation. However, the neuroectodermal origin and whether or not neuroectodermal differentiation of ESFT has a prognostic importance is still a matter of dispute [7–10]. Other histogenetic possibilities cannot be excluded because ESFT can also exhibit some epithelial and mesenchymal characteristics and can arise in organs not directly related to the neural crest (i.e. the kidney).

6. PROGNOSIS

The lack of prognostic factors obliges the use of nondifferentiated treatments for all patients, leading to overtreatment of those patients who could benefit from less toxic therapies. The reduction of delayed side effects is particularly important in this disease considering the young age of the patient and their long life expectancy. Several biological studies have been recently carried out on the identification of the peculiar features of ESFT, offering promising molecular targets to be potentially used for therapy. A better outcome for patients with localized tumors expressing the EWS/FLI-1 type 1 fusion has been reported [11–13]. The biological basis for the prognostic differences among the various fusion genes is unknown, but a recent report has shown a significant association with levels of expression of insulin-like growth factor (IGF-IR) and the proliferation rate [13,14]. Moreover, secondary cytogenetic and molecular alterations of ESFT (i.e. gains of chromosome 8 and 12, loss of chromosome 16, deletions at chromosome 1p36 and p53 mutations) have recently been associated with a worse prognosis [15–18], further indicating a possible association between genetic variants and clinical features. Recently, CD99 isoforms were reported to differentially dictate the fate of T cells [19] and preliminary in vitro results obtained from Partner 1 in collaboration with Partner 2 indicate that the two isoforms may have a differential biological role in affecting malignancy of ESFT. Finally, recent reports have shown that the expression of the NOVH protein, which belongs to the emerging CCN (Connective tissue growth factor, Cyr61/Cef10, nephroblastoma overexpressed gene) family of growth regulators [20,21], is related to a higher risk of developing metastases and an adverse clinical outcome in ESFT; preliminary data of Partner 1 (manuscript in preparation) also indicated an expression of erbB-2 in around 20–30% of cases. Evaluation of the prognostic role of erbB-2 requires larger studies. These encouraging but sporadic data should stimulate inclusion of these markers into retrospective and prospective studies.

7. THERAPY

In the current state of ESFT treatment there is a survival "plateau" (around 60% for patients with localized disease and 25% for high-risk groups) due to the lack of new drugs and toxicity that impedes more intense use of existing drugs [2,22,23]. The only exception is ET-743, a new marine compound with high effectiveness in sarcomas [24]. However, this is a conventional cytotoxic drug with the usual problems of high toxicity and development of drug resistance [25]. A new targeted drug (Imatinib/Glivec) that inhibits c-kit activity has been recently proposed due to expression of this receptor in ESFT [26]. However, its effectiveness was found to be controversial and generally low even in preclinical conditions, therefore diminishing initial enthusiasm [27,28]. The identification of new targets for innovative therapeutic strategies is, therefore, strongly needed for this tumor.

Progresses are generally hampered by the rarity of the disease (in Europe about 400 cases/year) implying a limited number of cases for effective research. Moreover, because ESFT is an orphan disease, no private company will develop new therapeutic tool and charge on itself the costs to conduct preclinical investigation.

8. SOLUTIONS AND AREAS FOR FURTHER DEVELOPMENT

With respect to the problem of toxicity, the project, by identifying the clinical relevance of a number of markers, may allow the differentiation of patients in terms of risk to recur. This will enable more aggressive treatments where these are justified, and avoid toxicity in cases where such treatments may be known to be unnecessary, with particularly significant consequences for the quality of life of the patients. Moreover, successful treatment of therapy-resistant patients requires new strategies. Indeed, there is desperate need for new therapeutic approaches in ESFT. Practically, most patients relapsing after front-line chemotherapy, that have used virtually all active drugs currently available, are suitable for inclusion in phase II studies on new targeted therapies. Thanks to the presence of different laboratories with an extremely specialized and differentiated experience on the subject, a thorough study of the preclinical effectiveness of new targeted therapeutic strategies will be performed with the aim of the identification of the Achilles' heel in this disease and the consequent development of a tailored biological therapy to be used in association with conventional chemotherapy.

By providing an organizational framework for collaboration the project will also allow multicentre collection and analysis of cases as well as suitable collaborative research to allow genetic studies for the screening of high-risk patients and patients responding differently to chemotherapy.

9. SPECIFIC SCIENTIFIC AND TECHNOLOGICAL OBJECTIVES

Previous studies, mainly conducted by partners of this project, have indicated in the presence of EWS/FLI-1 or EWS/ERG chimeric proteins, CD99, IGF-IR, NOVH and TTF1 as well as in responsiveness to tumor necrosis factor-related apoptosis-inducing ligand (TRAIL) peculiar features of ESFT [21,29–42]. The main objective of this project is to evaluate the prognostic and therapeutic relevance for each of these factors and any correlations among them, to provide the rationale for the design of new options for therapy to patients escaping conventional treatments. Accordingly, the prognostic impact for each of these factors, by the retrospective and prospective analysis of a large number of tissue samples collected in the different centres, and the preclinical effectiveness of new strategies specifically targeting some of these molecules will be defined.

Objective 1.   Impact of Molecular Studies on Prognosis in ESFT

Definition of subgroups of patients with different survival probabilities through the analysis of the expression of EWS/FLI-1 or EWS/ERG fusion transcripts, secondary cytogenetic and molecular alterations, CD99, IGF-IR, NOVH, erbB-2 and TTF1 in relation to prognosis. This allows stratification of therapy, which may improve the ability to cure high-risk patients and minimize toxicity to the low-risk patients.

Quantification

Definition of the relative risk for each patient. Achievable through:
1. Preparatory activities and prospective data collection
2. Prognostic relevance of selected promising predictive markers in classical Ewing's sarcoma versus peripheral neuroectodermal tumors
3. Identification of genetic markers and tissue array construction
4. Validation of ESFT arrays and Definition of new therapeutic strategies

Verification

Products
1. Standardization of experimental procedures:
   - Guidelines for histological diagnosis of ESFT
   - Guidelines for tissue-handling experimental procedures
   - Guidelines for the experimental procedures to define the preclinical profile of new targeted therapies
2. Tissue data bank
3. Standardization of diagnostic criteria
4. Definition of patient selection criteria
5. Comparison between biological findings and clinical outcome, thorough statistical analysis, will verify which factors have the higher prognostic impact in ESFT patients
6. Guidelines for the detection and the significance of the selected biomolecular markers of prognosis

Objective 2.   Impact of Targeted Therapeutic Approaches in the Treatment of ESFT

Definition of innovative therapeutic strategies.
   Identification of the most effective treatment modalities to be associated with conventional chemotherapy.

Quantification

The use of antisense strategies, dominant negative mutants, neutralizing or antagonist antibodies, and specific inhibitors of intracellular signalling pathway, in vitro and in preclinical studies, will provide information on the effectiveness of targeting possible therapeutic molecules (EWS/FLI-1, CD99, erbB-2 and IGF-IR). The development of a humanised anti-CD99 MAb, siRNA to EWS/FLI-1 and IGF-IR, delivered inside cells by using polycationic vectors, will represent new bullets for the treatment of ESFT. Simultaneous and sequential treatments in combination with chemotherapeutic drugs will help to clarify the practical usefulness of these approaches in clinical settings. The validity of a TRAIL-mediated immunotherapy in ESFT will be also verified.

It is achievable through:
1. Definition of the therapeutic value of EWS/FLI-1 and creation of therapeutic tools for clinical application
2. Therapeutic value of targeted therapies against erbB-2 and IGF-IR
3. Development of new therapeutic strategies targeting CD99 molecule
4. Anti-Ewing tumor activity of TRAIL
5. Preparation and preclinical assessment of chimeric CD99 Mab
6. Validation of ESFT arrays and Definition of new therapeutic strategies

Verification

Methods of in vitro verification: Analysis of the effects of specific treatments with respect to growth, apoptosis, motility and invasive ability of ESFT cells, also in association with conventional anticancer agents.

Methods of in vivo verification: Analysis of the effectiveness of these innovative strategies to inhibit the tumorigenic and metastatic ability of ESFT cells in athymic mice or NOD/scid mice, also in association with conventional anticancer agents.

Products
1. Synthetic and vectorized siRNA compounds specific to EWS/FLI-1
2. Soluble, cell permeable and vectorized scFv to EWS/FLI-1
3. ESFT cells with impaired expression of EWS/FLI-1
4. Pilot phase analysis of siRNA compounds specific to EWS/FLI-1
5. Pilot phase analysis of methylated antisense oligonucleotides
6. ESFT cells with impaired expression of IGF-IR
7. Pilot phase analysis of HerceptinTM preclinical activity
8. Synthetic and vectorized siRNA compounds specific to IGF-IR
9. Pilot phase analysis of IGF-IR targeted strategies
10. Cells expressing different isoforms of CD99 and mutated molecule
11. Genetic signature of CD99-induced caspase-independent apoptosis
12. CD99Fc and CD99-ligand constructs
13. Pilot phase analysis for TRAIL/Apo2L and etoposide.
14. ESFT cell lines expressing NFκB-VP22 fusion proteins

15. ESFT cell lines expressing hypermethylated caspase genes
16. Chimerized CD99 "O662" Mab
17. Pilot phase preclinical analysis of the combination of chimerized anti-CD99 MAb with conventional chemotherapy
18. Guidelines for the design of new therapies in ESFT

Objective 3. Development of New Molecular Markers and/or Genomic Profiles to be Used for the Evaluation of Tumor Progression and Response to Chemotherapy

Creation of an extensive ESFT tissue bank, including frozen and paraffin-embedded material standardized for diagnosis and procedure, and of tumor tissue arrays.

Creation of a hierarchical database with data related to each patient including treatment, outcome, pathology, and molecular features.

Analysis of gene expression profile of ESFT cells upon modulation of EWS/FLI-1, CD99 IGF-IR and NOVH expression or functions as well as of ESFT samples in relation to metastatic ability and response to chemotherapy will clarify which genes are related to ESFT tumor progression. Moreover, new possible markers to be used for diagnostic, prognostic or therapeutic purposes will be identified.

Construction of specific ESFT c-DNA and CGH microarrays as prototype.

Quantification

The ESFT tissue bank will be formed by 50–60% of all the European cases. A list of genes that are specifically overexpressed or underexpressed in ESFT in relation to outcome will be identified with microarray and tissue array techniques. Their analysis and the study of gene expression changes in ESFT upon modulation of EWS/FLI-1, CD99 and IGF-IR will define a limited number of genes, for whom functional studies will be performed. This multistep processes will endup with the definition of few genes (a number between 50 and 100 genes is expected) directly related to the pathogenesis and progression of this tumor.

Achievable through:
1. Identification of genetic markers and tissue array construction
2. Role of NOVH (CCN3) in the progression of ESFT
3. Construction of ESFT-specific microarrays

Verification

The ESFT tissue bank together with the connected database will represent an invaluable source of standardized material and information available for the European Community. Construction of ESFT-specific cDNA and/or CGH DNA microarrays containing the sequences from the known ESFT-associated target genes, as well as genes identified in ESFT cells modified for the expression of EWS/FLI-1, CD99, NOVH or IGF-IR. Construction of ESFT-specific tissue arrays by using 100–200 tissue samples. These arrays will represent useful tools for the screening of high-risk groups by assessing the clinical significance of specific genes in relation to ESFT malignancy and response to chemotherapy, and for improving the histological classification of ESFT.

Products
1. Construction of a comprehensive ESFT tumor tissue array
2. Vectors expressing the full-length nov protein
3. Vectors expressing the different domains of the nov protein
4. ESFT cells expressing the full length or the different domain of NOV
5. Antibodies against subportions of the NOV protein
6. Construction of ESFT-specific cDNA microarrays
7. Construction of ESFT-specific CGH microarrays

REFERENCES

1. Bacci, G., Toni, A., Avella, M., Manfrini, M., Sudanese, A., Ciaroni, D., Boriani, S., Emiliani, E., and Campanacci, M. Long-term results in 144 localized Ewing's sarcoma patients treated with combined therapy. *Cancer* 63: 1477–1486, 1988.
2. Bacci, G., Picci, P., Ferrari, S., Mercuri, M., Brach del Prever, A., Rosito, P., Barbieri, E., Tienghi, A., and Forni, C. Neoadjuvant chemotherapy for Ewing's sarcoma of bone. No benefit observed after adding ifosfamide and etoposide to vincristine, actinomycin, cyclophosphamide, and doxorubicin in the maintenance phase. Results of two sequential studies. *Cancer* 6: 1174–1183, 1998.
3. Craft, A., Cotterill, S., Malcolm, A., Spooner, D., Grimer, R., Souhami, R., Imeson, J., and Lewis, I. Ifosfamide-containing chemotherapy in Ewing's sarcoma: the Second United Kingdom Children's Cancer Study Group and the Medical Research Council Ewing's Tumor Study. *J Clin Oncol* 16: 3628–3633, 1998.
4. Kovar, H., Dworzak, M., Strehl, S., Schnell, E., Ambros, I.M., Ambros, P.F., and Gadner, H. Overexpression of the pseudoautosomal gene MIC2 in Ewing's sarcoma and peripheral primitive neuroectodermal tumor. *Oncogene* 45: 1067–1070, 1990.
5. Delattre, O., Zucman, J., Melot, T., Garau, X.S., Zucker, J.M., Lenoir, G.M., Ambros, P.F., Sheer, D., Turc-Carel, C., Triche, T.J., Aurias, A., and Thomas, G. The Ewing family of tumors — a subgroup of small-round-cell tumors defined by specific chimeric transcripts. *N Engl J Med* 331: 294–299, 1994.
6. Scotlandi, K., Serra, M., Manara, M.C., Benini, S., Sarti, M., Maurici, D., Lollini, P.L., Picci, P., Bertoni, F., and Baldini, N. Immunostaining of the p30/332MIC2 antigen and molecular detection of EWS rearrangements for the diagnosis of Ewing's sarcoma and peripheral neuroectodermal tumor. *Hum Pathol* 27: 408–416, 1996.
7. Noguera, R., Triche, T.J., Navarro, S., Tsokos, M., and Llombart-Bosch, A. Dynamic model of differentiation in Ewing's sarcoma cells. *Lab Invest* 62: 143–151, 1992.
8. Pagani, A., Fischer-Colbrie, R., Eder, U., Pellin, A., Llombart-Bosch, A., and Bussolati, G. Neural and mesenchymal differentiations in Ewing's sarcoma cell lines. Morphological, immunophenotypic, molecular biological and cytogenetic evidence. *Int J Cancer* 63: 738–743, 1995.
9. Terrier, Ph. Henry-Amar, M., Triche, T.J., Horowitz, M.E., Terrier-Lacomber, M.-J., Miser, J.S., Kinsella, T.J., Contesso, G., and Llombart-Bosch, A. Is neuro-ectodermal differentiation of Ewing's sarcoma of bone associated with an unfavourable prognosis? *Eur J Cancer* 31A: 307–314, 1995.

10. Parham, D.M., Hijazi, Y., Steinberg, S.M., Meyer, W.H., Horowitz, M., Tzen, C.Y., Wexle, L.H., and Tsokos, M. Neuroectodermal differentiation in Ewing's sarcoma family of tumors does not predict tumor behaviour. *Hum Pathol* 30: 911–918, 1999.
11. de Alava, E., Kawai, A., Healey, J.H., Fligman, I., Meyers, P.A., Huvos, A.G., Gerald, W.L., Jhanwar, S.C., Argani, P., Antonescu, C.R., Pardo-Mindan, F.J., Ginsberg, J., Womer, R., Lawlor, E.R., Wunder, J., Andrulis, I., Sorensen, P.H., Barr, F.G., and Ladanyi, M. EWS-FLI1 fusion transcript structure is an independent determinant of prognosis in Ewing's sarcoma. *J Clin Oncol* 16: 1248–1255, 1998.
12. Ginsberg, J.P., de Alava, E., Ladanyi, M., Wexler, L.H., Kovar, H., Paulussen, M., Zoubek, A., Dockhorn-Dworniczak, B., Juergens, H., Wunder, J.S., Andrulis, I.L., Malik, R., Sorensen, P.H.B., Womer, R.B., and Barr, F.G. EWS-FLI1 and EWS-ERG gene fusions are associated with similar clinical phenotypes in Ewing's sarcoma. *J Clin Oncol* 17: 1809–1814, 1999.
13. de Alava, E., Panizo, A., Antonescu, C.R., Huvos, A.G., Pardo-Mindan, F.J., Barr, F.G., and Ladanyi, M. Association of EWS/FLI1 type 1 fusion with lower proliferative rate in Ewing's sarcoma. *Am J Pathol* 156: 849–855, 2000.
14. Toretzky, J.A., Kalebic, T., Blakesley, V., Le Roith, D., and Helman, L.J. The insulin-like growth factor-I receptor is required for EWS/FLI-1 transformation of fibroblasts. *J Biol Chem* 272: 30822–30827, 1997.
15. Armengol, G., Tarkkanen, M., Virolainen, M., Forus, A., Valle, J., Bohling, T., Asko-Seljavaara, S., Blomqvist, C., Elomaa, I., Karaharju, E., Kivioja, A.H., Siimes, M.A., Tukiainen, E., Caballin, M.R., Myklebost, O., and Knuutila, S. Recurrent gains of 1q, 8 and 12 in the Ewing family of tumours by comparative genomic hybridization. *Br J Cancer* 75: 1403–1409, 1997.
16. Maurici, D., Perez-Atayde, A., Grier, H.E., Baldini, N., Serra, M., and Fletcher, J.A. Frequency and implications of chromosome 8 and 12 gains in Ewing sarcoma. *Cancer Genet Cytogenet* 100: 106–110, 1998.
17. Tarkkanen, M., Kiuru-Kuhlefelt, S., Blomqvist, C., Armengol, G., Bohling, T., Ekfors, T., Virolainen, M., Lindholm, P., Monge, O., Picci, P., Knuutila, S., and Elomaa, I. Clinical correlations of genetic changes by comparative genomic hybridization in Ewing sarcoma and related tumors. *Cancer Genet Cytogenet* 114: 35–41, 1999.
18. Hattinger, C.M., Rumpler, S., Strehl, S., Ambros, I.M, Zoubek, A., Potschger, U., Gadner, H., and Ambros, P.F. Prognostic impact of deletions at 1p36 and numerical aberrations in Ewing tumors. *Genes Chromosomes Cancer* 24: 243–54, 1999.
19. Alberti, I., Bernard, G., Rouquette-Jazdanian, A.K., Pelassy, C., Pourtein, M., Aussel, C., and Bernard, A. CD99 isoforms expression dictates T cell functional outcomes. *FASEB J* 16: 1946–1948, 2002.
20. Perbal, B. NOV (nephroblastoma overexpressed) and the CCN family of genes: structural and functional issues. *J Clin Pathol: Mol Pathol* 54: 57–79, 2001.
21. Manara, M.C., Perbal, B., Benini, S., Strammiello, R., Cerisano, V., Perdichizzi, S., Serra, M., Astolfi, A., Bertoni, B., Alami, J., Yeger, H., Picci, P., and Scotlandi, K. The expression of ccn3(nov) gene in musculoskeletal tumors. *Am J Pathol* 160: 849–859, 2002.
22. Craft, A., Cotterill, S., Malcolm, A., Spooner, D., Grimer, R., Souhami, R., Imeson, J., and Lewis, I. Ifosfamide-containing chemotherapy in Ewing's sarcoma: the Second United Kingdom Children's Cancer Study Group and the Medical Research Council Ewing's Tumor Study. *J Clin Oncol* 16: 3628–3633, 1998.
23. Paulussen, M., Ahrens, S., Craft, A.W., Dunst, J., Frohlich, B., Jabar, S., Rube, C., Winkelmann, W., Wissing, S., Zoubek, A., and Jurgens, H. Ewing's tumors with primary lung metastases: survival analysis of 114 (European Intergroup) Cooperative Ewing's Sarcoma Studies patients. *J Clin Oncol* 16: 3044–3052, 1998.
24. Demetri, G.D. ET-743: the US experience in sarcomas of soft tissues. *Anticancer Drugs* 13: S7–S9, 2002.
25. Puchalski, T.A., Ryan, D.P., Garcia-Carbonero, R., Demetri, G.D., Butkiewicz, L., Harmon, D., Seiden, M.V., Maki, R.G., Lopez-Lazaro, L., Jimeno, J., Guzman, C., and Supko, J.G.

Pharmacokinetics of ecteinascidin 743 administered as a 24-h continuous intravenous infusion to adult patients with soft tissue sarcomas: associations with clinical characteristics, pathophysiological variables and toxicity. *Cancer Chemother Pharmacol* 50: 309–319, 2002.
26. Landuzzi, L., De Giovanni, C., Nicoletti, G., Rossi, I., Ricci, C., Astolfi, A., Scotlandi, K., Serra, M., Vitale, L., Bagnara, G.P., Nanni, P., and Lollini, P.-L. The metastatic ability of Ewing's sarcoma cells is modulated by stem cell factor and by its receptor c-*kit*. *Am J Pathol* 157: 2123–2131, 2000.
27. Merchant, M.S., Woo, C.W., Mackall, C.L., and Thiele, C.J. Potential use of imatinib in Ewing's Sarcoma: evidence for in vitro and in vivo activity. *J Natl Cancer Inst* 94: 1673–1679, 2002.
28. Hotfilder, M., Lanvers, C., Jurgens, H., Boos, J., and Vormoor, J. c-KIT-expressing Ewing tumour cells are insensitive to imatinib mesylate (STI571). *Cancer Chemother Pharmacol* 50: 167–169, 2002.
29. Amann, G., Zoubek, A., Salzer-Kuntschik, M., Windhager, R., and Kovar, H. Relation of neurological marker expression and EWS gene fusion types in MIC2/CD99-positive tumors of the Ewing family. *Hum Pathol* 30: 1058–1064, 1999.
30. Aryee, D.N., Sommergruber, W., Muehlbacher, K., Dockhorn-Dworniczak, B., Zoubek, A., and Kovar, H. Variability in gene expression patterns of Ewing tumor cell lines differing in EWS-FLI1 fusion type. *Lab Invest* 80: 1833–1844, 2000.
31. Spahn, L., Petermann, R., Siligan, C., Schmid, J.A., Aryee, D.N., and Kovar, H. Interaction of the EWS NH2 terminus with BARD1 links the Ewing's sarcoma gene to a common tumor suppressor pathway. *Cancer Res* 62: 4583–4587, 2002.
32. Lambert, G., Bertrand, J.R., Fattal, E., Subra, F., Pinto-Alphandary, H., Malvy, C., Auclair, C., and Couvreur, P. EWS fli-1 antisense nanocapsules inhibits ewing sarcoma-related tumor in mice. *Biochem Biophys Res Commun* 279: 401–406, 2000.
33. van Valen, F., Winkelmann, W., and Jurgens, H. Type I and II insulin-like growth factor receptors and their function in human Ewing's sarcoma cells. *J Cancer Res Clin Oncol* 118: 269–275, 1992.
34. Scotlandi, K., Benini, S., Sarti, M., Lollini, P.-L., Maurici, D., Picci, P., Manara, M.C., and Baldini, N. Insulin-like growth factor I receptor-mediated circuit in Ewing's sarcoma/peripheral neuroectodermal tumor: a possible therapeutic target. *Cancer Res* 56: 4570–4574, 1996.
35. Scotlandi, K., Benini, S., Nanni, P., Lollini, P.-L., Nicoletti, G., Landuzzi, L., Serra, M., Manara, M.C., Picci, P., and Baldini, N. Blockage of insulin-like growth factor-I receptor inhibits the growth of Ewing's sarcoma in athymic mice. *Cancer Res* 58: 4127–4131, 1998.
36. Benini, S., Manara, M.C., Baldini, N., Cerisano, V., Massimo, S., Mercuri, M., Lollini, P.L., Nanni, P., Picci, P., and Scotlandi, K. Inhibition of insulin-like growth factor I receptor increases the antitumor activity of doxorubicin and vincristine against Ewing's sarcoma cells. *Clin Cancer Res* 7: 1790–1797, 2001.
37. Scotlandi, K., Maini, C., Manara, M.C., Benini, S., Cerisano, V., Strammiello, R., Baldini, N., Lollini, P.L., Nanni, P., Nicoletti, G., and Picci, P. Effectiveness of insulin-like growth factor I receptor antisense strategy against Ewing's sarcoma cells. *Cancer Gene Ther* 9: 296–307, 2002.
38. Sohn, H.W., Choi, E.Y., Kim, S.H., Lee, I., Chung, D.H., Sung, U.A., Hwang, D.H., Cho, S.S., Jun, B.H., Jang, J.J., Chi, J.G., and Park, S.H. Engagement of CD99 induces apoptosis through a calcineurin-independent pathway in Ewing's sarcoma cells. *Am J Pathol* 153: 1937–1945, 1998.
39. Scotlandi, K., Baldini, N., Cerisano, V., Manara, M.C., Benini, S., Serra, M., Lollini, P.-L., Nanni, P., Nicoletti, G., Bernard, G., Bernard, A., and Picci, P. CD99 engagement: an effective therapeutic strategy for Ewing tumors. *Cancer Res* 60: 5134–5142, 2000.
40. Bernard, G., Breittmayer, J.-P., de Matteis, M., Trampont, P., Hofman, P., Senik, A., and Bernard, A. Apoptosis of immature thymocytes mediated by E2/CD99. *J Immunol* 158: 2543–2550, 1997.
41. Van Valen, F., Fulda, S., Truckenbrod, B., Eckervogt, V., Sonnemann, J., Hillmann, A., Rödl, R., Hoffman, C., Winkelmann, W., Schäfer, L., Dockhorn-Dworniczak, B., Wessel, T, Boos, J.,

Debatin, K.-M., and Jürgens, H. Apoptotic responsiveness of the Ewing's sarcoma family of tumours to tumour necrosis factor-related apoptosis-inducing ligand (TRAIL). *Int J Cancer* 88: 252–259, 2000.
42. Van Valen, F., Fulda, S., Schäfer, K.L., Truckenbrod, B., Hotfilder, M., Poremba, C., Debatin, K.M., and Winkelmann, W. Selective and nonselective toxicity of TRAIL/Apo2L combined with chemotherapy in human bone tumour cells vs normal human cells. *Int J Cancer* 107: 929–940, 2003:

CHAPTER 2
TARGETED THERAPIES IN EWING'S SARCOMA

KATIA SCOTLANDI

"Growth Factors and Receptors", Oncologic Research Laboratory, Istituti Ortopedici Rizzoli, Via di Barbiano 1/10, 40126 Bologna, Italy

Abstract: Ewing's sarcoma, the second most common malignant bone tumor, is an extremely aggressive neoplasm, mainly occurring in children and adolescents. Ewing's sarcoma shows a low survival rate despite the adoption of multimodal treatments, including local control of the disease by surgery and/or radiotherapy and multidrug adjuvant chemotherapy. No new effective drugs have been recently described and proposed for sarcomas and, therefore, innovative treatment modalities are very welcome and needed. In this respect, two new entry sites for therapeutic intervention may derive from tailored therapies against the insulin-like growth factor receptor I (IGF-IR) or CD99, a cell surface transmembrane protein highly expressed in Ewing's sarcoma. Neutralizing IGF-IR functions was shown to significantly affect Ewing's sarcoma malignancy. However, it is only recently that new clinically applicable drugs targeting IGF-IR are available and represent a concrete opportunity. Engagement of CD99 induces massive apoptosis of Ewing's sarcoma cells through caspase-independent mechanisms and reduces their malignant potential. Since the apoptotic functions of this molecule are of potential clinical interest, the effects of a tailored therapy triggering CD99 were analyzed against Ewing's sarcoma local tumors and distal (lung and bone) metastases in athymic mice. The effects of targeted therapies against CD99 or IGF-IR were evaluated in combination with conventional chemotherapeutic agents to assess best drug–drug interactions and treatment schedule. Toxic effects of these tailored therapies were also considered to offer the necessary rationale for the application of possible forthcoming clinical trials.

Key words: Ewing's sarcoma, EWS/FLI-1, IGFRI, CD99

Correspondence to: Katia Scotlandi, Ph.D., Laboratorio di Ricerca Oncologica, Istituti Ortopedici Rizzoli, Via di Barbiano 1/10, 40136 Bologna, Italy; TEL (+39) 051-6366760; FAX (+39) 051-6366761; e-mail: katia.scotlandi@ior.it

1. INTRODUCTION

Ewing's sarcoma ranks second amongst bone tumors and is the most frequent solid tumor in children and adolescents. Ewing's sarcoma is characterized by the presence of specific chromosomal translocations, which produce *EWS/ets* gene rearrangements (in more than 95% of cases, the gene fusion is EWS/FLI-1, due to the t(11;22) (q24;q12), or EWS/ERG due to the t(21;22) (q22;q12)),[1] as well as the expression, at extremely high levels, of an antigen encoded by the *MIC2* gene (also known as CD99 or p30/32^{MIC2}).[2,3,4] From a clinical point of view, combination of chemotherapy associated with surgery or radiation therapy have become standard practice in the treatment of patients with Ewing's sarcoma and the use of these multimodal treatments have shifted the survival rate of Ewing's sarcoma patients with localized disease to 65–70% after 5 years.[5–9] However, the most recent improvements in the cure rate of these patients have been achieved by dose-intensification, therefore paying the price of severe toxicity and high rate of life-threatening late events, such as secondary malignancies. This poses serious quality of life issues due to the young age of the patients and their long life expectancy. In addition, 25% of patients have metastases at the time of diagnosis, and for this high-risk group the survival rate at 5 years is still as low as 20%. The lack of new effective drugs in the treatment of sarcomas together with important side effects of high dose regimens in young patients with a long life expectancy supports the need of innovative therapeutic strategies, including targeted therapies against molecules that appear to be critical for the pathogenesis and progression of Ewing's sarcoma. Discoveries in the last years have led to a better understanding of the mechanisms involved in the genesis of this neoplasm and allowed the identification of some biological targets.[10–14] Here we focused our attention on the clinical relevance of targeting the IGF-IR and CD99, two targets commonly expressed in Ewing's sarcoma.

2. INSULIN-LIKE GROWTH FACTOR RECEPTOR I

There is compelling evidence that the IGF and its receptors plays a major role in human neoplasia and interfering with the IGF-signaling system may be an attractive strategy for the treatment of some human cancer, including Ewing's sarcoma.[15,16]

IGF is an endocrine factor involved in metabolic control and normal growth, having crucial roles in many types of cancer cells. The biological functions of IGFs are initiated by their interaction with cell surface tyrosine kinase receptors, in particular the IGF-IR.[17,18] Upon ligand interaction with the IGF-IR α-subunit, tyrosine residues in the intracellular, membrane-bound β-subunit become autophosphorylated.[19] This enables docking and phosphorylation of the insulin receptor substrates (IRS) and Shc, thereby activating two important pathways mediating proliferation and survival, i.e., the phosphatydilinositol 3-kinase (PI3-K)/Akt and the mitogen-activated protein kinase (MAPK) pathways (for a review see[17,18]). IGF-IR serves several important functions, such as mitogenicity, growth in size of the cell, protection from apoptotic injuries, regulation of cell adhesion and cell motility, all well connected to normal and malignant physiology. However, IGF-IR is not unique for these functions, with one notable exception, that is its requirement for anchorage-independent growth, a property being well established to be unique for malignant cells. This property also implicates the function of IGF-IR in tumor progression, since the degree of anchorage independency reflects the level of malignancy. This means that metastasis had acquired more anchorage independency, and more IGF-IR dependency compared with primary tumors. It is therefore not surprising that IGF-IR emerged as a very promising target for cancer therapy. Respect to Ewing's sarcoma we have extensively demonstrated the crucial role of IGF/IGF-IR system in the pathogenesis of the neoplasia. In particular, IGF-IR-mediated loop is constantly present in Ewing's sarcoma and is a major autocrine circuit of this neoplasm.[20,21] In addition, Toretsky et al.[22] reported IGF-IR expression to be necessary for EWS/FLI-1-mediated transformation of primary fibroblasts. Thus, the IGF-IR pathway may be considered a good site for therapeutic intervention in EWS/FLI-1-mediated tumor proliferation. We confirmed this finding by demonstrating that impairment of IGF-IR by neutralizing antibody and suramin reduces growth and increases apoptosis of Ewing's sarcoma cells both in vitro and in vivo, and significantly decreases migration, invasion and metastatic spread to the lungs and bones.[11] It must be stressed that the "cures" observed in athymic mice were probably real cures, because the experiments were designed so that mice were treated for 10 days and then allowed to live out for at least 12 weeks after discontinuing the treatment. No toxicity from use of antibodies to the IGF-IR was observed. Similar data were also observed when we used an antisense oligonucleotide approach[12] or when we used dominant negative mutants to inhibit IGF-IR functions.[23] In all these conditions, the blockage of IGF-IR-mediated circuit was found to effectively reduce the tumorigenic and metastatic ability of Ewing's sarcoma cells in athymic mice and increase the effectiveness of conventional cytotoxic drugs.[24,25] The synergistic effects observed with doxorubicin and vincristine (VCR), likely due to the induction of apoptotic that follows IGF-IR impairment, is an important component of the anticancer effect of IGF-IR targeting and has obvious practical implications. In addition, disruption of IGF-IR has recently been observed to increase sensitivity to other targeted therapies, in particular trastuzumab that inhibits erbB-2.[26] Therefore, IGF-IR represents a valuable therapeutic approach against Ewing's sarcoma. However, approaches targeting IGF-IR still show limited possibilities for a prompt application in Ewing's sarcoma therapy. Despite the promising findings obtained in preclinical conditions, both antisense and dominant negatives have the problem, common to all plasmids, of an efficient delivery into animals whereas the large size of antibodies to IGF-IR restricts its access to tumor cells, particularly in central regions of solid tumors. Smaller fragments are currently being studied in an effort to improve access and uptake.[27] More promising is the search for small molecules that may inhibit the tyrosine kinase activity of the IGF-IR or its signaling pathways. The major advantage of this approach is that small molecules have a considerable higher bioavailability compared with antibodies, dominant-negative receptors and antisense oligonucleotides. We demonstrated that disruption of either the MEK/MAPK or the PI3-K modules has profound functional consequences in Ewing's sarcoma cells.[28] However, in the presence of IGF-IR, the inhibition of MEK/MAPK pathway showed additional advantages. The critical role of MEK/MAPK pathway in Ewing's sarcoma, also supported by the fact that interference with the constitutive activation of members of the MAPK signaling pathway impairs EWS/FLI-1-dependent transformation,[29] provides impetus for future studies testing the in vivo therapeutic value and the general toxicity of these signaling pathway-specific inhibitors. The problem with these inhibitors is related to the ubiquitous ness of their targets. Similarly, the problem of IGF-IR tyrosine kinase inhibitors is the very high homology between the tyrosine kinase domains of the IGF-IR and the insulin receptor, which is higher than 94%. Actually many of the developed IGF-IR tyrosine kinase inhibitors have also caused substantial inhibition of the Insulin Receptor. Such cross-reaction would probably cause diabetic reactions, which would prevent their clinical use. However, there is an interesting exception. Recently, a very selective small molecule (NVP-AEW541) inhibiting the IGF-IR tyrosine kinase and inhibiting tumor growth in animals by oral administration has been reported.[30] The effectiveness of this compound was tested also against sarcomas.[31] Ewing's sarcoma cells were generally found to be more sensitive to the effects of this drug compared with rhabdomyosarcoma and osteosarcoma, in agreement with the high dependency of this neoplasm to IGF-IR signaling. NVP-AEW541 induced a G1 cell cycle block in all cells tested, whereas apoptosis was observed only in those cells that show a high level of sensitivity. Concurrent exposure of cells to NVP-AEW541 and other chemotherapeutic agents resulted in positive interactions with VCR, actinomycin D (ACT-D) and ifosfamide (IFO), whereas subadditive effects were observed with doxorubicin and cisplatin. Accordingly, combined treatment with NVP-AEW541 and VCR significantly inhibited tumor growth of Ewing's sarcoma xenografts in nude mice. Therefore, results encourage inclusion of this drug in the treatment of patients with Ewing's sarcoma in combination with VCR, ACT-D and IFO, three major drugs in the treatment of sarcomas. Among the small molecule, the cyclolignans also deserves to be mentioned. They have also been shown to be potent inhibitors of IGF-IR and malignant growth.[32,33] The cyclolignan PPP revealed to be nontoxic in mice and very effective in inhibiting xenograft growth. In addition etoposide is a lignan derivative already used in cancer therapy. The evaluation of the effectiveness of these compounds in Ewing's sarcoma certainly deserves a try.

Finally, it should also be considered that IGF-signaling is also influenced by the IGF-binding proteins (IGFBPs) that modulate the bioavailability and bioactivity of the IGFs. The IGFBP family contains at least six high affinity members with variable functions and mechanisms of action. IGFBP-3 is the most abundant IGFBP in post-natal serum and has been shown to be a growth inhibitory, apoptosis inducing molecule, capable of acting through IGF-dependent and IGF-independent mechanisms.[34,35] In particular, IGFBP-3 in serum forms a 150 kDa complex with acid-labile subunit and IGF-I or IGF-II, thus prolonging their half-life and regulating the distribution of IGFs and their endocrine actions. Locally produced IGFBP-3 acts as an autocrine/paracrine regulator of IGFs. IGFBP-3 has affinities for IGFs that are equal to or stronger than those of the IGF receptors and, therefore, inhibits the IGFs by sequestration in the extracellular compartment and preventing their interaction with IGF-IR. In addition, IGF-independent actions, including growth inhibition, apoptosis, and sensitization to radiation and chemotherapeutic agents,[34,35] as well as successful in vivo treatment of cancer models with IGFBP-3[36,37] were reported. Thus, these experimental evidences together with epidemiological studies that correlated the level of IGFBP-3 with the risk to develop cancer,[38,40] indicate IGFBP-3 as an anticancer molecule with potential therapeutic relevance. We evaluated whether IGFBP-3 may be exploited for therapeutic applications in the treatment of Ewing's sarcoma. We observed a generally low expression of IGFBP-3 either in a panel of Ewing's sarcoma cell lines or clinical samples. The addition of rhIGFBP-3 to Ewing's sarcoma cells induced dose–dependent growth inhibition, in monolayer cultures, inhibited anchorage-independent growth, and led to significant reduction of motility. IGFBP-3 acts mainly through IGF-dependent mechanisms and the protein may therefore represent an alternative strategy to inhibit IGF-IR functions. We also observed a significant inhibition in the production or activity of the matrixmetalloprotease (MMP)-2 and 9 and of vascular endothelial growth factor (VEGF)-A after treatment with IGFBP-3. In addition, forced expression and production of IGFBP-3 in stable TC-71 Ewing's sarcoma cells, confirmed the effects observed with the exogenous protein, and reduced the number of lung and bone metastases in athymic mice. These findings further support the therapeutic value of disrupting IGF/IGF-IR system in Ewing's sarcoma and indicate the therapeutic potential of IGFBP-3. Unfortunately, there are currently no strategies available to deliver exogenous IGFBP-3 protein as an effective anticancer therapeutic.

In conclusion, experimental and preclinical investigations have provided clear and encouraging results about the attractiveness of IGF-IR as a therapeutic target in Ewing's sarcoma, which has demonstrated a peculiarly high level of sensitivity towards IGF-IR-disrupting approaches. We now have compounds, such as NVP-AEW541 or cyclolignan PPP that seemingly show the necessary requirements of specificity, drugability and preclinical effectiveness. Experiments are concordant in indicating that IGF-IR targeting has very little toxic effects. So, it is time to perform clinical studies and definitely evaluate the usefulness and risks of targeting IGF-IR as an option in Ewing's sarcoma treatment.

3. CD99

CD99 is a ubiquitous 32-kDa transmembrane protein encoded by the *MIC2* gene, which is located to the pseudoautosomal regions of both X and Y human chromosomes, and shares no homology with any known family of proteins with the exception of *XG* gene.[41-44] Broadly distributed on normal cells, CD99 is highly expressed on T- and B-lymphocytes.[45-47] Immunohistochemistry detected CD99 in several types of cancer, including lymphoblastic lymphoma/leukemia,[46] some rhabdomyosarcomas,[48] synovial sarcoma[49] and mesenchymal chondrosarcoma.[50] Antigen retrieval technology has been extending the number of tumors in which detection of the antigen was found. However, Ewing's sarcoma remains the tumor that, without exception, most consistently expresses CD99, a feature that allows differential diagnosis to distinguish Ewing's sarcoma from other types of small round cell tumors.[51] The functional role of CD99 is not well known. The ligand (s) for CD99 has not been identified, and most of our knowledge about its functions derives from triggering CD99 on hematopoietic cells by agonistic monoclonal antibodies (MAb). In normal cells, ligation of CD99 has been functionally implicated in cell adhesion and homotypic aggregation, apoptosis, cell migration, Th1 cell differentiation, activation and proliferation of mature T cells and upregulation and transport of several transmembrane proteins.[47,52-57] In pathological conditions, a role for CD99 has been indicated in Ewing's sarcoma cells where it delivers cell–cell adhesion and apoptotic signals,[58-60] whereas very little is known about the biological significance of CD99 expression in other malignancies. In Ewing's sarcoma, the engagement of CD99 induces massive apoptosis through caspase-independent mechanisms. The intracellular signaling pathway associated to CD99 is still poorly defined, but certainly involves actin polimerization. Findings obtained with selective inhibitors indicated that only actin cytoskeleton integrity was essential for cell–cell adhesion and apoptosis of Ewing's sarcoma cells. Indeed, CD99 stimulation induced actin repolymerization likely through zyxin, a cytoplasmic adherens junction protein found to play a role in the regulation of the actin cytoskeleton and to act as a shuttle protein for communications to nuclear compartment.[61] Over-expression of zyxin after CD99 ligation was confirmed by real time PCR and western blot. Treatment of Ewing's sarcoma cells with zyxin antisense oligonucleotides inhibited CD99-induced cell aggregation and apoptosis, suggesting a functional role for this protein. Thus, actin and zyxin appear as early signaling events driven by CD99 engagement. Downstream effectors remain, however, to be determined. Their identification may have relevant clinical consequences since they are likely more common targets than CD99 and being involved in apoptotic pathways they may attract the companies interest. Ewing's sarcoma is an orphan disease and no private company will develop new therapeutic tools. Understanding apoptosis regulatory events occurring within Ewing's sarcoma cells may reveal molecular targets of general interest or for which drugs are already available and this may make the difference for translating experimental evidences in clinical practice.

Due to its effects on apoptosis, CD99 triggering by agonistic MAbs has an obvious therapeutic interest. We explored the preclinical effectiveness of combined treatments with the anti-CD99 0662 MAb and DXR. Simultaneous administration of the drugs or the use of the anti-CD99 MAb before DXR induced synergistic in vitro growth-inhibitory effects. In athymic mice, the combination of the two agents was remarkably effective against local growth and metastases. Accordingly, a significant increase in the survival of mice was observed. Systemic delivery of the anti-CD99 0662 MAb significantly reduced the number of lung and bone metastases and increased the time of their appearance. Combined treatments completely abrogate the metastatic ability of TC-71 cells. Thus, the anti-CD99 0662 MAb could be a useful adjuvant to cytotoxic chemotherapy for the treatment of Ewing's sarcoma patients. Toxic effects appear to be minimal in the animals. We also explored possible toxic effects of anti-CD99 MAb against human blood cell precursors because CD99 is highly expressed on many hematopoietic cell types, including CD34-positive cells[46] and its engagement is able to induce apoptosis of double positive T lymphocytes.[47] The in vitro analysis of growth of the different blood cell populations after exposure of CD34+ cells to a dose of anti-CD99 MAb that is effective against Ewing's sarcoma cells excluded an important toxic effect at bone marrow level, further supporting the use of anti-CD99 MAb in the therapy of Ewing's sarcoma.

In conclusion, we provide evidence for the preclinical effectiveness of combined treatments with anti-CD99 MAb and DXR against Ewing's sarcoma local tumors and distal metastases in athymic mice and, by assessing the potential toxicity of anti-CD99-based therapies against normal bone marrow precursors, we offer the necessary rationale for the application of a tailored therapy with high clinical potentialities that could be advantageously used in association with conventional agents.

REFERENCES

1. Delattre O, Zucman J, Melot T, et al. The Ewing family of tumors – a subgroup of small-round-cell tumors defined by specific chimeric transcripts. *N Engl J Med* 331: 294–299 (1994).
2. Ambros IM, Ambros PF, Strehl S, et al. MIC2 is a specific marker for Ewing's sarcoma and peripheral primitive neuroectodermal tumors. Evidence for a common histogenesis of Ewing's sarcoma and peripheral primitive neuroectodermal tumors from MIC2 expression and specific chromosome aberration. *Cancer* 67: 1886–1893 (1991).
3. Kovar H, Dworzak M, Strehl S, et al. Overexpression of the pseudoautosomal gene MIC2 in Ewing's sarcoma and peripheral primitive neuroectodermal tumor. *Oncogene* 45: 1067–1070 (1990).
4. Noguera R, Triche TJ, Navarro S, et al. Dynamic model of differentiation in Ewing's sarcoma cells. *Lab Invest* 62: 143–151 (1992).
5. Bacci G, Toni A, Avella M, et al. Long-term results in 144 localized Ewing's sarcoma patients treated with combined therapy. *Cancer* 63: 1477–1486 (1988).
6. Burgert EO, Nesbit ME, Garnsey LA, et al. Multimodal therapy for the management of non-pelvic, localized Ewing's sarcoma of bone: intergroup study IESS-II. *J Clin Oncol* 8: 1514–1524 (1990).
7. Paulussen M, Ahrens S, Craft AW, et al. Ewing's tumors with primary lung metastases: survival analysis of 114 (European Intergroup) Cooperative Ewing's Sarcoma Studies patients. *J Clin Oncol* 16: 3044–3052 (1998).
8. Bacci G, Picci P, Ferrari S, et al. Neoadjuvant chemotherapy for Ewing's sarcoma of bone. No benefit observed after adding iphosphamide and etoposide to vincristine, actinomycin,

cyclophosphamide, and doxorubicin in the maintenance phase. Results of two sequential studies. *Cancer* 6: 1174–1183 (1998).
9. Craft A, Cotterill S, Malcolm A, et al. Ifosfamide-containing chemotherapy in Ewing's sarcoma: the Second United Kingdom Children's Cancer Study Group and the Medical Research Council Ewing's Tumor Study. *J Clin Oncol* 16: 3628–3633 (1998).
10. Kovar H, Aryee D, Zoubek A. The Ewing family of tumors and the search for the Achilles'heel. *Curr Opin Oncol* 11: 275–284 (1999).
11. Scotlandi K, Benini S, Nanni P, et al. Blockage of insulin-like growth factor-I receptor inhibits the growth of Ewing's sarcoma in athymic mice. *Cancer Res* 58: 4127–4131 (1998).
12. Scotlandi K, Maini C, Manara MC, et al. Effectiveness of Insulin-like growth factor receptor I antisense strategy against Ewing's sarcoma cells. *Cancer Gene Ther* 9: 296–307 (2002).
13. Scotlandi K, Baldini N, Cerisano V, et al. CD99 engagement: an effective therapeutic strategy for Ewing tumors. *Cancer Res* 60: 5134–5142 (2000).
14. Landuzzi L, De Giovanni C, Nicoletti G, et al. The metastatic ability of Ewing's sarcoma cells is modulated by stem cell factor and by its receptor c-kit. *Am J Pathol* 157: 1–9 (2000).
15. Scotlandi K, Benini S, Nanni P, Lollini P-L, Nicoletti G, Landuzzi L, Serra M, Manara MC, Picci P, Baldini N. Blockage of insulin-like growth factor-I receptor inhibits the growth of Ewing's sarcoma in athymic mice. *Cancer Res* 58: 4127–4131 (1998).
16. Scotlandi K, Maini C, Manara MC, Benini S, Serra M, Cerisano V, Strammiello R, Baldini N, Lollini P-L, Nicoletti G, Picci P. Effectiveness of insulin-like growth factor I receptor antisense strategy against Ewing's sarcoma cells. *Cancer Gene Ther* 9: 296–307 (2002).
17. LeRoith D, Roberts CT Jr, The insulin-like growth factor system and cancer. *Cancer Lett* 195: 127–137 (2003).
18. Baserga R. The contradictions of the insulin-like growth factor 1 receptor. *Oncogene* 19: 5574–5581 (2000).
19. Ullrich A, Gray A, Tam AW, Yang-Feng T, Tsubokawa M, Collins C, Henzel W, Le Bon T, Kathuria S, Chen E, et al. Insulin-like growth factor I receptor primary structure: comparison with insulin receptor suggests structural determinants that define functional specificity. *EMBO J* 5(10): 2503–2512 (1986).
20. Yee D, Favoni RE, Lebovic GS, et al. Insulin-like growth factor I expression by tumors of neuroectodermal origin with the t(11;22) chromosomal translocation. A potential autocrine growth factor. *J Clin Invest* 86: 1806–1814 (1990).
21. Scotlandi K, Benini S, Sarti M, et al. Insulin-like growth factor I receptor-mediated circuit in Ewing's sarcoma/Peripheral neuroectodermal tumor: a possible therapeutic target. *Cancer Res* 56: 4570–4574 (1996).
22. Toretsky JA, Kalebic T, Blakesley V, LeRoith D, Helman LJ. The Insulin-like growth factor-I receptor is required for EWS/FLI-1 transformation of fibroblasts. *J Biol Chem* 272: 30822–30827 (1997).
23. Scotlandi K, Avnet S, Benini S, Manara MC, Serra M, Cerisano V, Perdichizzi S, Lollini PL, De Giovanni C, Landuzzi L, Picci P. Expression of an IGF-I receptor dominant negative mutant induces apoptosis, inhibits tumorigenesis and enhances chemosensitivity in Ewing's sarcoma cells. *Int J Cancer* 101(1): 11–16 (2002).
24. Toretsky JA, Thakar M, Eskenazi AE, Frantz CN. Phosphoinositide 3-hydroxide kinase blockade enhances apoptosis in the Ewing's sarcoma family of tumors. *Cancer Res* 59: 5745–5750 (1999).
25. Benini S, Manara MC, Baldini N, Cerisano V, Massimo Serra, Mercuri M, Lollini PL, Nanni P, Picci P, Scotlandi K. Inhibition of insulin-like growth factor I receptor increases the antitumor activity of doxorubicin and vincristine against Ewing's sarcoma cells. *Clin Cancer Res* 7(6): 1790–1797 (2001).
26. Scotlandi K, Manara MC, Hattinger CM, Benini S, Perdichizzi S, Pasello M, Bacci G, Zanella L, Bertoni F, Picci P, Serra M. Prognostic and therapeutic relevance of HER2 expression in osteosarcoma and Ewing's sarcoma. *Eur J Cancer* 41(9): 1349–1361 (2005).

27. Sachdev D, Li SL, Hartell JS, Fujita-Yamaguchi Y, Miller JS, Yee D. A chimeric humanized single-chain antibody against the type I insulin-like growth factor (IGF) receptor renders breast cancer cells refractory to the mitogenic effects of IGF-I. *Cancer Res* 63(3): 627–635 (2003).
28. Benini S, Manara MC, Cerisano V, Perdichizzi S, Strammiello R, Serra M, Picci P, Scotlandi K. Contribution of MEK/MAPK and PI3-K signaling pathway to the malignant behavior of Ewing's sarcoma cells: therapeutic prospects. *Int J Cancer* 108(3): 358–366 (2004).
29. Silvany RE, Eliazer S, Wolff NC, Ilaria RL Jr, Interference with the constitutive activation of Erk1 and Erk2 impairs EWS/FLI-1-dependent transformation. *Oncogene* 19: 4523–4530 (2000).
30. Garcia-Echeverria C, Pearson MA, Marti A, et al. In vivo antitumor activity of NVP-AEW541-a novel, potent, and selective inhibitor of the IGF-IR kinase. *Cancer Cell* 5: 231–239 (2004).
31. Scotlandi K, Manara MC, Nicoletti G, Lollini PL, Lukas S, Benini S, Croci S, Perdichizzi S, Zambelli D, Serra M, Garcia-Echeverria C, Hofmann F, Picci P. Antitumor activity of the insulin-like growth factor-I receptor kinase inhibitor NVP-AEW541 in musculoskeletal tumors. *Cancer Res* 65(9): 3868–3876 (2005).
32. Girnita A, Girnita L, del Prete F, Bartolazzi A, Larsson O, Axelson M. Cyclolignans as inhibitors of the insulin-like growth factor-1 receptor and malignant cell growth. *Cancer Res* 64(1): 236–242 (2004).
33. Vasilcanu D, Girnita A, Girnita L, Vasilcanu R, Axelson M, Larsson O. The cyclolignan PPP induces activation loop-specific inhibition of tyrosine phosphorylation of the insulin-like growth factor-1 receptor. Link to the phosphatidyl inositol-3 kinase/Akt apoptotic pathway. *Oncogene* 23(47): 7854–7862 (2004).
34. Mohan S, Baylink DJ. IGF-binding proteins are multifunctional and act via IGF-dependent and -independent mechanisms. *J Endocrinol* 175: 19–31 (2002).
35. Ali O, Cohen P, Lee KW. Epidemiology and biology of Insulin-like growth factor binding protein-3 (IGFBP-3) as an anti-cancer molecule. *Horm Metab Res* 35: 726–733 (2003).
36. Lee HY, Moon H, Chun KH, Chand YS, Hassan K, Ji L, Lotan R, Khuri FR, Hong WK. Effects of insulin-like growth factor binding protein-3 and farnesyltransferase inhibitor SCH66336 on Akt expression and apoptosis in non-small-cell lung cancer cells. *J Natl Cancer Inst* 96: 1536–1548 (2004).
37. Kirman I, Poltoratskaia N, Sylla P, Whelan RL. Insulin-like growth factor-binding protein 3 inhibits growth of experimental colocarcinoma. *Surgery* 136: 205–209 (1999).
38. Yu H, Spitz MR, Mistry J, Gu J, Hong WK, Wu X. Plasma levels of insulin-like growth factor-I and lung cancer risk: a case-control analysis. *J Natl Cancer Inst* 91: 151–156 (2004).
39. Petridou E, Skalkidou A, Dessypris N, Moustaki M, Mantzoros C, Spanos E, Trichopoulos D. Insulin-like growth factor binding protein-3 predicts survival from acute childhood leukemia. *Oncology* 60: 252–257 (2001).
40. Renehan AG, Zwahlen M, Minder C, O'Dwyer ST, Shalet SM, Egger M. Insulin-like growth factor (IGF)-I, IGF binding protein-3, and cancer risk: systematic review and meta-regression analysis. *Lancet* 363: 1346–1353 (2004).
41. Levy R, Dilley J, Fox RI, Wamk R. A human thymus-leukemia antigen defined by hybridoma monoclonal antibodies. *Proc Natl Acad Sci USA* 76: 6552–6556 (1979).
42. Fouchet C, Gane P. A study of coregulation and tissue specificity of XG and MIC2 gene expression in eukaryotic cells. *Blood* 95: 18919–18926 (2000).
43. Petit C, Levilliers J, Wessenbach J. Physical Mapping of the human pseudo-autosomal region; comparison with genetic linkage map. *EMBO J* 7: 2369–2376 (1988).
44. Banting GS, Pym B, Darling SM, Goodfellow PN. The MIC2 gene product: epitope mapping and structural prediction analysis define an integral membrane protein. *Mol Immunol* 26: 181–188 (1989).
45. Dworzak MN, Fritsch G, Buchinger P, et al. Flow cytometric assessment of human MIC2 expression in bone marrow, thymus, and peripheral blood. *Blood* 83: 415–425 (1994).

46. Dworzak MN, Fritsch G, Fleischer C, et al. CD99 (MIC2) expression in paediatric B-lineage leukaemia/lymphoma reflects maturation-associated patterns of normal B-lymphopoiesis. *Br J Haematol* 105: 690–695 (1999).
47. Bernard G, Breittmayer J-P, de Matteis M, et al. Apoptosis of immature thymocytes mediated by E2/CD99. *J Immunol* 158: 2543–2550 (1997).
48. Ramani P, Rampling D, Link M. Immunocytochemical study of 12E7 in small round-cell tumors of childhood: an assessment of its sensitivity and specificity. *Histopathology* 23: 557–561 (1993).
49. Fisher C. Synovial sarcoma. *Ann Diagn Pathol* 2: 401–421 (1998).
50. Brown RE, Boyle JL. Mesenchymal chondrosarcoma: molecular characterization by a proteomic approach, with morphogenic and therapeutic implications. *Ann Clin Lab Sci* 33: 131–141 (2003).
51. Ambros IM, Ambros PF, Strehl S, et al. MIC2 is a specific marker for Ewing's sarcoma and peripheral primitive neuroectodermal tumors. Evidence for a common histogenesis of Ewing's sarcoma and peripheral primitive neuroectodermal tumors from MIC2 expression and specific chromosome aberration. *Cancer* 67: 1886–1893 (1991).
52. Alberti I, Bernard G, Rouquette-Jazdanian AK, et al. CD99 isoforms expression dictates T cell functional outcomes. *FASEB J* 16: 1946–1948 (2002).
53. Schenkel AR, Mamdouh Z, Chen X, Liebman RM, Muller WA. CD99 plays a major role in the migration of monocytes through endothelial junctions. *Nature Immunol* 3: 143–150 (2002).
54. Bernard G, Raimondi V, Alberti I, et al. CD99 (E2) up-regulates alpha4beta1-dependent T cell adhesion to inflamed vascular endothelium under flow conditions. *Eur J Immunol* 30: 3061–3065 (2000).
55. Wingett D, Forcier K, Nielson CP. A role for CD99 in T cell activation. *Cell Immunol* 193: 17–23 (1999).
56. Waclavicek M, Majdic O, Stulnig T, Berger M, Sunder-Plassmann R, Zlabinger GJ, Baumruker T, Stöckl J, Ebner C, Knapp W, Pickl WF. CD99 engagement on human peripheral blood T cells results in TCR/CD3-dependent cellular activation and allows for Th1-restricted cytokine production. *J Immunol* 161: 4671–4678 (1998).
57. Kim SH, Choi EY, Shin YK, Kim TJ, Chung DH, Chang SI, Kim NK, Park SH. Generation of cells with hodgkin's and Reed-Sternberg phenotype through downregulation of CD99 (Mic2). *Blood* 92: 4287–4295 (1998).
58. Sohn HW, Choi EY, Kim SH, et al. Engagement of CD99 induces apoptosis through a calcineurin-independent pathway in Ewing's sarcoma cells. *Am J Pathol* 153: 1937–1945 (1998).
59. Scotlandi K, Baldini N, Cerisano V, et al. CD99 engagement: an effective therapeutic strategy for Ewing tumors. *Cancer Res* 60: 5134–5142 (2000).
60. Cerisano V, Aalto Y, Perdichizzi S, et al. Molecular mechanisms of CD99-induced caspase independent-cell death and cell–cell adhesion in Ewing's sarcoma cells: Actin and Zyxin as key intracellular mediators. *Oncogene* 23: 5664–5674 (2004).
61. Wang Y, Gilmore TD. Zyxin and paxillin proteins: focal adhesion plaque LIM domain proteins go nuclear. *Biochim Biophys Acta* 1593(2–3): 115–120 (2003).

CHAPTER 3

THE CCN3 PROTEIN AND CANCER

BERNARD PERBAL

Laboratoire d'Oncologie Virale et Moléculaire, UFR de Biochimie, Université Paris 7 D. Diderot, Case 7048 – 2 Place Jussieu – 75,005 Paris, France

Abstract: A new family of cell growth and differentiation regulators has emerged in the past decade. These signaling proteins grouped under the CCN acronym, participate to fundamental biological functions, during normal development and in adulthood, from birth to death. Disregulation of their expression has been associated to tumorigenesis. Even though part of their physiological properties may be related to their capacity to bind several integrins, the CCN protein also interact with several other receptors and ligands that play critical roles in the regulation of cell signaling and communication. The multimodular structure of the CCN proteins provides the ground for the myriad of roles in which they participate, but it remains a challenge to those who wish to decipher the structure–function relationship that govern their multifunctional properties. The recent discovery of CCN variants whose expression is associated to the development of cancer raises fascinating questions regarding their role in the establishment and maintenance of the tumor state.

Identifying the pathways in which the CCN proteins act and establishing the role of these proteins in intercellular communication will constitute new promising avenues among the trends for 21st century in cancer research.

1. INTRODUCTION

The CCN acronym stands for a group of genes whose three founding members were isolated in the 1990s and were designated Cyr61, Ctgf, and Nov. Later, other proteins named ELM-1/WISP-1, rCop-1/WISP-2/CTGF-L, and WISP-3 were reported to share structural identity with CTGF, CYR61, and NOV [1–4]. The CCN family of genes presently consists of six members (Figure 3.1) and a unifying nomenclature was recently proposed for the CCN family [5], numbering the proteins CCN1–CCN6 in the order in which they were first described in the literature.

With the exception of CCN5, which lacks the CT module, all CCN proteins, share a common multimodular organization (Figure 3.2) with four structural

CHAPTER 3

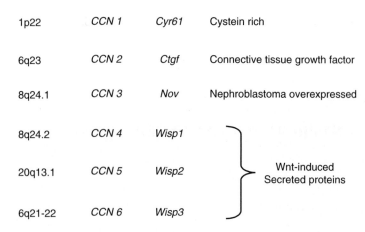

Figure 3.1. The CCN family of proteins. The figure shows the localization of CCN genes on human chromosomes. Original names of the genes are indicated on the right side.

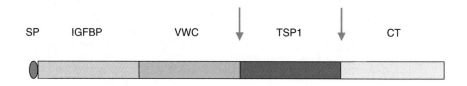

SP : Signal peptide

IGFBP : Insulin-like growth factor binding protein

VWC : Von Willebrand type C repeat

TSP1 : Thrombospondin type 1 repeat

CT : C-terminal cystin knot

Figure 3.2. Tetramodular structure of the CCN proteins. Each constitutive module is represented with the name of the protein families with which they share a partial identity. The arrows indicate sites of proteolytic cleavage.

modules (IGFBP, VWC, TSP1, and CT) resembling insulin-like growth factor binding proteins, Von Willebrand factor, thrombospondin 1, and a series of matricellular proteins and growth factors that contain a cystin knot motif. The high degree of structural identity and the conservation of 38 cystein residues in all CCN proteins but CCN5, suggested that they might play similar or redundant functions. However, the bulk of data that has been obtained in the recent years have challenged this view.

2. BIOLOGICAL FUNCTIONS OF CCN PROTEINS: A BRIEF OVERVIEW

The CCN proteins are now defined as a novel family of growth and differentiation regulators acting on a large variety of cell types (see recent reviews in [6]). Despite their similar structure, they show significant functional divergence and their biological functions are highly dependent upon the type of cells and the cellular context. Their multiple interactions with a variety of regulatory molecules [3,7,8] may account for the different functions attributed to CCN proteins.

During the embryonic life, the CCN proteins are expressed in derivatives of the three embryonic sheets: (1) ectoderm; (2) mesoderm; and (3) endoderm and they are implicated in the development of different tissues such as kidney, nervous system, muscle, bone marrow, cartilage, and bone [3].

The CCN proteins are known to participate during adulthood, in various biological processes including normal growth and development, angiogenesis, wound healing, tissue regeneration and uterine function, fibrosis in various tissues/organs, and inflammation such as arthritis and tumor growth [1–4, 8–11]. In the developing human embryo, the major site of expression for ccn3 is the nervous system, followed by the adrenal cortex. The urogenital system and smooth and striated muscles also express significant amount of ccn3 [3]. There are several documented examples showing that the expression of ccn3 is associated with differentiation of cartilage, muscle, kidney, and brain in normal situations [3].

The multiple biological functions of the CCN proteins are exerted through various signaling pathways (Figure 3.3) involving cell surface receptors such as integrins [2,11], 240-kDa protein [12,4], LRPs [13–15], Notch [16], as well as connexins [17,18] and calcium channels [19]. The regulation of intracellular calcium concentration by CCN3 and CCN2 has established these proteins as genuine signaling factors. CCN proteins have been proposed to participate in the formation of multifunctional regulatory complexes and to coordinate signaling pathways governing intracellular and intercellular communication required for the efficient control of cell growth and differentiation [20].

The existence of four potentially functional domains in these proteins raised fundamental questions about their contribution to the various biological properties of the CCN proteins. Although very little is known about the biological activities of individual domains, it is currently accepted that each of the four modules acts both independently and interdependently. The multimodular structure of the CCN proteins provides the basis for a wide range of interactions with different partners and a support for the variety of signaling pathways in which CCN proteins are acting.

Conflicting results have been obtained regarding the ability of Module I to bind IGF. Affinity-labeling studies suggested a weak IGF-binding affinity for CCN2 and CCN3 [21,22], whereas ligand-blotting experiments did not permit to confirm binding [23]. Recently, a chimearic protein has been constructed in

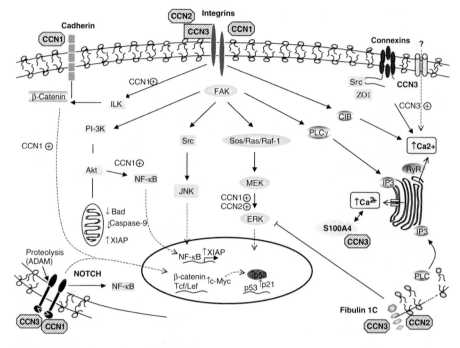

Figure 3.3. Signaling pathways in which CCN proteins are known to play a role.

which the aminoterminal domain of IGFBP-3 was substituted with the aminoterminal domain of CCN3. The CCN3-IGFBP-3 chimeara bound IGFs and inhibited IGF activity very weakly, similar to CCN3 itself (Yan et al., submitted for publication). Therefore, module I cannot replace the aminoterminal domain of IGFBP-3 in its IGF-binding function. However, preliminary observations suggest that this domain participate to binding-independent IGF signaling.

The second module of CCN2 was reported to bind BMP4 [24] and TGFbeta. Both modules 2 and 3 interact with various types of integrins. The interaction of CCN proteins with integrins being cell- and context-dependent. Module 3 itself binds LRP-1.

The CT module binds heparin sulphate proteoglycans (HSPGs) [25–27], and contains heparin-binding sequences and is sufficient to promote cell growth. The use of yeast two hybrid system and immunoprecipitation performed with specific antibodies raised against CCN3 permitted to establish that the CT module is also involved in the interaction of CCN3 with fibulin-1C, connexins, and notch1 [3]. Since module 4 is critical for the interaction of CCN proteins with their partners, one can predict that CCN5, which is lacking this module, should behave as a competitor for binding and modulator of CCN biological activities. This possibility is presently under current investigation.

3. CCN3 EXPRESSION IN TUMORS

At a cellular level, CCN proteins are involved in the control of adhesion, migration, proliferation, and survival, which are basic biological features that are known to be altered in cancer cells.

For example, CCN1, CCN2, CCN3, and CCN5 were found to support the adhesion of normal cells from various origins, such as fibroblasts, epithelial cells, stellate cells, platelets and monocytes, endothelial cells, and smooth muscle cells. The adhesion of these cells is directed by the interaction of CCN proteins with the various integrins expressed at the cell surface [6].

Differential responses are observed with respect to cell migration, proliferation, and survival. Whereas CCN1, CCN2, and CCN3 stimulate the migration of normal fibroblasts, vascular endothelial cells, vascular smooth muscle cells, and mesangial cells, CCN4 and CCN5 show inhibitory effects. The effects on cell proliferation are dependent upon the coordinate action of growth factors whose action is enhanced by the CCN proteins. From studies performed with both normal and tumor cells, it appears that CCN proteins show both pro- and antiapoptotic activities, with differential effects on cell survival.

The implication of CCN3 in the regulation of cell proliferation and differentiation stems from early studies that established this protein as a potential proto-oncogene with antiproliferative activity [28].

The analysis of avian nephroblastomas induced by myeloblastosis associated virus type 1 (MAV-1) [29,30] led to the identification of ccn3 as an integration site for MAV-1 in these tumors, which represent a unique model of the Wilms tumor [31]. The *ccn3* gene was first cloned from normal chicken embryo fibroblasts (CEF), and later from mouse and human. The xenopus and rat orthologs were isolated later (for review see [3]). The chicken *ccn3* gene shows the typical CCN organization, with five exons that encode from the 5' to the 3' end, a signal peptide, and the four prototypic structural domains that constitute the CCN proteins (Figure 3.4). Upon integration of the MAV proviral genome in the *ccn3* gene, a chimearic mRNA containing 91 nucleotides of viral origin at the 5' end was highly expressed in the tumor cells. As a result of MAV LTR integration within exon 2 of ccn3, the resulting 252 aminoacid proteins encoded by this mRNA was lacking the signal peptide and part of the IGFBP domain. The first ATG contained in this chimearic mRNA is localized at the beginning of exon 3.

Two main consequences can be foreseen as the result of ccn3 disruption by MAV: (1) the CCN3 protein expressed in these conditions is lacking the IGFBP domain and (2) the absence of signal peptide is expected to alter the fate of the protein.

By Northern hybridization, we could show that all avian nephroblastomas expressed high levels of ccn3 RNA as compared with the levels of message detected in normal kidney tissue of the same age (Figure 3.5). Interestingly, the levels of ccn3 were relatively high in the developing kidney tissue. We originally designated this gene *nov* for nephroblastoma overexpressed [28,31].

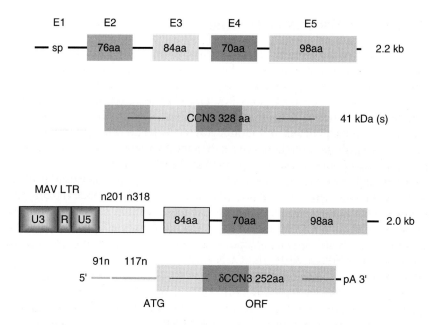

Figure 3.4. Organization of the ccn3 gene. The exon-intron organization of the chicken gene is represented on the top, with the corresponding putative CCN3 protein. The bottom part shows the remodeling of the ccn3 gene upon integration of the proviral MAV DNA. The structure of the chimearic mRNA resulting from the MAV integration is represented with the size and composition of the putative CCN3 protein expressed in the corresponding tumor. ATG, first initiation codon after the 5′ terminus of the chimearic RNA. ORF: open reading frame.

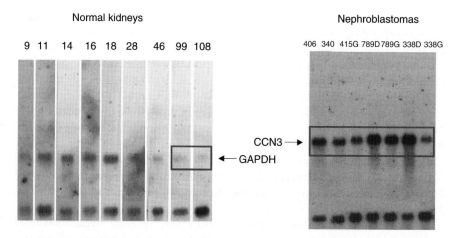

Figure 3.5. Overexpression of ccn3 in nephroblastomas. Northern blotting of RNA species extracted from normal developing kidneys (left panel) and MAV-induced nephroblastomas (right panel). Age in days is indicated above each sample. Gapdh: glyceraldehyde 3 phosphate dehydrogenase used as a quantitation marker.

In the light of the antiproliferative activity of CCN3 (see below) the overexpression of ccn3 in all avian nephroblastomas may appear parodoxical. However, two lines of evidence indicated that the involvement of ccn3 in kidney development might account for these observations. Firstly, studies performed with the avian system identified the blastemal cells committed to differentiation as targets for MAV-1, and established that ccn3 is not a common integration site for MAV in these tumors [32]. Secondly, the analysis of ccn3 expression in samples representing the whole panel of Wilms tumors established that ccn3 is not always overexpressed in these tumors, even though the levels of mRNA and proteins were altered in the tumors. These results indicated that in both systems, blastemal cells expressed high levels of ccn3 and that the quantities of CCN3 RNA and protein detected in the tumors were somehow related to the level of heterotypic differentiation in these tumors. More importantly, these results also indicated that the high expression of ccn3 detected in avian nephroblastomas was the result of a tumor-induced expansion of target blastemal cells that already express high levels of ccn3.

Recent results obtained with a MAV2 strain failed to detect any overexpression of ccn3 in the MAV2-induced nephroblastomas [33], an observation suggesting that either the type of viral strain or the route of injection used, or both can influence the outcome of the tumor cells.

In any case, these observations indicated that high expression of CCN3 may be restricted to certain types of nephroblastomas and led to the conclusion that the nov acronym being misleading it is preferable to use the new CCN nomenclature [5].

Because several examples of proviral insertional mutagenesis have been reported in the past, we have examined the biological activities of the full length CCN3 protein which was expressed in most tumors and of the truncated variant highly expressed in one of the tumors. As a first step in this approach, we have made use of the RCAS retroviral expression system, which is derived from RSV. When the recombinant viral strain expressing the full length ccn3 is used to infect CEFs, cells stop growing and detach (Figure 3.6) whereas infection of CEF with a retroviral construct expressing the aminotruncated CCN3 protein leads to morphological transformation (Figure 3.7). In the former case, the protein is normally secreted, while in the latter, the protein is not secreted. Therefore, we could draw two conclusions from these experiments: (1) the full length secreted protein showed potential growth inhibitory functions, (2) the truncation of the CCN3 resulted in a transforming capacity that might be revealed by the modification of addressing anticipated from the lack of signal peptide in the truncated protein.

To check whether the properties that we assigned to both forms were applying to a more general situation, we have first studied the effects of CCN3 on cell growth both in a wider context. We have also conducted studies aimed at identifying the molecular basis for the transforming potential of the aminotruncated CCN3 variant.

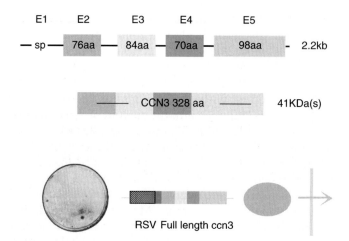

Figure 3.6. Inhibition of cell growth by RSV-CCN3. The full length ccn3 sequences have been cloned in a RCAS backbone and the recombinant virus used to infect chicken fibroblasts. Upon infection, cells die and detach from the plate. In this condition, the full length CCN3 protein is secreted.

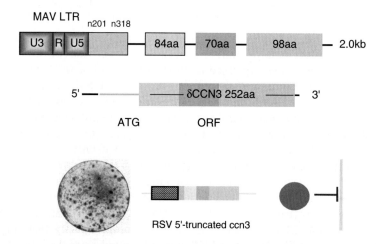

Figure 3.7. Morphological transformation induced by truncated CCN3. The coding sequence of the aminotruncated CCN3 protein expressed in nephroblastoma cells has been cloned in the RCAS vector. Upon infection of chicken embryo fibroblasts with the recombinant virus expressing the aminotruncated CCN3 protein, transformation foci are observed. In these conditions, the CCN3 protein lacking the signal peptide is not secreted.

4. ANTIPROLIFERATIVE ACTIVITY OF THE FULL LENGTH CCN3

In order to study the effects of CCN3 on cell proliferation we prepared a series of expression vectors in which the human full length ccn3 cDNA was inserted downstream to a pCMV promoter sequence, either in the sense or antisense orientation. These vectors were used to isolate stably transfected cells after G418, or blasticidin selection.

Since we had previously shown that in freshly explanted glioblastoma cells the level of CCN3 mRNA expression is inversely related to their tumorigenic potential, we used this system to assess the potential antiproliferative activity of ccn3. As previously reported [34] the expression of ccn3 in glioma cells resulted in a marked decrease in their growth rate (Figure 3.8). The inhibitory effect that resulted from ccn3 expression was increasing with time, therefore suggesting that larger quantities of CCN3 protein are produced in confluent cell cultures (Bleau et al., submitted for publication). A similar inhibition of cell growth was observed when the ccn3 expression vector was used to isolate stable transfectants derived from choriocarcinoma, Ewing's tumor cells [17,35]. More recently, results have been obtained that extend the negative effects of CCN3 to chronic myeloid leukemia cells and to normal cells of various origin (Fukanaga et al., submitted for publication; Buteau et al., submitted for publication; Bleau et al., submitted for publication).

To confirm that the effects observed did not result from the integration of the plasmid DNA in a particular area of the recipient genome, but were indeed due to the CCN3 protein, we used two different approaches. In the first one, we

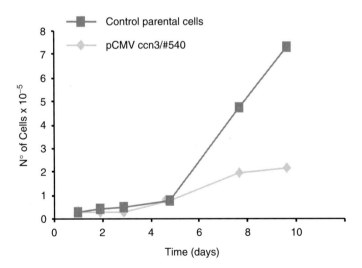

Figure 3.8. Antiproliferative effect of CCN3. Glioblastoma cells (G59) and ccn3-transfected derivatives (pCMV ccn3 540) have been grown for 10 days on plastic plates. To avoid deprivation in essential nutriments, the culture medium was changed every other day.

constructed vectors in which the expression of CCN3 was driven by an inducible promoter. This vector was used to transfect Ewing's sarcoma cells. The results obtained clearly indicated that the reduction of proliferation was observed only when the CCN3 protein expression was induced [35]. In the second approach, we made use of exogenous protein. The addition of either CCN3 protein purified from cell culture medium or recombinant GST-CCN3 protein produced in a bacterial system, also resulted in a marked decrease of cell proliferation. In some cases, the effects obtained with purified exogenous CCN3 proteins were even greater than those observed in transfection (Bleau et al., submitted for publication). A practical consequence of the growth inhibitory effects of CCN3 is that continuous passage of ccn3-transfected cells tends to eliminate cells that produce or select cells that produce lower levels.

Another consequence of ccn3 expression in tumor cells was exemplified by a dramatic reduction of the tumorigenic potential of transfected cells when assayed after injection in nude mice (Figure 3.9). In the case of Ewing's sarcoma, choriocarcinoma and glioblastoma cells, the expression of ccn3 did not permit the tumor cells to expand and give rise to large tumors. Only abortive growth was observed. In the case of glioblastomas, a decrease in the vascularization of the growing tumor was proposed to account for the inhibitory effect of CCN3 [34], even though this protein was shown to act as a proangiogenic factor when tested in a rat corneal micropocket implant assay [36]. It is interesting to point out that in all three cases, the implantation of tumor cells did not seem to be affected as much as their ability to expand.

Altogether, these results suggested that the CCN3 protein is devoted of antiproliferative activity and may interfere with tumor growth, i.e., acts as a tumor suppressor.

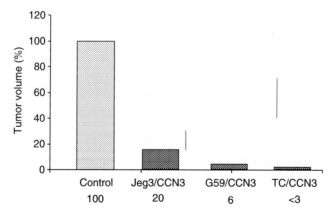

Figure 3.9. Antitumoral effect of CCN3. The ccn3-transfected cells derived from choriocarcinomas (JEG) glioblastomas (G59) and Ewing's tumor cell lines were injected in nude mice. The mean volume of tumors was measured as indicated.

In support of this conclusion, an elevated level of ccn3 expression was detected in tumors with favorable outcome and tumors showing a high degree of differentiation. However, in a few other cases, including Ewing's sarcoma cells, the expression of ccn3 was associated with a higher proliferative rate or bad prognostic (Figure 3.10).

In a study performed on a group of 45 patients, it was shown that CCN3 expression was associated with a higher risk of developing metastasis [35]. Indeed, none of the 15 patients whose primary tumor as negative for CCN3 developed metastasis after 10 years time. On the contrary, 50% of the patients who developed either lung or bone metastasis were shown to have primary tumors positive for CCN3 (Figure 3.11). In addition the expression of CCN3 in Ewing's tumor cells was conferring on Ewing's transfected cells the capacity to

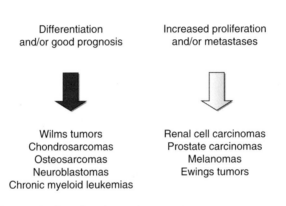

Figure 3.10. CCN3 expression in various human tumors.

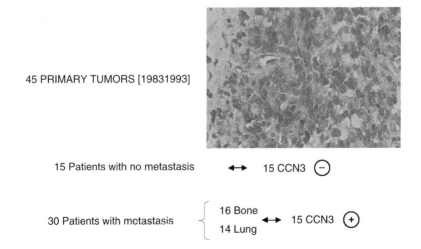

Figure 3.11. CCN3 is a marker of metastatic potential in Ewing's cells.

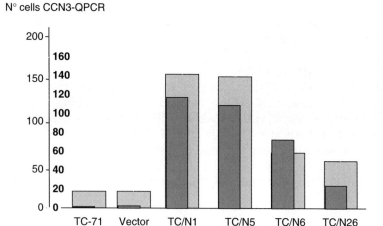

Figure 3.12. CCN3 increases Ewing's cells migration. The effect of ccn3 expression on migration (counts of cells) is shown to be directly proportional to the quantity of ccn3 mRNA detected by quantitative PCR analysis (qPCR). TC/N: independent stable transformants obtained after transfection of TC71 parental Ewing's cells with a ccn3 expression plasmid.

better migrate and invade matrigel [35]. For example, the increase of migratory capacity of ccn3-transfected Ewing's sarcoma cells was proportional to the amount of CCN3 expressed in cells (Figure 3.12).

Thus, it appears that CCN3 may act as a double edge sword in this system, on one hand, increasing the capacity of cells to invade and give rise to metastasis, and on the other hand decreasing their proliferation and tumorigenic potential. Although we do not have definitive explanation for these effects that may seem contradictory, we believe that these results allow to disconnect two sets of parameters that have been intuitively associated in the past. Ongoing experiments should permit to elucidate the relationships and contradictions that may exist between these two phenotypes.

It is interesting to point out that this situation is not restricted to the case of Ewing's cells since the expression of ccn3 is also associated to a higher metastatic potential in other tumors (unpublished results).

5. CCN3 PROTEINS

The use of a specific antibody (K19M) raised against a peptide contained in the C-terminal domain of CCN3 permitted us to detect CCN3-related proteins in different cellular compartments. As shown in Figure 3.13, when cells are untreated the staining for CCN3 is detected at the cell membrane and when cells are permeabilized, cytoplasmic staining is observed [37]. Following a gentle scraping off the cells that were attached on the plate, it is possible to detect the

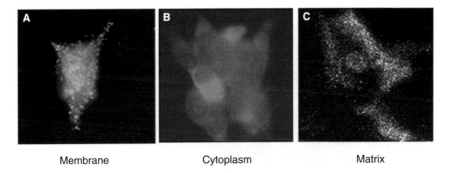

Figure 3.13. Subcellular localization of CCN3. Slides were stained with the K19M anti CCN3 antibody. Cells were not treated (left), permeabilized (center) and scraped (right). For experimental details see [35].

CCN3 protein in the form of a footprint indicating that CCN3 was indeed released in the extracellular matrix. The CCN3 protein is also detected in the cell culture supernatant as shown by western blotting analysis of conditioned medium. Similar observations were made for CCN1 and CCN2, which were detected both in the culture medium and at the cell surface [26,38].

Depending upon the amount of protein expressed by the cells, the detection of CCN3 can be performed either directly or after enrichment by affinity chromatography through heparin sepharose [23]. In the case of glioblastoma-transfected cells and adrenocarcinoma cells, the full length secreted CCN3 was detected as a doublet resulting from posttranslational modifications occurring soon after synthesis. In addition to this doublet of 50 kDa apparent molecular weight, an additional doublet was detected with an molecular weight of 32 kDa. Interestingly, CCN3 proteins with the same size were detected within the cellular lysates prepared from either ccn3-transfected glioma cells or adrenocarcinoma cells that produce CCN3 naturally (Figure 3.14).

Since truncated CCN proteins were also detected in several cases, the production of such variants raise fundamental questions regarding their potential biological functions.

It has been reported for several years that bioactive 10–20 kDa truncated variants containing different combinations of C-terminal modules are generated in utero through limited proteolysis of the full length CCN2 protein [39,40]. Inasmuch as these fragments were able to induce DNA synthesis and possessed heparin binging activities, one can expect that they might enhance or counteract the functions of the parental full length protein. Along this line, it is striking that the amount of CCN2-truncated variants is dependent upon progression in the uterine cycle.

At the same time, a truncated CCN3 isoform was identified in cell culture medium of insect cells infected with a ccn3-recombinant baculovirus. Quite

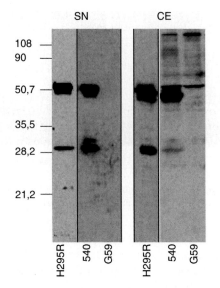

Figure 3.14. Western blot characterization of CCN3. Cell culture supernatant and cell lysates were analysed for the presence of CCN3 proteins. H295R: adrenocortical tumor cells, 540: glioblastoma transfected cells, G59 parental glioblastoma cells. SN cell culture supernatant. CE: cell lysate. For experimental details see [35].

interestingly, the CCN3-truncated form that is recovered from this cell culture supernatant shows the same aminoterminus as one of the CCN2 variants identified in pig uterine fluids, in spite of the slight differences in the primary sequences of these two proteins at the junction of domain 2 and 3. These observations therefore indicated that the proteolytic process that is involved in generating the two truncated forms is conserved and common to CCN2 and CCN3.

A 30 kDa aminotruncated CCN3 was also identified in the nucleus of two cancer cell lines that stained positive for CCN3 in immunofluorescence experiments using two different specific antibodies [3]. These observations raised the possibility that posttranslational processing of the full length CCN3 protein might produce shorter variants that would be addressed to the nucleus and raised several questions regarding (1) the mechanisms involved in nuclear addressing of the truncated proteins; (2) the biological activities of these variants; and (3) the process controlling this posttranslational processing.

The detection of aminotruncated CCN3 proteins in cancer cells was reminiscent of the situation encountered in the case of the MAV insertional mutagenesis that resulted in the production of a transforming aminotruncated CCN3 variant (see above).

Therefore, we hypothesized that the production of CCN3 with different subcellular localizations might be dependent upon the physiological state of the cells and that the mechanisms involved in the partition of these variants might be altered in cancer cells.

6. NUCLEAR ADDRESSING OF THE AMINOTRUNCATED CCN3 PROTEINS

The detection of nuclear CCN3-related proteins by immunofluorescence and the identification of a 30 kDa aminotrunctaed CCN3 variant in the nucleus of several cancer cells raised the possibility that the transforming activity of the MAV-truncated CCN3 protein might result from addressing to another cellular compartment and that aminotruncated variants generated by either genetic alterations or posttranslational processing might also show oncogenic properties.

In order to address this question, a series of plasmids were constructed in which various tags (GFP, myc, and 6HIS) have been added either at the carboxy- or at the amino-terminal end of ccn3 coding sequences representing different combinations of exons. These clones were engineered to produce proteins deprived of a signal peptide, in order to recapitulate the situation that was encountered in the case of MAV-induced disruption of CCN3.

The results that were obtained [41] indicated that the CCN3 variants lacking a signal peptide were addressed to the nucleus where they can repress transcription of a reporter gene. Considering that the expression of such a truncated CCN3 variant induced morphological transformation of CEFs, these observations suggested that repression of transcription might apply to genes with an antiproliferative activity.

Although preliminary attempts failed to identify specific targets for CCN3, nuclear addressing of the truncated forms is in agreement with previous reports which documented the binding of CCN3 CT domain to the PAI2 promoter, the interaction of CCN3 with subunit 7 of RNA polymerase II and the colocalization of CCN3 with the ICP4 protein in HSV1-infected cells (reviewed in [3]).

The uptake and intracellular transport of the CCN2 protein has also been reported and was suggested to play a role in the regulation of transcription [42].

The situation that we report for CCN proteins adds to the large number of secreted regulatory proteins among which FGF2, lactoferrin, epidermal growth factor, and IGFBP-3 that were detected in the nucleus and directly regulate transcription after internalization.

In this context, we proposed a model (Figure 3.15) in which the dual effect of CCN3 on proliferation is dependent upon the relative amount of secreted and intracellular isoforms. In the normal situation, the full length secreted CCN3 acts negatively on cell proliferation through its interactions with various receptors and ligands. In this conditions, the amount of truncated protein produced by posttranslational processing at internalization is antagonizing the inhibitory effects of CCN3 and can be used to trigger proliferation when needed. When the balance between full length and amino truncated CCN3 variants is disrupted, either at the level of posttranslational processing or through the abnormal production of a

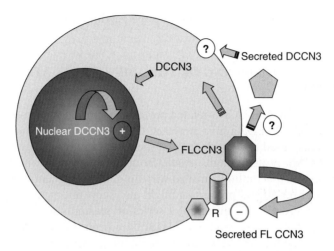

Figure 3.15. Schematic model for the dual activity of CCN3 proteins. The model shows the antiproliferative effects of the secreted full length CCN3 protein, which are transmitted through the binding to various receptors with which CCN3 was shown to interact. The aminotruncated CCN3 isoform generated by posttranslational proteolysis is thought to re-enter within the cell where it is addressed to the nucleus. The nuclear protein is shown to repress transcription and induce proliferation (Planque et al., submitted for publication).

truncated form (i.e., by viral insertion or any other genetic event), cells would respond to the constant positive effects of nuclear CCN3, a situation that provides the ground for the occurrence of further tumorigenic events.

A challenging question is to know whether this dual activity is also a feature of other CCN proteins for which variants have been described but are ignored by those who limit their attention to the properties of the secreted forms. The history of sciences is full of discoveries that have challenged our views and we should keep in mind these words from the Greek philosopher Heraclite: «the one who is seeking the Truth should be prepared to the unexpected, because it is difficult to find, and when one face it, it is often disconcerting.»

ACKNOWLEDGMENTS

The research performed in my laboratory has been funded by grants from Ligue Nationale contre le Cancer (Comité du Cher), Ministère de la Recherche et de l'Education Nationale, and the European PROTHETS project. I am grateful to Annick Perbal, Michel Le Rigoleur, and O. Julienne for help and financial support.

REFERENCES

1. Brigstock DR. The connective tissue growth factor/cysteine-rich 61/nephroblastoma overexpressed (CCN) family. *Endocr Rev* 20: 189–206 (1999).
2. Lau LF, Lam SC. The CCN family of angiogenic regulators, the integrin connection. *Exp Cell Res* 248: 44–57 (1999).

3. Perbal B. NOV (nephroblastoma overexpressed) and the CCN family of genes, structural and functional issues. *Mol Pathol* 54: 57–79 (2001).
4. Takigawa M. CTGF/Hcs24 as a multifunctional growth factor for fibroblasts, chondrocytes and vascular endothelial cells. *Drug News Perspect* 16(1): 11–21 (2003).
5. Brigstock DR, Goldschmeding R, Katsube KI, Lam SC, Lau LF, Lyons K, Naus C, Perbal B, Riser B, Takigawa M, Yeger H. Proposal for a unified CCN nomenclature. *Mol Pathol* 56(2): 127–128 (2003).
6. Perbal B, Takigawa M. *CCN Proteins. A new family of cell growth and differentiation regulators.* Imperial College Press, London, 2005.
7. Planque N, Perbal B. A structural approach to the role of CCN (CYR61/CTGF/NOV) proteins in tumourigenesis. *Cancer Cell Int* 3(1): 15 (2003).
8. Perbal B. CCN proteins: multifunctional signaling regulators. *Lancet* 363(9): 62–64 (2004).
9. Takigawa M, Nakanishi T, Kubota S, Nishida T. Role of CTGF/HCS24/ecogenin in skeletal growth control. *J Cell Physiol* 194(3): 256–266 (2003).
10. Brigstock DR. The CCN family: a new stimulus package. *J Endocrinol* 178(2): 169–175 (2003).
11. Gao R, Brigstock. Connective tissue growth factor (CCN2) induces adhesion of rat activated hepatic stellate cells by binding of its C-terminal domain to integrin alpha(v)beta(3) and heparin sulfate proteoglycan. *J Biol Chem* 279(10): 8848–8855. Epub 2003 Dec 17 (2004).
12. Nishida T, Nakanishi T, Shimo T, Asano M, Hattori T, Tamatani T, Tezuka K, Takigawa M. Demonstration of receptors specific for connective tissue growth factor on a human chondrocytic cell line (HCS-2/8). *Biochem Biophys Res Commun* 247: 905–909 (1998).
13. Gao R, Brigstock DR. Low density lipoprotein receptor-related protein (LRP) is a heparin-dependent adhesion receptor for connective tissue growth factor (CTGF) in rat activated hepatic stellate cells. *Hepatol Res* 27: 214–220 (2003).
14. Mercurio S, Latinkic B, Itasaki N, Krumlauf R, Smith JC. Connective-tissue growth factor modulates WNT signaling and interacts with the WNT receptor complex. *Development* 131: 2137–2147 (2004).
15. Segarini PR, Nesbitt JE, Li D, Hays LG, Yates JR 3rd, Carmichael DF. The low density lipoprotein receptor-related protein/alpha2-macroglobulin receptor is a receptor for connective tissue growth factor. *J Biol Chem* 276: 40659–40667 (2001).
16. Sakamoto K, Yamaguchi S, Ando R, Miyawaki A, Kabasawa Y, Takagi M, Li CL, Perbal B, Katsube K. The nephroblastoma overexpressed gene (NOV/ccn3) protein associates with Notch1 extracellular domain and inhibits myoblast differentiation via Notch signaling pathway. *J Biol Chem* 277: 29399–29405 (2002).
17. Gellhaus A, Dong X, Propson S, Maass K, Klein-Hitpass L, Kibschull M, Traub O, Willecke K, Perbal B, Lye SJ, Winterhager E. Connexin 43 interacts with NOV: a possible mechanism for negative regulation of cell growth in choriocarcinoma cells. *J Biol Chem* 279: 36931–36942 (2004).
18. Fu CT, Bechberger JF, Ozog MA, Perbal B, Naus CC. CCN3 (NOV) interacts with connexin43 in C6 glioma cells: possible mechanism of connexin-mediated growth suppression. *J Biol Chem* 279(35): 36943–36950. Epub 2004 Jun 21 (2004).
19. Li CL, Martinez V, He B, Lombet A, Perbal B. A role for CCN3 (NOV) in calcium signaling. *Mol Pathol* 55(4): 250–261 (2002).
20. Perbal B. Communication is the key. *Cell Commun Signal* 1(1): 3 (2003).
21. Kim HS, Nagalla SR, Oh Y, Wilson E, Roberts CT Jr, Rosenfeld RG. Identification of a family of low-affinity insulin-like growth factor binding proteins (IGFBPs): characterization of connective tissue growth factor as a member of the IGFBP superfamily. *Proc Natl Acad Sci USA* 94: 12981–12986 (1997).
22. Burren CP, Wilson EM, Hwa V, Oh Y, Rosenfeld RG. Binding properties and distribution of insulin-like growth factor binding protein-related protein 3 (IGFBP-rP3/NovH), an additional member of the IGFBP Superfamily. *J Clin Endocrinol Metab* 84: 1096–1103 (1999).
23. Chevalier G, Yeger H, Martinerie C, Laurent M, Alami J, Schofield PN, Perbal B. novH: differential expression in developing kidney and Wilm's tumors. *Am J Pathol* 152(6): 1563–1575 (1998).
24. Abreu JG, Ketpura NI, Reversade B, De Robertis EM. Connective-tissue growth factor (CTGF) modulates cell signaling by BMP and TGF-beta. *Nat Cell Biol* 4: 599–604 (2002).

25. Brigstock DR, Steffen CL, Kim GY, Vegunta RK, Diehl JR, Harding PA. Purification and characterization of novel heparin-binding growth factors in uterine-secretory fluids. *J Biol Chem* 272: 20275–20282 (1997).
26. Kireeva ML, Latinkic BV, Kolesnikova TV, Chen C-C, Yang GP, Abler AS, Lau LF. Cyr61 and Fisp12 are both signaling cell adhesion molecules: comparison of activities, metabolism, and localization during development. *Exp Cell Res* 233: 63–77 (1997).
27. Ball DK, Rachfal AW, Kemper SA, Brigstock DR. The heparin-binding 10 kDa fragment of connective tissue growth factor (CTGF) containing module 4 alone stimulates cell adhesion. *J Endocrinol* 176(2): R1–R7 (2003).
28. Joliot V, Martinerie C, Dambrine G, Plassiart G, Brisac M, Crochet J, Perbal B. Proviral rearrangements and overexpression of a new cellular gene (nov) in myeloblastosis-associated virus type 1-induced nephroblastomas. *Mol Cell Biol* 12(1): 10–21 (1992).
29. Perbal B, Lipsick JS, Svoboda J, Silva RF, Baluda MA. Biologically active proviral clone of myeloblastosis-associated virus type 1: implications for the genesis of avian myeloblastosis virus. *J Virol* 56(1): 240–244 (1985).
30. Soret J, Kryceve-Martinerie C, Crochet J, Perbal B. Transformation of Brown Leghorn chicken embryo fibroblasts by avian myeloblastosis virus proviral DNA. *J Virol* 55(1): 193–205 (1985).
31. Perbal B. Contribution of MAV-1-induced nephroblastoma to the study of genes involved in human Wilms' tumor development. *Crit Rev Oncog* 5(6): 589–613 (1994).
32. Li CL, Coullin P, Bernheim A, Joliot V, Auffray C, Zoroob R, Perbal B. Integration of myeloblastosis associated virus proviral sequences occurs in the vicinity of genes encoding signaling proteins and regulators of cell proliferation. *Cell Commun Signal* 4:1 (2006).
33. Pajer P, Pecenka V, Kralova J, Karafiat V, Prukova D, Zemanova Z, Kodet R, Dvorak M. Identification of potential human oncogenes by mapping the common viral integration sites in avian nephroblastoma. *Cancer Res* 66(1): 78–86 (2006).
34. Gupta N, Wang H, McLeod TL, Naus CC, Kyurkchiev S, Advani S, Yu J, Perbal B, Weichselbaum RR. Inhibition of glioma cell growth and tumorigenic potential by CCN3 (NOV). *Mol Pathol* 54(5): 293–299 (2001).
35. Benini S, Perbal B, Zambelli D, Colombo MP, Manara MC, Serra M, Parenza M, Martinez V, Picci P, Scotlandi K. In Ewing's sarcoma CCN3(NOV) inhibits proliferation while promoting migration and invasion of the same cell type. *Oncogene* 24(27): 4349–4361 (2005).
36. Lin CG, Leu SJ, Chen N, Tebeau CM, Lin SX, Yeung CY, Lau LF. CCN3 (NOV) is a novel angiogenic regulator of the CCN protein family. *J Biol Chem* 278(26): 24200–24208. Epub 2003 Apr 13 (2003).
37. Kyurkchiev S, Yeger H, Bleau AM, Perbal B. Potential cellular conformations of the CCN3(NOV) protein. *Cell Commun Signal* 2(1): 9 (2004).
38. Yang GP, Lau LF. Cyr61, product of a growth factor-inducible immediate early gene, is associated with the extracellular matrix and the cell surface. *Cell Growth Differ* 2(7): 351–357 (1991).
39. Brigstock DR, Steffen CL, Kim GY, Vegunta RK, Diehl JR, Harding PA. Purification and characterization of novel heparin-binding growth factors in uterine secretory fluids. Identification as heparin-regulated Mr 10,000 forms of connective tissue growth factor. *J Biol Chem* 272(32): 20275–20282 (1997).
40. Ball DK, Surveyor GA, Diehl JR, Steffen CL, Uzumcu M, Mirando MA, Brigstock DR. Characterization of 16- to 20-kilodalton (kDa) connective tissue growth factors (CTGFs) and demonstration of proteolytic activity for 38-kDa CTGF in pig uterine luminal flushings. *Biol Reprod* 59(4): 828–835 (1998).
41. Planque N, Long Li C, Saule S, Bleau AM, Perbal B. Related Articles, Links Nuclear addressing provides a clue for the transforming activity of amino-truncated CCN3 proteins. J Cell Biochem. 2006 Apr 5; [Epub ahead of print]
42. Wahab NA, Brinkman H, Mason RM. Uptake and intracellular transport of the connective tissue growth factor: a potential mode of action. *Biochem J* 359(Pt 1): 89–97 (2001).

CHAPTER 4

EWS-FLI1 IN EWING'S SARCOMA: REAL TARGETS AND COLLATERAL DAMAGE

JOZEF BAN, CHRISTINE SILIGAN, MICHAEL KREPPEL, DAVE ARYEE, AND HEINRICH KOVAR

Children's Cancer Research Institute, Kinderspitalgasse 9, A1190 Vienna, Austria

Abstract: Ewing's sarcoma family of tumors (ESFT) are a clinically and scientifically very demanding group of tumors in children and young adults with still unknown histogenesis. The rate-limiting oncogenic mutation in this disease has been identified as a chromosomal translocation, t(11;22)(q24;q12), that leads to the expression of a chimeric transcription factor, EWS-FLI1. We have studied the downstream pathway of EWS-FLI1 by a dual strategy including the isolation of direct target genes from ESFT chromatin and the monitoring of transcriptomic changes after silencing of EWS-FLI1 by RNA interference. This study has lead to the identification of several directly EWS-FLI1-regulated genes and the characterization of their genomic distribution. By comparing several ESFT cell lines, not only variation in overall gene expression patterns downstream of EWS-FLI1 was observed, but also differential regulation of directly EWS-FLI1-bound genes. Interestingly, there was variation between members of the same functional gene families. Studies on CD99, another diagnostic hallmark of ESFT, in relation to EWS-FLI1 provided additional evidence for context dependence of fusion protein function. Together, our study represents a first approach to the separation of essential molecular consequences from noise generated by the *EWS-FLI1* gene rearrangement in ESFT.

1. THE EWING'S SARCOMA FAMILY OF TUMORS: AN ENIGMA

Ewing's sarcoma (ES) is a challenging disease in many aspects. Although it predominantly arises in bone, there is no evidence for an osteoblastic histogenesis. It affects mainly children in puberty and young adults with a slight male preponderance but, so far, nothing is known about a hormonal component in the

Corresponding author: Heinrich Kovar, PhD. Phone: ++43 1 40470-4090; FAX: ++43 1 40470-7150; e-mail: heinrich.kovar@ccri.univie.ac.at

development of the disease. In the prechemotherapy era with surgery alone, almost all patients died from recurrence at distant sites due to a generally high metastatic potential of ES. Two thirds of patients present with localized disease and, with current multimodal treatment regimens, 60% of them can be made long-term survivors. In contrast, patients with clinically overt metastases at diagnosis have a dismal prognosis (<20% survival), which has almost not changed over the last 30 years. Thus, there appears to be an intrinsic biological difference between these two patient groups, which remains to be defined. In addition, the proportion of patients with localized but treatment-resistant disease has only slightly decreased over the last two decades, but significant progress has been achieved with respect to the prolongation of life in this group of patients. Consequently, while the number of early relapses has been successfully reduced in this cohort, an increasing number of late recurrences is observed, sometimes more than 5 or even 10 years after completion of therapy. However, despite strong efforts to identify prognostic markers for patients with localized disease, we are still unable to predict the course of the disease with confidence, and the nature of dormant tumor cells giving rise to late relapse is completely unknown.

Pathologists used to distinguish between less differentiated ES and peripheral primitive neuroectodermal tumor (pPNET) based on the absence or presence of scarce neuroglial marker expression, until the discovery in the early 1990s of two molecular traits common to both tumors; high CD99 expression[1] and a recurrent, highly specific gene rearrangement, *EWS-FLI1* or a related gene fusion.[2] Based on these common characteristics and since there appears to be no clinical difference between ES and pPNET with modern treatment protocols, they have been combined to form the ESFT.

2. EWS-FLI1, A CHIMERIC TRANSCRIPTION FACTOR WITH TISSUE-SPECIFIC TARGETS

The *EWS* gene aberration results from chromosomal translocation and comes in several flavors. 85% of ESFT show *EWS* rearrangements with *FLI1* with a high degree of genomic breakpoint variation resulting in the expression of chimeric EWS-FLI1 proteins with variable architecture. In the remaining cases of ESFT, *FLI1* is replaced by other members of the same gene family, all of them encoding ETS transcription factors, *ERG* in 10%, *E1AF*, *ETV1*, and *FEV* in less than 1% of cases, each.

The ES gene *EWS* encodes for an RNA polymerase II and, presumably, spliceosome-associated RNA-binding protein of largely unknown function. In the ESFT associated gene rearrangements, the COOH-terminal RNA-binding domain is replaced by the ETS transcription factor-derived DNA-binding domain, and early experimental data indicated that the resulting fusion protein has potent transcription factor activity itself.[3,4] It binds to GGAA/T consensus sequences through its ETS DNA-binding domain and, in vitro, drives reporter gene activity by the EWS NH_2-terminal domain. In response to ectopic EWS-FLI1 expression in mammalian cells, however, approximately equal numbers of genes are found to be repressed as activated.[5] Transcriptional repression by EWS-FLI1 cannot solely be attributed to indirect effects, since the *TGFβRII* promoter, e.g., has been convincingly demonstrated to be directly bound and suppressed by the fusion protein.[6] The mechanism of transcriptional suppression by EWS-ETS proteins, however, remains elusive.

In the classical NIH3T3 transformation assay, EWS-FLI1 is capable of inducing surface-independent growth[7] and, when transformed cells are transferred into nude mice, of giving rise to tumors with a small-cell morphology somehow resembling ES.[8] The tight association between *EWS-ETS* gene rearrangements and ESFT, and the potent oncogenic activity of the chimeric transcription factor EWS-FLI1 have led to the conclusion that the pathogenic properties of this protein are largely the consequence of altered transcription of EWS-FLI1 target genes. In order to identify these genes, several investigators have stably or inducibly expressed EWS-FLI1 in different cellular models and analysed associated changes in gene expression profiles. However, dependent on the model used, ectopic EWS-FLI1 elicited heterogeneous cellular responses ranging from cell cycle arrest or cell death, which was observed in the majority of primary cells, to transformation, and from blocked differentiation potential to trans-differentiation.[9,5,10,11,12] Consequently, the EWS-FLI1 imprint on the cells' transcriptomes varied between the models.[13] Several potential target genes have been isolated, mainly from murine cells, however, ESFT specific expression could not be confirmed but in rare cases.[14] These results may indicate that EWS-FLI1 function is cell type-specific and combinatorial.[15]

Since the tissue of origin for ESFT development is unknown, the second best choice to study EWS-FLI1 function in its authentic cellular context remains ESFT cells. Modern RNA interference-based technologies enable silencing of the fusion gene to study the consequences for ESFT cell biology and several studies have recently identified pathway components downstream of EWS-FLI1 in ESFT cell lines.[16,17,18]

3. IDENTIFYING A FUNCTIONAL HIERARCHY IN EWS-FLI1 ACTIVITY

So far, two methods of EWS-FLI1 silencing have been used: (1) transient transfection of synthetic small interfering (si)RNA and (2) stable expression of small hairpin (sh)RNA. Both approaches require incubation of cells with the silencing vector for a minimal period of at least 48 h and, therefore, do not allow to identify immediate early changes in gene expression associated with EWS-FLI1 knockdown. Thus, differential gene expression patterns between control and si/shRNA-transfected cells reflect intermediate endpoints characterizing the global consequences of switching off EWS-FLI1 in ESFT cells, but do not allow for discrimination between direct EWS-FLI1 targets and transcriptional consequences further downstream. In addition, EWS-FLI silencing results in growth arrest[17] and even cell death of ESFT cell lines,[16] as has also been demonstrated in earlier studies using antisense strategies,[19,20,21] making it difficult to discern between causes and consequences of cellular fate and altered gene expression profiles. Like with forced EWS-FLI1 expression in heterologous systems, silencing of EWS-FLI1 in ESFT cells results in the altered expression of hundreds of genes and it remains to be defined which of them are essential for tumorigenesis and which might be considered as "collateral damage".

As a first approach to solve this problem, we have established a dual target gene identification strategy (Figure 4.1). Using EWS-FLI1 specific antibodies, chromatin complexes containing endogenous EWS-FLI1 were isolated from cultured ESFT cells after chemical cross-linking, and genomic DNA bound to

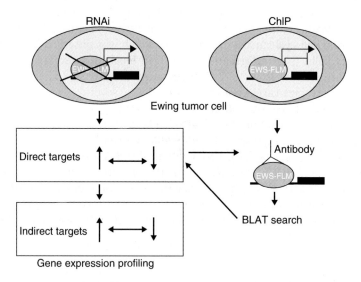

Figure 4.1. Strategy for the isolation of direct EWS-FLI1 target genes. Three different ESFT cell lines were treated with EWS-FLI1-specific shRNA expression vectors and subjected to comparative Affymetrix GeneChip-based gene expression analysis (left branch). Concomitantly, after chemical cross-linking, chromatin complexes were isolated from cultured ESFT cells using EWS- and FLI1-specific antibodies, genomic DNA was recovered from precipitates, directly cloned and sequenced, and assigned to the respective genomic regions from which they were derived by BLAST search (right branch). The expression of genes isolated by the chromatin precipitation approach was tested on Affymetrix arrays before and after EWS-FLI1 silencing to identify directly EWS-FLI1-regulated genes.[17]

the fusion protein was directly cloned. The expression of genes from which the identified directly EWS-FLI1-binding sequences were derived was then tested in several ESFT cell lines upon shRNA-mediated silencing of EWS-FLI1. The chromatin immunoprecipitation (ChIP) approach resulted in almost 40% gene-derived genomic sequences when EWS- and FLI1-specific antibodies were used as compared with about 10–20% in no antibody or irrelevant controls reminiscent of random precipitation (Figure 4.2). This twofold enrichment indicated that at least half of the genes precipitating with specific antibodies most likely represent true EWS-FLI1 targets. A total of 99 genes were identified among the cloned genomic fragments of which 33 were subjected to altered expression when EWS-FLI1 was silenced in the ESFT cell line from which they were derived.[17] Since only nine genes were hit more than once, we concluded that this list of putative direct EWS-FLI1 target genes was far from being complete.

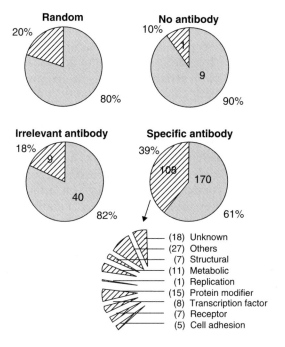

Figure 4.2. Chromatin immunoprecipitation results from the ESFT cell line STA-ET-7.2. The number of clones obtained from intergenic genomic regions (grey areas) and from genes (shaded areas) are indicated for precipitations performed with no antibody, an irrelevant antibody, and with EWS-FLI1-specific antibodies. The proportions of gene hits obtained as compared with the expected proportion in a random precipitation are indicated in percent of all clones obtained per assay. For EWS-FLI1-specific genes, functional annotations are separately shown.

4. LESSONS FROM THE COMBINED ChIP/shRNA STUDY OF DIRECT EWS-FLI1 TARGET GENES

Despite the only twofold enrichment in gene hits among ChIP clones when EWS- and FLI1-specific antibodies were used as compared with controls, several nonrandom observations support the specificity of our results: (1) there was a clear enrichment for genomic fragments from the upstream regions and, importantly, the first two introns of the cloned genes. Such a distribution is reminiscent of other transcription factors like p53. (2) The identified genes were not evenly distributed over the chromosomes but tended to cluster in certain genomic regions (specifically on chromosomes 3q and 9q) while intergenic clones followed a random distribution (Figure 4.3). Interestingly, there was also an overrepresentation of gene-derived ChIP clones from chromosome 8, which is frequently trisomic in ESFT, potentially contributing to increased gene dosage of EWS-FLI1 target genes. (3) With respect to a potential role of EWS-FLI1 in differentiation, it is noteworthy that more than 20% of genes identified by the ChIP approach have been reported to play a role in either neural differentiation or neural function. This finding supports results from EWS-FLI1 overexpression studies in NIH3T3 cells, human rhabdomyosarcoma, and neuroblastoma cell lines, where the fusion protein imposes features of a cholinergic neural expression pattern.[22,11,12] Together, and parallel to normal FLI1 function in the development of neural crest derived mesenchyme (reviewed in[15]), these data suggest that the partial neural differentiation of at least some ESFT is the consequence of the *EWS* gene rearrangement.

An important lesson for the biology of direct EWS-FLI1 target genes was learned from the extent of change after EWS-FLI1 silencing. In Affymetrix GeneChip analysis, approximately equal numbers of genes isolated from ESFT chromatin by EWS-FLI1 specific immunoprecipitation were found to be suppressed and activated, compatible with findings in heterologous EWS-FLI1 expression models. However, the majority of these putative directly EWS-FLI1-regulated genes showed a less than twofold induction or repression and did, therefore, not score among top EWS-FLI1-regulated genes on expression arrays. In fact, many of them might have been missed by isolated differential gene expression profiling since their level of change would have been considered below threshold. In contrast, top regulated genes after EWS-FLI1 silencing showed an up to 200-fold change in microarray analysis. Based on this finding, we hypothesize that EWS-FLI1 causes only modest alterations in gene dosage of several master regulatory genes, which, by a hierarchical signal amplification mechanism, result in pronounced transcriptomic changes further downstream.

5. VARIABILITY IN EWS-FLI1 TARGET GENE EXPRESSION IN ESFT

A recent RNAi study in the ESFT cell line A673 described EWS-FLI1 binding to and preferential induction of *IGFBP3* and convincingly demonstrated its role

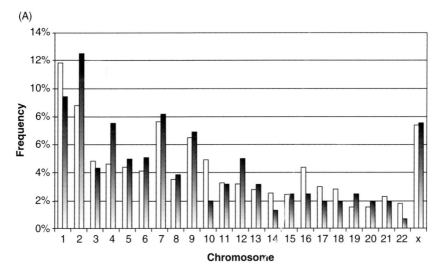

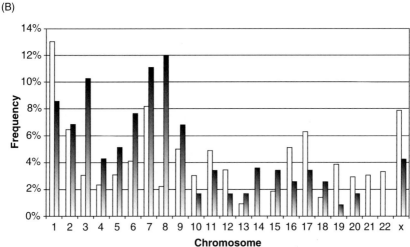

Figure 4.3. Chromosomal assignments of genomic fragments obtained by EWS-FLI1-specifc chromatin immunoprecipitation. White bars: expected distribution for random precipitation; grey bars: distribution obtained by precipitation of EWS-FLI1-containing chromatin complexes. (A) Distribution of EWS-FLI1-specific clones from intergenic regions, which largely followed the expected pattern of a random precipitation. (B) Distribution of EWS-FLI1-specific clones from genes, which showed a nonrandom distribution.

in ESFT apoptosis.[16] When studying several different ESFT cell lines with an EWS-FLI1-directed shRNA approach, we observed that the sets of genes affected by EWS-FLI1 suppression varied between cell lines. For *IGFBP* genes, several different family members were found to respond to EWS-FLI1 suppression in a cell line-specific way (Figure 4.4). Similarly, cyclin-dependent kinase

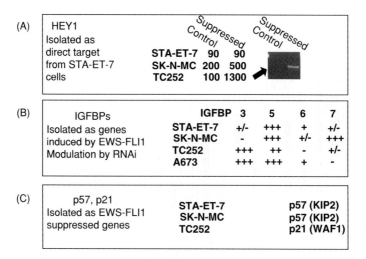

Figure 4.4. Suppression of EWS-FLI1 expression by RNA interference differentially affects the expression of functionally related gene family members in a cell line dependent way. Three examples are shown. (A) For the NOTCH pathway effector HEY1, Affymetrix gene expression values are shown for three different ESFT cell lines. Despite of binding of EWS-FLI1 to this gene in STA-ET-7.2 cells, it is neither expressed before nor after EWS-FLI1 silencing, while upon EWS-FLI1 knockdown there is moderate expression in SK-N-MC and strong induction in TC252 cells. This induction is also shown by RT-PCR. (B) For IGFBPs, relative induction upon EWS-FLI1 silencing is shown for STA-ET-7.2, SK-N-MC, and TC252 cells and compared with published data for A673.[16] (C) For cyclin-dependent kinase inhibitor genes p21 (WAF1) and p57 (KIP2) differential induction upon EWS-FLI1 silencing is shown for the cell lines STA-ET-7.2, SK-N-MC, TC252

inhibitor genes, which have previously been implicated in EWS-FLI1 function,[23,24] $CDKN1C^{p57/KIP2}$ and $CDKN1A^{p21/WAF1}$, were found to be alternatively activated in a cell line-dependent way. Interestingly, some genes from our ChIP-cloning approach, which were either not expressed at all or which remained unchanged when EWS-FLI1 was knocked down in the ESFT cell line, from which they were isolated, were significantly affected by the treatment in other ESFT cell lines. The NOTCH pathway gene *HEY1*, e.g., remained unexpressed in STA-ET-7.2 cells, the cell line used for ChIP, and was only weakly expressed in SK-N-MC, but was significantly induced upon EWS-FLI1 silencing in TC252 cells. Compatible with the findings for *IGFBP* and *CDKN1* family genes, other members of the *HEY* gene family replaced *HEY1* in ESFT cell lines VH64, WE68, and STA-ET-1 (Ban et al., manuscript in preparation). Closely related gene family members frequently serve similar functions but are expressed in either a tissue- or stage-specific manner during differentiation. It is, therefore, intriguing to speculate that ESFT originate from either different embryonal tissues or, more likely, from one tissue hit by the tumorigenic *EWS* gene rearrangement at different stages of differentiation. Further extending the panel of ESFT cell lines in EWS-FLI1 suppression studies and investigation of other gene families with developmentally regulated expression may uncover further biological heterogeneity including correlations with ES and pPNET phenotypes and possibly distinct clinical features.

6. COMMON PATHWAYS AFFECTED BY EWS-FLI1 IN ESFT

Pathway analysis of Affymetrix gene expression profiles in three ESFT cell lines treated in five independent experiments with three different EWS-FLI1-specific shRNAs revealed three major commonly affected pathways: (1) cell cycle regulation by CMYC and cyclin D1; (2) replication licensing; and (3) TGFβ signaling. With the exception of replication licensing, involvement of these pathways in ESFT biology has already previously been described.[3,23,25,26,27,6] Since CMYC is well-known to alter the expression of a high proportion of the human genome, its upregulation by EWS-FLI1, respectively its modulation by silencing of EWS-FLI1, may, at least in part, account for the multitude of genes affected by altered EWS-FLI1 expression. Are these three pathways sufficient to generate ESFT? Probably not. The majority of common EWS-FLI1 downstream genes cannot be assigned to these pathways and do not form large functionally related clusters. This is specifically true for the genes obtained by ChIP, which serve a multitude of functions with no apparent relation to each other. It will be a major challenge for the future to link these genes to downstream alterations in gene expression profiles in order to establish a hierarchical map of aberrant gene regulation initiated by the EWS-FLI1 fusion protein.

7. EWS-FLI1 AND CD99

The close association between the two diagnostic hallmarks of ESFT, the *EWS*-gene rearrangement, and constitutive high CD99 expression, raises the question whether the one is a consequence of the other. In fact, ectopic EWS-FLI1 expression in neuroblastoma and rhabdomyosarcoma cell lines has been shown to upregulate CD99 expression.[22,12] Interestingly however, knocking down EWS-FLI1 in several ESFT cell lines by RNA interference had no effect on the high CD99 expression level that is typical for ESFT ([16]and Ban et al., unpublished material). Since, among normal tissues, the highest CD99 levels are found in hematopoietic stem cells, CD99 positivity may be considered a hallmark of a stem cell-like ESFT progenitor, and forcing EWS-FLI1 expression in unrelated embryonal tumors may reactivate primitive stem cell features including CD99 expression. When testing primary ESFT samples on Affymetrix arrays, minor variability in high CD99 expression levels can be observed. Here, expression of the *KCMF1* gene shows an inverse correlation with CD99 RNA levels. Silencing of CD99 in several ESFT cell lines resulted in consistent induction of *KCMF1*[28] while there was no influence on this gene when EWS-FLI1 was switched-off, providing further evidence that EWS-FLI1 does not impact on the CD99 pathway in ESFT. However, silencing of CD99 influenced the in vitro growth properties of ESFT cells. Despite sustained

EWS-FLI1 expression, a prerequisite for continuous ESFT cell growth, ESFT cells showed significantly reduced clonogenicity in soft agar when CD99 was knocked down, resulting in less, smaller and more compact colonies than controls, suggesting that CD99 expression is a functionally important characteristic of ESFT supporting the growth promoting function of EWS-FLI1. In addition, it is involved in ESFT cell migration, at least in part by suppression of *KCMF1*.[28] Thus, CD99 acts as an important component of the cellular context that provides permissiveness to the oncogenic function of EWS-FLI1.

8. CONCLUSION

There is increasing evidence that EWS-FLI1 function is very sensitive to context-specific cellular factors,[15] as reflected by variation of gene expression profiles not only between different EWS-FLI1 transgenic cellular models, but also between different ESFT cell lines. In addition, there is a remarkable variability in the architecture of ESFT-associated EWS chimeric proteins. Despite this molecular diversity, there is much less variation in histopathological and clinical ESFT presentation and, so far, elucidation of the spectrum of molecular features of ESFT has not translated into a better prognostication of the disease. One of the many challenges of ESFT is to identify rate-limiting functional pathways, identify their essential components, which may vary between individual cases, and separate them from the noise of dysregulated gene expression that is generated as collateral damage by the activity of a mutated transcription factor, EWS-FLI1.

ACKNOWLEDGMENTS

This study was supported by grants from the Austrian Science Fund FWF (#16067-B04 and #18046-B12), the Austrian government (GEN-AU-Child, contract 200.071/2-VI/1/2001), and the European Commission (PROTHETS, contract LSHC-CT-2004-503036).

REFERENCES

1. H. Kovar, M. Dworzak, S. Strehl, E. Schnell, I.M. Ambros, P.F. Ambros, and H. Gadner. Overexpression of the pseudoautosomal gene MIC2 in Ewing's sarcoma and peripheral primitive neuroectodermal tumor. *Oncogene* 5: 1067–1070 (1990).
2. O. Delattre, J. Zucman, B. Plougastel, C. Desmaze, T. Melot, M. Peter, H. Kovar, I. Joubert, P. De Jong, G. Rouleau, et al. Gene fusion with an ETS DNA-binding domain caused by chromosome translocation in human tumours. *Arch Facial Plast Surg* 359: 162–165 (1992).
3. R.A. Bailly, R. Bosselut, J. Zucman, F. Cormier, O. Delattre, M. Roussel, G. Thomas, and J. Ghysdael. DNA-binding and transcriptional activation properties of the EWS-FLI-1 fusion protein resulting from the t(11;22) translocation in Ewing sarcoma. *Mol Cell Biol* 14: 3230–3241 (1994).
4. S.L. Lessnick, B.S. Braun, C.T. Denny, and W.A. May. Multiple domains mediate transformation by the Ewing's sarcoma EWS/FLI-1 fusion gene. *Oncogene* 10: 423–431 (1995).

5. B. Deneen and C.T. Denny. Loss of p16 pathways stabilizes EWS/FLI1 expression and complements EWS/FLI1 mediated transformation. *Oncogene* 20: 6731–6741 (2001).
6. K.B. Hahm, K. Cho, C. Lee, Y.H. Im, J. Chang, S.G. Choi, P.H. Sorensen, C.J. Thiele, and S.J. Kim. Repression of the gene encoding the TGF-beta type II receptor is a major target of the EWS-FLI1 oncoprotein. *Nat Genet* 23: 222–227 (1999).
7. W.A. May, M.L. Gishizky, S.L. Lessnick, L.B. Lunsford, B.C. Lewis, O. Delattre, J. Zucman, G. Thomas, and C.T. Denny. Ewing sarcoma 11;22 translocation produces a chimeric transcription factor that requires the DNA-binding domain encoded by FLI1 for transformation. *Proc Natl Acad Sci USA* 90: 5752–5756 (1993).
8. A.D. Thompson, M.A. Teitell, A. Arvand, and C.T. Denny. Divergent Ewing's sarcoma EWS/ETS fusions confer a common tumorigenic phenotype on NIH3T3 cells [In Process Citation]. *Oncogene* 18: 5506–5513 (1999).
9. S.L. Lessnick, C.S. Dacwag, and T.R. Golub. The Ewing's sarcoma oncoprotein EWS/FLI induces a p53-dependent growth arrest in primary human fibroblasts. *Cancer Cell* 1: 393–401 (2002).
10. E.C. Torchia, S. Jaishankar, and S.J. Baker. Ewing tumor fusion proteins block the differentiation of pluripotent marrow stromal cells. *Cancer Res* 63: 3464–3468 (2003).
11. M.A. Teitell, A.D. Thompson, P.H. Sorensen, H. Shimada, T.J. Triche, and C.T. Denny. EWS/ETS fusion genes induce epithelial and neuroectodermal differentiation in NIH 3T3 fibroblasts. *Lab Invest* 79: 1535–1543 (1999).
12. C.J. Rorie, V.D. Thomas, P. Chen, H.H. Pierce, J.P. O'Bryan, and B.E. Weissman. The Ews/Fli-1 fusion gene switches the differentiation program of neuroblastomas to Ewing sarcoma/peripheral primitive neuroectodermal tumors. *Cancer Res* 64: 1266–1277 (2004).
13. J.P. Zwerner, J. Guimbellot, and W.A. May. EWS/FLI function varies in different cellular backgrounds. *Exp Cell Res* 290: 414–419 (2003).
14. A. Arvand and C.T. Denny. Biology of EWS/ETS fusions in Ewing's family tumors. *Oncogene* 20: 5747–5754 (2001).
15. H. Kovar. Context matters: The hen or egg problem in Ewing's sarcoma. *Semin Cancer Biol* 15: 189–196 (2005).
16. A. Prieur, F. Tirode, P. Cohen, and O. Delattre. EWS/FLI-1 silencing and gene profiling of Ewing cells reveal downstream oncogenic pathways and a crucial role for repression of insulin-like growth factor binding protein 3. *Mol Cell Biol* 24: 7275–7283 (2004).
17. C. Siligan, J. Ban, R. Bachmaier, L. Spahn, M. Kreppel, K.L. Schaefer, C. Poremba, D.N. Aryee, and H. Kovar. EWS-FLI1 target genes recovered from Ewing's sarcoma chromatin. *Oncogene*. (2005).
18. S. Nozawa, T. Ohno, Y. Banno, T. Dohjima, K. Wakahara, D.G. Fan, and K. Shimizu. Inhibition of platelet-derived growth factor-induced cell growth signaling by a short interfering RNA for EWS-Fli1 via down-regulation of phospholipase D2 in Ewing sarcoma cells. *J Biol Chem* 280: 27544–27551 (2005).
19. H. Kovar, D.N. Aryee, G. Jug, C. Henockl, M. Schemper, O. Delattre, G. Thomas, and H. Gadner. EWS/FLI-1 antagonists induce growth inhibition of Ewing tumor cells in vitro. *Cell Growth Differ* 7: 429–437 (1996).
20. K. Tanaka, T. Iwakuma, K. Harimaya, H. Sato, and Y. Iwamoto. EWS-Fli1 antisense oligodeoxynucleotide inhibits proliferation of human Ewing's sarcoma and primitive neuroectodermal tumor cells. *J Clin Invest* 99: 239–247 (1997).
21. J.A. Toretsky, Y. Connell, L. Neckers, and N.K. Bhat. Inhibition of EWS-FLI-1 fusion protein with antisense oligodeoxynucleotides. *J Neurooncol* 31: 9–16 (1997).
22. S. Hu-Lieskovan, J. Zhang, L. Wu, H. Shimada, D.E. Schofield, and T.J. Triche. EWS-FLI1 fusion protein up-regulates critical genes in neural crest development and is responsible for the observed phenotype of Ewing's family of tumors. *Cancer Res* 65: 4633–4644 (2005).
23. L. Dauphinot, C. De Oliveira, T. Melot, N. Sevenet, V. Thomas, B.E. Weissman, and O. Delattre. Analysis of the expression of cell cycle regulators in Ewing cell lines: EWS-FLI-1 modulates p57KIP2 and c-Myc expression. *Oncogene* 20: 3258–3265 (2001).

24. F. Nakatani, K. Tanaka, R. Sakimura, Y. Matsumoto, T. Matsunobu, X. Li, M. Hanada, T. Okada, and Y. Iwamoto. Identification of p21WAF1/CIP1 as a direct target of EWS-Fli1 oncogenic fusion protein. *J Biol Chem* 278: 15105–15115 (2003).
25. Y. Matsumoto, K. Tanaka, F. Nakatani, T. Matsunobu, S. Matsuda, and Y. Iwamoto. Downregulation and forced expression of EWS-Fli1 fusion gene results in changes in the expression of G(1)regulatory genes. *Br J Cancer* 84: 768–775 (2001).
26. H. Nishimori, Y. Sasaki, K. Yoshida, H. Irifune, H. Zembutsu, T. Tanaka, T. Aoyama, T. Hosaka, S. Kawaguchi, T. Wada, J. Hata, J. Toguchida, Y. Nakamura, and T. Tokino. The Id2 gene is a novel target of transcriptional activation by EWS-ETS fusion proteins in Ewing family tumors. *Oncogene* 21: 8302–8309 (2002).
27. J. Zhang, S. Hu, D.E. Schofield, P.H. Sorensen, and T.J. Triche. Selective usage of D-Type cyclins by Ewing's tumors and rhabdomyosarcomas. *Cancer Res* 64: 6026–6034 (2004).
28. M. Kreppel, D.N. Aryee, K.L. Schaefer, G. Amann, R. Kofler, C. Poremba, and H. Kovar. Suppression of KCMF1 by constitutive high CD99 expression is involved in the migratory ability of Ewing's sarcoma cells. *Oncogene*. (2005).

CHAPTER 5

MOLECULAR KARYOTYPING IN SARCOMA DIAGNOSTICS AND RESEARCH

HANNA VAUHKONEN, SUVI SAVOLA, SIPPY KAUR, MARCELO L. LARRAMENDY, AND SAKARI KNUUTILA

Department of Pathology, Haartman Institute and HUSLAB, University of Helsinki and Helsinki University Hospital, Helsinki, Finland

Abstract: Conventional cytogenetic and molecular genetic studies have both clinical and biological significance in sarcomas. However, the resolution of these methods does not always suffice to screening of novel, specific genetic changes, such as small deletions, amplifications, and fusion genes. Tumor-specific chromosomal translocations revealed by cytogenetic and molecular methods play a decisive role in the differential diagnosis of sarcomas. The novel molecular karyotyping techniques have proven to be powerful in the screening of clinically and biologically relevant molecular changes in human neoplasias. A variety of platforms for molecular karyotyping is available, e.g., arrayed cDNA clones or oligonucleotides that can be used in microarray-based comparative genomic hybridization (CGH) and gene expression analysis. We review here the clinically most relevant cytogenetic and molecular changes in sarcomas and describe latest microarray techniques for screening of clinically relevant gene copy number and expression changes.

1. SARCOMAS AND GENETIC ALTERATIONS

Bone and soft tissue sarcomas are a heterogeneous group of rare mesenchymal tumors comprising only 1% of all human neoplasms. In most sarcoma types, the cell of origin remains undefined, but there is some evidence that tumor cells in Ewing's sarcoma are derived from neural crest cells [1]. Osteosarcoma, chondrosarcoma, and Ewing's sarcoma as the three most common forms of bone sarcoma, and the most prevalent soft tissue sarcomas are malignant fibrous histiocytoma, liposarcoma, leiomyosarcoma, and synovial sarcoma [2,3]. The

*Correspondence: SK or HV, Department of Pathology, Haartman Institute, POB 21 (Haartmaninkatu 3), FI-00014 University of Helsinki, Helsinki, Finland. Fax: +358-9-19126788; E-mail: hanna.vauhkonen@helsinki.fi

patients' age distribution varies according to the histological type of the tumor, but bone tumors frequently affect teenagers and young adults, whereas soft tissue sarcomas usually show a gradual increase of incidence with advancing age [3]. Although the etiology and pathogenesis of specific sarcomas are still largely unknown, in recent years considerable progress has been made in the genetic typing of solid tumors, including sarcomas. Based on cytogenetics, the sarcomas can be divided into two groups: (1) cancers with translocations (e.g., Ewing's sarcoma, desmoplastic round cell tumor, alveolar rhabdomyosarcoma, and synovial sarcoma) and (2) cancers with complex karyotypes without any uniform translocations (e.g., osteosarcoma, malignant fibrous histiocytoma, leiomyosarcoma, and chondrosarcoma). Small round cell tumors include histologically indistinguishable sarcomas, neuroblastomas, and lymphomas, which makes differential diagnosis difficult (for review, see Tarkkanen and Knuutila [4]). To ensure proper prognostic and therapeutic assessment, early and accurate diagnosis can be aided using cytogenetic and molecular methods.

Specific chromosomal translocations have proven to be useful in diagnostics and monitoring residual disease [4]. By coupling a transcriptional activation domain with a DNA-binding domain, the translocation results often in creation of novel fusion protein with oncogenic properties. Table 5.1 shows a summary of selected sarcoma types and their characteristic translocations. A classical example in the Ewing family of tumors is the t(11;22)(q24;q12) translocation, which leads to production of the EWS-FLI1 fusion protein [5]. This translocation is seen in about 85% of Ewing's sarcoma patients while the remaining 15% display other nonrandom gene rearrangements with *EWS* (for review see Burchill [6]), as shown in Table 5.1. Gains of chromosomes 8, 12, and 1q are detected secondary aberrations in Ewing tumors and a deletion at 9p21 (encompassing the *p14* and *p16* genes) has been suggested to be a marker for poor prognosis [7,8]. Although the role of secondary aberrations is not fully understood, they may have a role in tumor progression.

The sarcomas with complex karyotypes show numerous chromosomal aberrations, e.g., translocations, inversions, gains, losses, and amplifications of genetic material [9–11]. This limits the potential of conventional cytogenetic analysis, especially when mitotic cells are scarce and yield poor chromosome banding and morphology. Many specific aberrations can be detected by molecular methods, i.e., fluorescence and chromogenic in situ hybridization (FISH and CISH), PCR, CGH, and newly developed microarray-based techniques [4,12].

Clinical and diagnostic features often fail to classify the tumor in time. Rapid diagnosis is particularly crucial for sarcoma patients whose treatment strategy to eradicate the disease should be determined as the optimal combination of chemo-and/or radiotherapy together with extensive surgery. Detection of residual disease and distinct prognostic markers for recurrence or metastasis are also of clinical importance. Today novel molecular karyotyping methods can be used for high-resolution genome-wide screening for molecular signatures that assist us in tumor classification and offer new reliable prognostic parameters in

Table 5.1. Cytogenetic and molecular genetic features of selected sarcomas

Bone and soft tissue tumors	Translocation	Gene fusion	Other cytogenetic characteristics
Alveolar soft part sarcoma	t(X;17)(p11;q25)	*ASPL/TFE3*	–
Clear cell sarcoma	t(12;22)(q13;q12)	*EWS/ATF1*	+7, +8
Congenital fibrosarcoma/ Infantile fibrosarcoma	t(12;15)(p13;q25)	*ETV6/NTRK3*	+8, +11, +17, +20
Desmoplastic small round cell tumor	t(11;22)(p13;q12)	*EWS/WT1*	–
Dermatofibrosarcoma protuberans	t(17;22)(q22;q13)	*COILA1/PDGFB*	Amp (17q,22q)
Extraskeletal myxoid chondrosarcoma	t(9;22)(q22;q12) t(9;17)(q22;q11)	*NR4A3/EWS* *NR4A3/RBP56*	– –
Ewing sarcoma/PNET	t(11;22)(q24;q12) main	*FLI1/EWSR1*	+1q, +8, +12
	t(21;22)(q22;q12)	*ERG/EWSR1*	–
	t(7;22)(p22;q12)	*ETV1/EWSR1*	–
	t(17;22)(q21;q12)	*ETV4/EWSR1*	–
	t(2;22)(q33;q12)	*FEV/EWSR1*	–
Lipoma	t(3;12)(q28;q14)	*HMGIC/LPP*	Changes in 12q13–15, 6p and 13q
Liposarcoma, myxoid	t(12;16)(q13;p11) t(12;22)(q13;12)	*FUS/DDIT3* *EWS/DDIT3*	– –
Rhabdomyosarcoma, alveolar	t(2;13)(q35;q14) t(1;13)(p36;q14)	*PAX3/FKHR* *PAX7/FKHR*	12q13–15 amplification, MYCN amplification
Synovial sarcoma	t(X;18)(p11;q11) – –	*SS18/SSX1* *SS18/SSX2* *SS18/SSX4*	–3, +7, +8, +12 – –

diagnosis. In the following sections we will briefly describe the microarray-based molecular karyotyping methods and discuss its clinical potentials.

2. METHODOLOGICAL CONSIDERATIONS

2.1. Cytogenetics and molecular genetics

In addition to chromosomal translocations, alterations in DNA copy numbers are important in the progression of carcinogenesis, as reviewed by Knuutila [12]. The activation of cellular proto-oncogenes may occur by increase in the copy number of the corresponding genomic region, and the inactivation of tumor suppressor genes may occur by copy number decrease. CGH is a powerful method in genome-wide detection of copy number changes, e.g., gains, losses, and amplifications [13]. Differentially labeled tumor and normal tissue DNAs are hybridized competitively on a metaphase spread, and based on the tumor tissue copy numbers in relation to normal tissue, specific chromosomal regions appear as gained or lost. The tissue used for the analysis may be fresh or fresh-frozen tumor, or even paraffin-embedded fixed tumor material [14]. An illustration of the CGH methods is

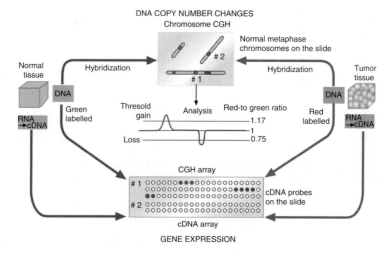

Figure 5.1. Comparative genomic hybridization (CGH) and gene expression analysis. DNA or RNA is extracted from tumor tissue and labeled with a fluorescent dye. Before labeling RNA can be converted to cDNA. A reference DNA or RNA with normal DNA copy number or expression level is labeled with another dye. The samples are combined and hybridized on a metaphase spread (e.g., chromosomal CGH) or on arrayed cDNA clones (e.g., array CGH, expression array). The resulting images are analyzed by compatible software, which shows over- and underrepresentation of tumor DNA or RNA as compared with the control sample. Thus, the results can be interpreted as gains or losses of genomic regions (DNA analysis) or as over- or underexpression of particular genes (RNA analysis).

shown in Figure 5.1. A comprehensive data compilation of copy number changes is publicly accessible at www.helsinki.fi/cmg/cgh_data.html [15]. Figure 5.2 shows recurrent amplicons seen in all sarcomas.

Given that the resolution of conventional chromosomal CGH is at the level of chromosomal bands (10–20 Mb), the method is not sufficient to detect specific amplification or deletion targets at gene level. Therefore microarray methods have been developed using BAC or cDNA-based targets [16,17]. In cDNA-based platforms, cDNA clones arrayed on a glass slide yield genome-wide gene-level information, as exemplified by our osteosarcoma array results [18]. Figure 5.3 shows a genome-wide copy number profile of an osteosarcoma sample, analyzed using cDNA array CGH. The latest technological advance arrays 30–80-mer oligonucleotides representing different exons of a specific gene or intergenic regions on a single platform [19,20]. The oligonucleotide-based CGH arrays give an average resolution of 35 kb (Human Genome CGH microarray Kit 44B, Agilent, Palo Alto, CA), which reveals aberrations not visible by other techniques, e.g., small deletions or amplifications. As shown in Figure 5.4, a sample with *MLL-ARG* fusion indicates that also apparently balanced translocations may harbor cryptic deletions [21].

The CGH techniques have certain limits, e.g., balanced translocations cannot be detected since the copy number levels are not altered. It has been argued

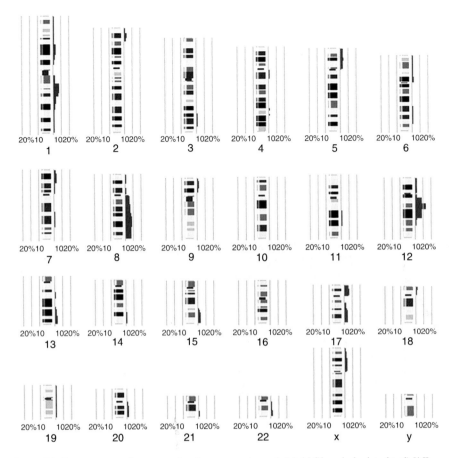

Figure 5.2. Recurrent amplicons seen in all sarcomas (www.helsinki.fi/cmg/cgh_data.html) [15].

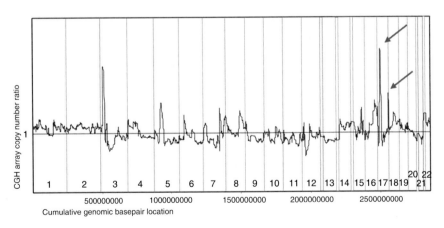

Figure 5.3. Example of a cDNA array CGH result of an osteosarcoma sample [18]. The cumulative base pair locations of the genome (1pter-Xqter) are indicated on the *X*-axis and the relative fluorescence ratios (tumor vs normal tissue) on the *Y*-axis. In addition to the well-known amplification at 17p11–13 (*arrow*), a novel region of amplification was found at 17q25 (*arrow*).

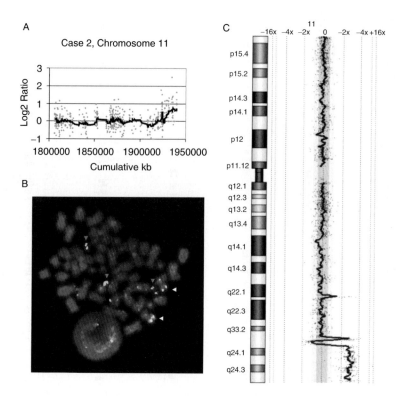

Figure 5.4. A deletion observed within a fusion gene in leukemia [21]. Panel A shows the results obtained using cDNA arrays and panel B shows the presence of the fusion gene with FISH (white arrowheads). The results shown in Panel C are produced using CGH on oligoarray platforms. The 1.8-Mb deletion containing approximately 30 genes (*arrow*) was not detected in conventional CGH or cDNA aCGH. The result is a fusion gene in chromosome 11q (*MLL-LARG*), which is thereafter amplified. The oligoarray results in Panel C are shown as Log2 ratios of fluorescence intensities with the corresponding genomic location in the ideogram. The losses are shown on the left side (negative Log2 values) and gains on the right side of the X-axis (positive Log2 values).

that CGH methods are less useful in detecting primary events, for instance translocations in Ewing's sarcoma. Instead, CGH is capable of revealing gains and losses, which are thought to be secondary events [7]. Whether apparently balanced translocations in sarcomas contain small regions of deletion or small amplifications has yet to be shown. But, the secondary changes are valuable in the prognostification of the disease. Applied to sarcoma research, the oligonucleotide-based platform will provide novel information of small copy number gains and losses not detected by other methods and it will possibly help to identify novel fusion genes that can be generated by minor aberrations, e.g., small deletions.

2.2. Gene expression analysis

The gene expression of a tissue sample is assessed by analyzing its mRNA content on arrays [22,23] (Figure 5.1), using either cDNA or oligonucleotide platforms. The observation of up- or downregulated genes in tumors supports diagnosis and prognosis and helps to specify targets for drug development. Furthermore, molecular signatures of sarcomas, which have been obtained using gene expression profiling, can be applied in diagnosis, prognosis, and treatment of these malignancies. Systems biology approach, integration of genome-wide gene expression data (coregulated transcripts) with gene function annotations (e.g., GO, KEGG, and Biocarta databases) and genomic sequence (regulatory DNA elements), provides knowledge of functional networks and activated signal transduction pathways in sarcomas. Holistic inferences of gene functionality and gene expression regulating machinery are essential in understanding carcinogenesis of sarcomas at molecular level. A variety of sarcomas have been subjected to expression analysis, e.g., osteosarcoma, synovial sarcoma, clear cell sarcoma, chondrosarcoma, and liposarcoma [24–29]. The main aim of these works has been to obtain gene expression signatures and novel marker genes for specific tumor types, aiding their separation from other solid tumors with similar appearance [30,31]. As an example of studies with clinical aims, expression profile-based clustering of Ewing's sarcomas [32] and leiomyosarcomas [33] has distinguished low-risk patients from patients with high risk of relapse or metastasis.

2.3. Laboratory practice

Preparation of samples requires special care because mRNA degrades at a rapid rate. Accordingly, expression analysis is generally not applicable on paraffin-embedded material. It should be further emphasized that expression analysis covers only a cross-section of the mRNA content at the time point of the tumor excision. In addition, rapidly degrading mRNA molecules, which can also be up- or downregulated, may not necessarily show the actual state of the transcriptome in the tumor. Besides, the heterogeneity of the tumor sample including, differential cancer cell clones, contaminating normal tissue or inflammatory cells may interfere with interpretation [34]. Still, the gene expression profiles of different macroscopic parts of a tumor have been shown to cluster together and to give a distinct profile to individual tumors [35].

2.4. Tissue microarrays

Tissue microarrays can be used to validate the clinical significance of targets found in aCGH and expression array studies [36]. A tissue microarray consists of a large number of tissue samples with known histology and patient history. The samples are embedded in a single paraffin block, of which sections are sliced to represent all of the samples. Then FISH, CISH, or immunohistochemistry is performed on the tissue sections to facilitate the association of molecular karyotyping results with the clinical parameters and the validation of specific targets [30]. The presence of translocations can be tested using breakapart probe-FISH [37,38], thus avoiding the need for immunohistochemistry or PCR. Engellau et al. [39] used 140 arrayed soft tissue sarcoma samples with different clinicopathological characteristics and found that immunopositivity of Ki-67, β-catenin, CD44, and Pgp has prognostic value for tumor metastasis. In another example, the prognostic significance of a novel amplicon in breast cancer was validated using 262 arrayed tissue samples [38]. The amplicon was found to reside close to a rearranged gene and to co-occur in the same samples with the rearrangement. Furthermore, the novel amplicon was confirmed to be associated with poor prognosis using FISH and breakapart probe-FISH on the tissue microarray [38]. Similarly, the results obtained from DNA copy number and expression studies of sarcoma samples could be tested on a tissue microarray with large sample populations to evaluate the clinical significance of the findings.

3. CLINICAL APPLICATIONS

Sarcomas are divided into prognostic subtypes according to histopathological features (e.g., tumor size, location, and number of mitotic cells). Specific translocations are determined by cytogenetic analysis, fusion gene transcripts by RT-PCR, and chromosomal gains or losses by FISH or conventional CGH. Table 5.1 gives an overview of sarcoma-specific translocations. However, these methods are not accurate enough to give sufficient gene-level information for diagnostic purposes, classification of the tumors, or prognostification. The molecular profiles generated by array-based methods have provided predictive markers for treatment response in breast cancer [40–42], molecular classification of tumors in lung adenocarcinomas [43] and diffuse large B-cell lymphomas [44], and markers for resistance of treatment in pediatric osteosarcomas [29]. The methodology enables also the detection of novel fusion genes [21], which may be useful in cancer diagnostics, and the detection of residual disease after primary tumor eradication [45]. However, further clinical trials should be conducted to validate these findings.

4. CONCLUSIONS AND FUTURE PROSPECTS

Molecular karyotyping methods include genome-wide techniques to assess the DNA copy-number and gene expression changes in tumor samples. The results provide distinct molecular signatures or single targets, which can be used for tumor classification, prognosis, and prediction of treatment response and occurrence of residual disease. Gene expression microarrays and systems biology tools can be used to identify novel drug targets, genes, which are actively transcribed in sarcomas, and signaling proteins and transcription factors, which promote sarcomagenesis.

REFERENCES

1. MS. Staege, C. Hutter, I. Neumann, S. Foja, UE. Hattenhorst, G. Hansen, D. Afar and SE. Burdach. DNA microarrays reveal relationship of Ewing family tumors to both endothelial and fetal neural crest-derived cells and define novel targets. *Cancer Res* 64: 8213–8321 (2004).
2. HC. Bauer, TA. Alvegard, O. Berlin, M. Erlanson, P. Gustafson, A. Kivioja, R. Klepp, T. Lehtinen, P. Lindholm, TR. Moller, A. Rydholm, G. Saeter, CS. Trovik, O. Wahlstrom and T. Wiklund. The Scandinavian Sarcoma Group Register. *Acta Orthop Scand Supp.* 285: 41–44 (1999).
3. CE. Herzog. Overview of sarcomas in the adolescent and young adult population. *J Pediatr Hematol Oncol* 27: 215–218 (2005).
4. M. Tarkkanen and S. Knuutila. The diagnostic use of cytogenetic and molecular genetic techniques in the assessment of small round cell tumours. *Current Diagnostic Pathology* 8: 338–348 (2002).
5. O. Delattre, J. Zucman, B. Plougastel, C. Desmaze, T. Melot, M. Peter, H. Kovar, I. Joubert, P. de Jong, G. Rouleau, et al. Gene fusion with an ETS DNA-binding domain caused by chromosome translocation in human tumours. *Nature* 359: 162–165 (1992).
6. SA. Burchill. Ewing's sarcoma: diagnostic, prognostic, and therapeutic implications of molecular abnormalities. *J Clin Pathol* 56: 96–102 (2003).
7. CM. Hattinger, U. Potschger, M. Tarkkanen, J. Squire, M. Zielenska, S. Kiuru-Kuhlefelt, L. Kager, P. Thorner, S. Knuutila, FK. Niggli, PF. Ambros, H. Gadner and DR. Betts. Prognostic impact of chromosomal aberrations in Ewing tumours. *Br J Cancer* 86: 1763–1769 (2002).
8. G. Wei, CR. Antonescu, E. de Alava, D. Leung, AG. Huvos, PA. Meyers, JH. Healey and M. Ladanyi. Prognostic impact of INK4A deletion in Ewing sarcoma. *Cancer* 89: 793–799 (2000).
9. M. Tarkkanen, A. Kaipainen, E. Karaharju, T. Bohling, J. Szymanska, H. Helio, A. Kivioja, I. Elomaa and S. Knuutila. Cytogenetic study of 249 consecutive patients examined for a bone tumor. *Cancer Genet Cytogenet* 68: 1–21 (1993).
10. J. Szymanska, M. Tarkkanen, T. Wiklund, M. Virolainen, C. Blomqvist, S. Asko-Seljavaara, E. Tukiainen, I. Elomaa and S. Knuutila. Gains and losses of DNA sequences in liposarcomas evaluated by comparative genomic hybridization. *Genes Chromosomes Cancer* 15: 89–94 (1996).
11. S. Kiuru-Kuhlefelt, W. El-Rifai, M. Sarlomo-Rikala, S. Knuutila and M. Miettinen. DNA copy number changes in alveolar soft part sarcoma: a comparative genomic hybridization study. *Mod Pathol* 11: 227–231 (1998).
12. S. Knuutila. Cytogenetics and molecular pathology in cancer diagnostics. *Ann Med.* 36: 162–171 (2004).
13. A. Kallioniemi, OP. Kallioniemi, D. Sudar, D. Rutovitz, JW. Gray, F. Waldman and D. Pinkel. Comparative genomic hybridization for molecular cytogenetic analysis of solid tumors. *Science* 258: 818–821 (1992).
14. M. Zielenska, P. Marrano, P. Thorner, J. Pei, B. Beheshti, M. Ho, J. Bayani, Y. Liu, BC. Sun, JA. Squire and XS. Hao. High-resolution cDNA microarray CGH mapping of genomic imbalances in osteosarcoma using formalin-fixed paraffin-embedded tissue. *Cytogenet Genome Res* 107: 77–82 (2004).
15. S. Knuutila, K. Autio and Y. Aalto. Online access to CGH data of DNA sequence copy number changes. *Am J Pathol* 157: 689 (2000).
16. JR. Pollack, CM. Perou, AA. Alizadeh, MB. Eisen, A. Pergamenschikov, CF. Williams, SS. Jeffrey, D. Botstein and PO. Brown. Genome-wide analysis of DNA copy-number changes using cDNA microarrays. *Nat Genet* 23: 41–46 (1999).
17. D. Pinkel, R. Segraves, D. Sudar, S. Clark, I. Poole, D. Kowbel, C. Collins, WL. Kuo, C. Chen, Y. Zhai, SH. Dairkee, BM. Ljung, JW. Gray and DG. Albertson. High resolution analysis of

DNA copy number variation using comparative genomic hybridization to microarrays. *Nat Genet* 20: 207–211 (1998).
18. J. Atiye, M. Wolf, S. Kaur, O. Monni, T. Böhling, A. Kivioja, É. Tas, M. Serra, M. Tarkkanen and S. Knuutila. Gene amplifications in osteosarcoma — CGH microarray analysis. *Genes Chromosomes Cancer* 42: 158–163 (2005).
19. C. Brennan, Y. Zhang, C. Leo, B. Feng, C. Cauwels, AJ. Aguirre, M. Kim, A. Protopopov and L. Chin. High-resolution global profiling of genomic alterations with long oligonucleotide microarray. *Cancer Res* 64: 4744–4748 (2004).
20. AJ. Aguirre, C. Brennan, G. Bailey, R. Sinha, B. Feng, C. Leo, Y. Zhang, J. Zhang, JD. Gans, N. Bardeesy, C. Cauwels, C. Cordon-Cardo, MS. Redston, RA. DePinho and L. Chin. High-resolution characterization of the pancreatic adenocarcinoma genome. *Proc Natl Acad Sci USA* 101: 9067–9072 (2004).
21. A. Tyybäkinoja, U. Saarinen-Pihkala, E. Elonen and S. Knuutila. Amplified, lost, and fused genes in 11q23-25 amplicon in acute myeloid leukemia -array-CGH study. *Genes Chromosomes Cancer*, 45: 257–264 (2006).
22. M. Schena, D. Shalon, R. Heller, A. Chai, PO. Brown and RW. Davis. Parallel human genome analysis: microarray-based expression monitoring of 1000 genes. *Proc Natl Acad Sci USA* 93: 10614–10619 (1996).
23. M. Schena, D. Shalon, RW. Davis and PO. Brown. Quantitative monitoring of gene expression patterns with a complementary DNA microarray. *Science* 270: 467–470 (1995).
24. S. Nagayama, T. Katagiri, T. Tsunoda, T. Hosaka, Y. Nakashima, N. Araki, K. Kusuzaki, T. Nakayama, T. Tsuboyama, T. Nakamura, M. Imamura, Y. Nakamura and J. Toguchida. Genome-wide analysis of gene expression in synovial sarcomas using a cDNA microarray. *Cancer Res* 62: 5859–5866 (2002).
25. KL. Schaefer, K. Brachwitz, DH. Wai, Y. Braun, R. Diallo, E. Korsching, M. Eisenacher, R. Voss, F. Van Valen, C. Baer, B. Selle, L. Spahn, SK. Liao, KA. Lee, PC. Hogendoorn, G. Reifenberger, HE. Gabbert and C. Poremba. Expression profiling of t(12;22) positive clear cell sarcoma of soft tissue cell lines reveals characteristic up-regulation of potential new marker genes including ERBB3. *Cancer Res* 64: 3395–3405 (2004).
26. KM. Skubitz, EY. Cheng, DR. Clohisy, RC. Thompson and AP. Skubitz. Differential gene expression in liposarcoma, lipoma, and adipose tissue. *Cancer Invest* 23: 105–118 (2005).
27. S. Subramanian, RB. West, RJ. Marinelli, TO. Nielsen, BP. Rubin, JR. Goldblum, RM. Patel, S. Zhu, K. Montgomery, TL. Ng, CL. Corless, MC. Heinrich and M. van de Rijn. The gene expression profile of extraskeletal myxoid chondrosarcoma. *J Pathol* 206: 433–444 (2005).
28. CM. Schorle, F. Finger, A. Zien, JA. Block, PM. Gebhard and T. Aigner. Phenotypic characterization of chondrosarcoma-derived cell lines. *Cancer Lett* 226: 143–54 (2005).
29. MB. Mintz, R. Sowers, KM. Brown, SC. Hilmer, B. Mazza, AG. Huvos, PA. Meyers, B. Lafleur, WS. McDonough, MM. Henry, KE. Ramsey, CR. Antonescu, W. Chen, JH. Healey, A. Daluski, ME. Berens, TJ. Macdonald, R. Gorlick and DA. Stephan. An expression signature classifies chemotherapy-resistant pediatric osteosarcoma. *Cancer Res* 65: 1748–1754 (2005).
30. TO. Nielsen, RB. West, SC. Linn, O. Alter, MA. Knowling, JX. O'Connell, S. Zhu, M. Fero, G. Sherlock, JR. Pollack, PO. Brown, D. Botstein and M. van de Rijn. Molecular characterisation of soft tissue tumours: a gene expression study. *Lancet* 359: 1301–1307 (2002).
31. M. van de Rijn and BP. Rubin. Gene expression studies on soft tissue tumors. *Am J Pathol* 161: 1531–1534 (2002).
32. A. Ohali, S. Avigad, R. Zaizov, R. Ophir, S. Horn-Saban, IJ. Cohen, I. Meller, Y. Kollender, J. Issakov and I. Yaniv. Prediction of high risk Ewing's sarcoma by gene expression profiling. *Oncogene* 23: 8997–9006 (2004).
33. YF. Lee, M. John, A. Falconer, S. Edwards, J. Clark, P. Flohr, T. Roe, R. Wang, J. Shipley, RJ. Grimer, DC. Mangham, JM. Thomas, C. Fisher, I. Judson and CS. Cooper. A gene expression signature associated with metastatic outcome in human leiomyosarcomas. *Cancer Res* 64: 7201–7204 (2004).

34. M. O'Sullivan, V. Budhraja, Y. Sadovsky and JD. Pfeifer. Tumor heterogeneity affects the precision of microarray analysis. *Diagn Mol Pathol* 14: 65–71 (2005).
35. P. Francis, J. Fernebro, P. Eden, A. Laurell, A. Rydholm, HA. Domanski, T. Breslin, C. Hegardt, A. Borg and M. Nilbert. Intratumor versus intertumor heterogeneity in gene expression profiles of soft-tissue sarcomas. *Genes Chromosomes Cancer* 43: 302–308 (2005).
36. J. Kononen, L. Bubendorf, A. Kallioniemi, M. Barlund, P. Schraml, S. Leighton, J. Torhorst, MJ. Mihatsch, G. Sauter and OP. Kallioniemi. Tissue microarrays for high-throughput molecular profiling of tumor specimens. *Nat Med* 4: 844–847 (1998).
37. J. Terry, TS. Barry, DE. Horsman, FD. Hsu, AM. Gown, DG. Huntsman and TO. Nielsen. Fluorescence in situ hybridization for the detection of t(X;18)(p11.2;q11.2) in a synovial sarcoma tissue microarray using a breakapart-style probe. *Diagn Mol Pathol* 14: 77–82 (2005).
38. LM. Prentice, A. Shadeo, VS. Lestou, MA. Miller, RJ. Deleeuw, N. Makretsov, D. Turbin, LA. Brown, N. Macpherson, E. Yorida, MC. Cheang, J. Bentley, S. Chia, TO. Nielsen, CB. Gilks, W. Lam and DG. Huntsman. NRG1 gene rearrangements in clinical breast cancer: identification of an adjacent novel amplicon associated with poor prognosis. *Oncogene* 24: 7281–7289 (2005).
39. J. Engellau, PO. Bendahl, A. Persson, HA. Domanski, M. Akerman, P. Gustafson, TA. Alvegard, M. Nilbert and A. Rydholm. Improved prognostication in soft tissue sarcoma: independent information from vascular invasion, necrosis, growth pattern, and immunostaining using whole-tumor sections and tissue microarrays. *Hum Pathol* 36: 994–1002 (2005).
40. LJ. van't Veer, H. Dai, MJ. van de Vijver, YD. He, AA. Hart, M. Mao, HL. Peterse, K. van der Kooy, MJ. Marton, AT. Witteveen, GJ. Schreiber, RM. Kerkhoven, C. Roberts, PS. Linsley, R. Bernards and SH. Friend. Gene expression profiling predicts clinical outcome of breast cancer. *Nature* 415: 530–536 (2002).
41. T. Sorlie, CM. Perou, R. Tibshirani, T. Aas, S. Geisler, H. Johnsen, T. Hastie, MB. Eisen, M. van de Rijn, SS. Jeffrey, T. Thorsen, H. Quist, JC. Matese, PO. Brown, D. Botstein, P. Eystein Lonning and AL. Borresen-Dale. Gene expression patterns of breast carcinomas distinguish tumor subclasses with clinical implications. *Proc Natl Acad Sci USA* 98: 10869–10874 (2001).
42. CM. Perou, T. Sorlie, MB. Eisen, M. van de Rijn, SS. Jeffrey, CA. Rees, JR. Pollack, DT. Ross, H. Johnsen, LA. Akslen, O. Fluge, A. Pergamenschikov, C. Williams, SX. Zhu, PE. Lonning, AL. Borresen-Dale, PO. Brown and D Botstein. Molecular portraits of human breast tumours. *Nature* 406: 747–752 (2000).
43. T. Shibata, S. Uryu, A. Kokubu, F. Hosoda, M. Ohki, T. Sakiyama, Y. Matsuno, R. Tsuchiya, Y. Kanai, T. Kondo, I. Imoto, J. Inazawa and S. Hirohashi. Genetic classification of lung adenocarcinoma based on array-based comparative genomic hybridization analysis: its association with clinicopathologic features. *Clin Cancer Res* 11: 6177–6185 (2005).
44. H. Tagawa, M. Suguro, S. Tsuzuki, K. Matsuo, S. Karnan, K. Ohshima, M. Okamoto, Y. Morishima, S. Nakamura and M. Seto. Comparison of genome profiles for identification of distinct subgroups of diffuse large B-cell lymphoma. *Blood* 106: 1770–1777 (2005).
45. S. Avigad, IJ. Cohen, J. Zilberstein, E. Liberzon, Y. Goshen, S. Ash, I. Meller, Y. Kollender, J. Issakov, R. Zaizov and I. Yaniv. The predictive potential of molecular detection in the nonmetastatic Ewing family of tumors. *Cancer* 100: 1053–1058 (2004).

CHAPTER 6

TUBAFROST: EUROPEAN VIRTUAL TUMOR TISSUE BANKING

P.H.J. RIEGMAN[1,*], M.H.A. OOMEN[1,*], W.N.M. DINJENS[1,*],
J.W. OOSTERHUIS[1,*], K.H. LAM[1,*], A. SPATZ[2,*], C. RATCLIFFE[3,*],
K. KNOX[3,*], R. MAGER[3,4,*], D. KERR[3,*], F. PEZZELLA[4,*], B. VAN
DAMME[5,*], M. VAN DE VIJVER[6,*], H. VAN BOVEN[6,*],
M.M. MORENTE[7,*], S. ALONSO[7,*], D. KERJASCHKI[8,*], J. PAMMER[8,*],
J.A. LÓPEZ-GUERRERO[9,*], A. LLOMBART-BOSCH[9,*], A. CARBONE[10,*],
A. GLOGHINI[10,*], I. TEODOROVIC[11,*], M. ISABELLE[11,*],
A. PASSIOUKOV[11,*], S. LEJEUNE[11,*], P. THERASSE[11,*],
AND E.-B. VAN VEEN[12,*]

*The TuBaFrost Consortium; [1]Erasmus MC, Dr Molewaterplein 50, 3015 GE Rotterdam, The Netherlands; [2]Institut Gustave Roussy, Rue Camille Desmoulins 39, 94,805 Villejuif, France; [3]National Translational Cancer Research Network, University of Oxford, Radcliffe Infirmary, Woodstock Road, Oxford OX2 6HE, United Kingdom; [4]Nuffield Department of Clinical Laboratory Sciences, University of Oxford, John Radcliffe Hospital, Headington Oxford OX3 9DU, United Kingdom; [5]U.Z-K.U.-Leuven, Herestraat 49, B-3000 Leuven, Belgium; [6]Netherlands Cancer Institute, Plesmanlaan 121, NL-1066 CX Amsterdam, The Netherlands; [7]Centro Nacional de Investigaciones Oncologicas, Melchor Fernández Almagro, 3. E-28029 Madrid, Spain; [8]Alllgemeines Krankenhaus, University of Vienna, Waeringer Guertel 18-20, 1090 Vienna, Austria; [9]Fundación Instituto Valenciano de Oncología, C/Profesor Beltran Baguena, 8+11, Valencia, Spain; [10]Centro di Riferimento Oncologico, Via Pedemontana Occidentale, 12, I-33081 Aviano (PN), Italy; [11]EORTC Data Center, Avenue E. Mounier 83, B-1200 Brussels, Belgium; [12]Medlaw Consult, Postbus 11,500, 2502 AM Den Haag, The Netherlands

Grants: European Commission 5th framework Quality of life and living resources QLRI-CT-2002-01551

Abstract: TuBaFrost is a consortium responsible for the task to create a virtual European human frozen tumor tissue bank, composed of high quality frozen tumor tissue collections with corresponding accurate diagnosis stored in European cancer centers

Corresponding author: P.H.J. Riegman Ph.D.; Head Erasmus MC Tissue Bank; Co-ordinator of the European Human Frozen Tumor Tissue Bank (TuBaFrost); Department of Pathology, Josephine Nefkens Institute Be 235b; Erasmus Medical Center, PO box 1738, 3000DR Rotterdam, The Netherlands; p.riegman@erasmusmc.nl http://www.tubafrost.org http://www.eur.nl/fgg/pathol/

and universities, searchable on the Internet, providing rules for access and use and a code of conduct to comply with the various legal and ethical regulations in European countries. Such infrastructure would enlarge tissue availability and accessibility in large amounts of specified or even rare tumor samples. Design of an infrastructure for European residual tissue banking with the described characteristics, clear focus points emerge that can be broken down in dedicated subjects: (1) standardization and quality assurance (QA) to avoid inter-institute quality variation; (2) law and ethics enabling exchange of tissue samples possible between institutes in the different European countries, where law and ethics are characterized by a strong variability; (3) rules for access, with sufficient incentives for collectors; (4) central database application containing innovations on search and selection procedures; (5) support when needed with histology images; and (6) Internet access to search and upload, with in addition a solid website giving proper information on the procedures, intentions and activities not only to the scientific community, but also to the general public.

One consortium decision, part of the incentives for collectors, had major impact on the infrastructure; custodianship over the tissues as well as the tissues stay with the collector institute. Resulting in specimens that are not given to an organization, taking decisions on participation of requests, but instead the local collected tissues stay very easy to access by the collector and allows autonomous negotiation between collector and requestor on cooperation, coauthorship in publication or compensation in costs. Thereby, improving availability of large amounts of high quality samples of a highly specified or rare tumor types and contact opportunities for cooperation with other institutes.

Key words: European virtual tumor tissue banking, TuBaFrost, bio-repositories, virtual microscope, access rules, ethics

1. INTRODUCTION

Recent years, an increasing demand for large quantities of high quality and well-documented fresh frozen tumor tissue has arisen, due to new queries available to address in translational cancer research. Thanks to rapid developments of high throughput techniques in medical molecular science. Human tissue collections form a wealthy source of scientific information, with the prerequisite that the tissues are well documented and well conserved [1–5].

1.1. Types of bio-repositories

Today, collections of human tissue can be found in a variety of forms collected for many different purposes. Tissue collections stored for medical research are stored in bio-repositories and upon the aim, three major categories are recognized in the area of medical research: (1) population-driven; (2) project-driven; and (3) systematic-driven banks. Organ or transplantation banks form a distinct class of tissue bank with their own regulations. Although they can be involved in medical research, these banks collect tissue for reuse in human beings and therefore need stricter forms of regulation. To avoid confusion on this point they will not be further discussed here. The population banks are collected to follow a group of people prospectively to characterize the developments of the group in samples collected at a certain time point [1,6,7]. The collections are often very large and clinical data, gathered over time, accompanying the samples is instrumental for the experiments that can be performed.

The project-driven collection is characterized by collections that are disease or organ based. Specialized research groups performing dedicated research on these organs or diseases collect the residual samples and most often not only have an interest in the retrospective, but also are actively collecting the follow up data. To set up such a valuable collection of population-based or project-driven banks takes many years of investment.

To avoid the long time involved in setting up these collections, the systematic collections, which is also the main subject of discussion in this chapter, forms an important alternative. It looks very similar to the project-driven collection, but is not limited by an organ type or disease. Both collections can be seen as disease-driven. Available residual tissues are collected and stored without the need of a specific aim at the time of collection. This way valuable time can be saved to start up medical research based on tissue samples. Data is also collected, but most often less detailed as in project-driven collections. This can also be done in later stages, when it is clear what data is specifically needed. Uniting systematic tissue banks is especially useful for those studies involving high numbers or highly specialized or rare types of tissue specimens, which can take a very long time to build even for systematic tissue banks of large institutes, can save even more valuable time.

1.2. Networks

Networks of systematic tumor tissue banks need to be developed in answer to the demand arisen due to the rapid developments of high throughput techniques [3,4]. TuBaFrost is a project funded by the European Commission in the fifth framework to set up a European human frozen tumor tissue bank network. A mission statement covering the aims of the project was formulated: create an innovating virtual European human frozen tumor tissue bank for the whole European scientific community composed of high quality frozen tumor tissue collections accompanied by a solid diagnosis stored in major cancer centers and universities in Europe, which can be easily consulted through the Internet, provided with rules for access, use of the tissues complete with a code of conduct to comply with the various legal and ethical regulations in European countries. To unite systematic banks of different institutes several important demands must be met.

2. STANDARD OPERATING PROCEDURES

Local methods to store and collect tissue vary from one institute to the other. In fact these methods are differing everywhere in the smallest details. These differences can lead to inter-institute experimental variability or intrinsic bias in results of experiments performed with specimen derived from different institutes. Therefore, minimal standard operating procedures (SOPs) have to be developed and QA measures have to be taken. The ideal form of standardization would be a totally equal operating procedure, however this would involve unnecessary costs that would have to be made for joining a network of tissue banks. Therefore, all activities needed for collecting, freezing, and storing high quality tissue were identified and subsequently weighed on their impact on the quality of the sample. Matters of high impact were described as SOP mandatory for implementation, whereas low impact matters were recommended. In addition, choices were described in methods where the sample would get equal quality in the end. These SOPs and QA needs to be implemented at all participating collecting institutes. For most institutes involved in the TuBaFrost project implementation of the SOPs and QA has led to the adaptation of used protocols. Many aspects in the local protocol do not form a threat to the quality or differences in the quality of the sample were not included for change. There stays however awareness for the best procedures, described in golden standards to which an institute must strive, e.g., mechanical freezers versus liquid nitrogen for long-term storage of tissue specimens. The most important measures are: a pathologist selects the fresh tissue that can be dissected from the surgical specimen, because he is responsible for the diagnosis and can judge which part of the tissue is needed for diagnostic purposes and what tissues are residual [7,8]. The lag time between surgical removal and snap freezing must not be longer than 30 min or otherwise be noted. Alarm systems, an adequate registration and labeling system for tracking the samples must be in place and sufficient backup systems must be operational. For QA, 1% of the new samples have to be randomly checked every year on RNA quality, accuracy of the position in the storage system, and database system and contents of the sample compared with the description. An SOP derived from the here developed SOP to avoid multicenter study intrinsic bias has been worked out for future EORTC multicenter studies.

3. RULES FOR ACCESS AND USE

The infrastructure must not form a burden for the participating collectors to join and at best have incentives for collectors at several fronts. The TuBaFrost consortium has for this reason rejected the idea of a review commission, which would judge all requests on their scientific impact. Instead the custodianship over the tissues stays with the local collector. The local collector takes the decision to participate in a request. This allows collections with high scientific value to be uploaded to the central database. The permission to use the tissues is now depending on the request proposal and on the conditions under which they will be issued. This permits the local collector to negotiate on participation, copublication, and compensation in costs. In addition, the local collector is able to issue the tissues for use in their own institute; they only need to update the central database on availability of the samples. Due to the variations in scientific value it might not only occur that the outcome of the negotiation between the requesting institutes differs, but it might even lead to differences in acceptation and rejection of requests. These aspects are all laid down in rules for access and use:

1. Participation in the European Human Frozen Tumor Tissue Bank Network as collector will only be possible for those European institutes that can contribute tissues (collectors), such as *Hospitals* or *existing networks of tumor banks*. These institutions must, in addition, be able to meet the demands set by the TuBaFrost consortium on minimum standards, protocols, and quality control.
2. Access as a user or requestor to the European Human Frozen Tumor Tissue Bank Network will only be possible after registration, which can be done if the following conditions are met:
   (a) The requestor must be involved as a researcher at any tumor research group, hospital, university, existing network of tumor banks, national cancer association, research center, collaborative group, European commission, associated countries and international associations and involved in medical research only.
   (b) The requestor must be familiar with and comply to the European Law and Ethics on residual tissue as described in D 7.1 and MS 7.1 Code of Conduct or if the law and ethics has changed the then actual situation.
   (c) Requestor must be able to give an exact description of all planned research involving the tissue samples.
   (d) The requestor must have permission of the local Medical Ethics Commission to execute the planned research.
   (e) The planned research needs to be of sufficient quality to spend the valuable tissue samples on, to be judged by the individual collectors, which are involved in the application.
   (f) Requestor must have sufficient expertise at their disposal to perform the planned research.
   (g) Access is limited to those who are prepared to make a reference to the European Human Frozen Tumor Tissue Bank and the involved collectors of the tissue in future publications based on the results obtained on the received tissues.

   Acknowledgment policy: If results obtained with tumor tissues from European Human Frozen Tumor Tissue Bank Network result in a publication, the following statement should be included in the Acknowledgments or Material and Methods section of the manuscript: "The tissue used in this publication was provided by TuBaFrost the European Human Frozen Tumor Tissue Bank". In case facilities were used from the collecting institute(s), beyond the sole activity of issuing tissue, which in addition have contributed to a publication, the persons involved need to be treated as coauthor of that publication.
   (h) The requestor will respect the evaluation of their proposal, which is judged by the local collectors involved in the request and is reflecting the

permission on tissues requested at that local institute. If permission is denied, the requestor can submit the request again to other Institutes involved in collecting samples. In addition, requests of collectors are considered more important over requests of sole requestors, therefore, in case of competitive interests collectors requests are considered first.
3. In order to participate in the European Human Frozen Tumor Tissue Bank Network collector institutes will have to comply with the following criteria:
    (a) All collector institutes must have availability and access to cancer patient specimens.
    (b) Collector institutes must have the availability of sufficient personnel (e.g., technicians and pathologists) and infrastructure capacity for the development and/or maintenance of a tumor bank, which meets the minimal standards and can collect tissue according to the protocols and rules set by the European Human Frozen Tumor Tissue Bank Network.
    (c) In order to participate in the European Human Frozen Tumor Tissue Bank Network all or potential collector institutes will establish a tissue bank, collect specimens and corresponding clinical data according to the standardized collection methods and policies of the European Human Frozen Tumor Tissue Bank Network in order to assure quality control of specimens and data.
        - The collector institutes will name a member of the institute, who is responsible for the scientific tasks of the tumor bank (macroscopic analysis, and selection and harvest of the surgery specimens). Those selected persons should demonstrate their interest in participating in the tumor bank and cancer research. He/she will be responsible for the functional aspects such as reception, processing and storing of the samples, quality controls, legal and ethical aspects, management of the information referring to each sample, and the distribution of the samples.
        - The collector institutes has a technician responsible for the processing, storing, cryopreservation, retrieval, and shipping of the samples.
        - The collector institutes will indicate the existence of a budget, a physical space and equipment to perform the activities of the tumor bank.
4. The collector institutes are responsible for the association of samples with a locally valid consent or at least an opting out system, where patients can object to the use of their tissues for research purposes, to ensure legal research use of TuBaFrost material by the requestor institutions.
5. The collector institutes will be required to accept and implement the common policies and procedures approved by the TuBaFrost consortium (or the Steering Committee).
6. The collector institutes will be required to accept cooperative action between the tumor banks of the European Human Frozen Tumor Tissue Bank Network.

The TuBaFrost infrastructure supports and facilitates different aspects of these complex tasks.

4. ETHICS AND LAW

TuBaFrost is a European Tissue bank and has to deal with a variety of laws and ethics in the different participating countries. It is absolutely key for translational cancer research in Europe to find a practical way to exchange tissues. The project proposal had already strong emphasis on this point. One of the end points was a code of conduct for exchange of tissues in Europe.

To come to a code of conduct for exchanging residual tissues in Europe first an inventory was made on the variation of laws of the participating countries, including some remarkable developments in others. From this a strategy was determined by looking at the possibilities. Harmonization was one that was considered, however this would lead to the strictest form of regulation, making exchange of tissue in Europe virtually impossible and too expensive for research purposes. Another alternative is to harmonize only the two countries involved in every exchange. This gave also a too complicated outcome. Therefore, instead of harmonizing attempts a solution was sought in the creation of a coordinating rule by using the principal of home country control and governance over the tissue samples.

If tissue may legitimately be used for a certain kind of research in the country where it was taken out, it may also be used for such research in the country where it is sent to in the context of a scientific program even if in that other country other regulations would apply for research with residual tissue taken out from patients under their jurisdiction.

This outcome results in an environment enabling tissue exchange possible in Europe, which keeps collection of tissue under local law legitimate for later tissue exchange. This principle is described in a code of conduct, which as a minimum requires an opt-out consent procedure. All institutes lacking consent procedures for collecting residual tissues for medical research and participating in the TuBaFrost network, need to give the patient the opportunity to object to the use of their residual tissue for medical experimental purposes. Therefore, an adequate system needs to be in place enabling the destruction of the tissues of the patient, after he or she objected to the use of their residual tissues in medical research. Furthermore, when tissue goes from one institute to another both parties involved must sign a tissue transfer agreement, which states the rights and obligations of both parties involved on a basic level. The tissue transfer agreement leaves room for negotiations between the recipient and collector on cooperation, copublication and/or a compensation in costs made for collecting freezing and retrieval of the samples.

5. CENTRAL DATABASE

The central database is much more than just a simple database where tissue records are stored [2,9]. It is a complete application that supports the visibility of tissues for the research community, whereas it not only facilitates the search for tissues needed for research, but also supports the request with decision making for the collector and the communication between requestors, collectors, and central office. The application has a hierarchic structure; all the different actors, that need to be registered, have on basis of their login permission to use facilities in the central database. The lowest is the requestor followed by the head requestor (one per institute), collector, head collector (one per institute), and the administrator or central office. Support and start of the communication process is done by automatically generated e-mails. The e-mail addresses need to be given upon registration. The e-mail is used to identify the institute especially when new actors are registering from the same institute. False e-mail addresses block further contact and communication will not take place. The data that is uploaded to the central database is coded locally under the responsibility of the local institute. This code is again changed when uploaded to identify the institute. Uploading can be done by single samples and by batch upload. The patient case data knows the following fields: (1) local patient case code; (2) histopathological diagnosis 1,2,3 (morphology); (3) stage (and TNM); (4) grade; (5) age at time of biopsy/sampling (years); (6) gender; and (7) clinical, trial involvement, medical history/treatment/complications/toxicity/concomitant disease/secondary tumors/laboratory data.

Tissue sample data: (1) local inventory tissue code; (2) site of tumor (topography); (3) tissue condition (tumor/nontumor/interface); (4) time elapsed after biopsy was performed and tissue sample was frozen; and (5) availability of tissue sample (available/issued/in use/terminated).

A requestor can search for tissues characterized by all fields recorded. He can select tissues for research and put them in a cryo-cart or basket, which is comparable with a shopping cart used in Internet shopping. The contents of this cryo-cart can be ordered after filling out a form, which is information required for the collector involved in the request to participate in the request or not and will be part of the request proposal send to the local collector. Digital images of the sample histology can be uploaded to support the selection process of the tissue specimens. Tissues with a controversial diagnosis can be supported by virtual microscopy.

6. VIRTUAL MICROSCOPY

Virtual microscopy is a technique that enables to digitize complete tissue slides on high magnification and store the information on disk. The user can approach the large digital image over a network connection (Internet) and demand arbitrarily magnifications and positions scanned by the VM. The image server will not send the whole picture to the user, but only small pictures to the user enough to fill the screen every time the user requests a certain magnification or position and have sufficient resolution to render a correct diagnosis as a virtual pathology slide [10,11]. The pictures need a large quantity of disc space depending on the magnification and the area that needs to be scanned. Pictures are on average

500 Mb using the 40× objective. Therefore, only tissues having a difficult diagnosis are supported with the virtual microscope. Scanning is done at the EORTC, where a Nanozoomer from Hamamatsu (Japan) has been installed for this purpose. It supports automatic slide feeding (210/session), imaging with a software chosen 20× or 40× objective and has fast scanning times. A viewer is made available for Internet approach of all interested parties. A link is the only thing that needs to be uploaded to the central database to make access possible to the virtual microscopic image.

7. INTERNET ACCESS

Requestors, collectors, and central office need to be able to contact the central database on a regular basis. In addition, there need to be a tool to show the general public, patient carer groups and new requestors and collectors in a transparent way what the TuBaFrost aims are and how these aims are to be reached. In addition, the way you can participate in the network must be completely clear. Therefore, an integrated website is needed to explain to all the actors mentioned, how TuBaFrost functions and what requirements need to be fulfilled to take part in this initiative. At www.tubafrost.org an experimental version is presented, but as soon as it is finished the final version will be put under this name. The experimental version was needed to discuss and develop the professional one. A company has developed the final website and central database application with knowledge on website design, because it is important for the first impression of the visitor of the site. The site reflects the professionalism of programming and needs to be programmed with the latest insights.

8. PERSPECTIVES

After EC funding the TuBaFrost network is implemented under the umbrella of the OECI (Organization of European Cancer Institutes) where it will be used by the OECI members as a tool for networked tumor tissue banking as an integral part of comprehensive cancer institutes.

TuBaFrost is just a start, but many connections need to be made between biorepositories, nationally as well as internationally. Discussions on a global level have even been started. The structure of the here presented infrastructure allows for project-driven and systematic tissue bank to join forces and therewith can come to higher numbers needed for significant results and verification or reach the needed critical mass to start research on rare cases. This is enabled due to the unique vision on custodianship over the collected samples. It is also tempting to speculate on the new scientific opportunities that would arise if population banks could join forces with the disease-driven banks preferably intermediated with cancer registries. Ways need to be found to let information find its way without harming the privacy of the patients involved.

REFERENCES

1. J.W. Oosterhuis, J.W. Coebergh, E.B. van Veen. Tumour banks: well-guarded treasures in the interest of patients. *Nat Rev Cancer* 3: 73–77 (2003).
2. K. Knox, D.J. Kerr. Establishing a national tissue bank for surgically harvested cancer tissue. *Br J Surg* 91: 134–136 (2004).
3. I. Teodorovic, P. Therasse, A. Spatz, et al. Human tissue research: EORTC recommendations on its practical consequences. *Eur J Cancer* 39: 2256–2263 (2003).
4. M.M. Morente. Tumour banks and oncological research. *Rev Oncol* 5, 63–64 (2003).
5. I. Hirtzlin, C. Dubreuil, N. Preaubert, et al. A EUROGENBANK Consortium. An empirical survey on biobanking of human genetic material and data in six EU countries. *Eur J Hum Genet* 11: 475–488 (2003).
6. M.T. Goodman, B.Y. Hernandez, S. Hewitt, et al. Tissues from population-based cancer registries: a novel approach to increasing research potential. *Hum Pathol* 36: 812–820 (2005).
7. W.E. Grizzle, R. Aamodt, K. Clausen, V. LiVolsi, T.G. Pretlow, S. Qualman. Providing human tissues for research: how to establish a program. *Arch Pathol Lab Med* 122: 1065–1076 (1998).
8. W.E. Grizzle, K.H. Woodruff, T.D. Trainer. The pathologist's role in the use of human tissues in research — legal, ethical, and other issues. *Arch Pathol Lab Med* 120: 909–912 (1996).
9. E.J. Kort, B. Campbell, J.H. Resau. A human tissue and data resource: an overview of opportunities, challenges, and development of a provider/researcher partnership model. *Comput Methods Programs Biomed* 70: 137–150 (2003).
10. S.P. Costello, D.J. Johnston, P.A. Dervan and D.G. O'Shea. The virtual pathology slide. A new Internet telemicroscopy tool for tracing the process of microscopic diagnosis and evaluating pathologist behaviour (abstract) *Arch Pathol Lab Med* 126: 781–802 (2002).
11. S.P. Costello, D.J. Johnston, P.A. Dervan, D.G. O'Shea. Evaluation of the virtual pathology slide: using breast needle core biopsy specimens. *Br J Cancer Suppl* 86: S34 (2002).

CHAPTER 7
VIRTUAL MICROSCOPY IN VIRTUAL TUMOR BANKING

M. ISABELLE[1], I. TEODOROVIC[1], J.W. OOSTERHUIS[2],
P.H.J. RIEGMAN[2], AND THE TUBAFROST CONSORTIUM[#]

[#]THE TUBAFROST CONSORTIUM: I. TEODOROVIC[1], M. ISABELLE[1],
A. PASSIOUKOV[1], S. LEJEUNE[1], P. THERASSE[1], P.H.J. RIEGMAN[2*],
W.N.M. DINJENS[2], J.W. OOSTERHUIS[2], K.H. LAM[2], M.H.A. OOMEN[2],
A. SPATZ[3], C. RATCLIFFE[4], K. KNOX[4], R. MAGER[4,5], D. KERR[4],
F.PEZZELLA[5], B. VAN DAMME[6], M. VAN DE VIJVER[7], H. VAN
BOVEN[7], M.M. MORENTE[8], DR. S. ALONSO[8], D. KERJASCHKI[9],
J. PAMMER[9], J.A. LÓPEZ-GUERRERO[10], A. LLOMBART-BOSCH[10],
A. CARBONE[11], A. GLOGHINI[11], AND E.-B. VAN VEEN[12]

[1]*EORTC Data Center, Avenue E. Mounier 83, B-1200 Brussels, Belgium*
[2]*Erasmus MC, Dr Molewaterplein 50, 3015 GE Rotterdam, The Netherlands*
[3]*Institut Gustave Roussy, Rue Camille Desmoulins 39, 94805 Villejuif, France*
[4]*National Translational Cancer Research Network, University of Oxford, Radcliffe Infirmary, Woodstock Road, Oxford OX2 6HE, United Kingdom*
[5]*Nuffield Department of Clinical Laboratory Sciences, University of Oxford, John Radcliffe Hospital, Headley Way, Oxford OX3 9DU, United Kingdom*
[6]*U.Z-K.U-Leuven, Herestraat 49, B-3000 Leuven, Belgium*
[7]*Netherlands Cancer Institute, Plesmanlaan 121, NL-1066 CX Amsterdam, The Netherlands*
[8]*Centro Nacional de Investigaciones Oncologicas, Melchor Fernández Almagro, 3. E-28029 Madrid, Spain*
[9]*Alllgemeines Krankenhaus, University of Vienna, Waeringer Guertel 18-20, 1090 Vienna, Austria*
[10]*Fundación Instituto Valenciano de Oncología, C/Profesor Beltran Baguena, 8+11, Valencia, Spain*
[11]*Centro di Riferimento Oncologico, Via Pedemontana Occidentale, 12, I-33081 Aviano, PN, Italy*
[12]*Medlaw Consult, Postbus 11500, 2502 AM Den Haag, The Netherlands*

* *Corresponding authors: P.H.J. Riegman Ph.D.; Head Erasmus MC Tissue Bank; Coordinator of the European Human Frozen Tumor Tissue Bank (TuBaFrost) and J.W. Oosterhuis M.D., Ph.D.; Professor of Pathology, and Head of the Department, Department of Pathology, Josephine Nefkens Institute Be 235b. Erasmus University Medical Center, PO box 1738, 3000DR Rotterdam, The Netherlands. p.riegman@erasmusmc.nl; http://www.tubafrost.org; http://www.eur.nl/fgg/pathol/; j.w.oosterhuis@erasmusmc.nl*

Grants: European Commission 5th framework Quality of life and living resources QLRI-CT-2002-01551

Abstract: Many systems have already been designed and successfully used for sharing histology images over large distances, without transfer of the original glass slides. Rapid evolution was seen when digital images could be transferred over the Internet. Nowadays, sophisticated virtual microscope systems can be acquired, with the capability to quickly scan large batches of glass slides at high magnification and compress and store the large images on disc, which subsequently can be consulted through the Internet. The images are stored on an image server, which can give simple, easy to transfer pictures to the user specifying a certain magnification on any position in the scan. This offers new opportunities in histology review, overcoming the necessity of the dynamic telepathology systems to have compatible software systems and microscopes and in addition, an adequate connection of sufficient bandwidth. Consulting the images now only requires an Internet connection and a computer with a high quality monitor. A system of complete pathology review supporting bio-repositories is described, based on the implementation of this technique in the European Human Frozen Tumor Tissue Bank (TuBaFrost).

Keywords: Virtual microscope, virtual tumor bank, virtual bio-repository, virtual tissue banking

1. INTRODUCTION

Ever since Rudolf Virchow published the first attempts to a systematic classification of tumors based on histology, light microscopy has been the mainstay of tumor diagnosis. Virtual microscopy is an important technological development that broadens the range of applications of light microscopy. Here we describe the uses of virtual microscopy in the context of the European virtual tumor bank TuBaFrost.

Telepathology is the practice of "pathology" at a distance, using telecommunications technology as a means to transfer images of gross morphology and histological slides between remote locations for the purposes of diagnosis, education, and research [1,2]. In a broader sense telepathology also includes remote discussion from pathologist to pathologist and from clinicians to pathologist, remote quality assurance involving pathologists and referral labs, and research collaboration between research teams [1–2].

Telepathology systems have been traditionally defined as either dynamic or static. Dynamic systems allow a telepathologist to view images transmitted in real time from a remote robotic microscope that permits complete control of the field of view and magnification [3–6]. Static (or store and forward) telepathology involves the capture and storage of images followed by transmission over the Internet through e-mail attachment, file transfer protocol, or a Web page, or distribution through CD-ROM. Dynamic hybrids also exist, which incorporate aspects of both technologies [6].

Today, telepathology systems are divided into three major types: (1) static image-based systems; (2) real-time systems; and (3) virtual slide systems. To overcome problems attributable to sampling bias and interpretation resulting from limited field selection, telepathologists must be able to navigate to any field of view, at magnifications comparable with that of a conventional microscope, using images of sufficient resolution to render a correct diagnosis [7–9]. To meet such criteria the Virtual Pathology Slide has been developed [10,11]. The Virtual Pathology Slide mimics the use of a microscope in both the stepwise increase in magnification and in lateral motion in the X and Y Cartesian directions. This permits a pathologist to navigate to any area on a slide, at any magnification, similar to a conventional microscope. This is discussed further in the next section.

Static image systems have the significant drawback in only being able to provide selected microscopic fields. Conversely, both real-time and virtual slide systems allow a consultant pathologist the opportunity to evaluate the entire microscopic slide. With real-time systems, the consultant actively operates a microscope located at a distant site, whereas virtual slide systems utilize an automated scanner that takes a virtual image of the entire slide, which can then be forwarded to another location.

While real-time and virtual slide systems appear ideal for telepathology, there are certain drawbacks to each. Real-time systems perform best on local area networks, but performance may suffer if employed during periods of high network traffic. The scanning of virtual slides can be a time-intensive operation requiring anywhere from minutes to hours to accurately scan a single slide. Also the large data size of the virtual slide means that a large data storage space is required as well as adequate archival and backup systems. However recent developments in virtual slide systems have resulted in a dramatic reduction of the time required for scanning a single slide. In addition, auto slide feeding technology has been developed in conjunction with these systems to allow batch slide scanning of a number of slides without the user having to manually feed the system a single slide one at a time. Developments in software applications and compression methods have allowed for a smaller data size compared with what was possible in the past.

The diversity in telepathology systems reflects growing technological expertise in this area and the increasing importance of telepathology in education, training, quality assurance, and teleconsultation [12–17]. Numerous pathology archives are accessible on the Internet, providing links to both educational and commercial telepathology websites. These offer access to either static or dynamic image delivery systems [3,4,15–22,23–25].

2. VIRTUAL MICROSCOPY

Virtual Microcopy (virtual slide system or Virtual Pathology Slide) is the technique of digitizing an entire glass microscope slide at the highest resolution to produce a "digital virtual microscope slide" with diagnostic image quality. This "digital virtual slide" can be used in conjunction with image processing software tools (both windows-based and browser-based) to view, manipulate, position, and specify the magnification of the image on screen as if using a regular microscope to view the original glass slide (see Figure 7.1). As the glass slide is now in a digital or virtual format, it is possible to use the image for: archiving, replication, transferring over networks, remote consultation, integration with other media types for educational use on the web or DVD, and integration into laboratory information systems and image analysis (reference article 'Nanozoomer/Medical Solutions user manual and website').

Traditional histopathology diagnosis uses the standard light microscope to observe prepared tissue sections on glass slides. For second opinion is required to send the glass slides by post to another pathologist for analysis under his/her microscope and then the slides usually have to be returned to the original pathologist. This method takes time, risks permanent damage to the glass slides and incurs mailing costs.

Classroom viewing of histology slides in an educational environment would require the availability of microscopes for the students, or an optical microscope with a projector operated by an instructor (and not the students themselves) and would require the students to be physically present in the same classroom at the same time.

However, because of Virtual Microscopy technology and development of the broadband internet connection it is now possible for these histology slide to be digitized, placed on an image server and made available online by a (secure) website.

Note: The whole virtual slide image is not sent to the user over the Internet. When a user selects a point and magnification on the virtual slide map, the image server will select that relevant field of view section from the full image stored on disk (on average 500 MB) and will send it to the user over the Internet. This method is performed every time the user requests another location or magnification. This means that all the pathologists involved in reviewing a case

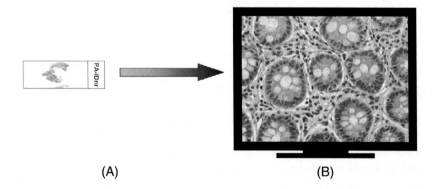

(A)   (B)

Figure 7.1. Virtual microscopy involves the digitization of an entire glass microscope slide (A) to produce a digital virtual microscope slide with diagnostic image quality (B).

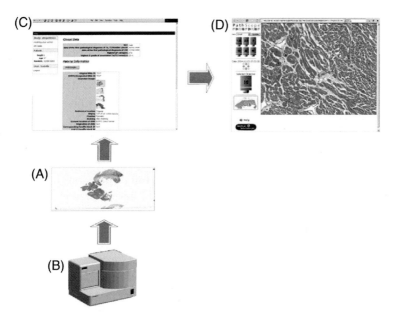

Figure 7.2. Virtual Slide Image (A) taken using the Virtual microscope system (B) can be accessed through the central database (C) using a website interface (D).

would be able to log into this website, access these images and be able to decide on their histopathological diagnosis of the case due to the diagnostic image quality of the virtual slide image (see Figure 7.2).

Students would also be able to access these images online and by using their computer can independently look at any image from a database containing thousands of slides. The viewing technology that is available with virtual microscopy technology allows the user to select magnifications and use UP/DOWN/LEFT/RIGHT arrow buttons to move the center of field of view. The viewing software can also be navigated by clicking on the map or field of view image.

3. APPLICATION OF TELEPATHOLOGY AND VIRTUAL MICROSCOPY IN ROUTINE PATHOLOGY

Current telepathology applications include intraoperative frozen sections services, routine surgical pathology services, second opinions, and subspecialty consultations. In this context, diagnostic accuracy of telepathology is comparable with that of conventional light microscopy for most diagnoses. Rapid and ultra-rapid virtual slide processors may further expand the range of telepathology applications. Next-generation digital imaging light microscopes may make virtual slide processing a routine laboratory tool [26].

Two kinds of virtual pathology laboratories are emerging: (1) those with distributed pathologists working in distributed (one or more) laboratories and (2) distributed pathologists working in a centralized laboratory. Both are under technical development. A virtual pathology institution (mode a) accepts a complete case with the patient's history, clinical findings, and virtual images for second opinion. The diagnostic responsibility is that of the requesting institution. The Internet serves as a platform for information transfer, and a central virtual slide server for coordination and performance of the diagnostic procedure. A group of pathologists is "on duty", or selects one member for a predefined duty period. The diagnostic statement of the pathologist(s) on duty is retransmitted to the virtual slide sender who maintains full responsibility. A centralized virtual pathology institution (mode b) depends upon the digitalization of a complete slide, and the transfer of the virtual slide to different pathologists working in one institution [27]. In order to acquire uniform high quality in pathologic diagnostics and to have fast spread of cancer research knowledge, a Virtual Pathology Institution group is under development in Italy. It is composed of distributed pathologists and distributed laboratories working in the seven Italian Cancer Institutes within the Alliance Against Cancer national project (ACC). Program implications of the ACC virtual pathology institution have included: (1) consensus sessions; (2) quality control; (3) education activities; (4) 2nd opinion; and (5) research. Virtual microscopy technology and Internet connection are used. Microscope slides images are digitized; virtual slides are then placed on a "central" image server and made available online through a website [4]. Since its implementation in 2003, the ACC virtual pathology institution has processed over 200 teleconsults, providing the communicating pathologists with diagnostic assistance on their most difficult cases. Today, the open access of the virtual microscopic images enables the user to send links of virtual microscopic images over e-mail. In addition, these links can be preset to a certain area and magnification of the image so allowing each pathologist to highlight interesting histology, whilst retaining the ability to view the complete histology of the scanned glass slide. Combining the use of other computer-aided communication allows for a large variety of new forms of applications in consulting.

4. VIRTUAL MICROSCOPE SYSTEM COMPONENTS

Figure 7.3 portrays a general setup for a virtual microscope and telepathology system. The glass slide is placed on the microscope stage and the digital camera (3-chip 24 bit) takes high quality "field of view" single images (usually at 20× objective, but 40× is possible) as the motorized stage moves the glass slide left-to-right and up and down. Eventually, a collection of "field of view images" are taken representing the entire glass slide. These images are saved to the PC that is connected to the digital camera and the image acquisition software installed on

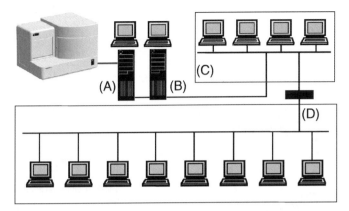

Figure 7.3. System setup for a virtual microscope and telepathology system. The virtual microscope network consists of a workstation (A) to operate the Virtual Microscope and an Image server (B) of large storage capacity and connected to the network. The network connection can involve a connection to the Intranet (C) enabling all institute computers with access to the image server to view the virtual microscopic images. It can also be connected to the Internet (D), enabling external users access to the images.

the PC will "stitch" together these field of view images to create an entire virtual slide image (between 100 MB and 2 GB in file size using JPEG image compression) depending on the area of the tissue slide and chosen magnification.

This virtual slide image can then be placed on the image file server and the server file address can be stored in a database under the tissue record. Client viewer software is installed on the web server so that when a user logs into the website and selects a tissue record, he/she can click on the low-objective slide image (taken, for instance, at 1/2× objective) which will load the client viewer software so that the user can view and navigate the matching virtual slide image that is stored on the image server (Figure 7.2 and Figure 7.4). The request is made to the image server, where the requested image is calculated and sent to the user as described in the section "What is a Virtual Microscope".

5. INVESTIGATION OF COMMERCIAL VIRTUAL MICROSCOPE SYSTEMS

During 2003–2004, as part of the TuBaFrost Project, 20 companies involved in virtual microscope systems around Europe were consulted and their systems evaluated.

From this evaluation the criteria for a good virtual microscope system were determined:
1. Good quality images (good resolution, focus, and sharpness)
2. Adequate range of objectives/magnification

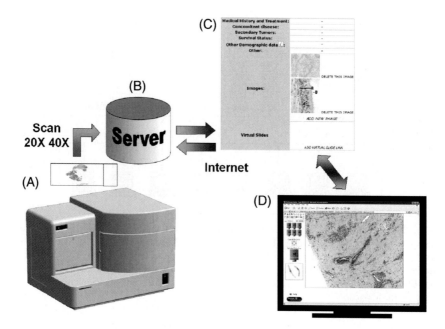

Figure 7.4. Once the virtual slide image is scanned (A), stored on the image server (B) and catalogued in the tissue database, users can access these images from the online tissue record (C) and discuss the slide image with colleagues (D).

3. Accurate focusing
4. Fast Scan speeds/low scan times
5. Best compression rates and low image sizes (but maintaining good quality images — e.g., JPEG image compression)
6. Easy integration into existing software, databases, and computer systems
7. Affordable system
8. Reliable

The components of a typical virtual microscope system are:
1. Microscope (standard or research) with XYZ motorized stage and controller (attached to PC), which includes focusing system.
2. 3-chip digital camera attached to a high performance PC with high resolution monitor.
3. High resolution flat-screen monitor attached directly to digital camera (for displaying current field of view).
4. Image acquisition software installed on PC that, through the motorized stage controller, allow the microscope to move across slide and collects frames of "field of views" and then stitches these field of view images together to make a complete virtual slide image of the complete glass slide.
5. Image file server to hold these compressed virtual slide images.

6. Web server to hold the web site files to provide an interface to the images and tissue information.
7. Pathology-Tissue database system that holds the image server file location for each virtual image within the tissue record.
8. Remote client viewing software installed on web server that works together with the tissue record database to allow pathologists and scientists to remotely access and navigate these virtual images.

Based on the criteria determined from the investigations and testing of some of the systems at that time, the Hamamatsu Nanozoomer Digital Pathology (NDP) system [28,29] was selected for the TuBaFrost Project. The hardware for this system is developed by Hamamatsu Photonics [28,29] and the software is developed by Medical Solution Plc [30,31].

6. APPLICATION OF VIRTUAL MICROSCOPY IN THE TUBAFROST NETWORK

The addition of the virtual microscope tool to the tissue bank model was primarily to assist when tissues with a difficult diagnosis are selected for experimental purposes. In this way, the user is aware of the exact constitution of the sample, so samples are not transported unnecessarily at cost to the requestor and samples are not unjustifiably unavailable. In addition, for very problematic cases a diagnosis review system can be easily established.

A central database was developed for the TuBaFrost network into which locally collected sample information is uploaded and is searchable by the scientific community in order to locate tissue material for their research projects. Associated with each tissue record in the central database is the diagnosis. In order to back up this diagnosis as well as provide images of the tumor sample for training, education, and reference, a virtual microscope system was employed within the network so that virtual slide images could be made from the tissue material and linked to the tissue samples entered into the central database.

Therefore, during the development of the TuBaFrost central database system, web software (provided by Medical Solution Plc) was integrated into the system to allow users to view these virtual slides online along with the tissue record and patient case information.

A large image server (1 TB = 1,000,000 MB) was integral to the NDP system, however since each image will be between 300 and 600 MB in file size the limiting factor is the storage capacity. The TuBaFrost consortium decided that it was not essential for all tumor samples to be supported with virtual slide images; support is only required for difficult to diagnose samples. In these cases, a local pathologist can determine the diagnosis by judging the virtual images. Less difficult cases can be supported by a normal representative image (0.2–1.0MB), which can be uploaded into the central database (see article Central Database in International Tumor Banking), whereas easily diagnosed cases need no image at all. In the future as storage media reduce in price, data capacity increases, and compression methods develop further, it will be possible to store more virtual slide images.

Pathologists associated with collecting institutes participating in the TuBaFrost network will select the slides required for virtual slide scanning (difficult cases) and send them to the Tumor Bank office at the EORTC in Brussels for digitization. The virtual slide images will be stored on an image server linked to the tissue records and patient case within the central database. After digitization is completed, the glass slides will be returned to the collector institute.

7. CONCLUSION AND FUTURE PROSPECTS

A relevant application for virtual slide technology is the documentation of tissue samples in tissue banks. This technology in the tissue bank can enhance traditional sample annotation, assist in ensuring remote tissue bank clients receive appropriate tissue for their research, and form the basis of quality assurance systems [32]. In this paper, the TuBaFrost methods for integration of whole slide imaging into the tissue bank workflow and information systems have been reported.

An additional application for this technology is the development of a virtual resource of histopathology images for educational purposes. Users of this resource can view any part of an entire specimen at any magnification within a standard web browser. The virtual slides can be supplemented with textual descriptions, but can also be viewed without diagnostic information for self-assessment of histopathology skills. Web-based virtual microscopy will probably become widely used at all levels in pathology teaching [33]. Virtual microscopy has significant advantages over real microscopy in education; it enables learning and has been favorably received by students and teachers alike [34].

In conclusion, as the acquisition of high-quality virtual slides is still a time-consuming task, the application of virtual microscopy in routine diagnostics cannot be recommended at this time. With the arrival of fast and easy to use virtual slide systems, we expect a significant change in telepathology application. In the near future, the digital (virtual) slide will conclude the automatic process from biopsy to digital archiving. As with biomedical technologies (MRT, PET, sonography, CT, and termography) histology may be integrated into the digital information systems for medical diagnostics, making these data available as part of the patient's electronic file.

ACKNOWLEDGMENTS

The European human frozen tumor tissue bank or TuBaFrost project was funded by the European Commission within the 5th framework of the division "Quality of life and living resources" under project number QLRI-CT-2002-01551, with the aid and commitment of the involved scientific officers

J. Namorado and M. Vidal. We thank Ms. K.K.M. de Wildt for her management support during the project.

REFERENCES

1. Wenstein RS. Prospects for telepathology (editorial). *Hum Pathol* 1986: 17: 433–434.
2. Wenstein RS, Bloom KJ, Rozek LS. Telepathology. Long-distance diagnosis. *Am J Clin Pathol* 1989 Apr, 91: S39–S42.
3. Armed Forces Institute of Pathology. Home page. URL: http://www.afip.org/ (accessed 2005 Apr 11).
4. Bacus Laboratories Inc. Microscope imaging and virtual microscopy products for pathology, research and medical education. URL: http://www.bacuslabs.com/ (accessed 2005 Apr 11).
5. Shimosato Y, Yagi Y, Yamagishi K, Mukai K, Hirohashi S, Matsumo T, et al. Experience and present status of telepathology in the National Cancer Center Hospital, Tokyo. *Zentralbl Pathol* 1992 Dec, 138: 413–417.
6. Leong FJW-M, Graham AK, Gahm G, McGee JO'D. Telepathology clinical utility and methodology. In: Lowe DG, Underwoord JCE (eds) *Recent advances in histopathology*. 18th edn. London: Churchill Livingstone, 1999, 217–240.
7. Halliday BE, Bhattacharyya AK, Graham AR, Davis JR, Leavitt SA, Nagle RB, et al. Diagnostic accuracy of an international static-imaging telepathology consultation service. *Hum Pathol* 1997, 28: 17–21.
8. Weinberg DS, Allaert FA, Dusserre P, et al. Telepathology diagnosis by means of still digital imagines: an international validation study. *Hum Pathol* 1996, 27: 111–118.
9. Demichelis F, Della Mea V, Forti S, Dalla Palma P, Beltrami CA. Digital storage of glass slides for quality assurance in histopathology and cytopatholgy. *J Telemed Telecare* 2002, 8: 138–142.
10. Costello SP, Johnston DJ, Dervan PA, O'Shea DG. The virtual pathology slide. A new internet telemicroscopy tool for tracing the process of microscopic diagnosis and evaluating pathologist behaviour (abstract). *Arch Pathol Lab Med* 2002, 126: 781–802.
11. Costello SP, Johnston DJ, Dervan PA, O'Shea DG. Evaluation of the virtual pathology slide: using breast needle core biopsy specimens. *Br J Cancer Suppl* 2002, 86: S34.
12. Brauchli K, Christen H, Haroske G, Meyer W, Kunze KD, Oberholzer M. Telemicroscopy by the Internet revisited. *J Pathol* 2002, 196: 238–243.
13. Wells CA, Sowter C. Telepathology: a diagnostic tool for the millennium? J Pathol 2002, 196: 351–355.
14. Petersen I, Wolf G, Roth K, Schluns K. Telepathology by the Internet. *J Pathol* 2000, 191: 8–14.
15. Afework A, Beynon MD, Bustamante F, Cho S, Demarzo A, Ferreira R, et al. Digital dynamic telepathology — the virtual microscope. In: Proceedings AMIA Symposium 1998. Bethesda, MD: American Medical Informatics Association 1998: 912–916.
16. Taylor RN, Gagnon M, Lange J, Lee T, Draut R, Kujawski E. CytoView. A prototype computer image-based Papanicolaou smear proficiency test. *Acta Cytol* 1999, 43: 1045–1051.
17. Barbareschi M, Demichelis F, Forti S, Dalla Palma P. Digital pathology: science fiction? *Int J Surg Pathol* 2000, 8: 261–263.
18. Telepathology City. Commercial solutions. URL: http://www.telepathologycity.com/systems.htm (accessed 2005 Apr 11).
19. Department of Pathology, University of Pittsburgh School of Medicine. On line case database. URL: http://path.upmc.edu/cases.html (accessed 2005 Apr 11).
20. The Johns Hopkins Medical Institutions, Department of Pathology, Division of Hematopathology. Flow cytometry tutorial. URL: http://162.129.103.34/leuk/toc.htm (accessed 2005 Apr 11).
21. University of Medicine and Dentistry of New Jersey, Department of Pathology. Bone tumors: tutorial for residents. URL: http://www.umdnj.edu/education/index.htm (accessed 2005 Apr 11).

22. Committee on Quality Assurance Training and Education of The European Federation of Cytology Societies. Home page. URL: http://crsg.ubc.kun.nl/quate/EFCS/EFCSindex.htm (accessed 2003 Apr 17).
23. Histokom, Telemedicine, telepathology, teleconsultation, telemicroscopy, teleendoscopy, telesonography, teledermatology. URL: http://www.ipe.unistuttgart.de/res/ip/histkome.html (accessed 2003 Apr 17).
24. Fairfield Imaging. Software for digital pathology. URL: http://www.fairfieldimaging.co.uk/ (accessed 2003 Apr 17).
25. Walter-Reed Army Medical Center. Pathology and area laboratory sciences: pathology to go! URL: http://www.wrmc.amedd.army.mil/departments/pathology/links.htm (accessed 2003 Apr 17).
26. Weinstein RS, Descour MR, Liang C, Bhattacharyya AK, Graham AR, Davis JR, Scott KM, Richter L, Krupinski EA, Szymus J, Kayser K, Dunn BE. Telepathology overview: from concept to implementation. *Hum Pathol* 2001, 32: 1283–1299.
27. Kayser K, Kayser G, Radziszowski D, Oehmann A. New developments in digital pathology: from telepathology to virtual pathology laboratory. *Stud Health Technol Inform* 2004, 105: 61–69.
28. Hamamatsu Nanozoomer Digital Pathology (NDP). URL: http://www.hpk.co.jp/Jpn/products/SYS/NDPJ.htm. Accessed 14th April 2005.
29. Hamamatsu Photonics to Market High-speed Digital Slide Scanner. URL: http://www.japan-corp.net/Article.Asp?Art_ID=9813. Accessed 14th April 2005.
30. Digital Pathology. Fairfield Imaging. URL: http://www.fairfield-imaging.co.uk/fairfield/frame.html. Accessed 14th April 2005.
31. Medical Solutions plc. Pathsight and Pathscope. URL: http://www.medical-solutions.co.uk/html/technical.htm. Accessed 14th April 2005.
32. Patel AA, Gupta R, Gilbertson J. Large scale implementation of whole slide imaging in a tissue bank. Medinfo 2004, 2004 (CD), 1801.
33. Lundin M, Lundin J, Helin H, Isola J. A digital atlas of breast histopathology: an application of web based virtual microscopy. *J Clin Pathol* 2004, 57: 1288–1291.
34. Lee SH. Virtual microscopy: applications to hematology. *Lab Hematol* 2005, 11: 38–45.

CHAPTER 8

HARMONIZING CANCER CONTROL IN EUROPE

ULRIK RINGBORG

Professor of Oncology, Karolinska Institute, Director of Cancer Center Karolinska, President, Organization of European Cancer Institutes

Cancer is an increasing global problem. According to WHO statistics [1,2] in 2000 we had 10.1 million new cases in the world. In 2020 we expect to have around 16 million new cases. During 2000 there were 6.2 million deaths due to cancer and in 2020 we expect around 10 million deaths unless we act more effectively. In 2000 there were 22.4 million people living with cancer. No doubt cancer diseases are causing an increasing problem worldwide and the problem is a fact also for Europe.

If we go to Sweden the incidence increase over the last decades has been 1.2% per year [3]. Of great interest is the prevalence trend. Sweden has had a prevalence increase of 3% per year [4]. An important part of the increase is caused by patients living longer with chronic disease. Thus, time-trends in survival show over time an improvement and a part of this survival benefit contributes together with outcome of prevention to decreased mortality. From an epidemiological point of view, we have to realize that the cancer problem is increasing. Even if outcome is improved we will have more patients with chronic disease living longer, which will have an impact on the allocation of resources in the health care.

There are pronounced inequalities in Europe. Mortality rates due to cancer are changing in a positive direction in most European countries, but not all. The decrease in mortality is also variable [5,6]. Estimation of survival after a cancer diagnosis shows pronounced differences if European countries are compared [7]. At the Conference organized by the Organization of European Cancer Institutes (OECI) in Piraeus, Greece, 2005 the problem of inequalities in Europe was discussed. It was concluded that there are pronounced differences, a problem which must be solved.

Corresponding author: Radiumhemmet, Karolinska University Hospital Solna, S-171 76 Stockholm, Sweden.

1. PROBLEMS IN EUROPEAN CANCER CARE AND RESEARCH

European cancer care is often criticized because it is fragmented. A unique feature of the cancer care is the complexity of needs of cancer patients. A fragmented health care organization is related to an insufficient global view of the patient. The cancer care is getting more complex over time due to new diagnostic tools and treatments. A serious criticism overall is a suboptimal implementation of new diagnostic and treatment methods.

European cancer research is also often criticized due to fragmentation. An important reason is the funding system in Europe, which does not stimulate enough collaboration and may even allow iteration of research. It is often said that the research has a limited effect on cancer care and prevention. This is probably correct, but mainly not dependent on the research. The limiting factor is often the health care, which suffers from an insufficient organization for implementation of new methods.

Nevertheless there are a number of research problems. There is a rapid expansion of knowledge in basic research. Due to the health care organization there are often restrictions for the clinical research. A bottleneck in the research process is the possibility to run clinical trials. Quite few cancer patients are involved in clinical trials. Further, there is a need to develop biologically driven clinical trials, which requires more specialized infrastructures not available in all hospitals. By tradition there is a gap between the basic and clinical research, a gap, which is preserved both by the funding mechanisms and education. Education is of great importance due to the rapid development in the basic research areas. It is therefore difficult for clinicians to follow this rapid development. It is also difficult for basic researchers to have insights in relevant clinical questions. There is a need to speed up the research process with the aim to shorten the time duration from important basic discoveries to implementation of new clinical knowledge based on these discoveries.

At present we are able to cure about 60% of patients with a cancer diagnosis. With a failure in about 40% of patients innovation is of great importance. Between a research result and a medical product there is a process named development. There may be significant differences if we compare outcome in a clinical trial on a selected group of patients and in routine care where all patients are treated. Therefore, new research information must be introduced into clinical practice in an organized way to make evaluation in clinical practice possible. The innovative potential of the cancer care organization is today far from optimal. This is demonstrated in a recently performed Pan-European comparison regarding patient access to different cancer drugs [8]. An analysis of introduction of anticancer drugs were compared in 19 European countries. Main differences between countries were demonstrated and overall, the introduction of new drugs, which has shown in clinical trials to be effective and are therefore registered, is a slow process. It is important to establish routines for introduction and clinical evaluation of new drugs.

It is also important to realize that this phenomenon is not restricted only to drugs. We often see unacceptable time durations when new diagnostic tools as well as new treatment modalities of other types will be implemented in clinical routine. The research potential in cell and molecular biology gives us enormous possibilities to improve cytology and pathology. Innovations in physics give us unique possibilities to develop imaging and radiation therapy. It is important to strengthen and more carefully analyse methodology behind innovation. If we are not able to use new research information for clinical implement them in an optimal way, it will be difficult to argue for more resources for cancer research.

Research financing is a problem for Europe. If we go to USA about seven times more public money is spent on cancer research [8,9]. Of interest is that the allocation of research money in USA is better organized. The majority of money is distributed to well functioning comprehensive cancer centers. In Europe we have fragmented systems both for research and funding. In Asia a number of research centers with tremendous financial support have grown up. To be competitive it is important for the European cancer research centers to find new ways to collaborate and we need more efficient ways to support funding. There are important advantages for research in Europe. There is an important intellectual capital. There are also health care organizations, which may be helpful to develop patient-oriented clinical research based on patient databases and biobanks. The strong emphasis on basic research in Europe will in future have an impact on the clinical research.

2. AIMS OF ACTIVITIES AGAINST CANCER

The overall aims of activities against cancer are reduction of mortality, reduction of morbidity, increase of survival, and increase of quality of life. The importance of quality of life is increasing. As earlier mentioned, improved treatments increase the survival of cancer patients, and their lives must be meaningful. Often, treatments are complicated by side effects.

With a global view of the patient, quality of life must be more in focus. In order to use all possibilities to reach the overall aims it is important to integrate care, prevention, research, development, and education. Prevention strategies should be used to a larger extend since it is often more efficient and more economic to prevent disease than wait until treatment is a necessity. Effects of the European code against cancer were recently demonstrated [10]. Encourage by the outcome a new programme has started [11]. It is also more efficient to treat early disease than advanced disease. We therefore need a rational balance when we compare resources for prevention and care. In order to optimize the clinical research integration between care and research is important and for innovation development must be visible. The integration of education is a necessity not only for the quality of care, but also for other cancer activities. An organization where care, prevention, research, development, and education are integrated is named comprehensive.

3. EUROPEAN ORGANIZATION OF EUROPEAN CANCER INSTITUTES FOCUS AT THE GLOBAL VIEW OF THE CANCER PATIENT AND THE CANCER ACTIVITIES

The OECI is a nongovernmental and nonprofit organization. Recently it became a European Economic Interest Grouping, which is a legal entity within the European Union. The Organization at present links more than 80 cancer institutes in Europe. OECI is working by developing a common vision of future European oncology and has a programme for the development of comprehensiveness by integrating care, prevention, research, development, and education. Important objectives for a cancer organization are quality, innovation, and evaluation. If quality aspects will be covered the global view of the cancer patient is important. With a global view a consequence is multidisciplinarity in the diagnosis and treatment of cancer patients. For innovation the research should aim at improving diagnostics, care, and prevention, i.e., translational research. Continuous evaluation gives us the information about areas, which should be developed.

For prevention and care we need a number of functions. Primary prevention is mainly a public issue, not primarily linked to the care system. The secondary prevention or early detection is on the other hand more linked to the care system. We have to make a distinction between prevention, which is population based and prevention focused at high-risk individuals. There are evidence-based programmes for population-based screening [12]. There is also an increasing prevention activity, both primary and secondary prevention, focused at high-risk individuals selected by diagnostic techniques in the health care system. For the care pathology, cytology, imaging, and laboratory medicine are important. Treatment is often multidisciplinary and includes surgical treatment, radiation therapy, and medical oncology. Important areas are psychosocial oncology and rehabilitation. Both for primary treatment but also palliation supportive care is a growing area and palliative oncology is another important function. Palliative oncology is often divided in an early phase where the cure is no longer possible but prolongation of survival is the aim. The late palliative phase is characterized by symptomatic treatments of different types, but without possibilities to prolong survival.

There are a number of research functions, which are important to improve the situation for cancer patients in the long run. To cover the translational research process we need laboratories for basic research as well as preclinical laboratory research. There must be an organization for clinical trials. The patient databases are getting more and more important as well as the biobanks. Combining work with biobanks and patient databases with different genomic techniques gives us extraordinary opportunities to develop molecular clinical epidemiology. Outcome research must be established for quality assessment and identification of needs in the care system.

Education activities in a cancer organization are of different types. Basic training should aim to educate medical students and nurses to get a global view of the cancer problem. A cancer organization must have resources for specialist training as well as continuous medical education. Research education is important and we have to stress the need to increase the number of physicians with a research education. Education of patients and relatives are getting more important in future and it is also mandatory to give decision makers of different types the relevant information for the future.

Obligatory characteristics of a cancer organization are the comprehensiveness. In several countries the comprehensive cancer centers are complete cancer hospitals. In other countries the clinics treating cancer patients are localized in regular hospitals, often in university hospitals. To be able to integrate research, development, and education with care and prevention we need the construction of virtual centers. In the countries where a comprehensive cancer center is a cancer hospital, a substantial number of patients are often treated at regular hospitals in the surrounding geographic area. In order to offer the patients innovative diagnostics and treatments but also to be able to include patients in clinical trials a close cooperation between the cancer center and surrounding clinics is necessary. It is also important that the cancer center has a contact with the university. For clinical departments localized at university hospitals it is important to develop a virtual organization, which will guarantee the comprehensiveness. There are advantages and disadvantages with both types of organizations, centers as cancer hospitals and coordinated cancer centers. With the concept of comprehensiveness in mind the necessary virtual organization should be established and formalized. If not, Europe will continue to decrease the competition potential compared with American oncology.

In future, oncology will to a larger extent require a closer collaboration between clinical and preclinical researchers in order to establish a successful translational research. It will also require the availability of detailed information about patients but also biological materials in order to stimulate biologically driven clinical trials and develop a more sophisticated clinical epidemiology. By time the different cancer diagnoses are split in to increasing number of subgroups. The development of molecular pathology will probably enlarge the number of subgroups and give us the possibility of treatment prediction. In this perspective, all centers will be incomplete regarding number of patients as well as advanced analytical methods. Therefore it is important to build cancer centers with a similar infrastructure and to stimulate collaboration between European centers.

4. ACCREDITATION OF EUROPEAN CANCER CENTERS

With the aim to establish a European future vision of oncology OECI has started a project for accreditation of cancer centers. The accreditation will be a voluntary process (self-assessment) to improve the infrastructure and quality of care. It will include consensus upon defined guidelines and compliance with guidelines will be the bases of labelisation, which includes an external assessment. There is a steering group with the following person: W van Harten (Amsterdam), R Otter (Groningen), U Ringborg (Stockholm). M Sagatchian (Villejuif), T Tursz (Villejuif) and D de Valeriola (Brussels). M Sagatchian is the chairperson for the project. The following persons are involved in the accreditation working group: A Llombart-Bosch (Valencia), C Lombardo (Brussels), M de Lena (Bari), B Maciejewski (Gliwice), Ch Mahler (Antwerp), Th Philip (Lyon) and G Storme (Brussels). There are two project managers, H. Hummel (Groningen) and N. Nabet (Villejuif).

The project has different parts. The first part will cover measurable characteristics of the cancer centers including different infrastructures. In order to describe quantitative aspects a questionnaire has been constructed and a second version of the questionnaire with about 300 questions is processed further in order to be validated. By the questionnaire it will be possible to establish a database of cancer centers and institutes in Europe. Data will be collected for care, prevention, research, and education.

For quality assessment a separate manual is under construction. There are already a manual from France and another one from the Netherlands. At present a work is ongoing to merge the two manuals in order to create a product, which can be used at the European level. The questionnaire and the manual will be harmonized in order to be complementary. With the accreditation methodology it will be possible to create a labelisation system for cancer centers/institutes in Europe with criteria for cancer units, cancer research institutes, cancer departments, cancer centers, and comprehensive cancer centers. Later on it will also be possible to identify criteria for excellence.

According to the time schedule the project will be finished within a 3-year period and the methodology will be available for the cancer centers/institutes. OECI is focusing on the global view of cancer activities, infrastructures, and organization. The OECI accreditation methodology will be a core. In future it will be extended to the specialized functions of a cancer center. It has already been agreed upon collaboration with EORTC and the different FECS organizations with the aim to create a unified European collaboration on accreditation with the aim to stimulate innovation of the European cancer activities with the focus on present and future cancer patients.

5. SUMMARY

Harmonizing cancer control in Europe should be a win situation not only for the cancer patients, but also for the different cancer research organizations. With the common vision of European oncology it will be possible to formulate strategic goals. One of these goals is the establishment of comprehensive cancer centers, either cancer hospitals or coordinated comprehensive cancer centers. It is important to identify the necessary structures needed to increase the innovation potential, and this includes not only care and diagnostic technologies but also prevention. We need to establish a harmonized evaluation procedure. Positive outcome measurements are important to demonstrate the effects of research on the treatment of our patients. For optimal innovation the establishment of translational research environments at the cancer centers are mandatory. In order to avoid fragmentation both in care and research, European collaboration must increase.

REFERENCES

1. Ferlay J, Bray F, Parkin DM, Pisani P. Cancer incidence and mortality Worldwide (IARC Cancer Bases no. 5). Globocan 2000, Lyon. IARC Press.
2. Stewart BW, Kleihues P. World Cancer Report. International Agency for Research on Cancer. WHO. 2003, Lyon.
3. Cancer Incidence in Sweden. The National Board of Health and Welfare, 2003.
4. Stenbeck M, Rosén M, Sparén P. Causes of increasing cancer prevalence in Sweden. *Lancet* 1999; 354: 1093–1094.
5. Levi F, Lucchini F, Negri E, Boyle P, La Vecchia C. Mortality from major cancer sites in the European Union 1955–1998. *Annals of Oncology* 2003; 14: 490–495.
6. Quinn MJ, d'Onofrio A, Möller R, Black R, et al. Cancer Mortality trends in the EU and acceding countries up to 2015. *Annals of Oncology* 2003; 14: 1148–1152.
7. Survival of Cancer Patients in Europe: the EUROCARE-3 Study. *Annals of Oncology* 2003; 14: Supplement 5.
8. Wilking N, Jönsson B. A pan-European comparison regarding patient access to cancer drugs. Karolinska institutet in collaboration with Stockholm School of Economics, Stockholm, Sweden, 2005.
9. Eckhouse S, Castanas E, Chieco-Bianchi L, Cinca S, et al. European Cancer Research Funding Survey. European Cancer Research Managers Forum, 2005.
10. Boyle P, d'Onofrio A, Maisonneuve P, Severi G, Robertson C, Tubiana M, Veronesi U. Measuring progress against cancer in Europe: has the 15% decline targeted for 2000 come about? *Annals of Oncology* 2003; 14: 1312–1325.
11. Boyle P, Autier P, Bartelink H, Baselga J, et al. European Code Against Cancer and scientific justification: third version (2003). *Annals of Oncology* 2003; 14: 973–1005.
12. Lynge E. Recommendations on cancer screening in the European Union. Advisory Committee on Cancer Prevention. *European Journal of Cancer* 2000; 36: 1473–1478.

CHAPTER 9

THE DIAGNOSIS OF CANCER: "FROM H&E TO MOLECULAR DIAGNOSIS AND BACK"

JOSE COSTA

Department of Pathology and Yale Cancer Center, Yale University School of Medicine, New Haven, Connecticut, USA

1. INTRODUCTION

The diagnosis of a patient enables us to predict the fate of the person and to decide the most effective course of treatment. This decision is based on knowledge about the natural history of the disease and the characteristics of the patient.

In order to gain the required knowledge about the spontaneous course of disease, it is important to classify diseases in distinct categories; only then can we begin to test the efficacy of different forms of therapy.

The classification of tumors has been driven by their site of origin combined with their histopathological appearance. Until the last 25 years of the 20th century, prognosis and therapy were essentially based on morphology (i.e., diagnosis, grading, and staging), but after 1975 the functional characteristics of the tumors begin to play a role in the classification of disease. Monoclonal antibodies put molecular entities at easy reach for histopathologists and different populations of cells difficult to distinguish on the basis of conventional stains can be dissected with relative ease and precision on frozen or paraffin-embedded sections. Concomitantly basic, mechanistic research on cancer cells began to unravel the molecular defects responsible for the emergence and maintenance of the malignant phenotype and this knowledgebase spawned the field of molecular diagnostics.

The capability to demonstrate the molecular alterations responsible for the cancerous nature of a cell or tissue provides a novel and very precise criterion for the diagnosis, prognosis, and classification of disease. Furthermore, in the instances when the molecular defect can somehow be repaired or neutralized, molecular diagnosis provides the indication for targeted molecular therapy. By defining the molecular alterations present in cancer, irrespective of the morphology, we can build a molecular nosology. In this fashion we can establish relationships between diseases that are very distant in terms of their manifestations. The efficacy of Gleevac to treat GIST, a tumor initially not targeted by the drug, provides a dramatic example of the power of this approach.

Rapid technological advances including sequencing, genotyping (SNP's), functional genomics (expression arrays), and the other "omics" provide us with "new microscopes" to analyze lesions found in tissues or molecular profiles in biological fluids. By correlating the patterns seen in very large quantitative data sets to the basic features of the disease we can further our knowledge and change classification and diagnosis.

2. THE TENSION

Today, two facts challenge the time-honored preeminence of histopathology in the diagnosis and classification of cancer: first objective and robust molecular tests can provide the answer to patient management questions that make a difference; second the "new microscopes" provide insights not previously attainable by pure morphology.

The tension is thus set between these two major approaches (morphology vs function at the molecular scale), but tension can serve to provide substantial strength if one understands how to resolve it. Calatrava, among others, has shown us how to build from tension!

3. RESOLUTION OF THE TENSION AT THE PRESENT TIME

The practice of oncologic pathology today has settled in a mode that can be characterized by the judicious use of molecular diagnostics to augment the resolutive power of anatomical pathology. This "balanced mode" brings progress both to surgical pathology and cytology.

Many diagnostic problems and clinically relevant questions in anatomical pathology can be resolved using molecular techniques. The most widely used involves comprehensive immunophenotyping of neoplasms. When the results are nonconclusive analysis of structural genetic alterations such as rearrangements, deletions or point mutations can "nail down" a diagnosis. Demonstration of chimeric transcripts by RT-PCR contributes to the precise classification of soft tissue and bone tumors when the phenotype is equivocal. Clonal markers serve as convenient indicators of common origin and establish the relationship or independence of metachronous disease. In some instances pathognomonic genetic lesions not only provide sound and robust diagnostic evidence but also, as is the case for oligodendrogliomas they contribute to guide therapy.

4. SOME POSSIBLE AVENUES FOR THE FUTURE

I like to suggest that the resolution of the tension between histopathological and molecular diagnosis, thus the "way back," can be found in Integrative Tumor

Biology. A discipline that uses the tools of computational biology to integrate very diverse data sets concerning tumors and their hosts. This approach not only furthers our understanding of tumors, but it also generates combinations of parameters that yield predictive, personalized diagnosis, and therapies.

Integrative Tumor Biology makes two contributions; first, it enables the treatment of complexity (e.g., systems biology/pathology); and second, it recognizes space (e.g., tissue architecture) as a powerful determinant of the biology of disease.

Systems biology integrates quantitative interaction of all the elementary parts constituting a whole, to examine and define the emergence of new properties that stem from these dynamic interactions.

Space can act as a strong shaping force on the biology of cancer. The work of Novak's group illustrates how the size of the patches can change the composition of the cells that populate the patch. Empirical data and simulation studies carried out in my laboratory indicate that the study of the dynamics among populations of mutated somatic cells reflects both the degree of genetic instability and the strength of selection.

There is yet another approach that may breathe life into morphology. I like to return to the value of morphology by asking the question: "can rigorous mathematical treatment of morphometric data reveal hidden structures in the data, and are the patterns in themselves informative and clinically relevant?"

Until now we have apprehended the phenotype, color, and shape, using our "onboard" computer. Our brain is wired to recognize patterns very efficiently, but has trouble building the pattern from the quantitative analysis of the parts. However it turns out that machine learning finds information encrypted in morphology, information that is hidden from and cannot be accessed by our natural powers.

The team headed by Saidi at Aureon Laboratories has presented evidence that morphometric parameters that can be extracted from tissue possess fractal structure, a structure that enables the distinction between normal and tumor state. Aureon's team has shown that automated image processing, data extraction, cleansing, and analysis can yield a diagnosis of normal versus tumor with specificity superior to 90%. It is possible to think that this line of work may produce automated algorithms that could extract predictive information from histopathology.

I would like to conclude by suggesting to you that the seamless integration of form and function at the cell, tissue and organ level, will be achieved by functional imaging. Here the in vivo study of form and function is truly fused and can be carried out in a dynamic fashion. Until that moment we will have to hobble along the best we can with some of the approaches here described.

CHAPTER 10
FROM MORPHOLOGICAL TO MOLECULAR DIAGNOSIS OF SOFT TISSUE TUMORS

MARKKU MIETTINEN

Department of Soft Tissue Pathology, Armed Forces Institute of Pathology, Washington, DC

Abstract: Cytogenetic discoveries of balanced translocations in soft tissue tumors have opened the way to molecular genetic definition of these translocations as gene fusions from the late 1980s. Many sarcomas are known to have such fusions, and the demonstration of the fusion transcripts in tumor tissue is of great value in specific diagnosis of synovial sarcoma (SYT-SSX), Ewing sarcoma (EWS-Fli1), clear cell sarcoma (EWS-ATF1), myxoid liposarcoma (FUS-CHOP), and other sarcomas. These translocations are believed to be disease-specific and pathogenetic forces, despite occasional observations to the contrary. Demonstration of SYT-SSX and EWS-ATF1 fusion assists in the diagnosis of synovial and clear cell sarcomas in unusual locations, such as the gastrointestinal tract, where these tumors occur with low frequency. Demonstration of sarcoma translocations and their fusion by different assays is well established; use of in situ hybridization is limited by availability of specific probes. In two exceptional instances, the same translocation and gene fusion occurs in two unrelated diseases: ETV6-NTRK fusion in infantile fibrosarcoma and secretory carcinoma of the breast, and ALK-TPM3 fusion in inflammatory myofibroblastic tumor and large cell anaplastic lymphoma. Thus, the target cell of the genetic change is an important factor to define the resulting disease. Activating mutations in two related receptor tyrosine kinases (RTKs), KIT, and platelet-derived growth factor receptor alpha (PDGFRA) is central to the pathogenesis of gastrointestinal stromal tumors (GISTs), and countering the mutational activation by specific tyrosine kinase inhibitors, such as Imatinib mesylate, is now standard treatment for metastatic GISTs. KIT exon 11 mutations (in frame deletions, point mutations, and duplications) occur in GISTs of all locations, whereas a characteristic exon 9 insertion–duplication AY502-503 is nearly specific for intestinal vs gastric tumors. In contrast, PDGFRA mutations are nearly specific for gastric GISTs, especially those with epithelioid morphology. Mutation type influences therapy responsiveness, but fortunately very few GISTs carry primarily Imatinib-resistant mutations. Secondary drug resistance acquired during Imatinib treatment based on new, Imatinib-resistant mutations is a major problem limiting treatment success. Loss of NF2 tumor suppressor gene in a biallelic fashion is believed to be central in the pathogenesis of neurofibromatosis 2 (NF2) associated and sporadic schwannomas and meningiomas. The mechanism includes nonsense or missense mutation in NF2 gene, and loss of the other NF2 allele as a part of losses in chromosome 22q. Schwannoma types may differ in their pathogenesis: gastrointestinal schwannomas lack NF2 changes suggesting a different pathogenesis. Intraneural and sclerosing perineuriomas display similar NF2 gene alterations as seen in meningioma, indicating a similar pathogenesis and molecular homology. Specific viral sequences of human herpesvirus 8 (HHV8) are diagnostic markers for Kaposi sarcoma (KS), and are absent in angiosarcoma. Despite discovery on simian virus SV40 sequences in mesothelioma as a possible pathogenetic factor, recent studies suggest that the presence of these sequences may be artifactual and based on common presence of some SV40 sequences as PCR contaminants.

Key words: Sarcoma, translocation, fusion transcript, gastrointestinal stromal tumor, mutation, KIT, PDGFRA, schwannoma, perineurioma, NF2, human herpesvirus 8

1. INTRODUCTION

The purpose of this chapter is to review examples of application of molecular pathology as a tool in the diagnosis of soft tissue tumors and understanding their pathogenesis, and discuss problems related to the application of molecular pathologic analysis. Examples of the different molecular changes include tumor translocations, mutational activation of oncogenes, tumor suppressor gene alterations, and presence of viral sequences.

2. TUMOR-SPECIFIC FUSION TRANSLOCATIONS IN SOFT TISSUE SARCOMAS

2.1. General comments

Cytogenetic studies form the mid-1980s and on revealed several recurrent translocations in soft tissue sarcomas, such as t(X;18) in synovial sarcoma, t(11;22) and others in Ewing family tumors, t(12;22) in clear cell sarcoma of tendons and aponeuroses, t(9;22) in extraskeletal myxoid chondrosarcoma, t(11;22) in desmoplastic small round cell tumor (DSRCT), t(2;13) in alveolar rhabdomyosarcoma, t(17;22) in dermatofibrosarcoma protuberance, and t(X;17) in alveolar soft part sarcoma, among others (Borden et al. 2003; Lasota 2003). Thus, typical translocations occur in most sarcomas that are composed of morphologically homogenous tumor cells with limited overall atypia. It is a great tribute to morphological pathologists that many of the earlier described tumor entities have been confirmed as distinct genetically defined entities with a specific translocation — in fact in every case, description of the cytogenetic changes was preceded by morphological definition of the tumor.

In contrast to many of the morphologically homogeneous tumors, no tumor-specific translocations are known for sarcoma types generally characterized by high degree of atypia, such as leiomyosarcoma, malignant peripheral nerve sheath tumor, pleomorphic liposarcoma, and malignant fibrous histiocytoma.

2.2. Molecular diagnosis of sarcoma translocations

Sarcoma translocations often lead to fusions between functional domains of two different genes, resulting in formation of pathologic fusion transcripts and ultimately fusion proteins many of which act as aberrant transcription factors and are key pathogenetic factors. Many of these genes encode for nucleic acid-binding nuclear proteins such as transcription factor, or other regulatory proteins.

Molecular diagnosis of sarcoma translocations is possible by PCR-based methods that detect gene fusions or their fusion transcripts. Although sensitive and often feasible in formalin-fixed and paraffin-embedded tissue, these methods may have a problem of false positive results due to cross contamination of DNA or cDNA templates, especially if nested amplification is used, because the sizes and sequences of the PCR-products are generally identical. Fusion transcript assays require recovery of RNA, because the breakpoints are either unknown or too variable to be practically detectable at the genomic DNA level by PCR-based methods. However, Southern blot analysis is feasible for detection of these gene rearrangements in genomic DNA from fresh or frozen tissue.

The fact that many sarcoma translocations involve the Ewing sarcoma (*EWS*) gene in 22q12 as one of the translocation partners, offers a unique approach for diagnosis by detection of the breakage of the *EWS* gene in a fluorescent in situ hybridization (FISH) assay by a "break apart probe" (Lee et al. 2005). Although this assay does not detect the specific fusion transcript and would not specifically identify the type of fusion, it can be very useful to distinguish tumors with EWS break from their mimics that do not have such a break, e.g., in the differential diagnosis of Ewing sarcoma and poorly differentiated synovial sarcoma, because the latter has normal configuration of the *EWS* gene.

Although in general, sarcoma translocations are disease specific, there are two intriguing examples of the same translocation and gene fusion being present in two unrelated tumor types. One of these examples is the ALK-TPM3 fusion, which has been detected in inflammatory myofibroblastic tumor and large cell anaplastic lymphoma (Lawrence et al. 2000; Cools et al. 2002). Another example is ETV6-NTRK3 fusion translocation in infantile fibrosarcoma that also occurs in secretory carcinoma of the breast, an uncommon variant of ductal breast carcinoma that has a predilection to young women (Knezevich et al. 1998; Tognon et al. 2002). These examples illustrate that a tumor translocation is not always disease specific and that the same molecular lesion can produce a different disease in a different target cell.

2.3. Synovial sarcoma

Synovial sarcoma fusions involving SYT-SSX1, SYT-SSX2, or very rarely SYT-SSX4 genes, are specific for this tumor. Biphasic variants typically involve

SSX1, whereas monophasic variants can have SSX1 or SSX2 type fusions. Although initially fusion involving SSX2 were believed to be associated with a better prognosis, more recent large studies cast some doubt on that (Ladanyi 2001; Ladanyi 2002; Guillou 2004).

Molecular diagnosis is useful in confirming a poorly differentiated synovial sarcoma, which morphologically and immunohistochemically can simulate other sarcomas, especially Ewing family tumors, often also being CD99 positive. The lack of epithelial markers in some of these tumors adds to diagnostic difficulty and need to perform molecular diagnostic studies.

Also, diagnosis of synovial sarcoma in unusual locations, such as stomach, may need support of an independent molecular diagnosis. Indeed, there are gastric tumors that are histologically, immunohistochemically, and molecularly identical with the peripheral synovial sarcomas. Apparently only 1 case of primary synovial sarcoma of stomach has been reported so far (Billings et al. 2000). We have identified six cases among tumors originally believed to be "gastric leiomyosarcomas." Long survival in some patients mitigates against metastatic nature of the tumor.

Minute synovial sarcoma of hands and feet <1 cm is a diagnosis that can meet skepticism of clinicians, because these tumors are typically clinically believed to be benign processes; we recently analyzed 21 such biphasic and monophasic tumors and demonstrated SYT-SSX fusions to support the unexpected diagnosis.

Some reports utilizing PCR-based detection have suggested that other tumors, especially malignant peripheral nerve sheath tumors, and even neurofibromas can also have SYT-SSX2 fusions (O'Sullivan et al. 2001). These reports have been met with skepticism and not confirmed by others. The apparent detection of synovial sarcoma fusions may have been result of differences in tumor classification, or false positive assay resulting from PCR template contamination. Convincing arguments against the occurrence of SYT-SSX fusion translocations in MPNST include lack of cytogenetic evidence for t(X;18) translocation in any of these tumors, and lack of reproducibility of such findings in other large series (Ladanyi et al. 2001).

What is synovial sarcoma, a sarcoma with truly epithelial differentiation: is it more closely related to a carcinoma or mesothelioma? If we look the expression of markers, synovial sarcoma epithelia has features of both: it often has calretinin and keratin 5 expression, similar to mesothelioma, although not WT1 (Miettinen et al. 2001a, b). In our experience, markers more typical of carcinomas, such as BerEp4 and occasionally CEA are also expressed, along with a complex array of epithelial mucins, such as MUC1, MUC2, MUC5A, and MUC6. Therefore, by expression of markers of epithelia and mesothelia, synovial sarcoma has unique hybrid features not present in any normal cell type.

2.4. Desmoplastic small round cell tumor

One of the more recent sarcoma entities, DSRCT, was discovered based on its distinct immunohistochemical profile by Rosai and his coworkers (Gerald et al. 1991). This intra-abdominal small round cell tumor in children and young adult was previously variably believed to be carcinoid tumor, neuroblastoma, Ewing sarcoma, or rhabdomyosarcoma. The tumor was discovered as a small round cell tumor with nested growth pattern with a desmoplastic stroma, and the tumor being simultaneously keratin-desmin and NSE-positive, showing multidirectional differentiation. Presentation in children and young adults as an intra-abdominal tumor made DSRCT also a cohesive clinicopathological entity (Gerald et al. 1991). Soon the t(11;22) translocation was discovered in, but it was found to be different from the t(11;22) of Ewing sarcoma by the breakpoint in chromosome 11 being in p13 in the short arm (and not in the q24 in the long arm, as in Ewing sarcoma). Soon followed identification of the breakpoints involving Wilms tumor gene (*WT1*) in 11p13 and *EWS* gene in 22q (Ladanyi et al. 1994); it took less than 3 years to proceed form the initial description of the entity to the discovery of its specific gene fusion and molecular pathogenesis. In this tumor, *WT1* gene (tumor suppressor gene in Wilms tumor) is highly expressed to be a diagnostic marker not present in other small round cell tumors (Barnoud et al. 2000).

Subsequent discovery of identical tumors in the pleural cavity (Parkash et al. 1995), and the fact that DSRCT shares a number of features, coexpression of keratins and desmin and WT1-expression with mesothelial cells, especially the fetal ones (van Muijen 1987), raises the intriguing possibility that this tumor typically occurring in the serous cavities, could be related to mesothelial cells and their tumors.

2.5. Clear cell sarcoma

Clear cell sarcoma of tendons and aponeuroses, originally described by Enzinger (1965), is a rare sarcoma of peripheral soft tissues usually occurring in young subjects in distal extremities, often in association with tendons. This tumor has some biologic relatedness to melanoma, although it differs from melanoma in some respects. Although gene expression profiles of clear cell sarcoma and melanoma have many similarities (Segal et al. 2003), the occurrence of typical translocation is specific for clear cell sarcoma.

The t(12;22) translocation with EWS-ATF1 fusion characterizes clear cell sarcoma. Intestinal clear cell sarcoma is an unusual mesenchymal tumor that occurs in both stomach and intestines. Some of these tumors contain osteoclastic giant cells, and in these, the presence of clear cell sarcoma translocation t(12;22) has been cytogenetically confirmed (Zambrano et al. 2003). In our experience, these tumors occur in a frequency of 1 for every 300 GI stromal tumors. They are S100-positive and KIT-negative tumors that in contrast with peripheral clear cell sarcoma often lack HMB45-positivity.

3. ACTIVATING RECEPTOR TYROSINE KINASE MUTATIONS: GASTROINTESTINAL STROMAL TUMOR AS AN EXAMPLE

3.1. Introduction

RTKs are key molecules in signal transductions pathways, where a growth factor signal initiates activation of the receptor and transmission of the activation to cytoplasm and nucleus by a number of downstream signaling proteins. There are approximately 60 RTKs of different types (Pawson et al. 2002). A receptor activated in many carcinomas is human epidermal growth factor receptor (HER). The best known activated receptors in soft tissue tumors are KIT and PDGFRA in GISTs.

3.2. Background on gastrointestinal stromal tumors

GISTs are the most common mesenchymal tumors of the GI tract; they typically express KIT, and comprise a great majority of tumors previously classified as gastrointestinal smooth muscle tumors. GISTs occur throughout the GI tract, most commonly in the stomach and small intestine. These tumors are currently the best known examples of soft tissue tumors with activating RTK mutations. This finding has a great clinical significance because of the availability of effective KIT inhibitor, Imatinib mesylate, as an effective treatment for metastatic and unresectable GISTs (Demetri et al. 2002). Unfortunately, this success has been tempered by common development of drug resistance, by acquired new mutations and rarely by KIT gene amplification over 1–2 years after institution of the therapy (Chen et al. 2005; Debiec-Rychter et al. 2005).

3.3. KIT and PDGFRA as oncogenes

KIT and PDGFRA are structurally and evolutionally closely related type III RTKs. Both are growth factor receptors that are the starting points for receptor-initiated cell signaling pathways. Under normal circumstances, these receptors are activated (= phosphorylated) by binding of their respective ligands (stem cell factor for KIT and PDGF alpha for PDGFA), and they then activate downstream target proteins by phosphorylation in a cascade-like manner. Activation of the signaling pathways leads to changes in cellular motility, prevention of apoptosis, and promotion of cell proliferation (Ronnstrand 2004). Activating mutations cause activation of KIT signaling independent of ligand binding, and similar mechanism is operational with many other oncogenes.

Activating mutations in KIT and PDGFRA are driving forces in GIST oncogenesis. This hypothesis has been confirmed at several levels approaching the fulfillment of criteria analogous to Koch's postulate for infections. First, most GISTs have either KIT or PDGFRA mutations, and second, introduction of mutation specific for GISTs in lymphoblastoid cell lines causes them to proliferate (Hirota et al. 1998). Third, transgenic mice with introduced GIST-type KIT mutations develop a GIST-tumor syndrome (Sommer et al. 2003; Rubin et al. 2005). The presence of KIT mutations in already very small tumors also supports the initiating role of mutation (Corless et al. 2002), although other factors are probably also involved. There are characteristic genomic losses in GIST, commonly those of copies of chromosome 14 and 22. Their specific role in pathogenesis is not yet clear (El-Rifai et al. 1996).

3.4. KIT mutations is GIST

Approximately 60–65% of GISTs have KIT mutations, and 15% have PDGFRA mutations that are mutually exclusive. KIT mutations are clustered in a small number of hotspots that include exon 11, exon 9, exon 13, and exon 17 in the order of frequency. PDGFRA mutations occur in homologous exon positions, including exons 12, 14, and 18; the latter is most commonly mutated in this gene. Common to these mutations is that they preserve the reading frame and produce a mutant protein that renders constitutional activation to KIT or PDGFRA. Most KIT mutants are responsive to Imatinib mesylate inhibitor treatment, but some, especially the rare exon 17 mutations are primarily resistant (Corless et al. 2004).

A great majority (>90%) of KIT mutations occur in exon 11 (juxtamembrane domain). The juxtamembrane domain mutations apparently disturb this alpha helical domain of KIT leading to constitutional KIT activation (Longley et al. 2001).

KIT exon 11 mutations include in-frame deletions most commonly involving the region of codons 555–560. Missense point mutations in the same range and almost exclusively involve codons 557, 559, and 560. Codon 576 in 3' part of exon 11 is less commonly involved. Both types of mutation occur in GISTs of all sites. In gastric GISTs, tumors with KIT exon 11 point mutations have a better prognosis than those with in-frame deletions (Miettinen et al. 2005a, b, c), but in our experience, small intestinal GISTs there seems to be no similar prognostic difference between tumor with KIT exon 11 deletions vs point mutations.

Internal tandem duplications (ITDs) that add genetic material typically occur in the 3' portion of exon 11, most commonly in gastric GISTs and very rarely in small intestinal tumors. Tumors with these mutations have a generally favorable behavior, and in the stomach, they often represent mitotically inactive and collagen-rich tumors (Lasota et al. 2003b).

KIT exon 9 mutation involving a two-codon insertion/duplication AY502-503 in the extracellular domain is a relatively rare but clinicopathologically distinctive KIT mutation. It occurs almost exclusively in intestinal vs gastric GISTs, and its frequency among the small intestinal GISTs is approximately 10% (Lasota et al. 2000; Lasota et al. 2003c; Antonescu et al. 2003). Tumors with this mutation are not morphologically separable from other small intestinal GISTs. Although tumors with this mutation generally have an unfavorable prognosis, their behavior is not statistically different from small intestinal tumors with other types of KIT mutations in our experience. Therefore, the adverse prognostic of the tumors with the exon 9 mutations is related to the more commonly adverse outcome of small intestinal (vs gastric) GISTs.

3.5. PDGFRA mutations in GIST

PDGFRA mutations occur almost exclusively in gastric and not in the intestinal GISTs. These mutations have a strong predilection to gastric GISTs with epithelioid morphology (tumors formerly often designated as leiomyoblastomas), and their overall frequency may be as high as 15% of all gastric GISTs (Heinrich et al. 2003; Corless et al. 2004; Lasota et al. 2004). Unfortunately, antibodies for consistent immunohistochemical demonstration of PDGFRA by immunohistochemistry are not currently available.

Most commonly, PDGFRA mutations occur in exon 18 of PDGFRA, and a great majority of these are point mutations D842V. Less commonly, they involve exon 12, where in-frame deletions similar to those in the homologous KIT exon 11 occur.

GISTs with PDGFRA mutations have a generally favorable prognosis, which partly relates to the fact that gastric GISTs are more favorable than the small intestinal ones. Exon 12 deletions are rare; the number of reported cases does not allow reliable comparison between them and other mutants (Lasota et al. 2004). It remains to be seen how the differential gene expression profiles between KIT and PDGFRA mutant GISTs is related to different behavior of these tumors (Antonescu et al. 2004, Subramanian et al. 2004).

There is a very small group of gastric GISTs (<1%) that have PDGFRA exon 14 mutations. These seem to be associated with better prognosis than expected based on size and mitotic rate parameters (Lasota et al. in press).

3.6. Unusual GIST subgroups that lack KIT and PDGFRA mutations

There are at least two clinicopathologically distinctive subgroups of GISTs that lack KIT and PDFGRA mutations: GISTs in children and neurofibromatosis1 patients. Thus, the pathogenesis of these GIST subgroups is currently unknown.

GISTs in children are very rare. In this age group, GISTs occur almost exclusively in the stomach, and are more common in girls. Prognosis is unpredictable, and there is often a long course of disease, with metastases developing over 10–20 years or more; even patients with liver metastases can have a long survival (Prakash et al. 2005; Miettinen et al. 2005a, b, c). A small number of childhood GISTs occurs in connection with Carney triad, a syndrome-combining GIST with pulmonary chondroma, paraganglioma, or both (Carney 1999).

Neurofibromatosis 1 is the most common autosomal dominant disorder, representing the most common tumor suppressor gene syndrome (Viskochil 2002).

Multiple neurofibromas are typical, and development of malignant peripheral nerve sheath tumors is one of the major complications.

Of some reason, NF1 patients commonly develop GISTs, with the risk of this occurrence estimated to be over100-fold (Andersson et al. 2005). In this patient population, the clinicopathological spectrum of GIST is distinctive: occurrence of multiple, often small GISTs typically in the small intestine. These patients more commonly have a good prognosis, but of some reason, those with duodenal GIST have more frequently progressive disease (Miettinen et al. in press). The NF1 patients show diffuse Cajal cell hyperplasia that most likely is a precursor stage for GISTs, somewhat similar to that seen in patients with familial GISTs or transgenic mice with constitutional KIT mutations.

4. NF2 TUMOR SUPPRESSOR GENE ALTERATIONS IN SCHWANN CELL AND RELATED TUMORS

4.1. General comments on tumor suppressor genes

A number of genes regulating cellular growth and proliferation have been found contributing to tumor pathogenesis upon their inactivation. Such inactivation typically is biallelic, and mechanism of inactivation includes most commonly allelic losses and mutations. Examples of tumor suppressor genes include retinoblastoma gene, neurofibromatosis 1, and *NF2* genes. The original concept was devised by Knudson (1971) on hereditary retinoblastoma as an epidemiologically based hypothesis before the actual retinoblastoma gene alterations were discovered.

4.2. Neurofibromatosis 2 gene

NF2 gene is located pericentromerically in the long arm of chromosome 22q12. It encodes for a cytoskeleton-associated protein merlin (also known as schwannomin) that regulates the growth of schwann cells and related cells (Xiao et al. 2003). Patients with the hereditary NF2 syndrome typically have an inactivating NF2 mutation, leading to loss of one fuctional allele. Tumors developing in NF2 syndrome: vestibular schwannomas, meningiomas, and pilocytic astrocytomas, usually have somatic allelic losses of the other copy of NF2 protein, leading to the loss of both NF2 alleles in the scheme of a recessive tumor suppressor gene requiring inactivation of both alleles for loss of function. Type of NF2 mutation with different degrees of NF2 alterations varying from total loss to alteration is a factor determining the severity of NF2 disease (Ruttledge et al. 1996).

4.3. NF2 alterations in schwannoma and meningioma

In addition to being a typical feature of the NF2 syndrome-associated schwannomas, peripheral schwannomas and meningiomas have similar alterations as somatic changes: e.g., allelic losses of chromosome 22q, including the

NF2 locus, and nonsense or missense mutations of NF2, rendering both NF2 loci nonfunctional (Baser et al. 2003).

4.4. Gastrointestinal schwannoma

Gastrointestinal schwannomas are rare, distinctive nerve sheath tumors specific to the GI tract. Especially in earlier days, these tumors were routinely confused with gastrointestinal stromal and smooth muscle tumors. However, these tumors are negative for KIT and positive for S100 protein. They usually occur in the stomach or colon in older adults, and have characteristic histological features, including peripheral lymphoid cuff, microtrabecular pattern, and focal nuclear atypia (Daimaru et al. 1988; Miettinen et al. 2001a, b). These tumors arise from the autonomic nervous system schwann cells in the walls of stomach or intestines. GI schwannomas have neither losses of NF2 allele nor NF2 mutations, indicating that their pathogenesis (and possibly proper classification, at least in terms of molecular pathology) is different form classical schwannomas (Lasota et al. 2003d).

4.5. Perineurioma: an NF2 mutant tumor

Perineurial cell tumors (perineuriomas) are rare nerve sheath tumors that feature perineurial cell differentiation. Perineurial cells are EMA-positive spindle cells that surround peripheral nerves and represent peripheral continuation of pia arachnoid meningeal cells. Perineuriomas are typically composed of slender, tapered spindle cells, in some cases forming onion skin-like formations, and in other cases forming meningioma-like patterns or trabecular infiltrates. One of their rare variants, intraneural perineurioma, is known to have losses in chromosome 22q, but NF2 mutations have not been analyzed (Emory et al. 1995; Brock et al. 2005). Sclerosing perineurioma is a distinctive, perhaps the most common variant of perineurioma that typically occurs in hands and fingers of young adults as a small nodule (Fetsch et al. 1997). Previously these tumors were considered fibromas or sclerosing glomus tumors.

Sclerosing perineuriomas demonstrate allelic losses in 22q, and they also have missense mutations in *NF2* gene, indicating that they have molecular homology with meningiomas (Lasota et al. 2001). However, similar mutations have not been found in other variants of perineuriomas suggesting genetic and perhaps conceptual heterogeneity.

5. VIRAL SEQUENCES IN SOFT TISSUE TUMORS

5.1. Kaposi sarcoma, angiosarcoma, and HHV8

The best documented example of viral presence and pathogenesis in human sarcoma is HHV8 in KS. This gammaherpesvirus occurs in endemic, as well as immunosuppression-associated KS, and it is more common in populations with higher frequency of this tumor, e.g., in Mediterranean region. HHV8 is believed to be a key pathogenetic factor for KS by interfering with host tumor suppressor mechanisms (Moore et al. 2004).

It has been somewhat contested whether HHV8 is truly specific for KS. Some PCR-based studies have revealed it in some angiosarcomas (McDonagh et al. 1996), whereas others did not find this virus in angiosarcoma or other non-Kaposi type vascular tumors (Lebbe et al. 1997; Lasota and Miettinen 1999). Cross-contamination of PCR templates is a significant problem in demonstration of viral sequences that are generally identical between positive specimens and can lead to overreporting of viral sequences in PCR-based studies.

Immunohistochemical demonstration of HHV8 viral proteins, especially the latent nuclear antigen-1, is a practical way to assess the presence of these viral sequences. Studies on vascular tumors have found consistent immunoreactivity in KS. However, no HHV8-immunoreactivity has been found in angiosarcoma, hemangioma, and other vascular tumors, indicating that HHV8 virus is generally not present in non-Kaposi type of vascular tumors and that its demonstration is useful for the differential diagnosis between KS and non-Kaposi type vascular tumors (Cheuk et al. 2004; Patel et al. 2004).

5.2. Simian virus SV40 in mesothelioma

Several investigators have reported on simian virus 40 (SV40) sequences in malignant mesothelioma, and its possible pathogenetic role has been suggested (Pass et al. 2004). On the other hand, a recent report detecting SV40 sequences as common SV40 laboratory contaminants, based on their presence in commonly used plasmids, has raised the question whether the significance of SV40 findings in mesothelioma has been overestimated (Lopez-Rios et al. 2004). The role of SV40 as an etiopathogenetic factor for mesothelioma has also been questioned based on lack of increased mesothelioma risk in patients who received an SV40-contaminated adenovirus vaccine (Rollison et al. 2004).

REFERENCES

Andersson J, Sihto H, Meis-Kindblom JM, Joensuu H, Nupponen N, Kindblom LG, 2005, NF1-associated gastrointestinal stromal tumors have unique clinical, phenotypic, and genotypic characteristics. *Am J Surg Pathol* 29: 1170–1176.

Antonescu CR, Sommer G, Sarran L, et al. 2003, Association of KIT exon 9 mutations in nongastric primary site and aggressive behavior: KIT mutation analysis and clinical correlates of 120 gastrointestinal stromal tumors. *Clin Cancer Res* 9: 3329–3337.

Antonescu CR, Viale A, Sarran L, et al. 2004, Gene expression in gastrointestinal stromal tumors is distinguished by KIT genotype and anatomic site. *Clin Cancer Res* 10: 3282–3290.

Barnoud R, Sabourin J, Pasquier D, et al. 2000, Immunohistochemical expression of WT1 by desmoplastic small round cell tumor. A comparative study with other small round cell tumors. *Am J Surg Pathol* 24: 830–836.

Baser ME, Evans DG, Gutmann DH, 2003, Neurofibromatosis 2. *Curr Opin Neurol* 16: 27–33.

Billings SD, Meisner LF, Cummings OW, Tejada E, 2000, Synovial sarcoma of the upper digestive tract: a report of two cases with demonstration of the X;18 translocation by fluorescence in situ hybridization. *Mod Pathol* 13: 68–76.

Borden EC, Baker LH, Bell RS, Branwell V, Demetri GD, Eisenberg BL, et al. 2003, Soft tissue sarcomas of the adults: state of the translational science. *Clin Cancer Res* 9: 1941–1956.

Brock JE, Perez-Atayde AR, Kozakevich HP, Richkind KE, Fletcher JA, Vargas SO, 2005, Cytogenetic aberrations in perineurioma: variation with subtype. *Am J Surg Pathol* 29: 1164–1169.

Carney JA, 1999, Gastric stromal sarcoma, pulmonary chondroma, and extra-adrenal paraganglioma (Carney triad): natural history, adrenocortical component, and possible familial occurrence. *Mayo Clin Proc* 74: 543–552.

Chen LL, Sabinpour M, Andtbacka RH, Patel SR, Feig BW, Macapinlac HA, Choi H, Wu EF, Frazier ML, Benjamin RS, 2005, Imatinib resistance in gastrointestinal stromal tumors. *Curr Oncol Rep* 7: 293–299.

Cheuk W, Wong KO, Wong CS, Dinkel JE, Ben-Dor D, Chan JK, 2004, Immunostaining for human herpesvirus 8 latent nuclear antigen-1 helps distinguish Kaposi sarcoma from its mimics. *Am J Clin Pathol* 121: 335–342.

Cools J, Wlodarska I, Somers R, Mentens N, Pedeutour F, Maes B, de Wolf-Peeters C, Pauwels P, Hagemeijer A, Marynen P, 2002, Identification of novel fusion partners of ALK, the anaplastic lymphoma kinase, in anaplastic large cell lymphoma and inflammatory myofibroblastic tumors. *Genes Chromosomes Cancer* 34: 354–362.

Corless CL, McGreevey L, Haley A, Town A, Heinrich MC, 2002, KIT mutations are common in incidental gastrointestinal stromal tumors one centimeter or less in size. *Am J Pathol* 160: 1567–1572.

Corless CL, Fletcher JA, Heinrich MC, 2004, Biology of gastrointestinal stromal tumors. *J Clin Oncol* 22: 3813–3825.

Daimaru Y, Kido H, Hashimoto H, Enjoji M, 1988, Benign schwannoma of the gastrointestinal tract: a clinicopathologic and immunohistochemical study. *Hum Pathol* 19: 257–264.

Debiec-Rychter M, Cools J, Dumez H, Sciot R, Stul M, Mentens N, Vranckx H, Wasag B, Prenen H, Roesel J, Hagemeijer A, van Osteroom A, Marynen P, 2005, Mechanisms of resistance to Imatinib mesylate in gastrointestinal stromal tumors and activity of the PKC412C inhibitor against imatinib-resistant mutants. *Gastroenterology* 128: 270–279.

Demetri G, van Mehren M, Blanke CD, 2002, Efficacy and safety of imatinib mesylate in advanced gastrointestinal stromal tumors. *N Engl J Med* 347: 472–480.

El-Rifai W, Sarlomo-Rikala M, Miettinen M, Knuutila S, Andersson LCA, 1996, DNA copy number losses in chromosome 14: an early change in gastrointestinal stromal tumors. *Cancer Res* 56: 3230–3233.

Emory TS, Scheithauer BW, Hirose T, Wood M, Onofrio BM, Jenkins RB, 1995, Intraneural perineurioma: a clonal neoplasm associated with abnormalities of chromosome 22. *Am J Clin Pathol* 103: 696–704.

Enzinger FM, 1965, Clear cell sarcoma of tendons and aponeuroses: an analysis of 21 cases. *Cancer* 18: 1163–1174.

Fetsch JF, Miettinen M, 1997, Sclerosing perineurioma. A clinicopathologic study of 19 cases of a distinctive soft tissue lesion with a predilection for the fingers and palms of young adults. *Am J Surg Pathol* 21: 1433–1442.

Gerald WL, Miller HK, Battifora H, Miettinen M, Rosai J, 1991, Intra-abdominal desmoplastic small round-cell tumor: report of 19 cases of a distinctive type of high-grade polyphenotypyic malignancy affecting young individuals. *Am J Surg Pathol* 15: 499–513.

Guillou L, Benhattar J, Bonichon F, Gallagher G, Terier P, stauffer E, de Saint Aubain Somerhausen N, Michels JJ, Jundt G, Vinvce DR, Taylor S, Genevay M, Collin F, Trassard M, Coindre JM, 2004, Histologic grade, but not SYT-SSX fusion type, is an important prognostic factor in patients with synovial sarcoma: a multicenter, retrospective analysis. *J Clin Oncol* 22: 4040–4050.

Heinrich MC, Corless CL, Duensing A, et al. 2003, PDGFRA activating mutations in gastrointestinal stromal tumors. *Science* 299: 708–710.
Hirota S, Isozaki K, Moriyama Y, et al. 1998, Gain-of-function mutations of c-kit in human gastrointestinal stromal tumors. *Science* 279: 577–580.
Knezevich SR, Mcfadden DE, Tao W, Lim JF, Sorensen PH, 1998, A novel ETV6-NTRK3 gene fusion in congenital fibrosarcoma. *Nat Genet* 18: 184–187.
Knudson AJ, 1971, Mutation and cancer: statistical study of retinoblastoma. *Proc Natl Acad Sci USA* 68: 820–823.
Ladanyi M, Gerald WL, 1994, Fusion of the EWS and WT1 genes in desmoplastic small round cell tumor. *Cancer Res* 54: 2387–2840.
Ladanyi M, 2001, Fusions of the SYT and SSX genes in synovial sarcoma. *Oncogene* 20: 5755–5762.
Ladanyi M, Woodruff JM, Scheithauer BW, Bridge JA, Barr FG, Golblum JR, Fisher C, Perez-Atayde A, Dal-cin P, Fletcher CD, Fletcher JA, 2001, *Mod Pathol* 14: 733–737.
Ladanyi M, Antonescu CR, Lung DH, Woodruff JM, Kawai A, Healey JH, Brennan MF, Bridge JA, Neff JR, Barr FG, Goldsmith JD, Brooks JS, Goldblum JR, Ali SZ, Shipley J, Cooper CS, Fisher C, Skytting B, Larsson O, 2002. Impact of SYT-SSX fusion type on the clinical behavior of synovial sarcoma: a multi-institutional retrospective study of 243 patients. *Cancer Res* 62: 135–140.
Lasota J, Miettinen M, 1999, Absence of Kaposi's sarcoma-associated virus (human herpesvirus 8) sequences in angiosarcoma. *Virchows Arch* 434: 51–56.
Lasota J, Wozniak A, Sarlomo-Rikala M, Rys J, Kordek R, Nassar A, Sobin LH, Miettinen M, 2000, Mutations in exons 9 and 13 of KIT gene are rare events in gastrointestinal stromal tumors. A study of two hundred cases. *Am J Pathol* 157: 1091–1095.
Lasota J, Fetsch JF, Wozniak A, Wasag B, Sciot R, Miettinen M, 2001, The neurofibromatosis type 2 (NF2) gene is mutated in perineurial cell tumors: a molecular genetic study of 8 cases. *Am J Pathol* 158: 1223–1229.
Lasota J, 2003, Genetics of soft tissue tumors. In: Miettinen M (ed.) *Diagnostic soft tissue Pathology*, Churchill Livingstone, New York.
Lasota J, Dansonka-Mieszkowska A, Stachura T, Schneider-Stock R, Kallajoki M, Steigen SE, Sarlomo-Rikala M, Bolze C, Kordek R, Roessner A, Stachura J, Miettinen M, 2003b, Gastrointestinal stromal tumors with internal tandem duplications in 3' end of KIT juxtamemberane domain occur predominantly in stomach and generally seem to have a favorable outcome. *Mod Pathol* 16: 1257–1264.
Lasota J, Kopczynski J, Sarlomo-Rikala M, Schneider-Stock R, Stachura T, Kordek R, Michal M, Bolze C, Roessner A, Stachura J, Miettinen M, 2003c, KIT 1530ins6 mutation defines a subset of predominantly malignant gastrointestinal stromal tumors of intestinal origin. *Hum Pathol* 34: 1306–1312.
Lasota J, Wasag B, Dansonka-Miszkowska A, Karcz D, Millward CL, Rys J, Stachura J, Sobin LH, Miettinen M, 2003d, Evaluation of NF2 and NF1 tumor suppressor genes in distinctive gastrointestinal nerve sheath tumors traditionally diagnosed as benign schwannomas: study of 20 cases. *Lab Invest* 83: 1361–1371.
Lasota J, Dansonka-Meiszkowska A, Sobin LH, Miettinen M, 2004, A great majority of GISTs with PDGFRA mutations represent gastric tumors with low or no malignant potential. *Lab Invest* 84: 874–883.
Lasota J, Stachura J, Miettinen M, 2006 GISTs with PDGFRA exon 14 mutations represent subset of clinically favorable gastric tumors with epithelioid morphology. Lab Invest. 86: 94–100.
Lawrence B, Perfez-Atayde A, Hibbard MK, Rubin BP, Dal Cin P, Pinkus JL, Xiao S, Yi ES, Fletcher CD, Fletcher JA, 2000, TPM3-ALK and TPM4-ALK oncogenes in inflammatory myofibroblastic tumors. *Am J Pathol* 157: 377–384.
Lebbe C, Pellet C, Flageul B, Sastre X, Avril MF, Bonvalet D, Morel P, Calvo F, 1997, Sequences of human herpesvirus 8 are not detected in various non-Kaposi sarcoma vascular lesions. *Arch Dermatol* 133: 919–920.

Lee J, Hopcus-Niccum DJ, Mulvihill JJ, Li S, 2005, Cytogenetic and molecular genetic studies of a variant of t(21;22), in (22;21)(q12;q21q22), with a deletion of the 3′ EWSR1 gene in a patient with Ewing sarcoma. *Cancer Genet Cytogenet* 159: 177–180.

Longley BJ, Reguera MJ, Ma Y, 2001, Classes of c-kit activating mutations: proposed mechanisms of action and implications for disease classification and therapy. *Leuk Res* 25: 71–76.

Lopez-Rios F, Illei PB, Rusch V, Ladanyi M, 2004, Evidence against role for SV40 infection in human mesotheliomas and high risk of false-positive PCR results owing to presence of SV40 sequences in common laboratory plasmids. *Lancet* 364: 1157–1166.

McDonagh DP, Liu J, Gaffey MJ, Layfield LJ, Azumi N, Traweek ST, 1996, Detection of Kaposi's sarcoma-associated herpesvirus-like DNA sequence in angiosarcoma. *Am J Pathol* 149: 1363–1368.

Miettinen M, Limon J, Niezabitowski A, Lasota J, 2001a, Calretinin expression in synovial sarcoma. Analysis of similarities and differences from malignant mesothelioma. *Am J Surg Pathol* 25: 610–617.

Miettinen M, Shekitka K, Sobin LH, 2001b, Schwannomas of the colon and rectum. Clinicopathologic and immunohistochemical study of 20 cases. *Am J Surg Pathol* 25: 846–855.

Miettinen M, Sobin LH, Lasota J, 2005a, Gastrointestinal stromal tumors of the stomach. A clinicopathologic, immunohistochemical, and molecular genetic study of 1765 cases with long term follow-up. *Am J Surg Pathol* 29: 52–68.

Miettinen M, Lasota J, Sobin LH, 2005b, Gastrointestinal stromal tumors of the stomach in children and young adults: a clinicopathologic, immunohistochemical and molecular genetic study of 44 cases with long-term follow-up and review of the literature. *Am J Surg Pathol* 29: 1373–1381.

Miettinen M, Fetsch JF, Sobin LH, Lasota J, 2006, Gastrointestinal stromal tumors in patients with neurofibromatosis 1. A clinicopathologic and molecular genetic study of 45 cases. *Am J Surg Pathol* 30: 90–96.

Moore PS, Chang Y, 2004, Kaposi's sarcoma-associated herpesvirus immunoevasion and tumorigenesis: two sides of the same coin? *Annu Rev Microbiol* 57: 609–639.

O'Sullivan MJ, Kyriakos M, Zhu X, Wick MR, Swanson PE, Dehner LP, Humphrey PA, Pfeifer JD, 2001, Malignant peripheral nerve sheath tumors with t(X;18). A pathologic and molecular genetic study. *Mod Pathol* 13: 1336–1346.

Parkash V, Gerald WL, Parma A, Miettinen M, Rosai J, 1995, Desmoplastic small round cell tumor of the pleura. *Am J Surg Pathol* 19: 659–665.

Pass HI, Bocchetta M, Carbone M, 2004, Evidence for and important role for SV40 in mesothelioma. *Thorac Surg Clin* 14: 489–495.

Patel RM, Goldglum JR, Hsi ED, 2004, Immunohistochemical detection of human herpesvirus 8 latent nuclear antigen-1 is useful in the diagnosis of Kaposi sarcoma. *Mod Pathol* 17: 456–460.

Pawson T, 2002, Regulation and targets of receptor tyrosine kinases. *Eur J Cancer* 38(Suppl. 5): S3–S10.

Prakash S, Sarran L, Socci N, deMatteo RP, Eisenstat J, Greco AM, Maki RG, Wexler LH, LaQuaglia MP, Besmer P, Antonescu CR, 2005, Gastrointestinal stromal tumors in children and young adults: a clinicopathologic, molecular, and genomic study of 15 cases and review of the literature. *J Pediatr Hematol Oncol* 27: 179–187.

Rollison DE, Page WF, Crawford H, Gridley G, Wacholder S, Martin J, Miller R, Engels EA, 2004, Case-control study of cancer among US army veterans exposed to simian virus 40-contaminated adenovirus vaccine. *Am J Epidemiol* 160: 317–324.

Ronnstrand L, 2004, Signal transduction via the stem cell factor receptor/c-Kit. *Cell Mol Life Sci* 61: 2535–2548.

Rubin BP, Antonescu CR, Scott-Browne JP, Comstock ML, Gu Y, Tanas MR, Ware CB, Woodell J, 2005, A knock-in mouse model of gastrointestinal stromal tumor harboring kit K641E. *Cancer Res* 65: 6631–6639.

Ruttledge MH, Andermann AA, Phelan CM, Vlaudio JO, Han FY, Chretien N, Rangaratnam S, MacCollin M, Short P, Parry D, Michels V, Riccardi VM, Weksberg R, Kitamura K, Bradburn

JM, Hall DB, Propping P, Rouleau GA, 1996, Type of mutation in the neurofibromatosis type 2 gene (NF2) frequently determines severity of disease. *Am J Hum Genet* 59: 331–342.

Segal NH, Pavlidis P, Noble WS, Antonescu CR, Viale A, Wasley UV, Busam K, Gallardo H, DeSantis D, Brennan MF, Cordon-Cardo C, Wolchok JD, Houghton AN, 2003, Classification of clear cell sarcoma as subtype of melanoma by genomic profiling. *J Clin Oncol* 21: 1775–1781.

Sommer G, Agosti V, Ehlers I, Rossi F, Corbacioglu S, Farkas J, Moore M, Manova K, Antonescu CR, Besmer P, 2003, Gastrointestinal stromal tumors in a mouse model by targeted mutation of the Kit receptor tyrosine kinase. *Proc Natl Acad Sci USA* 100: 6706–6711.

Subramanian S, West RB, Corless Cl, et al. 2004, Gastrointestinal stromal tumors (GISTs) with KIT and PDGFRA mutations have distinct gene expression profiles. *Oncogene* 23: 7780–7790.

Tognon C, Knezevich SR, Huntsman D, Roskelley CD, Melnyk N, Mathers JA, Becker L, Carneiro F, MacPherson N, Horsman D, Poremba C, Sorensen PH, 2002, Expression of the ETV6-NTRK3 gene fusion as a primary event in human secretory carcinoma. *Cancer Cell* 2: 367–376.

Van Muijen GN, Ruiter DJ, Warnaar SO, 1987, Coexpression of intermediate filament polypeptides in human fetal and adult tissues. *Lab Invest* 57: 359–369.

Viskochil D, 2002, Genetics of neurofibromatosis 1 and NF1 gene. *J Child Neurol* 17: 562–570.

Xiao GH, Chernoff J, Testa JR, 2003, NF2: the wizardry of merlin. *Genes Chromosomes Cancer* 38: 389–399.

Zambrano E, Reyes-Mugica M, Franchi A, Rosai J, 2003, An osteoclast-rich tumor of the gastrointestinal tract with features resembling clear cell sarcoma of soft parts: report of 6 cases of a GIST simulator. *Int J Surg Pathol* 11: 75–81.

CHAPTER 11
PREDICTION OF RESPONSE TO NEOADJUVANT CHEMOTHERAPY IN CARCINOMAS OF THE UPPER GASTROINTESTINAL TRACT

HEINZ HÖFLER[1], RUPERT LANGER[1], KATJA OTT[2], AND GISELA KELLER[1]

[1]*Institute of Pathology and* [2]*Department of Surgery, Technical University München, Trogerstr. 18, 81675 München, Germany*

Abstract: Multimodal treatment protocols are increasingly employed to improve the survival of patients with locally advanced adenocarcinomas of the upper gastrointestinal tract, however, only 30–40% per year of the patients respond to 5-FU and cisplatin-based neoadjuvant chemotherapy. The goal of our studies is the identification of reliable genetic markers, on the genomic DNA-level, mRNA, or protein level that could predict response of upper gastrointestinal carcinomas prior to neoadjuvant chemotherapy.

In esophageal carcinomas, a higher gene expression of methylenetetrahydrofolate reductase (MTHFR), an enzyme involved in folate metabolism, was more frequently found in responding patients. In addition high gene expression of *caldesmon* and of the two drug carrier proteins, MRP1 and MDR1 was associated with response to therapy. By performing a genome-wide profiling on the protein level in a small group of patients, new potential markers were identified, which have to be validated in ongoing studies.

In gastric carcinomas, mutations of the *p53* gene revealed no association with response or survival, but tumors with a high rate of loss of heterozygosity (LOH), determined by microsatellite analysis, showed a better response to a cisplatin-based chemotherapy. Analysis of expression of 5-FU-(e.g., *TS, DPD,* and *TP*) and cisplatin- (e.g., *ERCC1, ERCC4, GADD45A,* and *KU80*) related genes, demonstrated an association of *DPD* expression with response and survival. The combined consideration of *TP* and *GADD45* gene expression, showed the most obvious association with therapy response in this tumor.

Our studies point to promising markers with potential use for chemotherapy response prediction of adenocarcinomas of the upper gastrointestinal tract, but prospective studies for validation are necessary.

Key words: Neoadjuvant chemothepy, carcinomas, gastrointestinal tract, LOH

1. INTRODUCTION AND OBJECTIVE

Multimodal treatment protocols are increasingly employed to improve the survival of patients with locally advanced adenocarcinomas of the esophagus and stomach. Neoadjuvant chemotherapeutic treatment, mainly based on cisplatin and 5-FU, has been used since 1989 in several clinical trials and recently, a statistically significant improvement in respect to resectability, progression-free and overall survival in operable gastric and lower esophageal cancer has been demonstrated in a large randomized, controlled phase III trial (MAGIC trial) [1]. However, only 30–40% of the patients respond to therapy and the majority of patients undergo several month of toxic, expensive therapy without survival benefit. In particular, in the case of esophageal carcinomas, it has been shown that patients with nonresponding tumors seem to have an even worse prognosis than patients treated by surgery alone, which may be related to therapy-induced side effects, selection of chemotherapy-resistant, more aggressive tumor cells and delay of surgery [2]. Thus, the identification of reliable genetic markers that could predict response is highly demanding.

Several molecular markers had been investigated as potential response predictors. Thymidylate synthase as the target enzyme for 5-FU has been widely studied for 5-FU-containing regimens in gastrointestinal cancer, but the results are inconsistent [3,4,5]. Dihydropyrimidine dehydrogenase (DPD) and thymidine phosphorylase (TP) are two other important regulatory enzymes involved in the degradation of 5-FU, and low levels of DPD have been shown to be associated with response in gastric carcinoma [5,6], whereas conflicting results have been reported for TP.

The other major component used for the treatment of carcinomas of the upper gastrointestinal tract is cisplatin, which supposedly directly damages DNA. A significant association of the gene expression of the nucleotide excision enzyme *ERCC1*, which is involved in DNA repair, with response to neoadjuvant chemotherapy has been reported [4].

Other markers such as glutathione *S*-transferase, vascular endothelial growth factor and apoptosis-related genes have been such as *bcl-2*, *bax*, and *p53* have mostly been studied by immunohistochemistry, and the results have been inconclusive, so that no markers has been found to be clinically relevant at present [3,7].

Thus, the goal of our studies is to identify effective molecular markers for response prediction for patients with esophageal and gastric carcinomas treated by a neoadjuvant chemotherapy. We are using different strategies based on one side, on targeted approaches to characterize pretherapeutic biopsies for tumor-specific molecular alterations on the genomic DNA and mRNA-level. We also analyze constitutional genetic factors, e.g., DNA-polymorphisms in therapy-related genes. On the other side, we perform a genome-wide profiling on the protein level, to identify new marker proteins.

2. RESULTS

2.1. Characterization of pretherapeutic biopsies of esophageal carcinomas

2.1.1. Analysis of m-RNA expression of therapy-related genes

In this study, paraffin-embedded, formalin-fixed endoscopic esophageal tumor biopsies of 38 patients with locally advanced esophageal adenocarcinomas (Barrett's adenocarcinoma) were included. All patients underwent two cycles of cisplatin and fluorouracil (5-FU) therapy with or without additional paclitaxel followed by abdominothoracal esophagectomy. RNA expression levels of 5-FU metabolism-associated genes, *thymidylate synthase, TP, DPD, MTHFR, MAP7, ELF3*, as well as of platinum and taxane-related genes *Caldesmon, ERCC1, ERCC4, HER2-neu, GADD45* and multidrug resistance gene *MRP1* were determined using real-time RT-PCR. Expression levels were correlated with response to chemotherapy histopathologically assessed in surgically resected specimens.

The results demonstrated that the responding patients showed significantly higher pretherapeutic expression levels of *MTHFR* ($p = 0.012$), *Caldesmon* ($p = 0.016$), and *MRP1* ($p = 0.007$). In addition, patients with high pretherapeutic *MTHFR* and *MRP1* levels had a survival benefit after surgery ($p = 0.013$ and $p = 0.015$, respectively) [8]. Additionally, investigation of intratumoral heterogeneity of gene expression of relevant genes (e.g., *MTHFR, Caldesmon, Her2-neu, ERCC4*, and *MRP1*) — verified in nine untreated Barrett's adenocarcinomas by examination of five distinct tumor areas — revealed no significant heterogeneity in gene expression indicating that expression profiles obtained from biopsy material may yield a representative genetic expression profile of total tumor tissue [8].

Thus in conclusion, the results indicate that determination of mRNA levels of few genes may be useful for the prediction of the success of neoadjuvant chemotherapy in individual cancer patients with locally advanced Barrett's adenocarcinoma.

2.1.2. Differential quantitative ProteoTope analysis of fresh frozen biopsies

A comprehensive protein profiling approach, using the ProteoSys platform, has been performed until now for a small group of patients. Quantitative and qualitative protein expression analysis was performed using 2D ProteoTope techniques after radioactive labeling of the protein extract with I-125 and I-131. The results so far point to an interesting group of proteins, which may be associated with response. Validation of specific proteins by immunohistochemical analysis in a high number of cases is now part of ongoing studies.

2.2. Characterization of pretherapeutic biopsies of gastric carcinomas

2.2.1. Microsatellite analysis and p53 mutation analysis

We evaluated microsatellite instability (MSI) and LOH in 53 pretherapeutic gastric carcinoma biopsies using 11 microsatellite markers. The entire coding region of the p53 gene (exons 2–11) was analyzed for mutations by denaturing high-pressure liquid chromatography (DHPLC) and sequencing. P53 protein expression was evaluated by immunohistochemistry. Patients were treated with a cisplatin-based, neoadjuvant chemotherapy regimen. Therapy response was evaluated by CT scan, endoscopy, and endoluminal ultrasound [9,10].

We identified p53 mutations in 19 of the 53 (36%) analyzed tumors. No significant association with response or survival was found for p53 mutation or for p53 protein expression. Microsatellite instability (either MSI-H or MSI-L) did not show a correlation with response. With respect to LOH, LOH at chromosome 17p13 showed a significant association with therapy response ($p = 0.022$), but did not reach statistical significance in terms of patient survival. The global LOH rate, expressed as fractional allelic loss (FAL) was assessed and tumors were classified into tumors with a high (>0.5), a medium (>0.25–0.5), and a low (0–0.25) FAL-value. A statistically significant association of FAL with therapy response was found ($p = 0.003$), with a high FAL being related to therapy response.

Thus, a high level of chromosomal instability (high FAL-value) defines a subset of patients who are more likely to benefit from cisplatin-based neoadjuvant chemotherapy. p53 mutation status is not significantly associated with therapy response and is not a useful marker for response prediction [9,10].

2.2.2. Methylation analysis

We investigated the methylation profile of six genes, which are frequently methylated in gastric cancer (e.g., *14-3-3σ*, *E-cadherine*, *HPP1*, *Lysyl oxidase*, *MGMT*, and *p16*) for an association with response and survival in a set of 61 neoadjuvant-treated gastric cancer patients by bisulfite/methylation-specific PCR using the TaqMan system. Only 46% of the patients showed tumor-specific methylation signals in four or more genes. There was no significant correlation of response with global methylation status or with any of the genes alone. Patients with a low methylation status showed a tendency for response to therapy and patients with no or only one methylated gene demonstrated a statistically significant better survival ($p = 0.027$). This interesting finding raises the question if the use of inhibitors of DNA methylation and/or histone deacetylase inhibitors might represent a therapeutic alternative for gastric cancer patients demonstrating a high methylation status in their tumors [11].

2.2.4. Analysis of mRNA expression of therapy-related genes

For gastric carcinomas we performed gene expression analysis, focusing on genes related to the effects of 5-FU or cisplatin. Pretherapeutic, formalin-fixed and paraffin-embedded biopsies of 61 patients, who received a 5-FU and cisplatin-based chemotherapy were included. The expression of the 5-FU-related genes *TS*, *DPD* and *TP* and of the cisplatin-related genes *ERCC1*, *ERCC4*, *KU80* and *GADD45A* were analyzed by quantitative real-time PCR. The expression levels of single genes and of various combinations were tested for an association with response and overall survival [5]. High DPD levels were more frequently found in nonresponding patients and were associated with worse survival. *GADD45A* and *TP* levels demonstrated weak associations with response, but *GADD45A* expression correlated with survival. There was no association with response to TS expression, but tumors with a high TS level were associated with worse survival. The combination of *GADD45A* and *TP* revealed the strongest predictive impact. High expression values of *TP* and/or *GADD45A* were exclusively found in nonresponding patients ($p = 0.002$) and were associated with a significantly poorer survival ($p = 0.04$).

Thus, in conclusion, the combined gene expression levels of *TP* and *GADD45A* represent a new parameter to predict the clinical outcome after neoadjuvant chemotherapy in gastric cancer. The association of *DPD* expression with response and survival underlines a predominant role of *DPD* to predict 5-FU sensitivity. The association of *TS* expression levels with survival, but not with response, suggests an importance of this gene for tumor progression [5].

3. OUTLOOK

Although some of our studies point to promising markers with a potential use in chemotherapy response prediction for adenocarcinomas of the upper gastrointestinal tract, prospective studies for validation are necessary before they may be used in clinical practice. As chemotherapy response is considered to be highly complex, depending on tumor-specific characteristics as well as on the constitutional genetic makeup of the individual patient, integrative approaches for response prediction might be necessary. In addition the incorporation of early response evaluation by positron emission tomography (PET) for the therapeutic decision together with molecular markers, might result in superior sensitivity and specificity for a successful application of an individual therapy-strategy for patients with upper gastrointestinal malignancies.

REFERENCES

1. D. Cunnigham, W.H. Allum, S.P. Stenning, and S. Weeden, for the upper GI Cancer Clinical Study Group. Perioperative chemotherapy in operable gastric and lower oesophageal cancer: final results of a randomised, controlled trial (the MAGIC trial, ISRCTN 93793971). *ASCO Annual Meeting*, Abstract No. 4001 (2005).
2. J. Zacherl, A. Sendler, H.J. Stein, K. Ott, M. Feith, R. Jakesz, J.R. Siewert, and U. Fink. Current status of the neoadjuvant therapy for adenocarcinoma of the distal esophagus. *World J Surg* 27: 1067–1074 (2003).

3. K.H. Yeh, C.T. Shun, C.L. Chen, J.T. Lin, W.J. Lee, Y.C. Chen, and A.L. Cheng. High expression of thymidylate synthase is associated with the drug resistance of gastric carcinoma to high dose 5-fluorouracil-based systemic chemotherapy. *Cancer* 82: 1626–1631 (1998).
4. R. Metzger, C.G. Leichman, K.D. Danenberg, P.V. Danenberg, H.J. Lenz, K. Hayashi, S. Groshen, D. Salonga, H. Cohen, L. Laine, P. Crookes, H. Silberman, J. Baranda, B. Konda, and L. Leichman. ERCC1 mRNA levels complement thymidylate synthase mRNA levels in predicting response and survival for gastric cancer patients receiving combination cisplatin and fluorouracil chemotherapy. *J Clin Oncol* 16: 309–316 (1998).
5. R. Napieralski, K. Ott, M. Kremer, K. Specht, H. Vogelsang, K. Becker, M. Müller, F. Lordick, U. Fink, J.R. Siewert, H. Höfler, and G. Keller. Combined GADD45A and TP expression levels predict response and survival of neoadjuvant treated gastric cancer patients. *Clin Cancer Res* 11: 3025–3031 (2005).
6. Y. Ishikawa, T. Kubota, Y. Otani, M. Watanabe, T. Teramoto, K. Kumai, T. Takechi, H. Okabe, M. Fukushima, and M. Kitajima. Dihydropyrimidine dehydrogenase and messenger RNA levels in gastric cancer: possible predictor for sensitivity to 5-fluorouracil. *Jpn J Cancer Res* 91(1): 105–112 (2000).
7. N. Boku, K. Chin, K. Hosokawa, A. Ohtsu, H. Tajiri, S. Yosihida, T. Yamao, H. Kondo, K. Shirao, Y. Shimada, D. Saito, T. Hasebe, K. Mukai, S. Seki, H. Saito, and P. Johnston. Biological Markers as a predictor for response and prognosis of unresectable gastric cancer patients treated with 5-fluorouracil and cis-platinum. *Clin Cancer Res* 4: 1469–1474, (1998).
8. R. Langer, K. Specht, K. Becker, E. Philipp, M. Beckisch, M. Sarbia, R. Busch, M. Feith, H.J. Stein, J.R.Siewert, and H. Höfler. Prediction of response to neoadjuvant chemotherapy in Barrett`s carcinoma by quantitative gene expression analysis. *Clin Cancer Res* in press (2005).
9. T. Grundei, H. Vogelsang, K. Ott, J. Mueller M. Scholz, K. Becker, U. Fink, J.R. Siewert, H. Höfler H, and G. Keller. Loss of heterozygosity and microsatellite instability as predictive markers for neoadjuvant treatment in gastric carcinoma. *Clin Cancer Res* 6: 4782–4788 (2000).
10. K. Ott, H. Vogelsang, J. Mueller, K. Becker, M. Müller, U. Fink, J.R. Siewert, H. Höfler, and G. Keller. Chromosomal instability rather than p53 mutation is associated with response to neoadjuvant cisplatin-based chemotherapy in gastric cancer. *Clin Cancer Res* 9: 2307–2315 (2003).
11. R. Napieralski, K. Ott, M. Kremer, K. Becker, H. Vogelsang, F. Lordick, J.R. Siewert, H. Höfler, and G. Keller. Hypermethylation phenotype and association with response and survival in neoadjuvant treated gastric carcinoma, submitted.

CHAPTER 12

INTEGRATING THE DIAGNOSIS OF CHILDHOOD MALIGNANCIES

DOLORES LÓPEZ-TERRADA

Texas Children's Hospital and Baylor College of Medicine, Houston, Texas, USA

Abstract: Significant progress has been made in understanding the molecular basis of pediatric malignancies. Mechanisms of pediatric acute leukemia induction include hyperdiploidy, aberrant expression of proto-oncogenes, and activation of transcription factors or kinases by aberrant fusion genes. Molecular analysis of these alterations has facilitated the recognition of distinct groups with different sensitivity to therapy, and identified potential targets for antileukemic agents. Similar analysis of pediatric soft tissue and bone tumors also resulted in the identification of specific fusion genes, and their characterization has contributed greatly to understand their biology. Molecular assays for these rearrangements have become important tools in classifying these tumors, providing important prognostic data. However, the understanding of mechanisms involved in the pathogenesis of many other pediatric malignancies, including some embryonal tumors — believed to arise due to perturbation of the normal developmental program — is still vastly incomplete.

The Department of Pathology at Texas Children's Hospital is one of the Children's Oncology Group (COG) reference centers for pediatric liver tumors. We have been particularly interested in the biology of hepatoblastoma, the most common type of pediatric liver tumor. Although a number of cytogenetic and molecular abnormalities have been described for this type of embryonal tumor, its pathogenesis is still poorly understood. In an attempt to explore the role of different signaling pathways in this disease, we analyzed the expression patterns of different histologic subtypes of hepatoblastoma using cDNA microarray analysis, qualitative reverse transcription, polymerase chain reaction (QRT-PCR), and immunohistochemistry. Wnt signaling pathway, critical both in development and in neoplasia, appears to be particularly relevant in these tumors. Mutations of the β-catenin gene are present in over 90% of hepatoblastomas, leading to activating transcription of a number of target genes. The pattern of β-catenin expression and type of mutation in groups of tumors are crucial to understand the corresponding differences in their gene expression profiles. Our findings are consistent with a relationship between poor histologic phenotype and β-catenin activation, indicating the potential utility of targeted gene expression assays to identify molecular events related to the pathogenesis and prognosis of hepatoblastomas.

Integration of clinical, morphologic, phenotypic, cytogenetic, and molecular data has become the basis of novel prognostic prediction and therapeutic strategies in pediatric leukemia. Similarly, integration of new genetic and molecular data with clinical, and other diagnostic information will be crucial for accurate classification of pediatric tumors, risk stratification, and successful development of new therapies for pediatric oncologic patients.

1. INTRODUCTION

Significant progress has been achieved during the last few decades in understanding the molecular basis of numerous pediatric malignancies. In some instances this has resulted in the development of genetic and molecular tests that are being progressively integrated in the diagnosis and clinical management of these patients.

The pediatric and adult cancer disease spectrum is different, as it is the stem cell population targeted by mutations, the type and number of necessary mutations to induce a fully malignant phenotype, and the internal homeostatic environment of the host (developing vs a fully mature). All these result in a different approach to diagnose pediatric cancer, as many of these processes lack morphologic evidence of differentiation and are difficult to classify. Most of pediatric cancer patients are enrolled in cooperative group therapy protocols (90% of children in the USA), which are tailored to specific tumor types and subgroups, often requiring assessment of biologic tumor markers [1].

2. PEDIATRIC HEMATOPOIETIC MALIGNANCIES

The best example of how the application of newly gained biological knowledge in a malignancy type has resulted in improvements in diagnosis, classification and clinical management, is pediatric hematopoietic malignancies. True treatment success has been achieved in many pediatric acute leukemias and lymphomas, much more so than for adult hematopoietic malignancies. These differences in therapeutic success are probably due to a combination of factors, including biological differences of the neoplastic processes, host-dependent features and treatment strategies, and also a better understanding of normal hematopoietic development and of the molecular pathology of these malignancies [1,2].

2.1. Pediatric acute lymphoblastic leukemia

Pediatric acute lymphoblastic leukemia (ALL) is the most common malignancy in childhood, representing approximately 50% of all pediatric cancers. During the last decade a better understanding of normal hematopoietic development and of the molecular events involved in leukemic malignant transformation, has been achieved. As a result, significant improvements have occurred in our ability to diagnose, subclassify and treat these patients. During the last decade the survival rates have increased to close to 80% overall, as prognostic markers have been identified and risk adapted therapeutic protocols have been implemented. To the conventional methods of ALL diagnosis (morphology, cytochemistry, and immunophenotyping) a number of genetic and molecular diagnostic techniques have progressively become part of the diagnostic work-up and monitoring of ALL patients, including one or several of the following: conventional cytogenetics, southern blotting, polymerase chain reaction (PCR), fluorescence in situ hybridization (FISH), and less commonly spectral karyotyping (SKY), and comparative genomic hybridization (CGH) [3].

A small subset of important genetic abnormalities of variable frequency (Table 12.1) have been identified in pediatric ALL, and are presently being used in several clinical protocols for therapeutic stratification of these patients. The presence or absence of a number of chromosomal rearrangements and the resulting fusion genes are used as markers which, in combination with a number of other clinical parameters (immunophenotype, DNA index, age, white cell count, central nervous system or testicular involvement, and early response to therapy) serve to assess these patients standard, high or very high clinical risk, and to assign them to therapeutic protocols, accordingly [2].

Molecular testing in pediatric ALL fusion transcripts derived from chromosomal translocations has become critical for risk-stratification of patients and optimizing treatment strategies. Patient-specific transcripts can also be used as markers to monitor relapse or minimal residual disease.

Pediatric ALL has become the best example of integration of genetic and molecular biological markers in routine diagnosis, risk assessment, and monitoring of pediatric cancer patients [3,4]. In the last few years, cDNA microarray technology has been extensively utilized to identify distinct leukemia subtypes that can be defined exclusively on their expression profiles [5]. This newly gained knowledge has resulted in a number of challenges for the molecular testing of pediatric ALL patients, the first one being the potential need for individual molecular characterization of newly diagnosed leukemia patients. This will require the implementation of multiplexed diagnostic assays, sufficiently sensitive, efficient and cost-effective, requiring flexible new platforms to allow easy inclusion of new markers as well as potential of expansion into proteomics and pharmacogenomics testing.

Table 12.1. Genetic abnormalities and risk assessment in pediatric ALL

Cytogenetics	Fusion	Frequency (%)	Prognosis
t(9;22)(q34;q11.2)	*BCR/ABL*	3–4	Unfavorable
t(4;11)(q21;q23)	*AF4/MLL*	2–3	Unfavorable
t(1;19)(q23;p13.3)	*PBX/E2A*	6	Unfavorable
t(12;21)(p13;q22)	*TEL/AML1*	16–29	Favorable
Hyperdiploid >50	N/A	20–25	Favorable
Hypodiploid	N/A	5	Unfavorable

Our Department at Texas Children's Hospital is in the process of implementing a new ALL translocations Bead-based Assay (Ambion/Luminex). This technology, a combination of multiplex reverse transcription, polymerase chain reaction (RT-PCR) and detection of spectrally addressable fluorescent beads, allows the analysis of multiple targets in a single well in a 96 well format. This high throughput multiplexed assay is being implemented for the amplification and detection of six common leukemia-associated fusion transcripts in a single well, with a very short turnaround time (5 h) in a sensitive (reliably detects 1 fusion-transcript-carrying leukemia cell in 100–1,000 cells) and cost-effective manner [6].

2.2. Lymphomas of children and adolescents

Malignant lymphomas represent the third most common pediatric cancer, and between 8 and 10% of all pediatric malignancies. The four most common pediatric lymphoma subtypes are Burkitt lymphoma (40%), lymphoblastic lymphoma (30%), diffuse large B-cell lymphoma (20%), and anaplastic large cell lymphoma (ALCL, 10%) [7].

We will discuss briefly some of the most recent developments in the understanding of the biology of two subtypes of pediatric NHLs: ALCL and Burkitt lymphoma, particularly those most relevant to their diagnosis.

2.3. Anaplastic large cell lymphoma

ALCL, previously designated as "Ki-1 lymphoma" for its characteristic CD30 immunoreactivity, is a most commonly T-cell non-Hodgkin lymphoma, representing between 10 and 15% of the total pediatric NHLs [7]. This lymphoma subtype was initially associated with the presence of a t(2;5) translocation, which was cloned by Morris and colleagues, describing the genes involved in the resulting fusion gene: the nucleophosmin (*NPM*) gene, located on chromosome 5q35, and the anaplastic lymphoma kinase (*ALK*) gene, located on chromosome 2p23. Aberrant expression of ALK protein, a novel receptor tyrosine kinase of the insulin receptor family, is linked to tumorigenesis in ALK-positive ALCL, as well as in some other tumors [8,9,10]. A number of other variant translocations involving at least six other ALK partner genes have been recently described (Table 12.2).

Table 12.2. Anaplastic lymphoma kinase (ALK, 2p23) translocations in ALCL

Abnormality	Fusion gene
t(2;5)(p23;q35)	*ALK/NPM*
t(2;22)(p23;q11.2)	*ALK/CLCTL*
t(2;22)(p23;q11.2)	*ALK/MYH9*
t(1;2)(q35;p23)	*ALK/TPM3*
t(2;3)(p23;q21)	*ALK/TFG*
t(2;17)(p23;q25)	*ALK/ALO17*
inv(2)(q23q35)	*ALK/ATIC*

Histologic identification of the typical ALCL anaplastic morphology or any of its many morphological variants, in combination with phenotyping and characteristic CD30 immunoreactivity, are used to initially identify these lesions. Although a number of RT-PCR and FISH assays (*ALK*-specific FISH) are available to document translocations associated with ALCL, immunohistochemical detection of aberrant ALK protein expression is most commonly used [11,12,13].

2.4. Burkitt lymphoma

Histologic identification of the Burkitt lymphoma or L3 ALL blasts morphology, immunophenotyping, and correlation with the presence of *CMYC* oncogene (8q24.12–q24.13) rearrangements are usually required for accurate classification of these malignancies. Documenting these rearrangements can be either done by conventional cytogenetics, identifying the presence of the t(8;14) or variant translocations (t(2;8), t(8;22)), or *C-MYC* (8q24.13) oncogene rearrangements by FISH. More recently, the application of chromogenic in situ hybridization (CISH) probes to document *CMYC* oncogene rearrangements on tissue sections, touch imprints, and other cellular specimens, has allowed the simultaneous identification of the genetic rearrangement within their histologic context [14].

3. PEDIATRIC SOLID TUMORS

The majority of pediatric solid tumors (differently to adult malignancies, more commonly of epithelial origin) are either sarcomas or tumors of neuroectodermal origin. Many of these lesions are difficult to accurately classify by morphological means, as they are frequently poorly differentiated or undifferentiated neoplasms. Multiple diagnostic approaches are often required and ancillary tests are necessary to confirm the initial diagnostic and clinical impression in these patients.

Genetic abnormalities found in pediatric solid tumors include a number of chromosomal translocations (and other chromosomal rearrangements as deletion, inversions, and insertions) loss of tumor suppressor genes, amplification of oncogenes, abnormal methylation, genomic imprinting, defective DNA repair mechanisms, and telomerase activity. However, from the molecular diagnostic point of view, only those recurrent, well-characterized genetic changes rigorously correlated with specific tumor types and subtypes should be considered for clinical use.

No specific genetic rearrangements have been identified in a proportion of pediatric and adult sarcomas (Table 12.3). However, systematic cytogenetic analysis of pediatric solid tumors identified a number of other with tumor-specific chromosomal translocations (Table 12.4). Molecular cloning of these rearrangements resulted in the characterization of their breakpoints and the

Table 12.3. Sarcomas with complex karyotypes

Osteosarcoma
Rhabdomyosarcoma
   Embryonal
   Pleomorphic
Leiomyosarcoma
Fibrosarcoma
High-grade undifferentiated pleomorphic sarcoma
Chondrosarcoma
Liposarcoma
   Well differentiated/dedifferentiated
   Pleomorphic liposarcoma
Malignant peripheral nerve sheath tumor
Angiosarcoma

Table 12.4. Most common translocations and gene fusions in sarcomas

Tumor	Translocation	Fusion gene
Alveolar rhabdomyosarcoma	t(2;13)(q35;q14)	*PAX3-FKHR*
	t(1;13)(p36;q14)	*PAX7-FKHR*
Alveolar soft part sarcoma	t(X;17)(p11.2;q25)	*ASPL-TFE3*
Clear cell sarcoma	t(12;22)(q13;q12)	*ATF1-EWS*
Congenital fibrosarcoma and mesoblastic nephroma	t(12;15)(p13;q25)	*ETV6-NTRK3*
DFSP and giant cell fibroblastoma	t(17;22)(q22;q13)	*COL1A1-PDGFB*
Desmoplastic round cell tumor	t(11;22)(p13;q12)	*WT1-EWS*
Ewing and pPNET	t(11;22)(q24;q12)	*EWS-FLI1*
	t(21;22)(q22;q12)	*EWS-ERG*
	t(7;22)(p22;q12)	*EWS-ETV1*
	t(17;22)(q12;q12)	*EWS-E1AF*
	t(2;22)(q33;q12)	*FEV-EWS*
Inflammatory myofibroblastic tumor	t(2;19)(p23;p13.1)	*ALK-TPM4*
	t(1;2)(q22–23;p23)	*TPM3-ALK*
Myxoid chondrosarcoma, extraskeletal	t(9;22)(q22;q12)	*EWS-CHN(TEC)*
	t(9;17)(q22;q11)	*RBP56-CHN(TEC)*
	t(9;15)(q22;q21)	*TEC-TCF12*
Myxoid liposarcoma	t(12;16)(q13;p11)	*TLS(FUS)-CHOP*
	t(12;22)(q13;q12)	*EWS-CHOP*
Synovial sarcoma	t(X;18)(p11;q11)	*SYT/SSX1/2/4*

identification of fusion genes, chimeric transcripts, and proteins that result of these translocations. As a result of this, we have a better understanding of the molecular mechanisms involved in normal development in some cases, and malignant transformation of these tumors. Is also provided molecular pathologists with genetic markers for tumor classification and is impacted therapy which is becoming more risk-based and tumor specific. A number of translocation-specific novel therapeutic strategies are also being evaluated for these patients [15].

In addition of conventional cytogenetics and other karyotyping techniques (such as CGH and SKY) a number of molecular techniques, including RT-PCR, FISH and more recently CISH are clinically applied for the identification of tumor-specific chromosomal translocations, fusion genes, and fusion transcripts. The advantage of conventional cytogenetic analysis is the large amount of information provided by a complete karyotype, however it requires alive, dividing cells, its resolution is low, is expensive and slow. Both RT-PCR and FISH/CISH are affordable and fast tests, but noninformative if negative. RT-PCR allows an unlimited number of primer to be designed, but requires RNA obtained from the tissue tested. On the other hand, FISH/CISH probes are limited and test results may be difficult to interpret. Other conventional and ancillary techniques used for diagnosing pediatric solid tumors are included in Table 12.5. Probably the most important element to remember when a number of tests are incorporated for diagnosing these types of malignancies is that molecular testing should always be used in the context of other diagnostic tests and not used alone to arrive to a diagnosis [16].

The following section will mostly focus in tumor-defining chromosomal translocations identified in two pediatric sarcomas: (1) synovial sarcoma and (2) alveolar rhabdomyosarcoma (RMS). The most recent advances in the application of molecular diagnostic testing to these two sarcoma types, carrying characteristic, well-characterized chromosomal rearrangement with well-established diagnostic, therapeutic and prognostic relevance, will be discussed.

3.1. Synovial sarcoma

Synovial sarcoma is an aggressive, relatively common sarcoma (approximately 10% of all soft tissue sarcomas) of unknown histogenesis that affects children and young adults, with a slight male predominance [17,18]. The most common sites involved are the limbs, and particularly areas adjacent to joints, although they can arise in almost any area of the body (trunk, mediastinum, abdominal wall, head and neck, lung, and pleura) [19,20,21,22]. Histologically there are two

Table 12.5. Diagnostic methods for pediatric tumor diagnosis

Method	Application
Light microscopy	Mandatory in all cases
Immunohistochemistry	First-choice ancillary, widely used
RT-PCR	Most common Mol dx, now routine
FISH (CISH)	Supplanting CG in many cases
ISH	Specialized use (EBV)
Special stains	Still useful in some cases
Electron microscopy	Still widely used
Cytogenetics	Needed if no FISH probes available
Molecular CG: SKY and CGH	SKY useful in dx, CGH in LOH
Sequencing	Rarely useful (p53 mutations)

Adapted from Triche T, Hicks JM, and Sorensen PHB [16].

subtypes of synovial sarcoma: (1) monophasic, composed of a spindle cells component, or (2) biphasic, containing areas with variable degrees of epithelial differentiation, form carcinoma-like to "occult" subtle cases.

Cytogenetically both monophasic and biphasic synovial sarcomas share a recurrent reciprocal t(X;18)(p11.2;q11.2) translocation. This translocation fuses the *SYT* gene from chromosome 18q11 to either of three homologous genes at *Xp11*, *SSX1*, *SSX2* and rarely *SSX4* (only two cases have been described) [23,24]. The *SSX1* and *SSX2* genes encode closely related proteins (81% identity). The N-terminal portion of each SSX protein exhibits homology to the Kruppel-associated box (KRAB), a transcriptional repressor. Both the SYT-SSX1 and the SYT-SSX2 hybrid transcripts encode fusion proteins in which the C-terminal eight amino acids of the normal SYT protein have been replaced by 78 amino acids encoded by an *SSX* gene [25]. SYT and SSX proteins appear to be transcriptional regulators primarily through protein–protein interactions [26]. SYT acting as an activator of transcription and SSX as a repressor [27].

Cytogenetic studies on series of synovial sarcomas demonstrated a near-diploid karyotype in a majority of the cases with the t(X;18)(p11.2;q11.2) translocation as the sole cytogenetic abnormality present in approximately a third of synovial sarcomas [28]. Other chromosomal changes include numerical changes and no other recurrent structural abnormalities.

Molecular detection of SYT-SSX1 and two fusions has been demonstrated to be of tremendous clinical value. PCR analysis demonstrated the presence of SYT-SSX1 or SYT-SSX2 fusion transcripts in approximately 95% synovial sarcomas examined, indicating that the detection of these hybrid transcripts by PCR may represent a useful diagnostic method. Sequence analysis demonstrated further heterogeneity in the fusion transcripts with the formation of two distinct SYT-SSX1 fusion junctions and two distinct SYT-SSX2 fusion junctions. Coexisting SYT-SSX1 and SYT-SSX2 has been reported in 10% SYT-SSX-positive primary tumors [29].

Kawai et al. [30] found a relationship between the type of fusion transcript and the histologic subtype (*SYT-SSX1* associated mostly with biphasic, and *SYT-SSX2* with monophasic types) as well as with the prognosis, with a significantly better metastasis-free survival associated with the *SYT-SSX2* subtype. Skytting et al. [31] suggested that the base pair differences between the SSX transcripts may have biologic significance. The impact of the *SYT-SSX* fusion type on the clinical behavior of synovial sarcoma has since become a subject of scientific debate. Ladanyi et al. [32] found fusion type *the single most significant prognostic factor by multivariate analysis in patients with localized disease at diagnosis* for synovial sarcoma. However, a recently published European multicenter, retrospective analysis study found that is histologic grade, but not SYT-SSX fusion type, the most important prognostic factor determining these patients prognosis [33]. Further studies, with careful clinical, morphologic, and molecular correlation will be necessary to determine the significance of the molecular fusion type in synovial sarcoma.

3.2. Alveolar rhabdomyosarcoma

RMS, the most common soft tissue sarcoma in children, is a small round cell tumor of skeletal muscle histogenesis, thought to arise as a consequence of loss of growth control and differentiation of myogenic cells [34]. Their differential diagnosis often depends of the identification of rhabdomyoblasts or the detection of muscle-specific proteins in the tumor cells [35,36,37]. Histologically three main types of RMS can be identified: (1) botryoid; (2) embryonal; and (3) alveolar, associated with poorer prognosis. Separate categories have been established for undifferentiated sarcoma, anaplastic and sarcoma NOS (not otherwise specified) subtypes.

Cytogenetic analysis revealed chromosomal abnormalities, primary aneuploidies, in all subtypes of RMS. In the alveolar subtype two specific chromosomal translocations have been identified. The t(2;13)(q35;q14) translocation can be cytogenetically detected in approximately 60% of alveolar RMS [38,39,40]. This translocation juxtaposes the *Pax3* gene on 2q35, a transcription factor functional during early neuromuscular development, to *FKHR* gene (also known as *FoxO1A*) on 13q14. *FKHR* is a member of the forkhead family of transcription factors [41,42,43]. As a result of this translocation the 5' portion of the *Pax3* gene, including an intact DNA binding domain, is fused to *FKHR*, resulting in a chimeric transcript and protein containing the *Pax3* DNA-binding domain and the distal half of the fork head and C-terminal region of *FKHR*. A less common variant of the translocation, fusing *Pax7* gene (another member of the forkhead family of transcription factors located on 1p36) to *FKHR*, and resulting in a t(1;13)(p36;q14) translocation, has also been associated with alveolar RMS [44]. A third t(2;2)(q35;p23) translocation fusing *Pax3* to nuclear receptor co activator (*NCOA1*) gene [45] has been recently identified by gene expression profiling.

Gene expression profiling has demonstrated the activation of a myogenic transcription program by the *Pax3/FKHR* fusion oncogene and it is assumed that these unique fusion genes activate the transcription of downstream genes, ultimately responsible for the transformed phenotype seen in these tumors, however, the exact mechanism is still under investigation [46,47].

Molecular identification of these fusion transcripts, mostly using RT-PCR assays and FISH are helpful in diagnosing these tumors, particularly when limited diagnostic material available or when microscopic and immunohistochemical findings are equivocal [48,49,50,51,52,53]. In a recent COG study, Pax3–FKHR and PAX7–FKHR fusion transcripts were identified in a majority of alveolar RMSs analyzed (77%), with the first being almost twice as common as the *Pax7* fusion in these tumors [54]. *Pax7* fusion genes are more often associated with lesions in the extremities occurring in younger patients and have a better outcome than those carrying a *Pax3* fusion, representing another example of sarcomas with fusion gene variants of apparent clinical relevance. Barr et al. [55] have documented a true "fusion-negative" subset of alveolar RMS represented by genetically diverse subsets of tumors, including low-expressers of the common fusion genes, fusion variants with other genes and true negative case.

Expression of myogenin in RMS has been associated with the alveolar subtype and worse prognosis [56,57].

4. OTHER PEDIATRIC SOLID TUMORS

4.1. Pediatric liver tumors: hepatoblastoma

Neoplasms of epithelial origin represent only a minority of pediatric malignant solid tumors. Malignant tumors of the liver account approximately 1.1% of malignant childhood tumors in the USA with hepatoblastoma being the most common, particularly in early childhood [58].

Hepatoblastoma is an embryonal liver tumor, with an incidence that is still rising, of 0.5–1.5 cases per million children per year [58]. The reason(s) for the rising incidence are unclear, but may be due to combined effects of increasing survival rates of extreme prematurity, as well as exposures to environmental toxins in utero or early in life [59]. It affects children between 6 months and 3 years of age, with nearly 90% of hepatoblastomas seen in the first 5 years of life, with a distinct male predominance with a male to female ratio of 2:1.

Hepatoblastoma represents the most common type of primary pediatric liver malignancy, and accounting for just over 1% of all pediatric cancers. These tumors originate from immature liver precursor cells (hepatoblasts), and may recapitulate some aspects of the liver development. Histologically, all hepatoblastomas are composed of epithelial tissue, and about one third of them show also focal mesenchymal differentiation. The epithelial component can be further subdivided into four types: (1) pure fetal (31%); (2) embryonal (19%); (3) macrotrabecular (3%); and (4) small undifferentiated (3%), with most of the tumors showing a combination of more than one of these types.

Accepted staging systems include the conventional POG/CCG systems based on postsurgical status, and the International/SIOP system of presurgical staging [58]. The primary treatment of hepatoblastoma is surgical resection; however, chemotherapy plays an important role by increasing the number of tumors that are resectable. Both staging and histology play important roles in determining the prognosis in these patients. Virtually 100% of patients with Stage I (completely resected at diagnosis), favorable histology (i.e., pure fetal type), survive, while the survival rates of patients with Stage IV tumors (distant metastases at diagnosis) of all histologies is 0–27%. Patients with low stage tumors that demonstrate unfavorable histology (i.e., small cell), usually recur, and have poor overall prognosis [60]. With chemotherapy, there is an 85% survival rate for Stage III hepatoblastomas having diverse histology [58]. The characterization of gene expression profiles of tumor cells from the treatment failure cases versus the responsive tumors, as well as those of the tumor cells that remain viable after effective chemotherapy, is an ultimate goal of the longer-term project.

Some hepatoblastomas have been associated with constitutional genetic abnormalities, congenital malformations, and familial cancer syndromes, such as Beckwith-Wiedeman syndrome (BWS), familial polyposis coli (FAP), and rare cases of Prader-Willi and Li-Fraumeni syndrome, but most cases are sporadic [58,61]. Multiple cytogenetic and molecular cytogenetic abnormalities have been described especially in sporadic hepatoblastoma. These include: (1) extra copies of chromosomes 2q and 20; (2) frequent chromosome breaks at 1q12-q21 and 2q35-37; and (3) chromosomal CGH-based descriptions of gains of material on chromosomes 1, 2, 7, 8, 17, 20, and 22q as well as loss of 4q materials [58,62,63]. The clinical significance of such observations is unclear. They are, therefore not incorporated into any of the staging or grading systems. One of the long-term goals of this project is to correlate these molecular cytogenetic changes both with gene expression changes, and overall clinical behavior.

Activation of the canonical Wnt signaling pathway is a consistent finding in hepatoblastoma. The Wnt pathway has long been known to direct growth and patterning during embryonic development, as well as being one of the key signaling pathways regulating cell growth, motility, and differentiation [64,65]. This pathway has also emerged as a critical regulator of stem cells [66] integrally involved in tightly regulated self-renewal of stem and progenitor cells, as well as cancer development in the digestive (intestinal) epidermal and hematopoietic systems [67]. Nuclear β-catenin is the hallmark of Wnt signaling activation. β-catenin is an important scaffolding protein involved in both cellular adhesion and as the central regulator of canonical Wnt target gene transcription [68]. In quiescent cells, β-catenin is bound to the cellular membrane by E-cadherin and also to α-catenin, thus linking cellular adhesion to the cytoskeleton. Cytoplasmic β-catenin is maintained at a low level by a degradation complex including APC, Axin and Axin-2. This complex presents β-catenin to the CK1 and GSK3 kinases, resulting in phosphorylation of β-catenin at key serine and threonine residues in exon-3. This results in the recruitment of β-TrCP-containing E3 ubiquitin ligase to these phosphorylated residues and proteosomal degradation of cytoplasmic β-catenin. The ubiquitin ligase EBI also targets β-catenin for degradation in a p53-induced manner [69], thus linking DNA damage to inhibition of β-catenin-mediated transcription. Upon binding of Wnt ligand to the Frizzled/LRP corcceptor complex, the Dishevelled protein inhibits GSK3, allowing cytoplasmic accumulation and subsequent nuclear translocation of β-catenin. In the nucleus, β-catenin binds to BCL9 through the first four Arm repeats. β-catenin and BCL9 juxtapose Pygopus and the TCF/LEF transcription factors. The recently discovered BCL9-2 also links β-catenin to TCF/LEF [70]. The nuclear (TCF/LEF)/β-catenin/(BCL9/BCL9-2)/Pygopus complex facilitates the transcription of numerous canonical *Wnt* target genes, including those regulating diverse cellular activities such as proliferation (*CCND1*), apoptosis (*cMyc* and *survivan*), migration (*MMP-7* and *MMP-26*), growth factor ligands and receptors (*FGF18* and *MET*), Wnt transcription factors (*TCF-1* and *LEF1*) and inhibitors of Wnt canonical signaling (*Axin2, DKK1*, and *BTRC*) [71].

Mutations in the genes that constitute the canonical Wnt signaling pathway are frequently seen in adult cancers, including colorectal, liver, endometrial, prostate, thyroid, skin, and brain tumors [70]. Recent evidence also implicates this pathway in the development of childhood tumors (liver, kidney, brain, and pancreas) [77] that typically arise from germinal tissues, as opposed to the epithelial origin of the majority of adult cancer, most likely from a rapidly dividing stem cell. Nuclear translocation of β-catenin is present in 60–90% of hepatoblastomas [72,73] and has been associated with prognosis. Different types of *CTNNB1* gene-activating mutations, including point mutations in exon-3 and deletions either confined to exon-3 or extending from exon-3 to exon-4, have also been documented in hepatoblastomas with a frequency that varies from 15 to over 70% of cases, depending on the design of the primers used. Mutations of other components of the canonical Wnt pathway in hepatoblastoma include rare inactivating mutations in *APC, Axin,* and *Axin2* [74,75,76]. Although mutations in these members of the canonical Wnt pathway appear to be frequent in hepatoblastoma, a large correlation study including β-catenin status (mutation type), histological subtype and expression pattern of other components of this pathway and, most important, of its target genes, has not been previously reported.

Important information about the pathogenesis of childhood cancer has derived from the study of normal embryonic development. The similarities between growth and differentiation of cells and tissues and the dysregulation of these events in oncogenesis are evident at multiple levels. At the molecular level, this relationship has become substantial with the discovery that many proto-oncogenes encode components of signal transduction pathways that direct normal development, including the Wnt signaling pathway [77]. The hepatoblastoma progenitor undifferentiated cell of origin (hepatoblast) which equivalent during embryonic development (liver stem cell) is programmed to undergo massive proliferation, appears to be somehow locked in this uncontrolled proliferative state in these tumors.

Our Department of Pathology at Texas Children's Hospital and Baylor College of Medicine in Houston, Texas, is one of the two institutions designated for histological review of pediatric liver tumors by the COG in the USA. We have developed an increasing interest in studying the biology of hepatoblastoma and particularly in identifying biological differences that may explain behavior of histological subtypes and could be used as clinical markers.

We have studied 41 hepatoblastoma specimens and performed genomic profiling using BAC array CGH, gene expression profiling using Affymetrix chips, and QRT-PCR and are validating some of our data using immunohistochemistry, and in situ hybridization. We are particularly interested in the Wnt signaling pathway and the important role of its abnormal over activation may play in hepatoblastoma pathogenesis. One of our main objectives is to investigate additional mechanisms by which defects in regulation of the Wnt signaling pathway contribute to tumor progression in different subtypes of hepatoblastoma.

In order to investigate Wnt activation and hepatoblastoma we performed analysis of *CTNNB1* gene mutation status and β-catenin expression pattern in different subtypes and explored the possible contribution of additional canonical Wnt pathway molecules (epigenetic mechanisms) in Wnt signaling activation. We also analyzed the status of *Wnt/Tcf* target genes in hepatoblastoma.

We found markedly increased nuclear expression of β-catenin (particularly in the higher-grade tumor components) in hepatoblastomas with embryonal/small cell histology, and point mutations, deletions within or confined to the entire exon-3, (rarely no CTNNB1 mutations). By contrast, hepatoblastomas with pure fetal epithelial histology showed a predominantly membranous and only rare nuclear β-catenin expression, as well as predominantly large deletions including the entire exon-3 and most of exon-4 of *CTNNB1* gene. Our hypothesis is that when mutations are confined to exon-3 the BCL9-interaction domain is maintained, as well as the transcription of *Wnt* target genes, resulting in a more aggressive phenotype. Large deletions including the BCL9-interacting domain would not be as capable of facilitating canonical *Wnt* target gene expression (as wild type *CTNNB1*).

We studied other canonical and noncanonical Wnt pathway molecules by real-time QRT-PCR, using SYBR Green chemistry (value of the target gene calculated in relation to a control sample using $\Delta\Delta Ct$ method). This analysis demonstrated downregulation of canonical Wnt pathway genes, induction of upstream antagonists of canonical Wnt signaling, and induction of some non-canonical *Wnt* pathway genes in the majority of Hepatoblastomas.

We are now in the process of studying *Wnt/Tcf* target genes in hepatoblastoma (36 downstream targets, a subset of 14 involved in feedback regulation) by QRT-PCR, microarray analysis, and tissue immunohistochemistry in order to investigate the possibility of a hepatoblastoma subtype-specific Tcf target program that may explain clinical behavior and response to therapy.

We expect that a better understanding of hepatoblastoma biology, based on a systematic analysis of biologically important signaling pathways, and integration of this information with histology and other tumor markers, will lead to new biologically relevant and therapeutically significant tumor classification

5. CONCLUSIONS

The diagnosis of cancer and particularly, the diagnosis and prognostication of pediatric malignancies is undergoing tremendous challenges and improvements, as diagnostic pathology and cancer diagnosis are becoming part of the new molecular era. Classification of pediatric malignancies based on molecular markers and gene expression signatures has already begun. Pathologists will be playing a crucial translating role in assay development, clinical validation and diagnostic integration of new biological markers. This integration of new diagnostic information is already become clinically necessary to correctly diagnose, classify, and clinically stratify pediatric cancer patients. The integration of additional and more complex diagnostic parameters will become gradually more important in order to make optimal therapeutic decisions.

Abbreviations:
ALL: Acute Lymphoblastic Leukemia; PCR: Polymerase Chain Reaction; RT-PCR: Reverse Transcription, Polymerase Chain Reaction; FISH: Fluorescence in situ Hybridization; SKY: Spectral Karyotyping; CGH: Comparative Genomic Hybridization; CISH: Chromogenic In Situ Hybridization; NHL: non-Hodgkin lymphoma; RMS: Rhabdomyosarcoma; QRT-PCR: Quantitative Reverse Transcription Polymerase Chain Reaction; BAC: Bacterial Artificial Chromosomes.

REFERENCES

1. Gilliland, D.G. and M.S. Tallman, Focus on acute leukemias. *Cancer Cell* 1(5): 417–420 (2002).
2. Pui, C.H., M.V. Relling, and J.R. Downing, Acute lymphoblastic leukemia. *N Engl J Med* 350(15): 1535–1548 (2004).
3. Margolin, J.S., D.G. Poplack, Acute Lymphoblastic Leukemia, in: Pizzo, P.A. and Poplack, D.G. (eds) *Principles and Practice of Pediatric Oncology*. Lippincott Williams & Wilkins: Philadelphia, pp. 538–591 (2006).
4. Pui, C.H., D. Campana, and W.E. Evans, Childhood acute lymphoblastic leukaemia—current status and future perspectives. *Lancet Oncol* 2(10): 597–607 (2001).
5. Ross, M.E. et al., Classification of pediatric acute lymphoblastic leukemia by gene expression profiling. *Blood* 102(8): 2951–2959 (2003).
6. Wallace, J. et al., BARCODE-ALL: accelerated and cost-effective genetic risk stratification in acute leukemia using spectrally addressable liquid bead microarrays. *Leukemia* 17(7): 1411–1413 (2003).
7. Cairo, M.S. et al., Childhood and adolescent non-Hodgkin lymphoma: new insights in biology and critical challenges for the future. *Pediatr Blood Cancer* 45(6): 753–769 (2005).
8. Morris, S.W. et al., Fusion of a kinase gene, ALK, to a nucleolar protein gene, NPM, in non-Hodgkin's lymphoma. *Science* 263(5151): 1281–1284 (1994).
9. Bischof, D. et al., Role of the nucleophosmin (NPM) portion of the non-Hodgkin's lymphoma-associated NPM-anaplastic lymphoma kinase fusion protein in oncogenesis. *Mol Cell Biol* 17(4): 2312–2325 (1997).
10. Kutok, J.L. and J.C. Aster, Molecular biology of anaplastic lymphoma kinase-positive anaplastic large-cell lymphoma. *J Clin Oncol* 20(17): 3691–3702 (2002).
11. Simonitsch, I. et al., NPM/ALK gene fusion transcripts identify a distinct subgroup of null type Ki-1 positive anaplastic large cell lymphomas. *Br J Haematol* 92(4): 866–871 (1996).
12. Beylot-Barry, M. et al., Detection of t(2;5)(p23;q35) translocation by reverse transcriptase polymerase chain reaction and in situ hybridization in CD30-positive primary cutaneous lymphoma and lymphomatoid papulosis. *Am J Pathol* 149(2): 483–492 (1996).
13. Armstrong, G. et al., Early molecular detection of central nervous system relapse in a child with systemic anaplastic large cell lymphoma: case report and review of the literature. *Pediatr Blood Cancer* 44(4): 400–406 (2005).
14. Heerema, N.A. et al., State of the Art and Future Needs in Cytogenetic/Molecular Genetics/ Arrays in childhood lymphoma: summary report of workshop at the First International Symposium on childhood and adolescent non-Hodgkin lymphoma, April 9, 2003, New York City, NY. *Pediatr Blood Cancer* 45(5): 616–622 (2005).
15. Borden, E.C. et al., Soft tissue sarcomas of adults: state of the translational science. *Clin Cancer Res* 9(6): 1941–1956 (2003).

16. Triche, T., J.M. Hicks, and P.H.B. Sorensen, Diagnostic Pathology of Pediatric Malignancies, in: Pizzo, P.a.P.D. (ed.) *Principles and Practice of Pediatric Oncology.* Lippincott Williams & Wilkins: Philadelphia, pp. 185–236 (2006).
17. Enzinger, F.M. and S.W. Weiss, *Soft Tissue Tumors.* St Louis: Mosby (2001).
18. Fletcher, C., *Diagnostic Histopathology of Tumors*, Vol. 1. London: Churchill-Livingstone (2000).
19. Shmookler, B.M., F.M. Enzinger, and R.B. Brannon, Orofacial synovial sarcoma: a clinicopathologic study of 11 new cases and review of the literature. *Cancer* 50(2): 269–276(1982).
20. Shmookler, B.M., Retroperitoneal synovial sarcoma. A report of four cases. *Am J Clin Pathol* 77(6): 686–691 (1982).
21. Fetsch, J.F. and J.M. Meis, Synovial sarcoma of the abdominal wall. *Cancer* 72(2): 469–477 (1993).
22. Zeren, H. et al., Primary pulmonary sarcomas with features of monophasic synovial sarcoma: a clinicopathological, immunohistochemical, and ultrastructural study of 25 cases. *Hum Pathol* 26(5): 474–480 (1995).
23. Crew, A.J. et al., Fusion of SYT to two genes, SSX1 and SSX2, encoding proteins with homology to the Kruppel-associated box in human synovial sarcoma. *Embo J* 14(10): 2333–2340 (1995).
24. de Leeuw, B. et al., Identification of two alternative fusion genes, SYT-SSX1 and SYT-SSX2, in t(X;18)(p11.2;q11.2)-positive synovial sarcomas. *Hum Mol Genet* 4(6): 1097–1099 (1995).
25. Thaete, C. et al., Functional domains of the SYT and SYT-SSX synovial sarcoma translocation proteins and co-localization with the SNF protein BRM in the nucleus. *Hum Mol Genet* 8(4): 585–591 (1999).
26. Ladanyi, M., Fusions of the SYT and SSX genes in synovial sarcoma. *Oncogene* 20(40): 5755–5762 (2001).
27. dos Santos, N.R., D.R. de Bruijn, and A.G. van Kessel, Molecular mechanisms underlying human synovial sarcoma development. *Genes Chromosomes Cancer* 30(1): 1–14 (2001).
28. Sandberg, A.A. and J.A. Bridge, Updates on the cytogenetics and molecular genetics of bone and soft tissue tumors. Synovial sarcoma. *Cancer Genet Cytogenet* 133(1): 1–23 (2002).
29. Yang, K. et al., Co-existence of SYT-SSX1 and SYT-SSX2 fusions in synovial sarcomas. *Oncogene* 21(26): 4181–4190 (2002).
30. Kawai, A. et al., SYT-SSX gene fusion as a determinant of morphology and prognosis in synovial sarcoma. *N Engl J Med* 338(3): 153–160 (1998).
31. Skytting, B. et al., A novel fusion gene, SYT-SSX4, in synovial sarcoma. *J Natl Cancer Inst* 91(11): 974–975 (1999).
32. Ladanyi, M. et al., Impact of SYT-SSX fusion type on the clinical behavior of synovial sarcoma: a multi-institutional retrospective study of 243 patients. *Cancer Res* 62(1): 135–140 (2002).
33. Guillou, L. et al., Histologic grade, but not SYT-SSX fusion type, is an important prognostic factor in patients with synovial sarcoma: a multicenter, retrospective analysis. *J Clin Oncol* 22(20): 4040–4050 (2004).
34. Wexler, L., W.H. Meyer, and L.H. Helman, Rhabdomyosarcoma and undifferentiated sarcoma, in: Pizzo, P.a.P.D. (ed.) *Principles and Practice of Pediatric Oncology.* Lippincott Williams & Wilkins: Philadelphia, pp. 971–1002 (2006).
35. Dias, P. et al., Strong immunostaining for myogenin in rhabdomyosarcoma is significantly associated with tumors of the alveolar subclass. *Am J Pathol* 156(2): 399–408 (2000).
36. Qualman, S.J. et al., Intergroup Rhabdomyosarcoma Study: update for pathologists. *Pediatr Dev Pathol* 1(6): 550–561 (1998).
37. Parham, D.M., Pathologic classification of rhabdomyosarcomas and correlations with molecular studies. *Mod Pathol* 14(5): 506–514 (2001).
38. Whang-Peng, J. et al., Cytogenetic characterization of selected small round cell tumors of childhood. *Cancer Genet Cytogenet* 21(3): 185–208 (1986).
39. Whang-Peng, J. et al., Cytogenetic studies in subgroups of rhabdomyosarcoma. *Genes Chromosomes Cancer* 5(4): 299–310 (1992).

40. Shapiro, D.N., Valentine, M.B., Sublett, J.E., Sinclair, A.E., Tereba, A.M., Scheffer, H., Buys, C.H., Look, A.T., Chromosomal sublocalization of the 2;13 translocation breakpoint in alveolar rhabdomyosarcoma. *Genes Chromosomes Cancer* 4(3): 241–249 (1992).
41. Barr, F.G. et al., Rearrangement of the PAX3 paired box gene in the paediatric solid tumour alveolar rhabdomyosarcoma. *Nat Genet* 3(2): 113–117 (1993).
42. Shapiro, D.N. et al., Fusion of PAX3 to a member of the forkhead family of transcription factors in human alveolar rhabdomyosarcoma. *Cancer Res* 53(21): 5108–5112 (1993).
43. Galili, N. et al., Fusion of a fork head domain gene to PAX3 in the solid tumour alveolar rhabdomyosarcoma. *Nat Genet* 5(3): 230–235 (1993).
44. Biegel, J.A. et al., Detection of the t(2;13)(q35;q14) and PAX3-FKHR fusion in alveolar rhabdomyosarcoma by fluorescence in situ hybridization. *Genes Chromosomes Cancer* 12(3): 186–192 (1995).
45. Wachtel, M. et al., Gene expression signatures identify rhabdomyosarcoma subtypes and detect a novel t(2;2)(q35;p23) translocation fusing PAX3 to NCOA1. *Cancer Res* 64(16): 5539–5545 (2004).
46. Sublett, J.E., I.S. Jeon, and D.N. Shapiro, The alveolar rhabdomyosarcoma PAX3/FKHR fusion protein is a transcriptional activator. *Oncogene* 11(3): 545–552 (1993).
47. Bennicelli, J.L., R.H. Edwards, and F.G. Barr, Mechanism for transcriptional gain of function resulting from chromosomal translocation in alveolar rhabdomyosarcoma. *Proc Natl Acad Sci USA* 93(11): 5455–5459 (1996).
48. Barr, F.G., Q.B. Xiong, and K. Kelly, A consensus polymerase chain reaction-oligonucleotide hybridization approach for the detection of chromosomal translocations in pediatric bone and soft tissue sarcomas. *Am J Clin Pathol* 104(6): 627–633 (1995).
49. Biegel, J.A. et al., Chromosomal translocation t(1;13)(p36;q14) in a case of rhabdomyosarcoma. *Genes Chromosomes Cancer* 3(6): 483–484 (1991).
50. McManus, A.P. et al., Interphase fluorescence in situ hybridization detection of t(2;13)(q35;q14) in alveolar rhabdomyosarcoma—a diagnostic tool in minimally invasive biopsies. *J Pathol* 178(4): 410–414 (1996).
51. Anderson, J. et al., Amplification of the t(2; 13) and t(1; 13) translocations of alveolar rhabdomyosarcoma in small formalin-fixed biopsies using a modified reverse transcriptase polymerase chain reaction. *Am J Pathol* 150(2): 477–482 (1997).
52. Edwards, R.H. et al., Detection of gene fusions in rhabdomyosarcoma by reverse transcriptase-polymerase chain reaction assay of archival samples. *Diagn Mol Pathol* 6(2): 91–97 (1997).
53. Athale, U.H. et al., Use of reverse transcriptase polymerase chain reaction for diagnosis and staging of alveolar rhabdomyosarcoma, Ewing sarcoma family of tumors, and desmoplastic small round cell tumor. *J Pediatr Hematol Oncol* 23(2): 99–104 (2001).
54. Sorensen, P.H. et al., PAX3-FKHR and PAX7-FKHR gene fusions are prognostic indicators in alveolar rhabdomyosarcoma: a report from the children's oncology group. *J Clin Oncol* 20(11): 2672–2679 (2002).
55. Barr, F.G. et al., Genetic heterogeneity in the alveolar rhabdomyosarcoma subset without typical gene fusions. *Cancer Res* 62(16): 4704–4710 (2002).
56. Dias, P., Chen, B., Dilday, B., Palmer, H., Hosoi, H., Singh, S., Wu, C., Li, X., Thompson, J., Parham, D., Qualman, S., Houghton, P., Strong immunostaining for myogenin in rhabdomyosarcoma is significantly associated with tumors of the alveolar subclass. *Am J Pathol* 156(2): 399–408 (2000).
57. Hostein, I. et al., Rhabdomyosarcoma: value of myogenin expression analysis and molecular testing in diagnosing the alveolar subtype: an analysis of 109 paraffin-embedded specimens. *Cancer* 101(12): 2817–2824 (2004).
58. Mueller, B., D.H. Lopez-Terrada, and M.J. Finegold, Tumors of the liver, in: Pizzo, P.A.a.P.D.G. (ed.) *Principles and Practice of Pediatric Oncology*. Lippincott Williams & Wilkins: Philadelphia, pp. 887–905 (2006).
59. Buckley, J.D. et al., A case-control study of risk factors for hepatoblastoma. A report from the Childrens Cancer Study Group. *Cancer* 64(5): 1169–1176 (1989).

60. Haas, J.E., J.H. Feusner, and M.J. Finegold, Small cell undifferentiated histology in hepatoblastoma may be unfavorable. *Cancer* 92(12): 3130–3134 (2001).
61. Oda, H. et al., A mutational hot spot in the p53 gene is associated with hepatoblastomas. *Int J Cancer* 60(6): 786–790 (1995).
62. Ma, S.K. et al., Cytogenetic characterization of childhood hepatoblastoma. *Cancer Genet Cytogenet* 119(1): 32–36 (2000).
63. Weber, R.G. et al., Characterization of genomic alterations in hepatoblastomas. A role for gains on chromosomes 8q and 20 as predictors of poor outcome. *Am J Pathol* 157(2): 571–578 (2000).
64. Behrens, J. and B. Lustig, The Wnt connection to tumorigenesis. *Int J Dev Biol* 48(5–6): 477–487 (2004).
65. Lind, G.E. et al., A CpG island hypermethylation profile of primary colorectal carcinomas and colon cancer cell lines. *Mol Cancer* 3: 28 (2004).
66. Rattis, F.M., C. Voermans, and T. Reya, Wnt signaling in the stem cell niche. *Curr Opin Hematol* 11(2): 88–94 (2004).
67. Taketo, M.M., Shutting down Wnt signal-activated cancer. *Nat Genet* 36(4): 320–322 (2004).
68. Reya, T. and H. Clevers, Wnt signalling in stem cells and cancer. *Nature* 434(7035): 843–850 (2005).
69. Liu, J. et al., Siah-1 mediates a novel beta-catenin degradation pathway linking p53 to the adenomatous polyposis coli protein. *Mol Cell* 7(5): 927–936 (2001).
70. Brembeck, F.H. et al., Essential role of BCL9-2 in the switch between beta-catenin's adhesive and transcriptional functions. *Genes Dev* 18(18): 2225–2230 (2004).
71. Nusse, R., The Wnt Homepage, http://www.stanford.edu/~rnusse/wntwindow.html (2006).
72. Takayasu, H. et al., Frequent deletions and mutations of the beta-catenin gene are associated with overexpression of cyclin D1 and fibronectin and poorly differentiated histology in childhood hepatoblastoma. *Clin Cancer Res* 7(4): 901–908 (2001).
73. Park, W.S. et al., Nuclear localization of beta-catenin is an important prognostic factor in hepatoblastoma. *J Pathol* 193(4): 483–490 (2001).
74. Oda, H. et al., Somatic mutations of the APC gene in sporadic hepatoblastomas. *Cancer Res* 56(14): 3320–3323 (1996).
75. Taniguchi, K. et al., Mutational spectrum of beta-catenin, AXIN1, and AXIN2 in hepatocellular carcinomas and hepatoblastomas. *Oncogene* 21(31): 4863–4871 (2002).
76. Koch, A. et al., Mutations and elevated transcriptional activity of conductin (AXIN2) in hepatoblastomas. *J Pathol* 204(5): 546–554 (2004).
77. Koesters, R. and M. von Knebel Doeberitz, The Wnt signaling pathway in solid childhood tumors. *Cancer Lett* 198(2): 123–138 (2003).

CHAPTER 13

PRECLINICAL MODELS FOR CELL CYCLE-TARGETED THERAPIES

MARCOS MALUMBRES

Cell Division and Cancer Group, Centro Nacional de Investigaciones Oncológicas (CNIO), Madrid, Spain

Abstract: Deregulation of the cell cycle machinery is frequently associated to tumor development in most cell types. Many of these tumor-associated alterations result in the abnormal activation of cyclin-dependent protein kinases (Cdks) involved in the G1/S transition. Recent results from the genetic analysis of cyclins and Cdks in mouse models have raised some questions regarding the relevance of these molecules in cell cycle-targeted strategies for cancer therapy. The comprehensive evaluation of these biochemical and genetic data seems to be a necessary given the relevance of these and other new cell cycle kinases as targets for therapeutic approaches in cancer.

Key words: Cell cycle regulation, cyclin-dependent kinase, mouse models, kinase inhibitors, cancer therapy

1. INTRODUCTION

Cell cycle progression is controlled by diverse mechanisms including gene expression and protein phosphorylation or degradation. Progression through G1 mostly results in the expression of the genes required for the replication of the genome (S phase) and chromosome segregation in mitosis (M phase). Progression through these later phases requires the appropriate posttranslational modification of the existing proteins and degradation of these molecules when they are not required or are an obstacle for transition to the following phases. In the last few years, biochemical examination of G1/S Cdks, their regulators (cyclins and Cdk inhibitors, CKIs) and their substrates (mainly the retinoblastoma protein, pRb) has provided a general framework for the understanding of how mammalian cell cycle progression is regulated. During the G1 phase of the cell cycle, cells may decide whether to stay quiescent or to enter

S phase, where the genome is duplicated (Figure 13.1). Progression throughout the G1 phase is regulated by a complex mechanism involving at least Cdk4, Cdk6, and Cdk2 [1,2]. An additional kinase, Cdk3, can also function at this level although its physiological role is unclear [3,4]. During the G1 phase, cells receive and evaluate mitogenic and antiproliferative signals. Mitogenic signals, such as those emitted by growth factors, frequently result in the induction of D-type cyclins (cyclin D1, D2, or D3), which bind and activate Cdk4 and Cdk6 [1,2]. These kinases are then able to partially phosphorylate pRb and induce the synthesis of cyclin E (E1 and E2) which, in turn, binds and activates Cdk2. Cdk2–cyclin E complexes further phosphorylate the pRb protein canceling pRb-mediated repression of genes whose activities are necessary for S-phase entry. Further activity of Cdk2 bound to cyclin A is required to progress through S phase and to prepare cells for mitosis. Another member of the Cdk

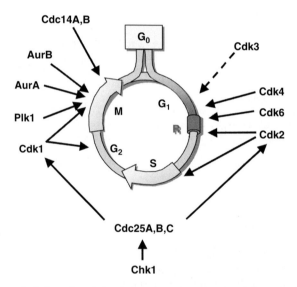

Figure 13.1. Control of the cell cycle by Cdks, other kinases, and phosphatases whose inhibition may have therapeutic potential in cancer. G1 cyclin-dependent kinases (Cdk) such as Cdk4, Cdk6, and Cdk2 (and possibly Cdk3) phosphorylate retinoblastoma proteins inducing progression throughout the G1 phase. "R" indicates the restriction point where cells become independent on sustained mitogenic signaling. G2/M progression is controlled by mitotic kinases such as Cdk1, Polo-like kinase (Plk1), and Aurora kinases (AurA, AurB, and perhaps AurC in some cell types; not shown). Some phosphatases are also involved in this regulation. Cdc25A, Cdc25B, and Cdc25C activate Cdks by removing inactivating phosphorylations. In some instances, these phosphatases are regulated by checkpoint kinases such as Chk1. Cdc14A and Cdc14B, on the other hand, are phosphatases that remove some phosphates previously introduced by mitotic kinases; an event required for the proper exit from mitosis.

family, Cdk1 (Cdc2) is involved in the control of the G2 and M phases [5]. The understanding of Cdk biology has opened a framework to understand cell cycle regulation and its deregulation in human cancer and to evaluate new therapeutic strategies.

Most proteins involved in Cdk regulation and some of their substrates have been linked to tumor development, underscoring the importance of these cell cycle regulators in maintaining appropriate proliferation rates [6]. D- and E-type cyclins are frequently overexpressed in human tumors, whereas Cdk inhibitors (such as p16INK4a, p21Cip, or p27Kip1) or Cdk substrates (such as pRb) are frequently considered as tumor suppressor proteins [6–8]. Genetic alteration of Cdk4 itself is more restricted although this gene is amplified and overexpressed in a wide variety of tumors and tumor cell lines. Some of them, mainly gliomas, sarcomas, breast tumors, and carcinomas of the uterine cervix display coamplification with the HDM2 locus as part of an amplicon located in human chromosome 12 (12q13–14) [8]. A point mutation in Cdk4 has also been described in spontaneous and familiar melanomas [9,10]. This mutation, substitution of Arg24 by Cys (R24C), leads to misregulation of the kinase activity by preventing binding of the INK4 family of cell cycle inhibitors without affecting the affinity of Cdk4 for cyclin D1. Similarly, Cdk6 is frequently overexpressed in hematopoietic malignancies, in some cases as a consequence of a translocation that places the Cdk6 locus under the control of strong promoters in these cells [5]. All these data have stimulated the design and development of small-molecule Cdk inhibitors as new drugs for cancer therapy (see below). However, the gap existing between biochemical/cellular models and human clinic has not been explored until the recent analysis of mice genetically engineered to express deficient (knock outs) or hyperactive (knock ins) alleles of some of these cell cycle regulators.

2. GENETIC ANALYSIS OF MAMMALIAN CDKS

Although the general functions of these Cdks during G1 have been extensively characterized in cultured cells, the exact role of each protein in vivo is obscured by the presence of multiple members of each family [11]. The generation of loss-of-function mutant mice for some of these proteins has provided us with important information regarding their individual roles. Lack of either Cdk4 or Cdk6 is compatible with embryonic life in the mouse and only results in alterations in specific cell types. Thus, only the pituitary gland, the beta cells of the endocrine pancreas, and the adipocytes seem to be severely affected by the lack of Cdk4 [12–15]. Similarly, the absence of Cdk6 only results in abnormalities in the hematopoietic compartment [16]. An obvious explanation for these phenotypes comes from the high structural and functional homology between Cdk4 and Cdk6. In vivo, lack of one of these cyclin D-dependent kinases might be compensated by the remaining one, at least in those cells where both proteins are expressed. Double Cdk4;Cdk6-deficient embryos display a more dramatic phenotype since they die during late embryonic development due to severe anemia [16]. These results demonstrate a partial compensation between these two Cdks, at least in the control of hematopoiesis. However, the combined lack of Cdk4 and Cdk6 does not result in defective G1/S transition or decreased proliferation in most other cell types even in late embryonic development [16]. A similar phenotype is observed in embryos lacking the three D-type cyclins [17] underscoring the strong functional correspondence between both group of proteins. The fact that D-type cyclins can also form complexes with Cdk2 [16] or Cdk5 [18] does not seem to be sufficient to rescue the lack of the canonical D-type Cdks, Cdk4, and Cdk6, although it could participate in the slightly increased survival of double Cdk4;Cdk6 mutant embryos (18.5 days of embryonic development versus 16.5 days in triple cyclin D1/D2/D3 mutant embryos). These D-type cyclin–Cdk2 complexes are also insufficient to rescue the defects induced by the deficiency in Cdk4 and Cdk6, since Cdk4;Cdk6 double knock out MEFs display an identical behavior to MEFs lacking the three D-type cyclins [16,17]. MEFs deficient in Cdk4 and Cdk6 or cells deficient in the three D-type cyclins display a decreased rate of proliferation in culture, accompanied by diminished phosphorylation of pRb and delayed expression of S-phase and M-phase markers such as E-type, A-type, and B-type cyclins [16,17]. The fact that all these cells are insensitive to the INK4 family of cell cycle inhibitors highlights the specificity of these inhibitors versus D-type cyclin–Cdk4/6 complexes. Yet, these MEFs are able to respond to mitogenic stimuli after serum starvation, indicating that cyclin D–Cdk4/6 complexes are not essential for mitogen-induced proliferation [16,17]. Whether this ability is dependent on Cdk2, other kinase or other kinase-independent pathways is still unknown [5].

The absence of Cdk2 is also compatible with life and only affects germ cell development [19,20]. In fact, the lack of mitotic phenotypes in the Cdk2-*null* mice has questioned the importance of this protein in several of the processes mentioned above such as DNA replication, centrosome maturation, and segregation and modulation of proteolysis [5]. Not only that; the differences between the targeted deletion of Cdk2 and those of the two E-type cyclins [21,22] underscore the functional differences between Cdk2 and these cyclins. Although the individual deficiency in one of these E-type cyclins has only minor effects in mouse development, combined deficiency in both cyclins results in embryonic lethality [21,22]. Cells without cyclin E1 and E2 are able to proliferate although they display a dramatic defect in the G0/G1 transition due to problems loading the Mcm proteins onto the pre-replication complexes [21], a defect not present in Cdk2-null cells [19,20]. Cdk2;Cdk6 double mutants do not display any synergistic phenotype [16]. Unfortunately, double Cdk2;Cdk4 mutants cannot be derived from the single knock out strains since both loci are located very closely in the mouse genome. Thus, generation of these double Cdk2;Cdk4 mutants will require additional recombination steps in ES cells.

The fact that all these three G1 Cdks can phosphorylate the pRb might suggest that these enzymes have overlapping roles. Although it has been reported that Cdk4/6 and Cdk2 phosphorylate different residues in pRb [23], the absence of one of these kinases could alter the affinities for specific sites. In fact, Cdk4 is quite effective in phosphorylating Cdk2-specific sites in Cdk2-depleted cells [24]. Not only these Cdks, but also Cdk1, Cdk3, and Cdk9 are able to phosphorylate pRb [11]. Alternatively, the control of pRb function might be modulated by the overall phosphorylation state rather than phosphorylation in specific sites, and a single Cdk could achieve the appropriate levels of phosphorylation. Compensatory roles are not so obvious for those substrates that seem to be specific for each of these G1 Cdks [5]. Thus, it has been recently described that Cdk4, but not Cdk2, phosphorylate and inactivate Smad3 proteins [25]. Similarly, Cdk2 (but not the other G1 Cdks) phosphorylates a variety of proteins involved in DNA replication, centrosome duplication and segregation and mitotic control [2,5]. In many cases, they have neither been reported as Cdk1 substrates, although given the promiscuity of these proteins [26] slight changes in affinities or subcellular localization could make these proteins available for phosphorylation by other kinases. Whether other Cdks (such as Cdk4 or even Cdk1) might be able to phosphorylate these proteins in the absence of Cdk2 is still unknown.

In a different scenario, one could assume that neither of these G1 Cdks nor their substrates are essential for cell cycle progression. In yeast, a single Cdk (most similar to human Cdk1) is sufficient for cell cycle progression [27]. In fact, Cdk1–cyclin B complexes, although initially responsible for the G2/M transition, might have cryptic S-phase-promoting abilities that could make the other Cdks dispensable for G1 transition [28]. Recently, it has also been observed that Cdk1 may also bind and be activated by E-type cyclins [29] raising the question whether this kinase may fully compensate for the lack of Cdk2 in the mitotic cell cycle.

3. THE G1/S TRANSITION AS A TARGET IN CANCER THERAPY

From human tumors and mouse models, it is now clear that misregulation of Cdk4/6 activity by either overexpression of D-type cyclins or loss of INK4 proteins almost invariably leads to hyperproliferative defects and eventually to tumor development. First, as stated above, many genetic mutations in human tumors affect Cdk4/6 regulation [6]. Second, knock-in mice expressing a deregulated Cdk4 develop a wide spectrum of spontaneous tumors confirming the central role of Cdk4 in the entry into the cell cycle [30,31]. Inactivating mutations in single INK4 inhibitors in the mouse, including p16^{INK4a} [32,33], p15^{INK4b} [34] and p18^{INK4c} [34,35], also result in tumor susceptibility. These observations, taken together, illustrate that activation of the Cdk4 pathway dramatically decreases the requirements that allow cells to enter the cell cycle and might participate in tumor development. These data have led to the design of therapeutic strategies for cancer treatment based on the inhibition of these G1 Cdks [6,36,37].

However, the fact that the absence of each of these G1 Cdks has no major effects on cell cycle progression has raised a note of caution in the design of small-molecule inhibitors against these kinases. The fact that cells can proliferate without these kinases, whatever this depends on lack of function or compensation by other proteins, could make G1 Cdk inhibitors less attractive for cancer therapy. Yet, the molecular effects resulting from the enzymatic inhibition of one protein might differ from the effects caused by the absence of that protein. Thus, lack of Cdk4 and Cdk6 favor the presence of complexes between Cdk2 and D-type cyclins [16] and these complexes could function to promote G1 progression. Small-molecule ATP analogs should provoke different results since they would allow the formation of inactive Cdk4/6–cyclin D complexes without redistribution of the cyclins to other kinase active complexes. The current technical requirements for the generation of mice deficient in multiple Cdks are preventing us from analyzing the effect of combined deficiency in multiple Cdk activities.

Second, the requirements for these kinases in tumor cells have not been characterized in vivo yet. Even though normal cells do not require these proteins, tumor cells that have increased proliferation rates could be more sensitive to the lack of these G1 kinases. In cultured cells, some tumor cell types are insensitive to the lack of Cdk2 but require Cdk4 [24]. These results also suggest that future design of small-molecule Cdk inhibitors could benefit if specificity towards a particular Cdk is reduced. In general, one could assume that drugs inhibiting all cell cycle Cdks could be more effective than specific ones. In fact, most of the Cdk inhibitors that have shown some activity in clinical trials have reduced specificity [38]. Obviously, inhibition of many proteins might produce unexpected or out-of-control results unless we have a detailed characterization of the function of these proteins in vivo.

The information we have obtained in the last years regarding the Cdk family and their involvement in cell cycle regulation and tumor development may help us to design new therapeutic strategies based upon cell cycle modulation. In addition to Cdks, many other cell cycle kinases and even phosphatases are now being considered as therapy targets (Figure 13.1). Among the kinases, there is a intensive effort to investigate some mitotic proteins such as Aurora or Polo-like kinases (Plk) and some other molecules involved in the different cell cycle checkpoints, such as Chk1 [39–41]. The therapeutic potential of cell cycle phosphatases has been explored mainly for the Cdc25 family [42] but might also apply to other molecules such as Cdc14. Other proteins involved in the spindle dynamics and the mitotic checkpoint [43] are also closely observed for their therapeutic use. Our experience with Cdks might suggest that these efforts should be accompanied by a parallel analysis of the roles of these molecules in vivo using genetically engineered mouse models. Hopefully, cancer patients will benefit from this knowledge if we are able to put together what we are learning from biochemical approaches, human tumors, clinical trials, and the genetic analysis of these proteins in preclinical models.

ACKNOWLEDGMENTS

I thank M. Barbacid for continuous support and the members of my laboratory and his group for helpful discussions. Work in my laboratory is supported by the Spanish Ministry of Science (BMC2003-06098 and GEN2001-4856-C13-09), Comunidad de Madrid (08.1/0045.1/2003), Fundación Ramón Areces, Fundación La Caixa and the Spanish Association Against Cancer (AECC).

REFERENCES

1. C.J. Sherr. Cancer cell cycles revisited. *Cancer Res* 60: 3689–3695 (2000).
2. C.J. Sherr and J.M. Roberts. CDK inhibitors: positive and negative regulators of G1-phase progression. *Genes Dev* 13: 1501–1512 (1999).
3. X. Ye, C. Zhu, and J.W. Harper. A premature-termination mutation in the *Mus musculus* cyclin-dependent kinase 3 gene. *Proc Natl Acad Sci USA* 98: 1682–1686 (2001).
4. S. Ren and B.J. Rollins. Cyclin C/Cdk3 Promotes Rb-dependent G0 Exit. *Cell* 117: 239–251 (2004).
5. M. Malumbres and M. Barbacid. Mammalian cyclin-dependent kinases. *Trends Biochem Sci* 30: 630–641 (2005).
6. M. Malumbres and M. Barbacid. To cycle or not to cycle: a critical decision in cancer. *Nat Rev Cancer* 1: 222–231 (2001).
7. M. Ruas and G. Peters. The p16INK4a/CDKN2A tumor suppressor and its relatives. *Biochim Biophys Acta* 1378: F115–177 (1998).
8. S. Ortega, M. Malumbres, and M. Barbacid. Cyclin D-dependent kinases, INK4 inhibitors and cancer. *Biochim Biophys Acta* 1602: 73–87 (2002).
9. T. Wolfel, M. Hauer, J. Schneider, M. Serrano, C. Wolfel, E. Klehmann-Hieb, E. De Plaen, T. Hankeln, K.H. Meyer zum Buschenfelde, and D. Beach. A p16Ink4a — insensitive Cdk4 mutant targeted by cytolytic T lymphocytes in a human melanoma. *Science* 269: 1281–1284 (1995).
10. L. Zuo, J. Weger, Q. Yang, A.M. Goldstein, M.A. Tucker, G.J. Walker, N. Hayward, and N.C. Dracopoli. Germline mutations in the p16INK4a binding domain of CDK4 in familial melanoma. *Nat Genet* 12: 97–99 (1996).
11. M. Malumbres. Revisiting the "Cdk-centric" view of the mammalian cell cycle. *Cell Cycle* 4: 206–210 (2005).
12. S.G. Rane, P. Dubus, R.V. Mettus, E.J. Galbreath, G. Boden, E.P. Reddy, and M. Barbacid. Loss of Cdk4 expression causes insulin-deficient diabetes and Cdk4 activation results in β-cell hyperplasia. *Nat Genet* 22: 44–52 (1999).
13. T. Tsutsui, B. Hesabi, D.S. Moons, P.P. Pandolfi, K.S. Hansel, A. Koff, and H. Kiyokawa. Targeted disruption of Cdk4 delays cell cycle entry with enhanced p27Kip1 activity. *Mol Cell Biol* 19: 7011–7019 (1999).
14. J. Martin, S.L. Hunt, P. Dubus, R. Sotillo, F. Nehme-Pelluard, M.A. Magnuson, A.F. Parlow, M. Malumbres, S. Ortega, and M. Barbacid. Genetic rescue of Cdk4 null mice restores pancreatic beta-cell proliferation but not homeostatic cell number. *Oncogene* 22: 5261–5269 (2003).
15. A. Abella, P. Dubus, M. Malumbres, S.G. Rane, H. Kiyokawa, A. Sicard, F. Vignon, D. Langin, M. Barbacid, and L. Fajas. Cdk4 promotes adipogenesis through PPARgamma activation. *Cell Metab* 2: 239–249 (2005).
16. M. Malumbres, R. Sotillo, D. Santamaria, J. Galan, A. Cerezo, S. Ortega, P. Dubus, and M. Barbacid. Mammalian cells cycle without the D-type cyclin-dependent kinases Cdk4 and Cdk6. *Cell* 118: 493–504 (2004).
17. K. Kozar, M.A. Ciemerych, V.I. Rebel, H. Shigematsu, A. Zagozdzon, E. Sicinska, Y. Geng, Q. Yu, S. Bhattacharya, R.T. Bronson, K. Akashi, and P. Sinicski. Mouse development and cell proliferation in the absence of D-cyclins. *Cell* 118: 477–491 (2004).

18. S. Kesavapany, B.S. Li, N. Amin, Y.L. Zheng, P. Grant, and H.C. Pant. Neuronal cyclin-dependent kinase 5: role in nervous system function and its specific inhibition by the Cdk5 inhibitory peptide. *Biochim Biophys Acta* 1697: 143–153 (2004).
19. S. Ortega, I. Prieto, J. Odajima, A. Martín, P. Dubus, R. Sotillo, J.L. Barbero, M. Malumbres, and M. Barbacid. Cyclin-dependent kinase 2 is essential for meiosis but not for mitotic cell division in mice. *Nat Genet* 35: 25–31 (2003).
20. C. Berthet, E. Aleem, V. Coppola, L. Tessarollo, and P. Kaldis. Cdk2 knockout mice are viable. *Curr Biol* 13: 1775–1785 (2003).
21. Y. Geng, Q. Yu, E. Sicinska, M. Das, J.E. Schneider, S. Bhattacharya, W.M. Rideout, R.T. Bronson, H. Gardner, and P. Sicinski. Cyclin E ablation in the mouse. *Cell* 114: 431–443 (2003).
22. T. Parisi, A.R. Beck, N. Rougier, T. McNeil, L. Lucian, Z. Werb, and B. Amati. Cyclins E1 and E2 are required for endoreplication in placental trophoblast giant cells. *EMBO J* 22: 4794–4803 (2003).
23. M. Kitagawa, H. Higashi, H.K. Jung, I. Suzuki-Takahashi, M. Ikeda, K. Tamai, J. Kato, K. Segawa, E. Yoshida, S. Nishimura, and Y. Taya. The consensus motif for phosphorylation by cyclin D1-Cdk4 is different from that for phosphorylation by cyclin A/E-Cdk2. *EMBO J* 15: 7060–7069 (1996).
24. O. Tetsu and F. McCormick. Proliferation of cancer cells despite CDK2 inhibition. *Cancer Cell* 3: 233–245 (2003).
25. I. Matsuura, N.G. Denissova, G. Wang, D. He, J. Long, and F. Liu. Cyclin-dependent kinases regulate the antiproliferative function of Smads. *Nature* 430: 226–231 (2004).
26. J.A. Ubersax, E.L. Woodbury, P.N. Quang, M. Paraz, J.D. Blethrow, K. Shah, K.M. Shokat, and D.O. Morgan. Targets of the cyclin-dependent kinase Cdk1. *Nature* 425: 859–864 (2003).
27. D.L. Fisher and P. Nurse. A single fission yeast mitotic cyclin B-p34cdc2 kinase promotes both S-phase and mitosis in the absence of G1 cyclins. *EMBO J* 15: 850–860 (1996).
28. J.D. Moore, J.A. Kirk, and T. Hunt. Unmasking the S-phase-promoting potential of cyclin B1. *Science* 300: 987–990 (2003).
29. E. Aleem, H. Kiyokawa, and P. Kaldis. Cdc2-cyclin E complexes regulate the G1/S phase transition. *Nat Cell Biol* 7: 831–836 (2005).
30. R. Sotillo, P. Dubus, J. Martin, E. de la Cueva, S. Ortega, M. Malumbres, and M. Barbacid. Wide spectrum of tumors in knock in mice carrying a Cdk4 protein insensitive to INK4 inhibitors. *EMBO J* 20: 6637–6647 (2001).
31. R. Sotillo, J.F. García, S. Ortega, J. Martín, P. Dubus, M. Barbacid, and M. Malumbres. Invasive melanoma in Cdk4 targeted mice. *Proc Natl Acad Sci USA* 98: 13312–13317 (2001).
32. P. Krimpenfort, K.C. Quon, W.J. Mooi, A. Loonstra, and A. Berns. Loss of p16Ink4a confers susceptibility to metastatic melanoma in mice. *Nature* 413: 83–86 (2001).
33. N.E. Sharpless, N. Bardeesy, K.H. Lee, D. Carrasco, D.H. Castrillon, A.J. Aguirre, E.A. Wu, J.W. Horner, and R.A. DePinho. Loss of p16^{INK4a} with retention of p19^{ARF} predisposes to tumorigenesis. *Nature* 413: 86–91 (2001).
34. E. Latres, M. Malumbres, R. Sotillo, J. Martin, S. Ortega, J. Martin-Caballero, J.M. Flores, C. Cordon-Cardo, and M. Barbacid. Limited overlapping roles of p15^{INK4b} and p18^{INK4c} cell cycle inhibitors in proliferation and tumorigenesis. *EMBO J* 19: 3496–3506 (2000).
35. D.S. Franklin, V.L. Godfrey, D.A. O'Brien, C. Deng, and Y. Xiong. Cdk inhibitors p18INK4c and p27Kip1 mediate two separate pathways to collaborative suppress pituitary tumorigenesis. *Genes Dev* 12: 2899–2911 (1998).
36. A.M. Senderowicz. Small-molecule cyclin-dependent kinase modulators. *Oncogene* 22: 6609–6620 (2003).
37. M. Knockaert, P. Greengard, and L. Meijer. Pharmacological inhibitors of cyclin-dependent kinases. *Trends Pharmacol Sci* 23: 417–425 (2002).
38. A.M. Senderowicz. Inhibitors of cyclin-dependent kinase modulators for cancer therapy. *Prog Drug Res* 63: 183–206 (2005).
39. P.D. Andrews. Aurora kinases: shining lights on the therapeutic horizon? *Oncogene* 24: 5005–5015 (2005).

40. S. Blagden and J. de Bono. Drugging cell cycle kinases in cancer therapy. *Curr Drug Targets* 6: 325–335 (2005).
41. F. Eckerdt, J. Yuan, and K. Strebhardt. Polo-like kinases and oncogenesis. *Oncogene* 24: 267–276 (2005).
42. K. Kristjansdottir and J. Rudolph. Cdc25 phosphatases and cancer. *Chem Biol* 11: 1043–1051 (2004).
43. G.J. Kops, B.A. Weaver, and D.W. Cleveland. On the road to cancer: aneuploidy and the mitotic checkpoint. *Nat Rev Cancer* 5: 773–785 (2005).

CHAPTER 14

WWOX, A CHROMOSOMAL FRAGILE SITE GENE AND ITS ROLE IN CANCER

D. RAMOS[1] AND C.M. ALDAZ[2]

[1]*Department of Pathology, Medical School, University of Valencia, Valencia, Spain;*
[2]*Department of Carcinogenesis, The University of Texas, M.D. Anderson Cancer Center, Texas, USA*

Abstract: Allelic imbalances affecting the long arm of chromosome 16 have been extensively reported in the literature as common abnormalities observed in various carcinoma types. As a result of loss of heterozygosity (LOH) studies in breast cancer, we delimited a genomic area within chromosome 16 that demonstrated the highest frequency of abnormalities. This led us to the identification and cloning of *WWOX*, a candidate tumor suppressor gene (TSG) that spans a fragile region of DNA located at 16q23.3–24.1 (FRA16D: the second most active common chromosomal fragile site in the human genome). This gene encodes a protein that contains two WW domains responsible of protein–protein interactions and a short-chain dehydrogenase (SDR) domain likely involved in sex steroid metabolism. Protein–protein interactions of WWOX with other peptides that act as apoptotic regulators as well as nuclear transcription factors have been described. We and other groups have studied the expression of WWOX in multiple tumor types in hormonally and nonhormonally regulated organs. In these studies, a significant correlation of loss of WWOX protein expression, with sex steroid hormone receptors expression and patient outcome, has been demonstrated. Reinsertion of the *WWOX* gene in WWOX-deficient tumorigenic cancer cell lines has shown a dramatic decrease of tumor growth in vivo, while inhibition of anchorage independent growth was observed in vitro. Further studies are necessary to elucidate the exact biological role of WWOX as a suppressor of tumor growth.

Key words: FRA16D, *WWOX* gene, tumor suppression, carcinogenesis

1. THE DISCOVERY OF WWOX

LOH affecting several chromosomes and especially affecting the long arm of chromosome 16 has been extensively observed in breast cancer (Sato et al., 1990; Cleton-Jansen et al., 1994; Aldaz et al., 1995). To better understand the natural history of breast cancer progression, we decided first to study comparatively the allelotypic profile of in situ and invasive ductal breast carcinomas by means of microsatellite length polymorphisms obtained from microdissected paraffin-embedded samples (Aldaz et at., 1995).

From this first investigation, allelic imbalances and losses of genomic material affecting chromosome 16q were considered to be an early event in breast carcinogenesis given that these genomic abnormalities appear in a high percentage of in situ ductal carcinomas of the mammary gland. Subsequent results from a high-resolution LOH analysis confirmed our previous observations and allowed us to accurately define a region localized on the long arm of chromosome 16 that was highly affected by genomic losses. This genomic area spanned the chromosomal region flanked by the STS markers D16S515 and D16S504, with the most affected locus being D16S518 at chromosome 16q23.3–24.1 (Cheng et al., 1996). The observed high frequency of LOH at preinvasive stages of the disease suggested to us that a candidate TSG may reside in the aforementioned chromosomal region, and, that this TSG could play an important role in breast carcinogenesis (Figure 14.1).

As in the case of breast carcinomas, similar findings affecting the long arm of chromosome 16 have also been described in prostate and hepatocellular carcinomas (Carter et al., 1990; Nishida et al., 1992).

The above-mentioned studies led to the cloning of the putative TSG using classical positional cloning strategies. Thus, we were the first group to report the cloning and identification of the gene spanning the common chromosomal

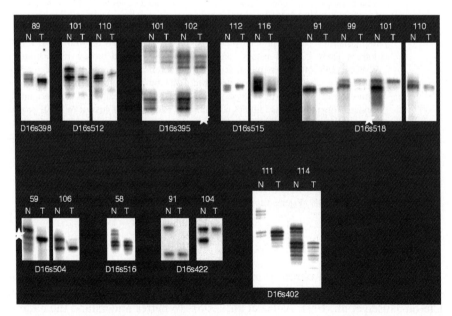

Figure 14.1. Loss of heterozygosity assessed at different SDS markers (chromosome 16q21). The most affected loci are referred with an asterisk. N: normal tissue. T: breast tumors.

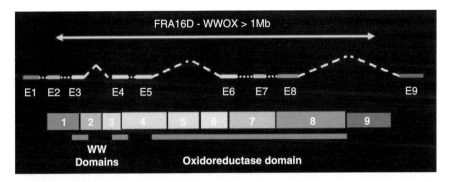

Figure 14.2. WWOX gene and protein are shown. *WWOX* gene presents a total of nine exons and encodes the WWOX protein, which shows structurally two main domains: the WW domain and the oxidoreductase domain.

fragile site FRA16D (Bednarek et al., 2000). In that study, we utilized shotgun BAC genomic sequencing as well as isolation and analysis of expressed sequenced tags (ESTs) specifically mapped to the genomic area of interest. We named the cloned gene *WWOX* because it encoded a protein, which contains two tryptophan (WW) domains coupled to a central domain with high homology to the SDR/reductase family of enzymes. The *WWOX* gene is composed of a total of nine exons, and it encodes a 414-amino acid protein (Figure 14.2). This is one of the largest known genes within the human genome, spanning over 1.1 MB; in fact, intron 8 is more than 400,000 bp in length. Various multiple myeloma translocation breakpoints have been already described mapping to intronic regions of WWOX (Chesi et al., 1998). This confirmed is candidacy as the *FRA16D* gene target.

2. WWOX PROTEIN DESCRIPTION

The specific amino acid sequence of the WWOX protein revealed a high homology to the SDR family of proteins, which encompasses a wide variety of similar enzymes. Northern blot and protein expression studies showed that WWOX had highly variable expression among breast cancer cell lines. In fact, some highly aggressive breast cancer lines were absolutely devoid of WWOX expression (e.g., MDA-MB 435 and MDA-MB 231) while other cell lines expressed high levels of WWOX protein, in particular the estrogen responsive breast cancer line MCF7. In normal tissues, high levels of WWOX expression were detected in hormonally regulated tissues such as testis, ovary, and prostate gland. Interestingly, this expression profile along with the presence of the SDR/reductase domain and specific amino acid features suggest a role for WWOX in steroid metabolism (Bednarek et al., 2000).

Other groups confirmed our original observations and, furthermore, they demonstrated deletion of *WWOX* gene in ovarian and lung carcinoma cell lines (Paige et al., 2001).

In in vitro studies, we observed inhibition of tumor cells growth in soft agar (anchorage independence) when increasing WWOX expression in MDA-MB 435 breast cancer cells (deficient in this gene). The same results were obtained with other breast cancer cell lines such as MDA-MB 431 and T47D. We also observed a dramatic inhibition of in vivo tumorigenicity by WWOX expression in MDA-MB 435 and 231 breast cancer cells, when these cell lines were injected into the mammary fat pad of nude mice (Bednarek et al., 2001). These observations strongly supported for the first time the original hypothesis that WWOX behaves as a suppressor of tumor growth. Similarly, Gabra and coworkers (Imperial College of London, UK) observed inhibition of tumorigenicity of the ovarian cancer cell line Peo 1, homozygously deleted for WWOX (personal communication). Confirming our original observations, very recently it was also showed inhibition of lung cancer cells growth both in vivo and in vitro when WWOX expression is restored (Fabbri et al., 2005).

We also detected the common occurrence of aberrant alternatively spliced WWOX transcripts with deletions of exons 5–8 or 6–8 in various carcinoma cell lines, multiple myeloma cell lines, and primary breast carcinomas as well. However, these aberrant mRNA forms of WWOX were undetectable in normal tissues. Interestingly, these aberrant variants when expressed as GFP fusion proteins, showed an abnormal nuclear localization instead of the wild-type WWOX protein that localizes to the Golgi apparatus (Bednarek et al., 2001). Even though these alternatively spliced forms are detectable at the mRNA level it is unclear whether they form viable protein since we failed to detect any smaller protein products in Western blot analyses. Nevertheless, it is clear that abnormalities occur at the transcriptional or splicing level affecting the *WWOX* gene, and, therefore, their potential relevance in carcinogenesis remains to be determined.

3. BIOLOGICAL/BIOCHEMICAL FUNCTIONS OF WWOX (WWOX INTERACTING PROTEINS)

As previously mentioned, the *WWOX* gene encodes a 46 kDa peptide containing two WW (tryptophans) domains involved in protein–protein interactions, and a short-chain oxidoreductase domain probably related to sex-steroid metabolism (Figure 14.2). High-density protein arrays were probed with a P^{32} radiolabeled portion of the WWOX protein spanning the WW domains in order to identify potential WWOX interacting proteins. As a result, one unknown peptide and, more interestingly, four known proteins resulted in a specific binding to WWOX: COTE1, WBP-1, SIMPLE/PIG7, and NF-KB AP (Ludes-Meyers et al., 2004). These finding were corroborated with GST fusion proteins pool down experiments and western blot analyses. Interestingly, the "small membrane protein of the lysosome/late endosome" (SIMPLE) (also known as p53-induced gene 7 or PIG7) protein harbors two PPSY motifs that specifically interact with the first WW motif of WWOX.

In MCF7 breast cancer cells, the first WW domain of WWOX (WWOX-1) is required for the physical interaction with endogenously expressed SIMPLE. Both proteins, WWOX and SIMPLE partially colocalize to the perinuclear compartment of the cell (Golgi zone) as determined in this cell line and using immunofluorescence staining methods: anti-WWOX, anti-Golgi 58K, anti-SIMPLE, and Mito-Tracker were used, fluorescent images captured and compared to determine colocalization or lack thereof (Ludes-Meyers et al., 2004). Notwithstanding, the specific role of protein–protein interaction between WWOX and SIMPLE/PIG7 still remains unknown. It is worth mentioning, however, that, Moriwaki et al. (2001) speculated that SIMPLE might be involved in apoptotic responses. SIMPLE has been recently reported as an activator of NF-KB-dependent gene transcription and this opens the possibility that WWOX could potentially be involved in transcriptional regulation processes. Other investigators have identified other potential WWOX interacting proteins such as P73 (a P53 homolog). Using co-immunoprecipitation and confocal microscopy analysis, it was shown that WWOX expression triggers redistribution of nuclear P73 to the cell cytoplasm, and, hence, it was suggested that WWOX could suppress p73 transcriptional activity by means of sequestration in the cytoplasmic compartment. Furthermore, it was suggested that cytoplasmic levels of P73 might determine the proapoptotic activity of WWOX. However, this study was not able to demonstrate under the same conditions a P53 binding to WWOX (Aqeilan et al., 2004a, b) even though such interaction has been previously described in another report (Chang et al., *J Biol Chem* Oct 10, 2005). Recently, Aqeilan and coworkers described yet another potential interaction for WWOX. They suggested that WWOX binds and antagonizes the biological function of YES-associated protein (YAP) through its first WW domain, and competes for the interaction with different targets, such as ErbB-4, regulating its transcriptional activity (Aqeilan et al., 2005). These same authors reported the physical and functional interaction between WWOX and AP-2 gamma transcription factor. In this study, WWOX would act as a truly TSG sequestering this transcriptional factor within the cytoplasm and inhibiting its biological functions (Aqeilan et al., 2004a, b). It must be mentioned however, that most of these studies were performed utilizing either pull-down experiments or quite artificial conditions such as transfections that overexpress the genes of interest and only under such experimental conditions the co-immunoprecipitation of WWOX with the multiple potential candidates was shown. Thus, it remains to be determined which if any of the described protein–protein interactions truly occur under physiological conditions in normal cells.

4. WWOX EXPRESSION IS DOWNREGULATED BY EXPOSURE TO ENVIRONMENTAL CARCINOGENS

As previously mentioned, *WWOX* is the gene spanning the second most active fragile site in the human genome at 16q23. Common chromosomal fragile sites are exposed by exposure to replication inhibitors such as aphidicolin. The frequency of gaps and breaks that occur after aphidicolin treatment varies in the general population and appears to be determined by a combination of genetics and exposure to environmental carcinogens (Huebner and Croce, 2001). For instance it has been shown that smokers show a higher level of expression of breaks at chromosomal fragile sites than nonsmokers.

Loss of fragile site gene expression in cancer has been mostly attributed to allelic deletion, chromosomal translocation, or gene inactivation as a result of exposure to environmental carcinogens. In addition, epigenetic means of fragile site gene silencing has also been demonstrated (Huebner and Croce, 2001). However, no information was available on the regulation of expression of fragile site genes in cells following DNA damage by environmental carcinogens, some of which do not induce chromosomal fragility. Thus, we recently study the expression patterns of WWOX and FHIT (the gene target of FRA3B) following exposure to DNA damaging agents such as, ultraviolet (UV) light, benzo[a]pyrene diol epoxide (BPDE), and ionizing radiation (IR), three ubiquitous environmental carcinogens. UV and BPDE induce bulky DNA adducts while IR induces predominantly DNA strand breaks.

Interestingly, we observed that UV and BPDE, but not IR, exposure downregulate the expression of both fragile site genes (Figure 14.3).

As compared with control, WWOX and FHIT mRNA levels dramatically decreased by 24 h in the UV-irradiated samples. At the moderate UV-C doses tested cell number decreased about twofold at 24 h postirradiation without a significant decrease in cell viability (not shown). Similar to UV, treatment of MCF7 cells with BPDE, another environmental carcinogen which induces DNA bulky adduct formation, also caused a dramatic decrease of WWOX and FHIT transcripts at 24 h post-BPDE treatment. Next we asked whether IR, which

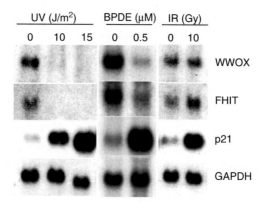

Figure 14.3. UV and BPDE treatment but not IR downregulate expression of fragile site genes, *WWOX* and *FHIT*. MCF7 cells were exposed to UV, BPDE, and IR at the indicated doses. Northern blots of total RNA (20 μg per lane) were hybridized with probes for the indicated genes.

causes DNA double strand breaks, also caused a similar effect on WWOX and FHIT expression. To test this, MCF7 cells were irradiated with IR at a dose (10 Gy) approximately equitoxic to 10 J/m^2 of UV-C radiation. As expected, the p21 mRNA level increased as a consequence of DNA damage by the three treatment protocols. Interestingly, we observed that in contrast to the other two carcinogens the IR treatment did not cause a significant reduction of WWOX or FHIT transcripts. We further determine that the observed effects of UV and BPDE were independent of p53 status since the same results were obtained using the SAOS-2 p53 null cell line (not shown). A decrease in WWOX protein levels however was only observed upon repeated UV irradiation, i.e., upon multiple cycles of UV treatment (Thavatiru et al., 2005).

We also determined that treatment with caffeine was able to abrogate most of the observed downregulation of WWOX and FHIT expression. Caffeine is known to block the S-phase checkpoint response. Therefore, the signaling pathways triggered at the S-phase checkpoint may play a role in the observed potential transcriptional downregulation.

The observed difference in the expression response at the WWOX and FHIT loci to damage induced by IR or UV is very intriguing. Both DNA damaging agents produce a response that includes the phosphorylation and stabilization of p53, the activation of MAPK pathways, dramatic changes in transcription of damage response genes, triggering one or more cell cycle checkpoints. However, these damage responses are mediated by different pathways. The ATM kinase pathway is primarily responsible for the IR-dependent response, whereas the related ATR kinase pathway is activated by UV. Similarly, there is also some evidence that BPDE-induced S-phase arrest is also mediated by the ATR kinase pathway. Alternatively, it is possible to speculate that the observed differences may be a consequence of the type of DNA damage produced by the different agents. As stated above, IR produces primarily double strand breaks and thymine glycols, the latter being a nondistortive DNA lesion. UV, on the other hand, produces cyclobutane pyrimidine dimers and 6-4 photoproducts both of which are classified as "bulky" lesions. The major BPDE-DNA adduct, formed by addition of this PAH to the exocyclic amino group of deoxyguanine, is also quite bulky, and has recently been shown to induce large kinks in adducted DNA. These bulky adducts represent strong blocks to both transcription and replication, and could directly affect the expression of genes such as FHIT and WWOX, which produce extremely long transcripts.

The decrease in WWOX and FHIT expression that we observed upon treatment with the DNA adduct inducing agents has the potential to be a mechanism of relevance for loss of expression of fragile site genes, in addition to the more classical described gross genomic abnormalities. Thus, one could imagine a scenario in which one allele of a fragile site gene could be inactivated by deletion or fragile site induction while the other allele could be subject to transcriptional downregulation by environmental carcinogens as observed in our studies, leading to complete silencing of the gene. We further hypothesize that such a continuous (transcriptional) silencing of fragile site associated putative TSGs such as *WWOX* and *FHIT* by protracted exposure to environmental carcinogens may play a significant role in the initiation and development of cancer.

5. WWOX EXPRESSION IN DIFFERENT TUMOR TYPES

By means of immunohistochemistry we have studied comparatively the level of expression of the WWOX protein in multiple neoplasias. In a first study we analyzed WWOX protein expression in a series of human breast carcinomas, correlating with diverse clinico-pathological characteristics.

In this study, normal breast samples showed all intense immunostaining against WWOX, whereas 69 out of 203 invasive breast carcinoma samples (34%) did not express WWOX, and an additional 26% showed a weak expression of WWOX protein. Interestingly, we also found a statistically significant correlation between loss of WWOX protein expression and loss of estrogen receptor alpha expression ($p = 0.003$), which clearly reinforces the hypothesis that WWOX may be involved in sex steroid metabolism (Nunez et al., 2005a, b) (Figure 14.4, left panels).

In a second study, we explored the role of WWOX expression in ovarian cancer (Figure 14.4, right panels). Similarly, loss of WWOX protein immunoexpression was observed in about a third of ovarian carcinomas in

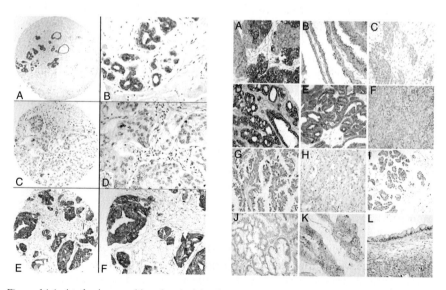

Figure 14.4. At the immunohistochemical level, WWOX protein is highly expressed in normal breast epithelium (A–B) while a variable levels of WWOX protein expression associated with ER status was demonstrated in human breast carcinomas (left panel, C–F). Similarly variable WWOX protein expression was observed in ovarian carcinomas, with expression varying significantly among different histotypes of ovarian neoplasms (right panel, serous carcinomas A–C, endometroid carcinomas D–F, clear cell carcinomas G–I, and mucinous carcinomas J–L).

general, however it was most commonly associated with specific ovarian carcinomas histotypes such as clear cell and mucinous ovarian neoplasms (Figure 14.4 G-I and J-L, respectively). A significant correlation between the loss of WWOX expression, clinical stage IV (FIGO), negative progesterone receptor status, and a shorter overall survival was also observed (Nunez et al., 2005a, b).

A potential role for the loss of WWOX has been also reported for human liver carcinomas (Park et al., 2005). Similarly, Pimenta et al. (2006) reported recently a study focused on the expression of the *WWOX* gene in a series of oral squamous cell carcinomas. In this study, the presence of abnormalities in mRNA transcription of WWOX strongly correlates with a remarkable reduction in WWOX protein expression in these tumor types, which may then contribute to the carcinogenetic process of oral cancer (Pimenta et al., 2005).

In close accordance with this last study, interesting preliminary data from the clinico-pathological evaluation of WWOX protein expression are currently being analyzed in a series of bladder cancer patients in collaboration with the Department of Pathology at the University of Valencia (Spain). In this series of bladder cancer samples, a remarkable reduction in WWOX protein expression has been found at the immunohistochemical level. The observed decrease in WWOX protein expression was quite dramatic in high-grade urothelial carcinomas cases with prominent squamous metaplasia as well as in the case of primary squamous cell carcinomas of the urinary bladder. In these neoplasms with a varying amount of a squamous cell component, the total absence of WWOX protein expression was remarkable. Nevertheless, a larger series of cases is currently under study to better define the role of loss of WWOX expression in bladder cancer development (Ramos et al., 2005).

In conclusion, *WWOX* behaves as TSG both in vivo and in vitro. Studies, of WWOX expression in a multiplicity of neoplasias from our laboratory and others support its candidacy as a bonafide TSG. Further studies and models are needed to fully understand the physiological role of WWOX in normal cells and the mechanistic implications that would favor the selection of cells that have lost its expression during cancer development.

ACKNOWLEDGMENTS

We are thankful to the many investigators from the Aldaz Lab that participated in the original work here reviewed: Drs. Andrzej Bednarek, John Ludes-Meyers, María Inés Núñez, and Elangovan Thavathiru.

These studies were supported by NCI grant RO1 CA 102444.

REFERENCES

Aldaz CM, Chen T, Sahin A, Cunningham J, Bondy M. Comparative allelotype of in situ and invasive human breast cancer: high frequency of microsatellite instability in lobular breast carcinomas. *Cancer Res* 55: 3976–3981, 1995.

Aqeilan RI, Donati V, Palamarchuk A, et al. WW domain-containing proteins, WWOX and YAP, compete for interaction with ErbB-4 and modulate its transcriptional function. *Cancer Res* 65: 6764–6772, 2005.

Aqeilan RI, Pekarsky Y, Herrero JJ, et al. Functional association between WWOX tumor suppressor protein and p73, a p53 homolog. *PNAS* 13: 4401–4406, 2004a.

Aqeilan RI, Palamarchuk A, Weigel RJ, et al. Physical and functional interactions between the WWOX tumor suppressor protein and the AP-2gamma transcription factor. *Cancer Res* 64: 8256–8261, 2004b.

Bednarek AK, Laflin KJ, Daniel RL, Liao Q, Hawkins KA, Aldaz CM. WWOX, a novel WW domain-containing protein mapping to human chromosome 16q23.3–24.1, a region frequently affected in breast cancer. *Cancer Res* 60: 2140–2145, 2000.

Bednarek AK, Keck-Waggoner CL, Daniel RL, Laflin KJ, Bergsagel PL, Kiguchi K, Brenner AJ, Aldaz Cm. WWOX, the *FRA16D* gene, behaves as a suppressor of tumor growth. *Cancer Res* 61: 8068–8073, 2001.

Carter BS, Ewing CM, Ward WS, et al. Allelic loss of chromosomes 16q and 10q in human prostate cancer. *Proc Natl Acad Sci USA* 87: 8751–8755, 1990.

Chang NS, Doherty J, Ensign A, Schultz L, Hsu LJ, Hong Q. WOX1 is essential for tumor necrosis factor-, UV light-, staurosporine-, and P53-mediated cell death, and its tyrosine 33-phosphorylated form binds and stabilizes serine 46-phosphorylated P53. *J Biol Chem* 280: 43100–8, 2005.

Chen T, Sahin A, Aldaz CM. Deletion map of chromosome 16q in ductal carcinoma in situ of the breast: refining a putative tumor suppressor gene region. *Cancer Res* 56: 5605–9, 1996.

Chesi M, Bergsagel PL, Shonukan OO, Martelli ML, Brents LA, Chen T, Schrock E, Ried T, Kuehl WM. Frequent dysregulation of the c-maf proto-oncogene at 16q23 by translocation to an Ig locus multiple myeloma. *Blood* 91: 4457–63, 1998.

Cleton-Jansen A, Moerland E, Kuipers-Dijkshoorn N, et al. At least two different regions are involved in allelic imbalance on chromosome arm 16q in breast cancer. *Cancer* 9: 101–107, 1994.

Fabbri M, Iliopoulos D, Trapasso F, et al. *WWOX* gene restoration prevents lung cancer growth in vitro and in vivo. *PNAS* 102: 15611–15616, 2005.

Huebner K, Croce CM. FRA3B and other common fragile sites. The weakest links. *Nat Rev Cancer* 1: 214–221, 2001.

Ludes-Meyers JH, Bednarek AK, Popescu NC, Bedford M, Aldaz CM. WWOX, the common chromosomal fragile site, *FRA16D*, cancer gene. *Cytogenet Genome Res* 100: 101–110, 2003.

Ludes-Meyers JH, Kil H, Bednarek AK, Drake J, Bedford MT, Aldaz CM. WWOX binds the specific praline-rich ligand PPXY: identification of candidate interacting proteins. *Oncogene* 23: 5049–5055, 2004.

Moriwaki Y, Begum NA, Kobayashi M, Matsumoto M, Toyoshima K, Seya T. Mycobacterium bovis Bacillus Calmette-Guerin and its cell wall complex induce a novel lysosomal membrane protein, SIMPLE, that bridges the missing link between lipopolysaccharide and p53-inducible gene, LITAF (*PIG7*), and estrogen-inducible gene, *EET-1*. *J Biol Chem* 276: 23065–23076, 2001.

Nishida N, Fukuda Y, Kokuryu H, et al. Accumulation of allelic loss on arms of chromosomes 13q, 16q and 17p in the advances stages of human hepatocellular carcinoma. *Int J Cancer* 51: 862–868, 1992.

Nunez MI, Ludes-Meyers J, Abba MC, Kil H, Abbey NW, Page RE, Sahin A, Klein-Szanto AJP, Aldaz CM. Frequent loss of WWOX expression in breast cancer: correlation with estrogen receptor status. *Breast Cancer Res Treat* 89: 99–105, 2005a.

Nunez MI, Rosen DG, Ludes-Meyers J, Abba MC, Kil H, Page RE, Klein-Szanto AJP, Godwin AK, Liu J, Mills GB, Aldaz CM. WWOX protein expression varies among ovarian carcinoma histotypes and correlates with less favorable outcome. *BMC Cancer* 5: 64, 2005b.

Paige AJ, Taylor KJ, Taylor C, et al. WWOX: a candidate tumor suppressor gene involved in multiple tumor types. *Proc Natl Acad Sci USA* 25; 98(20):11417–11422, 2001.

Park SW, Ludes-Meyers J, Zimonjic DB, Durkin ME, Popescu NC, Aldaz CM. Frequent downregulation and loss of *WWOX* gene expression in human hepatocellular carcinoma. *Br J Cancer* 91: 753–759, 2004.

Pimenta FJ, Gomes DA, Perdigao PF, Barbosa AA, Romano-Silva MA, Gomez MV, Aldaz CM, De Marco L, Gomez RS. Characterization of the tumor suppressor gene *WWOX* in primary human oral squamous cell carcinomas. *Int J Cancer* 118: 1154–58, 2006.

Ramos D, López-Guerrero JA, Rubio J, Martínez-Rodríguez M, Solsona-Narbón E, Almenar S, Aldaz CM, Llombart-Bosch A. WWOX protein expression in a series of bladder carcinomas. (Communication). II International Symposium on Cancer: New Trends in for the 21st century. November 12–15. Valencia. Spain, 2005.

Sato T, Tanigami A, Yamakawa K, et al. Allelotype of breast cancer: cumulative allele losses promote tumor progression in primary breast cancer. *Cancer Res* 50: 7184–7189, 1990.

Thavatiru E, Ludes-Meyers J, McLeod MC, Aldaz CM. Expression of common chromosomal fragile site genes, WWOX/FRA16D and FHIT/FRA3B is downregulated by exposure to environmental carcinogens, UV, and BPDE but not by IR. *Mol Carcinog* 44(3): 174–182, 2005.

CHAPTER 15

FROM GENOME TO PROTEOME IN TUMOR PROFILING: MOLECULAR EVENTS IN COLORECTAL CANCER GENESIS

JENS K. HABERMANN[1,2], UWE J. ROBLICK[1,3], MADHVI UPENDER[2], THOMAS RIED[2], AND GERT AUER[3]

[1]*Laboratory for Surgical Research, Department of Surgery, University Hospital Schleswig-Holstein, Campus Lübeck, Lübeck, Germany;* [2]*Genetics Department, National Cancer Institute, NIH, Bethesda, MD, USA;* [3]*Unit of Cancer Proteomics, Cancer Center Karolinska, Karolinska University Hospital Solna, Stockholm, Sweden*

Abstract: Biomedical research has advanced rapidly in recent years with the sequencing of the human genome and the availability of technologies such as global gene and protein expression profiling using different chip platforms. However, this progress has not yet been transferred to the bedside. While detection of cancer at early stages is critical for curative treatment interventions, efficient diagnostic and therapeutic markers for the majority of malignancies still seem to be lacking. Comprehensive tumor profiling has therefore become a field of intensive research aiming at identifying biomarkers relevant for improved diagnostics and therapeutics. This chapter will demonstrate a genomic and proteomic approach while focusing on tumor profiling during colorectal cancer development.

1. COLORECTAL CANCER

Colorectal cancer is one of the most common malignancies in the world.[1] While the 5-year disease-free survival rate for early stage tumors (UICC stage I) exceeds 90%, this percentage is reduced to 63% in advanced stage carcinomas (UICC stage III).[2] Therefore, detection of cancer at an early stage is critical for curative treatment interventions and utilization or application of tools and methodologies for early cancer detection can directly result in improving patient survival rates. In current clinical practice, screening for cancer and preinvasive polyps of the colorectum is based on clinical examination, detection of fecal occult blood (FOBT), and sigmoidoscopy or colonoscopy.[3] The successful implementation of these screening procedures has contributed to a reduction in the mortality of colorectal carcinomas.[4] However, despite these screening programs, about 70% of carcinomas are detected at advanced tumor stages (UICC III/IV) presenting poor patient prognosis.

The lifetime risk for the development of colorectal carcinomas is considerably increased in patients with ulcerative colitis (UC).[5] Ulcerative colitis can therefore be considered a bona fide premalignant condition leading to the recommendation that patients with UC should participate in surveillance programs to screen for early signs of malignancy.[6] However, reliable surveillance is difficult and 50% of the detected malignancies are already at an advanced tumor stage.[7-9]

2. GENOMICS

2.1. Genomic aneuploidy and its role in tumorigenesis

When the first quantitative measurements of the DNA content of cancer cells were performed, aneuploidy was defined as a variation in nuclear DNA content in the population of cancer cells within a tumor.[10] Since then, aneuploidy has been observed as a consistent genetic alteration of the cancer genome of different tumor entities.[11-13] In addition, aneuploidy seems to precede the manifestation of malignancy: Löfberg et al. reported aneuploid biopsies in 25% of ulcerative colitis patients with a high risk for colorectal cancer development at least once during 10 years of observation.[14] In other studies, aneuploidy has been repeatedly observed by flow cytometry even in nondysplastic mucosa of ulcerative colitis patients.[15] By means of image cytometry, we could detect highly aneuploid epithelial cell populations scattered over the colon and rectum in premalignant biopsies of eight patients with an ulcerative colitis-associated colorectal carcinoma (UCC).[16] These aneuploid lesions could be observed up to 11 years prior to the final cancer diagnosis (average 7.8 years). They were found in macro- and microscopically unsuspicious mucosa, could even be detected in regenerative epithelium, and were not related to dysplasia. This DNA aneuploidy occurred more frequently in biopsies of patients with a subsequent carcinoma (75%) than compared with biopsies from ulcerative colitis patients without a subsequent colorectal carcinoma (14%, $p = 0.006$). The carcinoma samples of the eight UCC patients also exhibited highly aneuploid DNA distribution patterns. The common conclusion of these studies strongly supports the hypothesis that genomic instability, represented by nuclear DNA aneuploidy, could initiate the process of malignant transformation in colitis as an early event. DNA aneuploidy would therefore indicate an increased risk of progression to invasive properties in genetically unstable cells. However, aneuploidy may be reversible over time once cells are not longer exposed to the inducing agent or carcinogen.[17,18] Thus, it is reasonable to suggest that the genomic instability reflected by aneuploidy has to be followed by multiple cellular alterations in order to reach malignant properties. One of the decisive steps in this transformational process is the ability of genomically altered cells to proliferate, which is compulsory for clonal expansion.[19] Interestingly, immunohistochemical expression of the proliferation marker Cyclin A was significantly correlated to aneuploidy in biopsies of the patients with a subsequent carcinoma ($r = 0.791$).

2.2. Chromosomal aneuploidies as tumor-specific patterns

With increased resolution of cytogenetic techniques, such as chromosome banding, comparative genomic hybridization (CGH), spectral karyotyping (SKY), and multicolor fluorescence in situ hybridization (FISH), it has become clear that in addition to nuclear aneuploidy, specific nonrandom chromosomal imbalances (heretofore referred to as chromosomal aneuploidy) exist.[20-23] Indeed, despite genetic instability in cancer genomes, cancer cell populations as a whole display a surprisingly conserved, tumor-specific pattern of genomic imbalances.[13,24,25] At early steps in the sequence of malignant transformation during human tumorigenesis, e.g., in preinvasive dysplastic lesions, chromosomal aneuploidies can be the first detectable genetic aberration found.[26-29] This suggests that there is both an initial requirement for the acquisition of specific chromosomal aneuploidies and a requirement for the maintenance of these imbalances despite genomic and chromosomal instability. This would be consistent with continuous selective pressure to retain a specific pattern of chromosomal copy number changes in the majority of tumor cells.[13,30-32] Additionally, in cell culture model systems in which cells are exposed to different carcinogens, chromosomal aneuploidy is the earliest detectable genomic aberration.[33,34] The conservation of these tumor-specific patterns of chromosomal aneuploidies suggests that they play a fundamental biological role in tumorigenesis.

The progression of colorectal cancer is defined by the sequential acquisition of genetic alterations.[35] At the cytogenetic level, many of these aberrations can be visualized as specific chromosomal gains and losses. These aneuploidies result in a recurrent pattern of genomic imbalances, which is specific and conserved for these tumors.[36] For instance, one of the earliest acquired genetic abnormalities during colorectal tumorigenesis are copy number gains of chromosome 7.[32] These trisomies can already be observed in benign polyps, and can emerge in otherwise stable, diploid genomes. At later stages, e.g., in high-grade adenomas or in invasive carcinomas, additional specific cytogenetic abnormalities become common, such as gains of chromosome and chromosome arms 8q, 13, and 20q, and losses that map to 8p, 17p, and 18q. For a comprehensive summary see the "Mitelman Database of Chromosome Aberrations in Cancer" at http://cgap.nci.nih.gov/Chromosomes/Mitelman. These chromosomal aneuploidies are accompanied by specific mutations in oncogenes and tumor suppressor genes, including ras, APC, and p53.[37] It is therefore well established that both, chromosomal aneuploidies and specific gene mutations, are required for tumorigenesis.

Detection of aneuploid lesions in ulcerative colitis patients seems to indicate imminent carcinogenesis with faithful progression to UCC. Recent reports have provided evidence that genomic aneuploidy in UCC is associated with chromosomal aneuploidies.[38, 39, 40, 41, 42] Unlike sporadic colorectal tumors, UCCs do not follow the adenoma–carcinoma sequence, and the sequential acquisition of chromosomal aneuploidy and gene mutations is less well established. It was therefore interesting to investigate whether the pattern of chromosomal gains and losses in UCC are similar to that described in sporadic carcinomas. This would indicate that the final distribution of genomic imbalances is the product of continuous selection, and that this distribution is independent of whether a carcinoma occurs spontaneously or as a result of, for example, chronic inflammation. We therefore determined the degree of genomic instability by DNA image cytometry and CGH for 23 UCC specimens.[43] All 23 UCC specimens revealed highly aneuploid DNA distribution patterns of the nuclear DNA content, independent of the tumor stage. CGH analysis could identify chromosomal imbalances as follows: the most common DNA gains were mapped to chromosomes or chromosome arms 20q (84% of all cases), 7 (74%), 8q (74%), 13q (74%), 11p and 12 (both 42%), 5p and 18p (both 37%), and 17q (31%). Recurrent losses occurred on 8p (58%), 18q (47%), and 5q (26%).[43] These results show that chromosomal imbalances observed in UCC mainly cluster on the same chromosomes as described for sporadic colorectal cancer. For instance, Ried et al. reported DNA gains that frequently mapped to chromosomes or chromosome arms 7, 8q, 13q, and 20 in sporadic colorectal carcinomas.[36] However, it also becomes clear that sporadic colorectal carcinomas have fewer genomic imbalances than UCCs (Figure 15.1). Additional significant differences exist that characterize UCCs in contrast to sporadic carcinomas: our previous analyses of sporadic colorectal carcinomas revealed an average number of DNA copy alterations (ANCA), calculated as the number of chromosomal copy number changes divided by the number of cases, of 5.6, which was elevated to 13.3 in UCC. This number exceeds that observed in primary liver metastases from colorectal carcinomas, for which the ANCA had been determined to be 11.7.[44] This high degree of genomic instability is also supported by measurements of the nuclear DNA content, which invariably revealed nuclear aneuploidy. We also observed a large number of localized high-level copy number increases (amplifications). Amplifications have been described as a reflection of advanced disease and poor prognosis in other malignancies.[45] Some of the amplifications occurred in regions known to be affected in colorectal carcinomas, such as chromosome arms 6p, 8q, 13q, 17q, and 20q, and for which the target genes are either known or likely candidates have been identified (http://www.helsinki.fi/cmg/cgh_data.html). For instance, the frequent gain of chromosome 8 and amplifications that map to band 8q24 target the *MYC* oncogene.

The CGH profile for UCC as presented here, dominated by overall gains and numerous amplifications, is in concordance with the relatively high ANCA value and severe aneuploidy observed in the majority of all 23 UCCs. In comparison, sporadic colon carcinomas show aneuploidy in only 70–80% of the cases, combined with an overall lower ANCA value. The surprisingly high level

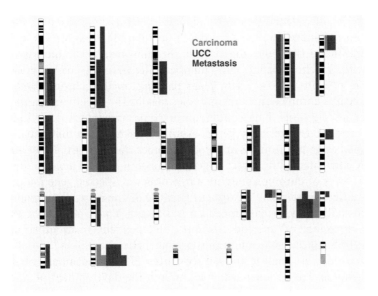

Figure 15.1. Comparison of genomic imbalances in sporadic colorectal carcinoma (SCC) ($n = 16$), UCC ($n = 19$), and liver metastasis of SCC ($n = 16$). The number of alterations per chromosome has been normalized to ten cases within each tumor type.

of ANCA values in UCC could be a reflection of a generally increased genetic instability in UCC, due to the long latency of inflammatory disease before overt tumors develop; however, the data presented here and in the literature clearly indicate that the tumor cell population as an entity selects for a distribution of genomic imbalances that is similar to sporadic carcinomas. Therefore, the tissue origin of the tumor cell, and not the mode of tumor induction, seems to define the similarity between sporadic colorectal cancers and UCC. This is in striking contrast to hereditary colorectal carcinomas arising in the background of mismatch repair deficiency, where neither aneuploidy nor specific chromosomal imbalances are observed.[46, 47]

In this respect, the positive correlation of high ANCA values and the occurrence of amplifications with elevated cyclin A expression ($p = 0.04$) indicate a high proliferative activity combined with genomic instability, necessary features for tumor growth and clonal expansion. This finding thus increases the value of cyclin A as an independent prognostic marker for carcinoma development in ulcerative colitis.

3. TRANSCRIPTOMICS

A rather well-defined correlation of tumor phenotype and genotype has been established, mainly through the application of molecular and molecular cytogenetic techniques to study sequential changes during tumorigenesis. However, it remains less clear how genomic aneuploidy and chromosomal imbalances impact on the transcriptome. One could postulate that expression levels of all transcriptional active genes on trisomic chromosomes would increase in accordance with the chromosome copy number. Alternatively, changing the expression level of only one or a few genes residing on that chromosome through tumor-specific chromosomal aneuploidies may be the selective advantage necessary for tumorigenesis. This would require the permanent transcriptional silencing of most of the resident genes. Another formal possibility that must be entertained is that chromosomal copy number changes are either neutral or inversely correlated with respect to gene expression levels. This would mean that gains or losses of chromosomes are a byproduct of specific gene mutations and may not offer any selective advantage. Because of the many chromosomal aberrations usually found in cancer cells, it is difficult, if not impossible, to identify the consequences of specific trisomies, independent from other coexisting genomic imbalances, gene mutations, or epigenetic alterations.[48]

Methodology to analyze the consequences of chromosomal imbalances in tumor genomes has become available through the development of microarray-based gene expression profiling. This method has been first described by Schena et al. and enables one the simultaneous analysis of thousands of genes for their gene expression.[49] Despite the exponential increase in the number of publications describing microarray experiments, only a few reports have attempted to specifically address the question regarding the immediate consequences of chromosomal aneuploidies vis-a-vis the dysregulation of the cellular transcriptome. These reports came to quite different conclusions and none of them attempt to address this question in the clinical setting of colorectal carcinogenesis.[44, 50-53] In addition, a comprehensive exploration of how global alterations of the cellular transcriptome might correlate with sequential steps of cellular transformation from normal mucosa by adenoma and carcinoma up to distant metastases has not been described. Such analyses, however, could reveal potential candidate genes for improved prognostics, diagnostics and therapeutics.

3.1. Immediate consequences of genomic imbalances on the transcriptome

Chromosomal aneuploidies are not only observed in sporadic and UCCs but in essentially all sporadic carcinomas. These aneuploidies result in tumor-specific patterns of genomic imbalances that are acquired early during tumorigenesis.[13, 24, 54] For instance, one of the earliest genetic abnormalities observed in the development of sporadic colorectal tumors is trisomy of chromosome 7.[32] Usually, once acquired, these specific imbalances are maintained despite ongoing chromosomal instability.[55] It is therefore reasonable to assume that continuous selective pressure for the maintenance of established genomic imbalances exists in cancer genomes. It is not known how genomic imbalances affect chromosome-specific gene expression patterns in particular and how chromosomal aneuploidy dysregulates the genetic equilibrium of cells in general. To model specific chromosomal aneuploidies in cancer cells and dissect the immediate consequences of genomic imbalances on the transcriptome, we set up an experimental model system in which the only genetic alteration between parental and derived cell lines is an extra copy of a single chromosome: we generated derivatives of the diploid yet mismatch repair-deficient colorectal cancer cell line, DLD1, and immortalized cytogenetically normal human mammary epithelial cell line (hTERT-HME) using microcell-mediated chromosome transfer to introduce extra copies of neomycin-tagged chromosomes 3, 7, and 13.[56] FISH with chromosome-specific probes confirmed the maintenance of extra copies of these chromosomes under neomycin-selective cell culture conditions. In addition, SKY was performed to determine whether the chromosome transfer process induced secondary karyotypic changes. With the exception of loss of the Y chromosome in all DLD1 + 3 cells analyzed and in a small percentage of DLD1 + 7 cells, all four derivative lines maintained the diploid karyotype of the parental cell line and contained the additional copy of the introduced chromosome. These results were also confirmed by CGH. The global consequences on gene expression levels were analyzed using cDNA arrays. Our results show that regardless of chromosome or cell type, chromosomal trisomies resulted in a significant increase in the average transcriptional activity of the trisomic chromosome ($p < 0.001$) (Figure 15.2).

Several important conclusions regarding the impact of chromosomal aneuploidy on cellular transcription levels can be drawn from our analysis. First, alterations in the copy number of whole chromosomes resulted in an increase in average gene expression for that chromosome. The average increase in gene expression (1.21) however, was lower than the average increase of genomic copy number (1.44). These results were consistent with results from similar analyses of aneuploid colorectal, pancreatic, and renal cancer derived cell lines in which we observed a trend, indicating that indeed chromosomal aneuploidies correlate with global transcription levels.[25, 44, 51] Second, chromosomes not observed to be aneuploid in particular tumor types (i.e., chromosome 3 in colorectal tumors) also had increased transcriptional activity when placed into that cellular environment. Thus, their presence is not neutral with respect to the transcriptome. Third, aneuploidy not only affects gene expression levels on the chromosomes present in increased copy numbers, but a substantial number of genes residing on other chromosomes significantly increased or decreased apparently in a stochastic manner. The influence of chromosomal aneuploidy on the expression level of individual genes was examined by considering only those genes whose expression ratios were >2.0 (upregulated) or <0.5 (downregulated) when compared with the parental cell line. Strikingly, none of the genes were affected in common among any of the four cell lines. This observation is of course consistent with known mechanisms of gene regulation (e.g., activator and suppressor proteins, signaling pathways) and the fact that genes residing in a given pathway are for the most part distributed throughout the genome on different chromosomes. Three groups have analyzed the consequences of constitutional chromosomal trisomies on

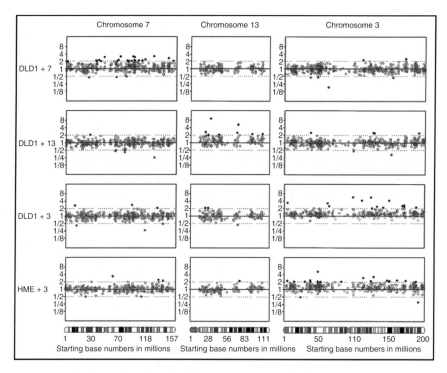

Figure 15.2. Global gene expression profiles. Each scatter plot displays all genes and their corresponding normalized ratio values along the length of each chromosome. Values in open circles represent ratio values between 0.5 and 2.0. Dark dots represent expression ratios ≥2.0 or ≤0.5.

transcriptional activity in noncancerous fetal cells and in a mouse model of human trisomy 21, and attained similar conclusions as us.[57, 58, 59] These studies concluded that the average gene expression of trisomic chromosomes is clearly increased, although this was not due to high-level upregulation of only a few specific genes. However, expression levels of multiple genes throughout the genome were dysregulated. Normal cells with constitutional chromosomal aneuploidies (or segmental duplications) cannot tolerate trisomy of >4.3% of the genome.[60, 61] However, this limit is not merely a reflection of the DNA content because multiple copies of an inactivated X-chromosome can be tolerated. Therefore, this limit is more likely to be imposed by global disturbance of the transcriptome as a consequence of genomic imbalances. This hypothesis is supported by the recent identification of differential average expression levels of specific chromosomes. For instance, the average gene expression levels and gene density of chromosomes 13, 18, and 21 are lower than those of smaller chromosomes, yet trisomy of only these chromosomes is compatible with life in noncancerous cells.[61] It is interesting to speculate that one of the specific features of emerging cancer cells, which would differ from nontransformed cells that carry constitutional trisomies, would be the ability to exceed this transcriptional threshold during the multiple steps required for tumorigenesis. This global dysregulation of the transcriptome of cancers of epithelial origin may also reflect on our ability for therapeutic intervention: although the consequences of a simple chromosomal translocation, such as the *BCR/ABL* fusion in chronic myelogenic leukemia, can be successfully targeted with an inhibitor of the resulting tyrosine kinase activity such as Gleevec, the normalization of the complex dysregulation of transcriptional activity in carcinomas requires a more general, less specific, and hence more complex interference.[62]

4. PROTEOMICS

The term proteome was first defined in 1994 and denotes the entirety of proteins expressed by the genome. Proteomics is thus understood as the consecutive step following genomics. Proteomics techniques have rapidly evolved and are now widely applied to monitor disease-specific alterations.[63]

4.1. Two-dimensional gel electrophoresis

In proteome research, two-dimensional gel electrophoresis (2-DE) is still the cornerstone separation technique for complex protein mixtures[64]. The 2-DE approach allows large-scale screening of the protein components of normal and disease cells[64]. We used quantitative 2-D SDS PAGE (pH 4–7) to analyze protein-based expression profiles for sporadic and hereditary, i.e., familial adenomatous polyposis (FAP), colorectal cancer samples. The 2-D gels were stained using silver, coomassie, or sypro ruby, images were scanned and digitally compared using PDQuest analysis software version 7.3. This is a powerful software offering automated spot detecting and matching function and integrates a statistical software package. Using this technique it is possible to highlight proteins that are differentially abundant in one state versus another (e.g., tumor vs normal).[65]

4.2. Matrix-assisted laser desorption ionization mass spectrometry

Within the matrix-assisted laser desorption ionization (MALDI) technique, matrix and sample are cocrystallized on the MALDI plate and irradiated with a laser pulse.[66] The matrix absorbs the energy and acts as an intermediary for the codesorption and ionization of sample and matrix. The ions are accelerated in an electrical field and enter a field free drift tube. The mass-related time of flight is detected and the analog signal converted and digitalized. The experimentally generated masses are compared with a set of mass profiles in a protein database, e.g., SwissProt, ExPASy, or UniProt. The most similar pattern determines the protein "hit." The tighter the mass tolerance, the more stringent is the identification.

The subsequent application of the 2-DE and mass spectrometry technique could show that tumor-specific quantitative or qualitative changes of protein patterns are indeed discernable. We performed a detailed analysis to identify sequential alterations of the proteome that define the transformation of normal epithelium and the progression from adenomas to invasive disease. We have analyzed tissue samples from 15 patients, including the mucosa–adenoma–carcinoma sequence from individual patients.[67] We determined the degree of genomic instability during carcinogenesis by measuring DNA contents and assessed protein expression levels by means of 2-DE and subsequent mass spectrometry. The 2-DE revealed a total of 112 polypeptide spots that showed an at least twofold differential expression between the four stages of carcinogenesis.[67] A total of 72 of these polypeptides could be characterized by mass spectrometry and 46 of those were exclusively overexpressed in tumors and metastases. Unsupervised principal component analysis allowed separation of adenomas, carcinomas, and metastases based on protein expression profiles. Interestingly, two dysplastic polyp samples did not conformingly cluster in their cohort and were closer located to the malignant samples (Figure 15.3). Both polyps revealed aneuploid DNA distribution patterns, indicating an increased malignancy potential.

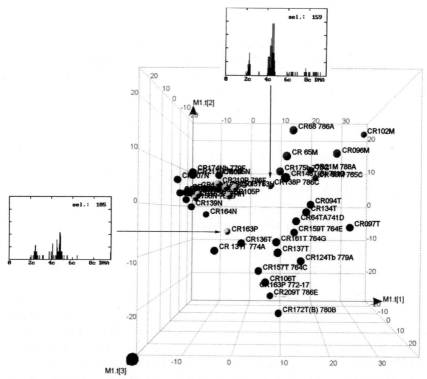

Figure 15.3. PCA plot of the protein expression data of all match set members, with the normal cohort (green), polyps (yellow), tumors (blue), and metastases (red). The *arrows* highlight two polyps that were outliers, showed aneuploidy in the DNA measurement (shown in the histograms) and clustered closer to the tumors in the three-dimensional space.

FAP also termed adenomatous polyposis coli (APC) is an autosomal dominant inherited disease with a germ line mutation of the APC gene on 5q21. In spite of this specific genetic alteration early diagnosis in young patients without polyposis onset and lack of a family history can be difficult and finally lethal. Thus, there is a need for a better understanding of the disease process on the molecular level in order to be able to introduce more sensitive diagnostic procedures. Proteomics is a multifaceted approach to study various aspects of protein expression. While DNA constitutes the "information archive of the genome," proteins actually serve as the functional effectors of cellular processes. We analyzed protein expression to elucidate pathways and networks involved in the pathogenesis of FAP coli, its associated carcinomas, and in comparison with the sporadic form of the disease.

Protein measurements were performed on 47 samples gained from 15 different patients. Proteins were separated by 2-DE revealing 1950 separated proteins. Qualitative and quantitative differences in expression levels between normal epithelium, adenoma, and carcinoma both in FAP and sporadic colon cancer were compared and statistically evaluated. In addition, collecting "triplets" from the same patient (normal, adenoma, and carcinoma) made also an intra- and inter-patient comparison possible. We found 17 proteins that showed quantitative changes between normal mucosa in FAP and sporadic normal mucosa with a false discovery rate (FDR) less than 10%. Furthermore, qualitative analysis discovered 47 proteins present in all FAP mucosa specimens and absent in the sporadic normal mucosa. Comparing FAP polyps with sporadic colonic polyps we found 49 polypeptides that are present in the FAP samples and absent in all sporadic polyps. One protein was found to be present in the sporadic polyps only. We also found 66 proteins whose absence/presence pattern coincides with the FAP/sporadic cancer grouping. The data obtained on the protein expression level make it possible to diagnose the FAP disease already in the "normally" appearing colorectal mucosa.

4.3. Surface-enhanced Laser Desorption Ionization

One particularly intriguing possibility develops if tumor-specific changes could be detected with noninvasive, cost-efficient formats, for instance, by detection of disease-specific markers in the peripheral blood. However, the use of single serum markers, e.g., carcinoembryonic antigen, has so far failed to deliver markers of high sensitivity and specificity for colon cancer and most other tumors.[68, 69] Comprehensive serum proteome profiling for such tumor-specific markers has therefore become a field of intensive research.[70] A particular promising technique for serum proteome screening is based on surface-enhanced laser desorption ionization time-of-flight (SELDI-TOF) mass spectrometry. A major advantage of surface-enhanced laser desorption ionization (SELDI) is that complex protein mixtures can be directly analyzed by mass spectrometry without any prior separation and purification. SELDI-TOF utilizes chromatographic surfaces that retain proteins from a complex sample mixture according to their specific properties (e.g., hydrophobicity and charge), with the molecular weights of the retained proteins then being measured by TOF mass spectrometry.[70] Microliter quantities of serum are directly applied to chips and the bound proteins are treated and analyzed by mass spectrometry. The mass spectra patterns obtained for different samples reflect the protein and peptide contents of these samples. Protein identification itself needs to be performed in an additional analysis step.[71-73] The reliability and reproducibility have been proven even if variation coefficients of 8–10% indicate the need for technical repeats.[72, 73] SELDI-TOF mass spectrometry is particularly well suited to evaluate low-molecular proteins (0.525 kDa) and is, as such, complementary to the 2-DE approach.

The identification of SELDI-based protein profiles and the subsequent protein identification of features that allow the distinction between malignancy-related and normal sera would be highly beneficial.

5. SUMMARY

Genomic aneuploidy occurs early and is commonly found in precancerous biopsies of ulcerative colitis patients who subsequently develop a UCC. The assessment of DNA ploidy could therefore become a basic element in future surveillance programs in ulcerative colitis. The complementary detection of increased cyclin A expression in aneuploid lesions — indicating clonal expansion — seems to be the most powerful combination to predict imminent malignant transformation for an individual patient. Moreover, genomic aneuploidy in UCC tumors correlates with specific chromosomal gains and losses, which, in turn, are associated with increased cyclin A levels. The overall pattern of specific chromosomal aberrations in UCC tumors is similar to that seen in sporadic colorectal carcinomas. The predominance of specific chromosomal aneuploidies in colorectal cancers also affects the transcriptome of cancer cells. In a well-defined model system we could observe that the introduction of an extra copy of a given chromosome increases very specifically the overall average expression of genes on the trisomic chromosomes. Additionally, a large number of genes on diploid chromosomes were also significantly increased, revealing a more complex global transcriptional dysregulation. In addition, increasing genomic instability and a recurrent pattern of chromosomal aberrations are accompanied by distinct protein expression patterns that correlate with subsequent stages of colorectal cancer progression. The identified proteins underwent extensive posttranslational modifications, thus multiplying the transcriptional dysregulation. Analyzing protein-based expression profiles for sporadic and hereditary colorectal cancer samples allowed the detection of a distinct protein expression pattern that seems to be characteristic for FAP-diseased tissue.

6. FUTURE PERSPECTIVES

The evolving technique of array CGH allows the identification of DNA copy number changes of, ideally, individual genes and thus enables an increasing resolution compared with conventional CGH, which is performed on more or less condensed chromosomes. One particular application, however, could be the analysis of amplicons identified in sporadic and UCCs. Custom designed arrays that contain genes located on distinct amplified chromosome segments would enable to select for individual genes rather than chromosomal segments that are highly amplified. Such genes could be used as marker genes for colorectal malignancy. Their diagnostic and prognostic potential could be tested with gene-specific probes by means of multicolor FISH in premalignant lesions (e.g., ulcerative colitis biopsies and adenomas). Thus, the combined analysis of DNA ploidy measurements and colorectal cancer-specific Multi-FISH probes in premalignant colorectal lesions could profoundly improve individual risk assessment for imminent colorectal cancer development. However, the design of customized arrays that contain the amplified genes might provide a more rapid and high throughput screening approach compared with Multi-FISH. The application of array CGH would also allow a direct correlation how single gene copy number changes influence the transcriptional equilibrium. The employment of comprehensive gene and protein expression profiling in subsequent stages of colorectal cancer progression allowed the identification of genes and proteins that now warrant further validation by RNA interference (RNAi) experiments, in order to prove their potential for gene and protein expression tailored individualized diagnostic, prognostic, and therapeutic approaches. The combined evaluation of ploidy status, amplification of disease and stage-specific gene probes, and gene and protein expression patterns in clinical tissue samples should be utilized in prospective studies to corroborate their value for improved diagnostics, prognostication, and identification of therapeutic targets. Proteomics-based antibody panels, ELISA tests and strategies combining surface-mediated protein enrichment with direct mass spectrometric quantification and identification of putative markers are thus promising ways for future cancer diagnosis. The field of proteomics promises accurate staging and, it is hoped, individualized prognosis and treatment tailoring in the not-so-distant future. The detection of malignancy related proteins in the serum might provide a rapid, sensitive, and specific screening method for colorectal malignancies even for early disease stages.

REFERENCES

1. Parkin, D.M., Pisani, P. and Ferlay, J. Global cancer statistics. *CA Cancer J Clin* 49: 33–64 (1999).
2. O'Connell, J.B., Maggard, M.A. and Ko, C.Y. Colon cancer survival rates with the new American Joint Committee on Cancer sixth edition staging. *J Natl Cancer Inst* 96: 1420–1425 (2004).

3. Mak, T., Lalloo, F., Evans, D.G. and Hill, J. Molecular stool screening for colorectal cancer. *Br J Surg* 91: 790–800 (2004).
4. Fleischer, D.E., et al. Detection and surveillance of colorectal cancer. *JAMA* 261: 580–585 (1989).
5. Bernstein, C.N., Blanchard, J.F., Rawsthorne, P. and Collins, M.T. Population-based case control study of seroprevalence of *Mycobacterium paratuberculosis* in patients with Crohn's disease and ulcerative colitis. *J Clin Microbiol* 42: 1129–1135 (2004).
6. Daperno, M., et al. The role of endoscopy in inflammatory bowel disease. *Eur Rev Med Pharmacol Sci* 8: 209–214 (2004).
7. Eaden, J., Abrams, K., McKay, H., Denley, H. and Mayberry, J. Inter-observer variation between general and specialist gastrointestinal pathologists when grading dysplasia in ulcerative colitis. *J Pathol* 194: 152–157 (2001).
8. Lynch, D.A., Lobo, A.J., Sobala, G.M., Dixon, M.F. and Axon, A.T. Failure of colonoscopic surveillance in ulcerative colitis. *Gut* 34: 1075–1080 (1993).
9. Riddell, R.H. How reliable/valid is dysplasia in identifying at-risk patients with ulcerative colitis? *J Gastrointest Surg* 2: 314–317 (1998).
10. Caspersson, T.O. Quantitative tumor cytochemistry—G.H.A. Clowes Memorial Lecture. *Cancer Res* 39: 2341–2345 (1979).
11. Duesberg, P., Rausch, C., Rasnick, D. and Hehlmann, R. Genetic instability of cancer cells is proportional to their degree of aneuploidy. *Proc Natl Acad Sci USA* 95: 13692–13697 (1998).
12. Lengauer, C., Kinzler, K.W. and Vogelstein, B. Genetic instabilities in human cancers. *Nature* 396: 643–649 (1998).
13. Ried, T., Heselmeyer-Haddad, K., Blegen, H., Schrock, E. and Auer, G. Genomic changes defining the genesis, progression, and malignancy potential in solid human tumors: a phenotype/genotype correlation. *Genes Chromosomes Cancer* 25: 195–204 (1999).
14. Lofberg, R., Brostrom, O., Karlen, P., Ost, A. and Tribukait, B. DNA aneuploidy in ulcerative colitis: reproducibility, topographic distribution, and relation to dysplasia. *Gastroenterology* 102: 1149–1154 (1992).
15. Rubin, C.E., et al. DNA aneuploidy in colonic biopsies predicts future development of dysplasia in ulcerative colitis. *Gastroenterology* 103: 1611–1620 (1992).
16. Habermann, J., et al. Ulcerative colitis and colorectal carcinoma: DNA-profile, laminin-5 gamma2 chain and cyclin A expression as early markers for risk assessment. *Scand J Gastroenterol* 36: 751–758 (2001).
17. Ono, J., et al. Reversibility of 20-methylcholanthrene-induced bronchial cell atypia in dogs. *Cancer* 54: 1030–1037 (1984).
18. Auer, G., et al. Reversibility of bronchial cell atypia. *Cancer Res* 42: 4241–4247 (1982).
19. Wang, A., et al. Different expression patterns of cyclins A, D1 and E in human colorectal cancer. *J Cancer Res Clin Oncol* 122: 122–126 (1996).
20. Caspersson, T., Zech, L., Johansson, C. and Modest, E.J. Identification of human chromosomes by DNA-binding fluorescent agents. *Chromosoma* 30: 215–227 (1970).
21. Kallioniemi, A., et al. Comparative genomic hybridization for molecular cytogenetic analysis of solid tumors. *Science* 258: 818–821 (1992).
22. Schrock, E., et al. Multicolor spectral karyotyping of human chromosomes. *Science* 273: 494–497 (1996).
23. Speicher, M.R., Gwyn Ballard, S. and Ward, D.C. Karyotyping human chromosomes by combinatorial multi-fluor FISH. *Nat Genet* 12: 368–375 (1996).
24. Knuutila, S., et al. DNA copy number amplifications in human neoplasms: review of comparative genomic hybridization studies. *Am J Pathol* 152: 1107–1123 (1998).
25. Forozan, F., Karhu, R., Kononen, J., Kallioniemi, A. and Kallioniemi, O.P. Genome screening by comparative genomic hybridization. *Trends Genet* 13: 405–409 (1997).

26. Hittelman, W.N. Genetic instability in epithelial tissues at risk for cancer. *Ann NY Acad Sci* 952: 1–12 (2001).
27. Hopman, A.H., et al. In situ hybridization as a tool to study numerical chromosome aberrations in solid bladder tumors. *Histochemistry* 89: 307–316 (1988).
28. Heselmeyer, K., et al. Gain of chromosome 3q defines the transition from severe dysplasia to invasive carcinoma of the uterine cervix. *Proc Natl Acad Sci USA* 93: 479–484 (1996).
29. Solinas-Toldo, S., et al. Mapping of chromosomal imbalances in pancreatic carcinoma by comparative genomic hybridization. *Cancer Res* 56 : 3803–3807 (1996).
30. Nowak, M.A., et al. The role of chromosomal instability in tumor initiation. *Proc Natl Acad Sci USA* 99: 16226–16231 (2002).
31. Desper, R., et al. Distance-based reconstruction of tree models for oncogenesis. *J Comput Biol* 7: 789–803 (2000).
32. Bomme, L., et al. Clonal karyotypic abnormalities in colorectal adenomas: clues to the early genetic events in the adenoma-carcinoma sequence. *Genes Chromosomes Cancer* 10: 190–196 (1994).
33. Barrett, J.C., Oshimura, M., Tanaka, N. and Tsutsui, T. Role of aneuploidy in early and late stages of neoplastic progression of Syrian hamster embryo cells in culture. *Basic Life Sci* 36: 523–538 (1985).
34. Oshimura, M. and Barrett, J.C. Chemically induced aneuploidy in mammalian cells: mechanisms and biological significance in cancer. *Environ Mutagen* 8: 129–159 (1986).
35. Fearon, E.R. and Vogelstein, B. A genetic model for colorectal tumorigenesis. *Cell* 61 : 759–767 (1990).
36. Ried, T., et al. Comparative genomic hybridization reveals a specific pattern of chromosomal gains and losses during the genesis of colorectal tumors. *Genes Chromosomes Cancer* 15: 234–245 (1996).
37. Vogelstein, B. and Kinzler, K.W. Cancer genes and the pathways they control. *Nat Med* 10: 789–799 (2004).
38. Kern, S.E., et al. Molecular genetic profiles of colitis-associated neoplasms. *Gastroenterology* 107: 420–428 (1994).
39. Holzmann, K., et al. Comparison of flow cytometry and histology with mutational screening for p53 and Ki-ras mutations in surveillance of patients with long-standing ulcerative colitis. *Scand J Gastroenterol* 36: 1320–1326 (2001).
40. Willenbucher, R.F., Zelman, S.J., Ferrell, L.D., Moore, D.H. II and Waldman, F.M. Chromosomal alterations in ulcerative colitis-related neoplastic progression. *Gastroenterology* 113: 791–801 (1997).
41. Loeb, K.R. and Loeb, L.A. Genetic instability and the mutator phenotype. Studies in ulcerative colitis. *Am J Pathol* 154: 1621–1626 (1999).
42. Aust, D.E., et al. Chromosomal alterations in ulcerative colitis-related and sporadic colorectal cancers by comparative genomic hybridization. *Hum Pathol* 31: 109–114 (2000).
43. Habermann, J.K., et al. Pronounced chromosomal instability and multiple gene amplifications characterize ulcerative colitis-associated colorectal carcinomas. *Cancer Genet Cytogenet* 147: 9–17 (2003).
44. Platzer, P., et al. Silence of chromosomal amplifications in colon cancer. *Cancer Res* 62 : 1134–1138 (2002).
45. Blegen, H., et al. Genetic instability promotes the acquisition of chromosomal imbalances in T1b and T1c breast adenocarcinomas. *Anal Cell Pathol* 22: 123–131 (2001).
46. Schlegel, J., et al. Comparative genomic in situ hybridization of colon carcinomas with replication error. *Cancer Res* 55 : 6002–6005 (1995).
47. Ghadimi, B.M., et al. Centrosome amplification and instability occurs exclusively in aneuploid, but not in diploid colorectal cancer cell lines, and correlates with numerical chromosomal aberrations. *Genes Chromosomes Cancer* 27: 183–190 (2000).

48. Matzke, M.A., Mette, M.F., Kanno, T. and Matzke, A.J. Does the intrinsic instability of aneuploid genomes have a causal role in cancer? *Trends Genet* 19: 253–256 (2003).
49. Schena, M., et al. Parallel human genome analysis: microarray-based expression monitoring of 1000 genes. *Proc Natl Acad Sci USA* 93: 10614–10619 (1996).
50. Virtaneva, K., et al. Expression profiling reveals fundamental biological differences in acute myeloid leukemia with isolated trisomy 8 and normal cytogenetics. *Proc Natl Acad Sci USA* 98: 1124–1129 (2001).
51. Phillips, J.L., et al. The consequences of chromosomal aneuploidy on gene expression profiles in a cell line model for prostate carcinogenesis. *Cancer Res* 61 : 8143–8149 (2001).
52. Pollack, J.R., et al. Microarray analysis reveals a major direct role of DNA copy number alteration in the transcriptional program of human breast tumors. *Proc Natl Acad Sci USA* 99: 12963–12968 (2002).
53. Hyman, E., et al. Impact of DNA amplification on gene expression patterns in breast cancer. *Cancer Res* 62 : 6240–6245 (2002).
54. Tirkkonen, M., et al. Molecular cytogenetics of primary breast cancer by CGH. *Genes Chromosomes Cancer* 21: 177–184 (1998).
55. Roschke, A.V., Stover, K., Tonon, G., Schaffer, A.A. and Kirsch, I.R. Stable karyotypes in epithelial cancer cell lines despite high rates of ongoing structural and numerical chromosomal instability. *Neoplasia* 4: 19–31 (2002).
56. Upender, M.B., et al. Chromosome transfer induced aneuploidy results in complex dysregulation of the cellular transcriptome in immortalized and cancer cells. *Cancer Res* 64: 6941–9 (2004).
57. Saran, N.G., Pletcher, M.T., Natale, J.E., Cheng, Y. and Reeves, R.H. Global disruption of the cerebellar transcriptome in a Down syndrome mouse model. *Hum Mol Genet* 12: 2013–2019 (2003).
58. Mao, R., Zielke, C.L., Zielke, H.R. and Pevsner, J. Global up-regulation of chromosome 21 gene expression in the developing Down syndrome brain. *Genomics* 81: 457–467 (2003).
59. FitzPatrick, D.R., et al. Transcriptome analysis of human autosomal trisomy. *Hum Mol Genet* 11: 3249–3256 (2002).
60. Brewer, C., Holloway, S., Zawalnyski, P., Schinzel, A. and FitzPatrick, D.A. chromosomal duplication map of malformations: regions of suspected haplo- and triplolethality—and tolerance of segmental aneuploidy—in humans. *Am J Hum Genet* 64: 1702–1708 (1999).
61. Caron, H., et al. The human transcriptome map: clustering of highly expressed genes in chromosomal domains. *Science* 291: 1289–1292 (2001).
62. Druker, B.J. Imatinib alone and in combination for chronic myeloid leukemia. *Semin Hematol* 40: 50–58 (2003).
63. Misek, D.E., Imafuku, Y. and Hanash, S.M. Application of proteomic technologies to tumor analysis. *Pharmacogenomics* 5: 1129–1137 (2004).
64. Gorg, A., Weiss, W. and Dunn, M.J. Current two-dimensional electrophoresis technology for proteomics. *Proteomics* 4: 3665–3685 (2004).
65. Oppermann, M., et al. Identification of foetal brain proteins by two-dimensional gel electrophoresis and mass spectrometry comparison of samples from individuals with or without chromosome 21 trisomy. *Eur J Biochem* 267: 4713–4719 (2000).
66. Veenstra, T.D., Prieto, D.A. and Conrads, T.P. Proteomic patterns for early cancer detection. *Drug Discov Today* 9: 889–897 (2004).
67. Roblick, U.J., et al. Sequential proteome alterations during genesis and progression of colon cancer. *Cell Mol Life Sci* 61: 1246–1255 (2004).
68. Macdonald, J.S. Carcinoembryonic antigen screening: pros and cons. *Semin Oncol* 26: 556–560 (1999).
69. Srivastava, S., Verma, M. and Henson, D.E. Biomarkers for early detection of colon cancer. *Clin Cancer Res* 7: 1118–1126 (2001).

70. Conrads, T.P., Hood, B.L., Issaq, H.J. and Veenstra, T.D. Proteomic patterns as a diagnostic tool for early-stage cancer: a review of its progress to a clinically relevant tool. *Mol Diagn* 8: 77–85 (2004).
71. Issaq, H.J., Conrads, T.P., Prieto, D.A., Tirumalai, R. and Veenstra, T.D. SELDI-TOF MS for diagnostic proteomics. *Anal Chem* 75: 148A–155A (2003).
72. Cordingley, H.C., et al. Multifactorial screening design and analysis of SELDI-TOF ProteinChip array optimization experiments. *Biotechniques* 34: 364–365, 368–373 (2003).
73. Semmes, O.J., et al. Evaluation of serum protein profiling by surface-enhanced laser desorption/ionization time-of-flight mass spectrometry for the detection of prostate cancer: I. Assessment of platform reproducibility. *Clin Chem* 51: 102–112 (2005).

CHAPTER 16

EFFECT OF HYPOXIA ON THE TUMOR PHENOTYPE: THE NEUROBLASTOMA AND BREAST CANCER MODELS

LINDA HOLMQUIST, TOBIAS LÖFSTEDT, AND SVEN PÅHLMAN

Department of Laboratory Medicine, Molecular Medicine, Lund University, University Hospital MAS, Malmö, Sweden

Abstract: The tumor oxygenation status associates with aggressive behavior. Oxygen shortage, hypoxia, is a major driving force behind tumor vascularization, and hypoxia enhances mutational rate, metastatic spread, and resistance to radiation and chemotherapy. We recently discovered that hypoxia promotes dedifferentiation of neuroblastoma and breast carcinoma cells and development of stem cell-like features. In both these tumor forms there is a correlation between low differentiation stage and poor outcome, and we conclude that the dedifferentiating effect of lowered oxygen adds to the aggressive phenotype induced by hypoxia. With neuroblastoma and breast carcinoma as human tumor model systems, we have addressed questions related to hypoxia-induced molecular mechanisms governing malignant behavior of tumor cells, with emphasis on differentiation and growth control. By global gene expression analyses we are currently screening for gene products exclusively expressed or modified in hypoxic cells with the aim to use them as targets for treatment.

1. INTRODUCTION

Cancer is a multigenetic disease, and the tumor origin determines, at least in part, which genes become affected. In total, some 100 different genes have been shown to be somatically deranged as a consequence of the selective pressure acting on tumor cells [1]. Even within a specific subgroup of tumors there is a considerable genetic variability. This is one explanation of the tremendous

Correspondence to Sven Påhlman, Department of Laboratory Medicine, Molecular Medicine, University Hospital MAS, Entrance, 78, SE-205 02 Malmö, Sweden. Phone: +46-40337403: Fax: +46-40337322; e-mail: sven.pahlman@med.lu.se

phenotypic heterogeneity seen in major tumor forms such as breast, colon, and prostate cancer and major pediatric cancers such as neuroblastoma. In the clinical setting, the complexity of cancer cell behavior has become even more evident during the last decades with the realization that not only tumor cells proper determine tumor aggressiveness. Our expanded knowledge of the pathophysiology of tumors shows that the interplay between the genetically unstable and altered tumor cells and the diploid, genetically stable stromal and blood vessel cells most profoundly determines the behavior of tumors and patient outcome. It is well established that tumor stroma affects the metastatic process and is involved in raised interstitial tumor pressure [2], and neovascularization with recruitment of vascular endothelial cells is a necessary and limiting factor in growth of solid tumors [3].

The tumor oxygenation status is tightly linked to aggressive behavior, in part explained by the facts that hypoxia is the major driving force behind tumor vascularization [4,5], and that hypoxia enhances the development of other hallmarks of aggressive tumor phenotypes [6], e.g., high mutational rate [7], metastatic spread [8], lowered pH, and resistance to treatment by radiation or cytotoxic drugs [9]. It is well established that growth of a solid tumor over a size of a few millimeter requires vascularization of the tumor. Tumor hypoxia and expression of vascular endothelial growth factor (VEGF) by stabilization and activation of the hypoxia inducible transcription factors (HIFs) are major players in the formation of new blood vessels, and consequently VEGF signaling and HIF activity are targets for novel treatment strategies aiming at blocking the angiogenic process, and thus tumor growth [5]. In addition, we recently discovered that low tumor oxygen levels lead to reduced expression of differentiation lineage specific genes and to the development of stem cell-like phenotypes, as demonstrated in neuroblastoma and breast cancer [10,11]. We conclude that this effect of hypoxia has a direct bearing on tumor aggressiveness, as tumors with immature features are more aggressive than corresponding, differentiated tumors, which is particularly true in the case of neuroblastoma and breast cancer [12–14]. Thus, in order to understand central issues like phenotypic heterogeneity, metastatic growth, angiogenesis, drug sensibility, and drug resistance, tumor physiological phenomena such as hypoxia, low pH, and high interstitial pressure has to be taken into account.

2. THE CELLULAR RESPONSE TO HYPOXIA

2.1. Hypoxia-inducible factors

All organisms from bacteria to humans have mechanisms for maintaining O_2 homeostasis in order to survive. A low availability of oxygen, hypoxia, results in cellular responses which in vertebrates improves oxygenation and viability through induction of angiogenesis, increase in glycolytic metabolism to raise energy production and upregulation of genes involved in cell survival/apoptosis

[5]. As we know today, the most important proteins governing the adaptive responses to hypoxia in mammals are the hypoxia inducible factors. HIF is a heterodimer consisting of an α-subunit (HIF-1α, HIF-2α, or HIF-3α), which is stabilized and activated with decreasing oxygen levels, and an oxygen-independent β-subunit also known as aryl hydrocarbon receptor nuclear translocator (ARNT) [15]. Both subunits contain basic helix-loop-helix (bHLH) and PAS domains (an acronym referring to the PER, ARNT, and SIM proteins, in which this motif first was discovered) mediating DNA binding and protein heterodimerization. Under hypoxic conditions (frequently 1% O_2 in experimental systems), the HIF-1α subunit translocates within minutes, independently of ARNT, to the nucleus by nuclear localization signals allowing rapid transcriptional responses to lowered intracellular oxygen levels [16,17]. Upon reoxygenation, HIF-1α protein is rapidly reduced with a half-life of <5 min and usually becomes undetectable under normoxic conditions. HIF-1/ARNT heterodimers bind to specific DNA sequences termed "hypoxia response elements" (HREs) present in promoters or enhancers of HIF target genes [18]. Importantly, binding of HIF to HREs and subsequent activation of genes depends not only on the amount of HIF complex, but also on coactivators such as CBP/p300 and posttranslational modifications [19,20].

Of the HIF proteins, HIF-1α was first described and has also been most studied. Two other proteins, HIF-2α and HIF-3α, were later discovered and share several characteristics with HIF-1α such as stabilization under hypoxia, heterodimerization with ARNT, DNA binding, and gene transactivation. Of the two proteins, HIF-2α, which was first named endothelial PAS protein (EPAS-1), has the highest similarity with HIF-1α [21]. HIF-2α is expressed in a complementary but not overlapping pattern to HIF-1α under systemic hypoxia. HIF-3α also shares considerable homology with HIF-1α and HIF-2α, but lacks the C-terminal transactivation domain and could thus act as a suppressor of the HIF pathway [22]. Recently, a dominant negative regulator of the HIF-α subunits named inhibitory PAS (IPAS) domain protein was also identified as a splice variant of the HIF-3α locus [23,24]. Moreover, discovery of sequence homologues of ARNT [25] that may have distinct physiological roles together with the HIF-α proteins further complicates the HIF network and the hypoxic signaling response.

Both HIF-1α and HIF-2α are essential for normal development. *HIF-1α$^{-/-}$* mice exhibit gross morphological aberrations, and die at approximately embryonal day 11 (E11) due to neural tube defects and cardiovascular malformations, reflecting the importance of HIF-1α during development [26,27]. The greater severity of the *HIF-1α$^{-/-}$* embryonic defects compared with ARNT-deficient mice [28] indicates that HIF-1α dimerizes with other partners such as ARNT2 [25]. Deficiency of HIF-2α has a more restricted effect on development. During embryogenesis HIF-2α is mainly expressed in vascular structures, but also in different parts of the developing sympathetic nervous system (SNS), responsible for catecholamine production prior to birth. *HIF-2α$^{-/-}$* mice also die at

E9.5–E13.5 of heart failure as a result of deficient catecholamine production, or of severe vascular defects [21,29]. However, the *HIF-2α*[-/-] phenotype appears to strongly depend on the genetic background, since defective catecholamine synthesis is not exhibited in all *HIF-2α* knockout strains [29,30].

Adaptation to hypoxia is critical also for cancer cells, implicating involvement of HIF proteins in tumorigenesis. Mouse hepatoma cell lines with mutated ARNT form much smaller tumors that express only low levels of VEGF and do not become highly vascularized [31]. Several studies have also associated HIF-1α with human cancer progression [32,33]. Histological analyses have shown that increased levels of HIF-1α correlate with poor prognosis and resistance to therapy in various solid tumors [34,35]. Overexpression of HIF-1α has also been associated with cell proliferation in several major tumor types, such as colon, breast, lung, skin, ovarian, prostate, and renal carcinomas [35]. The increased HIF-1α activity in cancers can result from intratumoral hypoxia as well as being a consequence of genetic alterations (i.e., by oncogene activation or tumor suppressor inactivation) or stimulation by growth factors [5]. In either case, higher HIF-1α activity leads to upregulation of genes involved in many aspects of cancer progression, including metabolic adaptation, resistance to apoptosis, increased angiogenesis, and metastasis. HIF-2α expression is also increased in a variety of human tumors, including neuroblastomas as will be further discussed in this contribution, but has mostly been associated with stromal cells [34,36]. Compared with bona fide cancer cells, stromal cells might mediate a different response to hypoxia.

2.2. Regulation of HIF stability and function

At hypoxia the HIFs are stabilized due to inhibited protein degradation, whereas at full access to oxygen the HIFs are degraded by the proteasomal pathway. The prolyl hydroxylases (PHD) are key regulators of HIF-stabilization. Using oxygen as substrate, PHDs hydroxylate conserved proline residues in the oxygen-dependent degradation domains (ODD) of the HIF-α subunits [37,38]. The abundance of PHD proteins and their affinities for oxygen are main regulators of the capacity of the PHDs to hydroxylate and thus promote degradation of HIFs. Hydroxylated prolines are recognized by the von Hippel-Lindau tumor suppressor protein (pVHL), which associates with the ubiquitin E3-ligase and forms a complex with HIFs resulting in proteasomal HIF degradation [39]. There are three identified PHDs that recognize HIF; PHD1, PHD2, and PHD3. PHD2 and PHD3 mRNA have been shown to be upregulated in many cell systems at hypoxia, also in neuroblastoma cells (to be published). The accumulation of PHD proteins at hypoxia is part of a negative feedback loop, which is responsible for enhanced degradation of the HIF-α subunits after reoxygenation. Cells transfected with siRNA against the PHDs show prolonged HIF-α stability after reoxygenation [40,41]. Furthermore, PHD2 seems to have a preference for HIF-1α, while PHD3 appears to target HIF-2α for hydroxylation

[42,43]. Both *PHD2* and *PHD3* are known to be HIF-1α target genes, which supports the idea of a negative feedback loop [44,45]. However, our data show a fast and robust, but transient, induction of HIF-1α protein levels at hypoxia (1% O_2) in neuroblastoma cells, whereas HIF-2α reveals a prolonged protein induction pattern at hypoxia [46,47]. The fact that HIF-1α drives the transcription of PHD2 and PHD3 at hypoxia indicates that the HIF-1α negative feedback loop affects HIF-1α protein levels also at low oxygen levels. Studies using siRNA further revealed that PHDs retain their functional activity at oxygen conditions as low as 2% oxygen, indicating a role for PHDs in HIF-regulation at low oxygen pressures [48]. In neuroblastoma cells we have shown that HIF-2α mRNA levels, but not HIF-1α mRNA levels increase at prolonged (72 h) hypoxia [49] and Uchida et al. [50] have seen similar patterns in lung epithelial cells. In summary, our current view is that the mechanisms by which HIF-α proteins accumulate differ between HIF-1α and HIF-2α. At hypoxia the PHDs possibly exert their effect primarily on HIF-1α and not HIF-2α, which could be an explanation for persistent HIF-2α protein levels at prolonged hypoxia in conjunction with increased HIF-2α mRNA levels.

In addition to the PHDs, an oxygen-dependent asparagyl hydroxylase factor inhibiting HIF (FIH) is involved in the regulation of HIF-α transcriptional activity. Under normoxic conditions, FIH hydroxylates an asparagine residue located in the C-terminal activation domain (CAD) of HIF-α. This post-translational modification, which is inhibited by hypoxia, reduces interaction with the CBP/p300 coactivators and results in a decreased ability of HIFs to transactivate their target genes [51].

3. HYPOXIA DEDIFFERENTIATES TUMOR CELLS

3.1. The neuroblastoma model

Neuroblastoma is a childhood tumor derived from immature SNS or SNS precursor cells, and based on differentiation marker gene expression analysis most neuroblastomas express a neuronal phenotype [52]. However, a subset of neuroblastomas contains tumor cells that have undergone a spontaneous neuronal-to-neuroendocrine/chromaffin lineage shift in areas adjacent to zones of tumor necrosis [52,53]. With the discovery of the HIFs, and in particular HIF-2α that is selectively expressed in the developing SNS, including neuroendocrine cells [10,21], we hypothesized that the differentiation lineage shift seen in these neuroblastomas might be hypoxia- and HIF-2α–driven. We did indeed detect HIF-2α and HIF-1α protein in these neuroendocrinely differentiated tumors in the zones of tumor cells surrounding necrotic areas, albeit HIF-2α protein was abundant also in tumor cells close to blood vessels, and thus presumably well oxygenized tumour areas [46]. However, when we experimentally addressed the question of whether hypoxia promotes neuroendocrine differentiation using a panel of human neuroblastoma cell lines grown under hypoxic conditions, our data suggested that hypoxic neuroblastoma cells, rather than adopting a neuroendocrine phenotype, lost their differentiated characteristics. We arrived at that conclusion based on expression analysis of neuronal differentiation markers, which became downregulated at the same time as HIF-2α and HIF-1α proteins were stabilized [10]. Thus, growth at low oxygen pressure (1% O_2) resulted in decreased expression of neuronal/neuroendocrine marker genes (*Chromogranin A* and *B*, *NPY*, *GAP43*, *dHAND*, and *HASH-1*) and neurofilament (*NEF3*), while genes linked to the decision of neural crest cells to develop into a sympathetic precursor cell (*KIT*, *ID2*, *NOTCH1*, and *HES1*) were upregulated [10,46,49]. Global gene expression analysis of hypoxic neuroblastoma cells confirmed these data and highlighted a number of additional genes supporting our conclusion that hypoxia drives neuroblastoma cells towards an immature phenotype [49].

As neuroblastomas are derived from the SNS, which in turn stems from the neural crest [54], we concluded that hypoxic neuroblastoma cells gain stem cell-like neural crest characteristics. Hedborg et al. have addressed the same question and also analyzed the hypoxic phenotype of cultured neuroblastoma cells [55]. They came to the conclusion that hypoxic neuroblastoma cells in culture go through a neuronal-to-neuroendocrine transition, similar to the in vivo situation in some neuroblastomas. However, their conclusions are based on a limited number of cell lines and marker genes, and importantly, the only established neuroendocrine marker investigated, *chromogranin A* (except for two bona fide hypoxia-driven genes, *IGF-2* and *TH* [10,56,57]) did not increase in their hypoxic cells. Thus, our and Hedborg's data are largely in agreement, but we come to different conclusions. As our conclusion is based on a substantial number of cell lines, on global gene expression data, and analyses of several recognized neuronal and neuroendocrine markers, we are confident that human neuroblastoma cells of established cell lines do not differentiate toward a neuroendocrine phenotype at hypoxic (1% O_2) growth conditions. Recently, it was reported that the nerve growth factor receptor TrkA exists in a truncated, NGF-independent, constitutively activated form in neural stem cells and in some neuroblastomas [58]. Treatment of neuroblastoma cells with the hypoxia mimetic cobalt chloride induced the expression of the truncated, stem cell-associated form of TrkA, supporting our findings that hypoxia pushes neuroblastoma cells to a stem cell-like phenotype, although these authors never directly tested the effect of hypoxia.

Clearly, cultured hypoxic neuroblastoma cells do not differentiate into a neuroendocrine lineage, and their response to hypoxia do not reiterate the neuronal-to-neuroendocrine lineage shift seen in a subset of clinical neuroblastomas. In retrospect, this finding is perhaps not so surprising as most neuroblastoma specimens do not show this lineage shift [12,52]. In general, neuroblastoma cell lines are established from highly aggressive, immature tumors, while those neuroblastomas showing neuroendocrine features appear to be more differentiated. Therefore, one might argue that established neuroblastoma cell lines most likely do not represent neuroblastoma cells with a capacity to differentiate into a neuroendocrine lineage. From a clinical as well as a tumor biological perspective, the dedifferentiating effect of hypoxia on tumor cells, highlighted by the study of high stage neuroblastomas, is probably a far more important observation.

3.2. The breast carcinoma model

The effect of hypoxia on tumor cell differentiation is not restricted to neuroblastoma and experimental tumor models, which we demonstrated in ductal breast carcinoma in situ [11]. In these tumor lesions, hypoxic cells, surrounding the necrotic zones and expressing HIF-1α protein, are morphologically immature by standard clinical histopathological criteria compared with well-oxygenized cells close to the basal membrane layer surrounding the tumor duct. Interestingly, hypoxic tumor cells had high expression of the breast epithelial stem cell marker, cytokeratin 19, and in estrogen receptor (ER) positive tumor lesions, the ER expression was downregulated in hypoxic cells ([11], Kronblad et al., to be published), presumably as part of a hypoxia-induced dedifferentiation process. We postulate that hypoxia-induced tumor cell dedifferentiation is one mechanism by which hypoxia contributes to the selection of a malignant tumor phenotype, as poor differentiation correlates positively to adverse outcome in both breast cancer and neuroblastoma [12–14].

4. MOLECULAR MECHANISMS OF HYPOXIA-INDUCED DEDIFFERENTIATION

The sequential molecular steps involved in the conversion of migrating neural crest cells to sympathetic precursors and finally to differentiated nonmigrating sympathetic neuroblasts, or SNS chromaffin cells, are far from known. However, genes shown to be important for lineage determination of SNS precursor cells were found to be affected by hypoxic treatment in neuroblastoma [10]. Hypoxia downregulated members of the bHLH transcription factor family, such as the neuronal markers *HASH-1* and *dHAND* while their counteracting HLH factors, the ID proteins, became upregulated. Furthermore, the dimerization partner for HASH-1 and dHAND, the E-protein E2-2 was also downregulated by hypoxia [49]. As also *NOTCH-1* and its downstream-regulated genes, including *HES-1*, were activated by hypoxia, a causal link between the NOTCH/HES pathway and *HASH-1* downregulation might exist and contribute to the dedifferentiated phenotype [10,49]. These results taken together suggest a contributing molecular mechanism by which hypoxic neuroblastoma cells dedifferentiate (Figure 16.1). A more direct involvement of HIF proteins in this process achieved further support when we demonstrated that hypoxic *ID2* expression was regulated by HIF-1 [59]. In that study the expression of another ID member, *ID1*, was also increased by hypoxia. The hypoxia-induced ID proteins could play a significant role in the initial phase of dedifferentiation by sequestering E-proteins such as

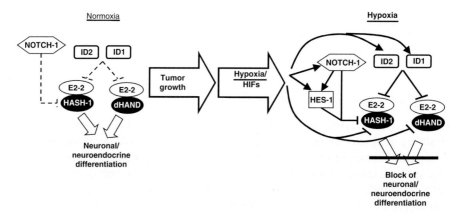

Figure 16.1. Molecular mechanisms involved in dedifferentiation of hypoxic neuroblastoma cells. Under normoxic conditions, the ID proteins are expressed at moderate levels allowing functionality of E-proteins (E2-2) in complex with tissue-specific bHLH proteins (i.e., dHAND and HASH-1) leading to expression of neuronal marker genes. In expanding solid tumors, regions of intratumoral hypoxia are formed where hypoxia-inducible factors (HIFs) are stabilized and functional. In the hypoxic microenvironment, the tissue-specific transcription factors and their dimerization partners are downregulated with a concomitant HIF-induced expression of *ID2* (and potentially *ID1*). These processes together with a hypoxia-induced activation of the NOTCH/HES signaling network result in downregulation of neuronal and neuroendocrine markers and development of a less differentiated phenotype. Based on results published in [10,49,59].

E2-2, and thereby inhibiting dimerization with HASH-1 and dHAND and their DNA-binding capacities. Reduced expression of neuronal bHLH transcription factors, together with inhibited bHLH function by ID proteins as a response to hypoxia would have profound effects on genes regulated by HASH-1 and dHAND. The fact that *ID* genes are downregulated upon induced differentiation in neuroblastoma [60], and are reduced during differentiation of several other cell types [61,62], further implicates their role in hypoxic dedifferentiation of neuroblastoma and likely other tumors as well. Since upregulation of *ID1* and *ID2* also was found in hypoxic breast cancer cells [59], and cells expressing HIF-1α in lesions of ductal breast carcinoma in situ were morphologically dedifferentiated [11], a possible connection between the ID proteins and breast tissue-specific bHLH factors may exist and be affected by hypoxia.

The involvement of HIF proteins, directly or indirectly, in hypoxic gene transcription of differentiation-related genes has not yet been investigated on a full scale in neuroblastoma or in other tumor types. It is likely, although not certain, that genes rapidly induced by hypoxia are direct HIF-target genes since HIF-1α and HIF-2α proteins become instantaneously stabilized and active in response to lowered oxygen in almost all cell types, including neuroblastoma [10,16].

5. CLINICAL HETEROGENEITY AND STABILITY OF THE HYPOXIC PHENOTYPE

Most solid tumor forms show extensive phenotypic heterogeneity both among tumors of the same diagnosis group and within a given tumor, as revealed by histopathological examination. As hypoxia has profound effects on the phenotype of any cell type including tumor cells, oxygen shortage contributes substantially to tumor heterogeneity. The hypoxic response is similar in most cells, i.e., increased expression of classical hypoxia/HIF-driven genes like *VEGF*, *GLUT-1* and *GLUT-3* and genes coding for glycolytic enzymes. Microarray analysis based on 27,000 genes and ESTs of seven different human neuroblastoma cell lines grown at 21 or 1% O_2, revealed an overall, uniform hypoxic response with induction of well-established hypoxia-driven genes like those mentioned above (Fredlund, Ovenberger, and Påhlman, to be published). A more detailed analysis disclosed a considerably more complex picture. While only four genes where up- or down-regulated twofold or more in all seven cell lines, the expression of as many as 7,000 genes were changed twofold or more in at least one of the seven cell lines. We could not show that the response in one or a few cell lines stood out against the others, which would offer an explanation why so few genes were uniformly regulated in all tested cell lines. Instead we have data showing that the discrepancies are explained by an unpredictable lack of or reduced hypoxic response of individual genes in one or two cell lines and that these cell lines differ from one gene to another. For example, in one neuroblastoma cell line the glucose transporter gene *GLUT-1* was downregulated when *GLUT-3* was upregulated by hypoxia. In another cell line, the opposite pattern was seen, while in the remaining five cell lines both *GLUT* genes were upregulated as expected [63], since both genes have been shown to be HIF-1α-driven [5]. Thus, for unknown reasons, the hypoxic response in at least human neuroblastoma cells is not as coherent as one might have expected, and translated to the clinical situation, adverse responses to low oxygen levels of subsets of tumor cells within a given tumor, will add to the complexity of the tumor phenotype.

In a growing tumor, the oxygenation level is not static but changes with the formation of new, and collapse of old tumor blood vessels. We therefore tested how persistent the immature features of hypoxic neuroblastoma cells are upon reoxygenation, as this will reflect the situation of an extravasating cancer cell encountering well-oxygenated blood. Expression analysis of a selection of early neural crest markers as well as later sympathetic markers revealed that the dedifferentiated hypoxic neuroblastoma phenotype persisted for up to 24 h. This was the situation irrespective of whether hypoxic neuroblastoma cells were reoxygenated to an atmospheric (21% O_2) and not very physiological condition, or to a more physiological tissue oxygen pressure (5%). However, the hypoxic phenotype was reversible after 72 h of reoxygenation (Figure 16.2). Intermittent (cycles of 1 and 21% O_2), as well as long-term (12 days) hypoxia reinforced the

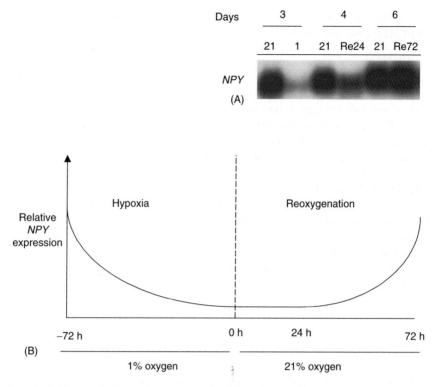

Figure 16.2. Phenotypic changes in reoxygenated (Re) hypoxic neuroblastoma cells, as exemplified by the expression of the neuronal marker, *neuropeptide Y (NPY)*. (A) Northern blot analysis of *NPY* in KCN-69n human neuroblastoma cells grown at 1% O_2 (1) for 3 days followed by reoxygenation at 21% O_2 (21) for 24 or 72 h. (B) Graph illustrating the relative expression of *NPY* (and other neuronal marker genes) in reoxygenated neuroblastoma cells. Adapted from [47].

hypoxic phenotype but it was still reversible after 72 h of reoxygenation at 21% O_2 [47]. In agreement with the findings of Hedborg et al., we conclude that the hypoxic response is reversible upon reoxygenation.

6. TARGETING THE HYPOXIC PHENOTYPE

The hypoxic phenotype is principally tumor-specific and one overriding aim of our ongoing hypoxia project is to identify genes that are strongly induced by hypoxia (and hypoglycemia) and to use these genes as targets for treatment. We are specifically focusing on genes that are coding for cell surface proteins. An ideal candidate gene would have limited expression in normoxic tumor cells, as well as in normal tissues. Given the dedifferentiating effect of hypoxia on tumor cells, we are currently searching our microarray databases for hypoxia-inducible genes normally expressed early during development, with initial focus on growth factor receptors, membrane channels, and adhesion molecules. As an alternative strategy, as well as an important step to verify that changes in gene expression identified by microarray analysis are indeed correlating to altered protein levels, protein fractions from normoxic and hypoxic tumor cells are analyzed by 2-D gel electrophoresis followed by mass spectroscopy identification of hypoxia-induced proteins or protein modifications induced by hypoxia. If we identify proteins that are more or less uniquely expressed or modified in hypoxic cells, the idea is to generate antibodies and to test their selectivity for hypoxic tumor cells both in vitro and in animal tumor models.

7. CONCLUSIONS

The observation that hypoxia appears to be a general mechanism by which tumors develop an immature phenotype is conceptually novel and highlights a new aspect of the aggressiveness of hypoxic tumors. The hypoxic phenotype, being essentially tumor-specific, is an attractive target for treatment. As outlined here, uniquely or highly overexpressed genes/proteins in hypoxic cells might successfully be explored as antibody targets, and might target the tumor stem cell compartment generally believed to be resistant to treatment.

ACKNOWLEDGMENTS

This work was supported by grants from the Swedish Cancer Society, the Children Cancer Foundation of Sweden, HKH Kronprinsessan Lovisas Förening för Barnasjukvård, Hans von Kantzows Stiftelse, and the research funds of Malmö University Hospital and Lund University Hospital.

REFERENCES

1. D. Hanahan and R.A. Weinberg. The hallmarks of cancer. *Cell* 100(1): 57–70. (2000).
2. C.H. Heldin, K. Rubin, K. Pietras, and A. Östman. High interstitial fluid pressure — an obstacle in cancer therapy. *Nat Rev Cancer* 4(10): 806–813 (2004).
3. J. Folkman. Angiogenesis in cancer, vascular, rheumatoid and other disease. *Nat Med.* 1(1): 27–31 (1995).
4. P. Carmeliet and R.K. Jain. Angiogenesis in cancer and other diseases. *Nature* 407(6801): 249–257 (2000).
5. G.L. Semenza. Targeting HIF-1 for cancer therapy. *Nat Rev Cancer* 3(10): 721–732 (2003).
6. M. Hockel, K. Schlenger, B. Aral, M. Mitze, U. Schaffer, and P. Vaupel. Association between tumor hypoxia and malignant progression in advanced cancer of the uterine cervix. *Cancer Res* 56(19): 4509–4515 (1996).
7. T.Y. Reynolds, S. Rockwell, and P.M. Glazer. Genetic instability induced by the tumor microenvironment. *Cancer Res* 56(24): 5754–5757 (1996).
8. D.M. Brizel, S.P. Scully, J.M. Harrelson, L.J. Layfield, J.M. Bean, L.R. Prosnitz, and M.W. Dewhirst. Tumor oxygenation predicts for the likelihood of distant metastases in human soft tissue sarcoma. *Cancer Res* 56(5): 941–943 (1996).
9. B.A. Teicher. Hypoxia and drug resistance. *Cancer Metastasis Rev* 13(2): 139–168 (1994).

10. A. Jögi, I. Øra, H. Nilsson, A. Lindeheim, Y. Makino, L. Poellinger, H. Axelson, and S. Påhlman. Hypoxia alters gene expression in human neuroblastoma cells toward an immature and neural crest-like phenotype. *Proc Natl Acad Sci USA* 99(10): 7021–7026 (2002).
11. K. Helczynska, A. Kronblad, A. Jögi, E. Nilsson, S. Beckman, G. Landberg, and S. Påhlman. Hypoxia promotes a dedifferentiated phenotype in ductal breast carcinoma in situ. *Cancer Res* 63(7): 1441–1444 (2003).
12. F. Hedborg, C. Bjelfman, P. Sparen, B. Sandstedt, and S. Påhlman. Biochemical evidence for a mature phenotype in morphologically poorly differentiated neuroblastomas with a favourable outcome. *Eur J Cancer* 31(4): 435–443 (1995).
13. J.S. Wei, B.T. Greer, F. Westermann, S.M. Steinberg, C.G. Son, Q.R. Chen, C.C. Whiteford, S. Bilke, A.L. Krasnoselsky, N. Cenacchi, D. Catchpoole, F. Berthold, M. Schwab, and J. Khan. Prediction of clinical outcome using gene expression profiling and artificial neural networks for patients with neuroblastoma. *Cancer Res* 64(19): 6883–6891 (2004).
14. R. Holland, J.L. Peterse, R.R. Millis, V. Eusebi, D. Faverly, M.J. Van De Vijver, and B. Zafrani. Ductal carcinoma in situ: a proposal for a new classification. *Semin Diagn Pathol.* 11(3): 167–180 (1994).
15. G.L. Wang, B.H. Jiang, E.A. Rue, and G.L. Semenza. Hypoxia-inducible factor 1 is a basic-helix-loop-helix-pas heterodimer regulated by cellular O_2 tension. *Proc Natl Acad Sci USA* 92(12): 5510–5514 (1995).
16. U.R. Jewell, I. Kvietikova, A. Scheid, C. Bauer, R.H. Wenger, and M. Gassmann. Induction of HIF-1α in response to hypoxia is instantaneous. *FASEB J* 15(7): 1312–1314 (2001).
17. P.J. Kallio, K. Okamoto, S. O'brien, P. Carrero, Y. Makino, H. Tanaka, and L. Poellinger. Signal transduction in hypoxic cells: inducible nuclear translocation and recruitment of the CBP/p300 coactivator by the hypoxia-inducible factor-1α. *EMBO J* 17(22): 6573–6586 (1998).
18. B.H. Jiang, E. Rue, G.L. Wang, R. Roe, and G.L. Semenza. Dimerization, DNA binding, and transactivation properties of hypoxia-inducible factor 1. *J Biol Chem* 271(30): 17771–17778 (1996).
19. Z. Arany, L.E. Huang, R. Eckner, S. Bhattacharya, C. Jiang, M.A. Goldberg, H.F. Bunn, and D.M. Livingston. An essential role for p300/CBP in the cellular response to hypoxia. *Proc Natl Acad Sci USA* 93(23): 12969–12973 (1996).
20. D.E. Richard, E. Berra, E. Gothie, D. Roux, and J. Pouyssegur. p42/p44 mitogen-activated protein kinases phosphorylate hypoxia-inducible factor 1α (HIF-1α) and enhance the transcriptional activity of HIF-1. *J Biol Chem* 274(46): 32631–32637 (1999).
21. H. Tian, R.E. Hammer, A.M. Matsumoto, D.W. Russell, and S.L. Mcknight. The hypoxia-responsive transcription factor EPAS1 is essential for catecholamine homeostasis and protection against heart failure during embryonic development. *Genes Dev* 12(21): 3320–3324 (1998).
22. Y.Z. Gu, S.M. Moran, J.B. Hogenesch, L. Wartman, and C.A. Bradfield. Molecular characterization and chromosomal localization of a third α-class hypoxia inducible factor subunit, HIF-3α. *Gene Expr* 7(3): 205–213 (1998).
23. Y. Makino, R. Cao, K. Svensson, G. Bertilsson, M. Asman, H. Tanaka, Y. Cao, A. Berkenstam, and L. Poellinger. Inhibitory PAS domain protein is a negative regulator of hypoxia-inducible gene expression. *Nature* 414(6863): 550–554 (2001).
24. Y. Makino, A. Kanopka, W.J. Wilson, H. Tanaka, and L. Poellinger. Inhibitory PAS domain protein (IPAS) is a hypoxia-inducible splicing variant of the hypoxia-inducible factor-3α locus. *J Biol Chem* 277(36): 32405–32408 (2002).
25. K. Hirose, M. Morita, M. Ema, J. Mimura, H. Hamada, H. Fujii, Y. Saijo, O. Gotoh, K. Sogawa, and Y. Fujii-Kuriyama. cDNA cloning and tissue-specific expression of a novel basic helix-loop-helix/pas factor (ARNT2) with close sequence similarity to the aryl hydrocarbon receptor nuclear translocator (ARNT). *Mol Cell Biol* 16(4): 1706–1713 (1996).
26. N.V. Iyer, L.E. Kotch, F. Agani, S.W. Leung, E. Laughner, R.H. Wenger, M. Gassmann, J.D. Gearhart, A.M. Lawler, A.Y. Yu, and G.L. Semenza. Cellular and developmental control of O_2 homeostasis by hypoxia-inducible factor 1 α. *Genes Dev* 12(2): 149–162 (1998).

27. H.E. Ryan, J. Lo, and R.S. Johnson. HIF-1α is required for solid tumor formation and embryonic vascularization. *EMBO J* 17(11): 3005–3015. (1998).
28. E. Maltepe, J.V. Schmidt, D. Baunoch, C.A. Bradfield, and M.C. Simon. Abnormal angiogenesis and responses to glucose and oxygen deprivation in mice lacking the protein ARNT. *Nature* 386(6623): 403–407 (1997).
29. J. Peng, L. Zhang, L. Drysdale, and G.H. Fong. The transcription factor EPAS-1/hypoxia-inducible factor 2α plays an important role in vascular remodeling. *Proc Natl Acad Sci USA* 97(15): 8386–8391 (2000).
30. M. Scortegagna, K. Ding, Y. Oktay, A. Gaur, F. Thurmond, L.J. Yan, B.T. Marck, A.M. Matsumoto, J.M. Shelton, J.A. Richardson, M.J. Bennett, and J.A. Garcia. Multiple organ pathology, metabolic abnormalities and impaired homeostasis of reactive oxygen species in EPAS1$^{-/-}$ mice. *Nat Genet* 35(4): 331–340 (2003).
31. P.H. Maxwell, G.U. Dachs, J.M. Gleadle, L.G. Nicholls, A.L. Harris, I.J. Stratford, O. Hankinson, C.W. Pugh, and P.J. Ratcliffe. Hypoxia-inducible factor-1 modulates gene expression in solid tumors and influences both angiogenesis and tumor growth. *Proc Natl Acad Sci USA* 94(15): 8104–8109 (1997).
32. B. Krishnamachary, S. Berg-Dixon, B. Kelly, F. Agani, D. Feldser, G. Ferreira, N. Iyer, J. Larusch, B. Pak, P. Taghavi, and G.L. Semenza. Regulation of colon carcinoma cell invasion by hypoxia-inducible factor 1. *Cancer Res* 63(5): 1138–1143 (2003).
33. S. Pennacchietti, P. Michieli, M. Galluzzo, M. Mazzone, S. Giordano, and P.M. Comoglio. Hypoxia promotes invasive growth by transcriptional activation of the met protooncogene. *Cancer Cell* 3(4): 347–361 (2003).
34. K.L. Talks, H. Turley, K.C. Gatter, P.H. Maxwell, C.W. Pugh, P.J. Ratcliffe, and A.L. Harris. The expression and distribution of the hypoxia-inducible factors HIF-1α and HIF-2α in normal human tissues, cancers, and tumor-associated macrophages. *Am J Pathol* 157(2): 411–421 (2000).
35. H. Zhong, A.M. De Marzo, E. Laughner, M. Lim, D.A. Hilton, D. Zagzag, P. Buechler, W.B. Isaacs, G.L. Semenza, and J.W. Simons. Overexpression of hypoxia-inducible factor 1α in common human cancers and their metastases. *Cancer Res.* 59(22): 5830–5835 (1999).
36. R.D. Leek, K.L. Talks, F. Pezzella, H. Turley, L. Campo, N.S. Brown, R. Bicknell, M. Taylor, K.C. Gatter, and A.L. Harris. Relation of hypoxia-inducible factor-2 α (HIF-2 α) expression in tumor-infiltrative macrophages to tumor angiogenesis and the oxidative thymidine phosphorylase pathway in human breast cancer. *Cancer Res* 62(5): 1326–1329 (2002).
37. M. Ivan, K. Kondo, H. Yang, W. Kim, J. Valiando, M. Ohh, A. Salic, J.M. Asara, W.S. Lane, and W.G. Kaelin Jr. HIFα targeted for VHL-mediated destruction by proline hydroxylation: implications for O_2 sensing. *Science* 292(5516): 464–468 (2001).
38. P. Jaakkola, D.R. Mole, Y.M. Tian, M.I. Wilson, J. Gielbert, S.J. Gaskell, A. Kriegsheim, H.F. Hebestreit, M. Mukherji, C.J. Schofield, P.H. Maxwell, C.W. Pugh, and P.J. Ratcliffe. Targeting of HIF-α to the von Hippel-Lindau ubiquitylation complex by O_2-regulated prolyl hydroxylation. *Science* 292(5516): 468–472 (2001).
39. N. Masson and P.J. Ratcliffe. Hif prolyl and asparaginyl hydroxylases in the biological response to intracellular O_2 levels. *J Cell Sci* 116(Pt 15): 3041–3049 (2003).
40. C.J. Schofield and P.J. Ratcliffe. Signalling hypoxia by HIF hydroxylases. *Biochem Biophys Res Commun*. BBRC 338(1): 617–626 (2005).
41. K. Hirota and G.L. Semenza. Regulation of hypoxia-inducible factor 1 by prolyl and asparaginyl hydroxylases. *Biochem Biophys Res Commun*. BBRC 338(1): 610–616 (2005).
42. R.J. Appelhoff, Y.M. Tian, R.R. Raval, H. Turley, A.L. Harris, C.W. Pugh, P.J. Ratcliffe, and J.M. Gleadle. Differential function of the prolyl hydroxylases PHD1, PHD2, and PHD3 in the regulation of hypoxia-inducible factor. *J Biol Chem* 279(37): 38458–38465 (2004).
43. O. Aprelikova, G.V. Chandramouli, M. Wood, J.R. Vasselli, J. Riss, J.K. Maranchie, W.M. Linehan, and J.C. Barrett. Regulation of HIF prolyl hydroxylases by hypoxia-inducible factors. *J Cell Biochem* 92(3): 491–501 (2004).

44. N. Pescador, Y. Cuevas, S. Naranjo, M. Alcaide, D. Villar, M.O. Landazuri, and L. Del Peso. Identification of a functional hypoxia-responsive element that regulates the expression of the EGL nine homologue 3 (EGLN3/PHD3) gene. *Biochem J* 390(1): 189–197 (2005).
45. E. Metzen, D.P. Stiehl, K. Doege, J.H. Marxsen, T. Hellwig-Burgel, and W. Jelkmann. Regulation of the prolyl hydroxylase domain protein 2 (PHD2/EGLN-1) gene: identification of a functional hypoxia-responsive element. *Biochem J* 387(3): 711–717 (2005).
46. H. Nilsson, A. Jögi, S. Beckman, A.L. Harris, L. Poellinger, and S. Påhlman. HIF-2α expression in human fetal paraganglia and neuroblastoma: relation to sympathetic differentiation, glucose deficiency, and hypoxia. *Exp Cell Res* 303(2): 447–456 (2005).
47. L. Holmquist, A. Jögi, and S. Påhlman. Phenotypic persistence after reoxygenation of hypoxic neuroblastoma cells. *Int J Cancer* 116(2): 218–225 (2005).
48. A.C. Epstein, J.M. Gleadle, L.A. Mcneill, K.S. Hewitson, J. O'rourke, D.R. Mole, M. Mukherji, E. Metzen, M.I. Wilson, A. Dhanda, Y.M. Tian, N. Masson, D.L. Hamilton, P. Jaakkola, R. Barstead, J. Hodgkin, P.H. Maxwell, C.W. Pugh, C.J. Schofield, and P.J. Ratcliffe. C. Elegans EGL-9 and mammalian homologs define a family of dioxygenases that regulate HIF by prolyl hydroxylation. *Cell* 107(1): 43–54 (2001).
49. A. Jögi, J. Vallon-Christersson, L. Holmquist, H. Axelson, A. Borg, and S. Påhlman. Human neuroblastoma cells exposed to hypoxia: induction of genes associated with growth, survival, and aggressive behavior. *Exp Cell Res* 295(2): 469–487 (2004).
50. T. Uchida, F. Rossignol, M.A. Matthay, R. Mounier, S. Couette, E. Clottes, and C. Clerici. Prolonged hypoxia differentially regulates hypoxia-inducible factor (HIF)-1α and HIF-2α expression in lung epithelial cells: implication of natural antisense HIF-1α. *J Biol Chem* 279(15): 14871–14878 (2004).
51. D. Lando, J.J. Gorman, M.L. Whitelaw, and D.J. Peet. Oxygen-dependent regulation of hypoxia-inducible factors by prolyl and asparaginyl hydroxylation. *Eur J Biochem* 270(5): 781–790 (2003).
52. J.C. Hoehner, C. Gestblom, F. Hedborg, B. Sandstedt, L. Olsen, and S. Påhlman. A developmental model of neuroblastoma: differentiating stroma-poor tumors' progress along an extraadrenal chromaffin lineage. *Lab Invest* 75(5): 659–675 (1996).
53. C. Gestblom, J.C. Hoehner, F. Hedborg, B. Sandstedt, and S. Påhlman. In vivo spontaneous neuronal to neuroendocrine lineage conversion in a subset of neuroblastomas. *Am J Pathol* 150(1): 107–117 (1997).
54. S. Påhlman and F. Hedborg. Development of the neural crest and sympathetic nervous system, in: G.M. Brodeur, T. Sawada, Y. Tsuchida and P.A. Voute (eds) *Neuroblastoma*. Elsevier Science, Amsterdam, 2000, pp. 9–19.
55. F. Hedborg, E. Ulleras, L. Grimelius, E. Wassberg, P.H. Maxwell, B. Hero, F. Berthold, F. Schilling, D. Harms, B. Sandstedt, and G. Franklin. Evidence for hypoxia-induced neuronal-to-chromaffin metaplasia in neuroblastoma. *FASEB J* 17(6): 598–609 (2003).
56. K.W. Kim, S.K. Bae, O.H. Lee, M.H. Bae, M.J. Lee, and B.C. Park. Insulin-like growth factor II induced by hypoxia may contribute to angiogenesis of human hepatocellular carcinoma. *Cancer Res* 58(2): 348–351 (1998).
57. M.L. Norris and D.E. Millhorn. Hypoxia-induced protein binding to O_2-responsive sequences on the tyrosine hydroxylase gene. *J Biol Chem* 270(40): 23774–23779 (1995).
58. A. Tacconelli, A.R. Farina, L. Cappabianca, G. Desantis, A. Tessitore, A. Vetuschi, R. Sferra, N. Rucci, B. Argenti, I. Screpanti, A. Gulino, and A.R. Mackay. TrkA alternative splicing: a regulated tumor-promoting switch in human neuroblastoma. *Cancer Cell* 6(4): 347–360 (2004).
59. T. Löfstedt, A. Jögi, M. Sigvardsson, K. Gradin, L. Poellinger, S. Påhlman, and H. Axelson. Induction of ID2 expression by hypoxia-inducible factor-1: a role in dedifferentiation of hypoxic neuroblastoma cells. *J Biol Chem* 279(38): 39223–39231 (2004).
60. A. Jögi, P. Persson, A. Grynfeld, S. Påhlman, and H. Axelson. Modulation of basic helix-loop-helix transcription complex formation by Id proteins during neuronal differentiation. *J Biol Chem* 277(11): 9118–9126 (2002).

61. R. Benezra, R.L. Davis, D. Lockshon, D.L. Turner, and H. Weintraub. The protein Id: a negative regulator of helix-loop-helix DNA binding proteins. *Cell* 61(1): 49–59 (1990).
62. X.H. Sun, N.G. Copeland, N.A. Jenkins, and D. Baltimore. Id proteins Id1 and Id2 selectively inhibit DNA binding by one class of helix-loop-helix proteins. *Mol Cell Biol* 11(11): 5603–5611 (1991).
63. H. Axelson, E. Fredlund, M. Ovenberger, G. Landberg, and S. Påhlman. Hypoxia-induced dedifferentiation of tumor cells — a mechanism behind heterogeneity and aggressiveness of solid tumors. *Semin Cell Dev Biol* 16(4–5): 554–563 (2005).

CHAPTER 17

METHYLATION PATTERNS AND CHEMOSENSITIVITY IN NSCLC

JOSÉ LUIS RAMÍREZ[1], M. FERNANDA SALAZAR[1], JALAJ GUPTA[1], JOSÉ MIGUEL SÁNCHEZ[1], MIQUEL TARON[1], MARIA SANCHEZ-RONCO[2], VICENTE ALBEROLA[3], AND RAMON DE LAS PEÑAS[4]

[1]*Catalan Institute of Oncology, Hospital Germans Trias i Pujol, Badalona, Spain;* [2]*Autonomous University of Madrid, Madrid, Spain;* [3]*Hospital Arnau de Vilanova, Valencia, Spain;* [4]*Hospital de Castellon, Castellon, Spain*

Abstract: Survival in advanced non-small-cell lung cancer (NSCLC) patients treated with platinum-based chemotherapy is rather variable. Methylation-dependent transcriptional silencing of *14-3-3σ*, a major G2/M checkpoint control gene, could be a predictor of longer survival. A sensitive methylation-specific polymerase chain reaction assay was used to evaluate 14-3-3σ methylation status in pretreatment serum DNA obtained from 115 cisplatin-plus-gemcitabine-treated advanced NSCLC patients. 14-3-3σ methylation was observed in all histologic types in 39 patients (34%). After a median follow-up of 9.8 months, median survival was significantly longer in the methylation-positive group (15.1 vs 9.8 months; $P = 0.004$). Median time to progression was 8 months in the methylation-positive group, and 6.3 months in the methylation-negative group ($P = 0.027$ by the log-rank test). A multivariate Cox regression model identified only 14-3-3σ methylation status and ECOG performance status (PS) as independent prognostic factors for survival. In an exploratory analysis, median survival for 22 methylation-positive responders has not been reached, while it was 11.3 months for 29 methylation-negative responders ($P = 0.001$). Methylation of 14-3-3σ is a new independent prognostic factor for survival in NSCLC patients receiving platinum-based chemotherapy. It can be reliably and conveniently detected in the serum, thus obviating the need for tumor tissue analysis.

Key words: 14-3-3σ, aberrant methylation, serum circulating DNA, NSCLC, cisplatin, gemcitabine

1. DNA METHYLATION: GENERALITIES

DNA methylation is one of the most common epigenetic events in mammalian genome. Recent studies show that it plays an important role in many biology events. Its influence in human carcinogenesis has been reported a few years ago [1].

Human genome contains regions of unmethylated segments interspersed by methylated regions. Small regions denominated CpG islands, ranging from 0.5 to 5.0 kb and occurring on average every 100 kb, are unmethylated. Because DNA methylation is a chemical modification that consists in the addition of a methyl group at the carbon 5 position of the cytosine ring in the sequence context 5′CG 3′. CpG islands are targeted at these genome modifications. This process is carried out by a group of enzymes known as DNA methyltransferases (DNMTs) [2]. As CpG islands are often associated with the promoter regions, methylation in a gene promoter region generally correlates with silencing of the gene [3].

In recent years, a high number of groups have extensively mapped an increasing number of gene CpG islands that are aberrantly hypermethylated in cancer. Genes can only be transcribed if three criteria are satisfied: (1) appropriate transcription factors are available; (2) the histones are acetylated and unmethylated; and (3) the cytosines in the CpG island remain unmethylated. However, in some exceptional cases, methylated DNA have been found in imprinted genes, on the X chromosome in women, in germ line-specific genes, and tissue-specific genes [4]. Genomic parental imprinting is restricted to one allele, either the paternal or maternal variant, by gene silencing due to selective DNA hypermethylation. A similar process occurs during X inactivation in women; the CpG island of all genes on one X chromosome are methylated, thus rendering the genes inactive and avoiding redundancy. Although DNA methylation is not widely used for regulating normal gene expression, there are important cases in which it plays an important role.

But among these important steps, hypermethylation of CpG islands also contributes to the carcinogenesis process. This is due to the fact that cancer cells exhibit two opposing changes in their pattern of DNA methylation: an overall decrease in DNA methylation in the bulky chromosome with an increasing methylation of CpG islands. These two effects, developed at the same time, play important roles in the tumoral process.

CpG hypermethylation can be an important topic in cancer if we consider the frequency of the process, the nature of some genes involved, the similar selective advantages conveyed by these changes, and by mutations in coding regions. Moreover, data suggest that epigenetic and genetic events interact in tumor progression [2]. These genes are involved in cell cycle regulation (p16, p15, and Rb), in DNA repair (BRCA1 and MGMT), apoptosis (DAPK and TMS1), drug resistance, detoxification, differentiation, angiogenesis, and metastasis.

1.1. Principal pathways affected by DNA methylation

1.1.1. Cell cycle

Hypermethylation of the cell cycle inhibitor p16INK41, very common in many tumors, enables cancer cells to escape senescence and begin to proliferate.

1.1.2. DNA repair

DNA methylation plays a major role in many repair pathways. The consequences of aberrant DNA methylation of repair pathways include: (1) microsatellite instability in sporadic colorectal cancer due to the silencing of the DNA mismatch repair gene *hMLH1*; (2) mutations in K-ras; and (3) p53 caused by hypermethylation of *MGMT* gene, preventing thus the removal of groups at the 06 position of guanine, hypermethylation [5,6].

2. NSCLC AND ABERRANT DNA METHYLATION

NSCLC is a heterogeneous disease with a remarkable variation in survival time amongst individual patients treated with the same approach. The platinum-based cytotoxic doublets have resulted in a high response rate and prolonged survival at 1 year. A large randomized clinical trial demonstrates the equivalence of four platinum-based cytotoxic doublets (with either paclitaxel, docetaxel, or gemcitabine) with a median survival of 7.9 months. This study shows a survival rate of 33% at 1 year and 11% at 2 years, among patients with good PS [7]. In a multivariate analysis, many negative survival prognostic factors in advanced NSCLC patients have been analyzed, in order to identify those that can be used to predict survival [8]. PS 1 was found to predict a worse outcome than PS 0, while patients with PS 2 had a high rate of serious adverse events and a very short survival [7,8].

A more recently randomized study of 1,218 patients reported a survival of 11 months in stage IIIB–IV patients [9]. However, there are no clinical parameters to understand the difference in survival amongst patients with advanced disease [8]. A better understanding of the alteration of gene functions in cancer could provide a breakthrough to help explain these survival variations.

In lung cancer, investigation has demonstrated that more than 40 genes seem to be altered by these epigenetic events [1].

3. DETECTING ABERRANT PATTERNS OF METHYLATION IN CIRCULATING DNA

Free DNA can be detected in different body fluids, e.g., urine, synovial fluid, pancreatic duct secretions, sputum, and serum/plasma. This DNA circulating in serum/plasma is found in small amounts in healthy controls, but in cancer patients higher concentrations of DNA are present. The presence of nucleic acids in plasma or serum of cancer patients has been recognized since the 1970s [10], when Leon et al. reported the presence of DNA in the serum of patients with diverse neoplastic diseases. Tumor cells released small quantities of doubles strand of DNA fragments into the circulation, with higher levels in patients with metastases [11]. In 1989, Stroun et al. [12] using a technique based on decreased strand stability of cancer cell DNA, recognized that part of the plasma DNA had the origin in cancer cells. After 5 years, Sorensoen et al. [13] reported the detection of tumor-derived mutated K-ras DNA sequences in human blood from pancreatic cancer patients. In many tumors, like head and neck [14], small-cell lung [15], and NSCLCs [16,17], the same microsatellite alterations detected in tumor were also found in plasma or serum DNA.

However, one interesting conclusion from most of these studies is the fact that it is possible to find alterations in serum/plasma in the vast majority of cancer patients. This finding opened a lot of possibilities for the use of this serum/plasma as a tool for the detection of potential molecular markers to avoid invasive tools in order to obtain tumoral tissue [18].

4. METHODOLOGY: MSP ANALYSIS

On the basis of this data, using a sensitive MSP technique for abnormal promoter hypermethylation was possible to detect an aberrant promoter hypermethylation of tumor suppressor genes in serum from NSCLC [18].

Methylation-specific PCR (MSP) procedure uses bisulphite-modified genomic DNA. Bisulphite modification converts cytosine to uracil at an efficiency $\geq 99\%$; but if the cytosine is methylated, it is not converted. The DNA template can then be amplified for specific genes that could be methylated through the design of methylation-specific primers. Therefore, with the MSP assay small amounts of DNA template (~50–200 ng) can be used to detect gene-specific promoter hypermethylation in DNA recovered from frozen or paraffin-embedded fixed tumors as well as biological fluids such as sputum and plasma [19]. Because primers are designed in order to recognize only methylated or unmethylated alleles, contaminating normal tissue does not interfere with the ability to detect methylation. Moreover, unlike mutation screens that survey numerous exons to detect gene dysfunction (e.g., in *TP53*), the MSP approach uses one primer set to assay a common genomic region where the detection of methylation is associated with loss of gene function. The MSP method is a sensitive assay (1 in 10^4) for detecting methylation in the presence of contaminating normal tissue or cells. The MSP assay can also be conducted directly on tissue sections (in situ MSP) to identify the clonality of the gene silencing in tumors and premalignant lesions [20]. Recent studies demonstrated the sensitivity of the MSP procedure to detect one methylated allele in ~50,000 unmethylated alleles by incorporating a nested PCR approach [21]. In this analysis, the use of nested MSP assay first amplifies the CpG island for the gene being evaluated without preference for methylated or unmethylated alleles, and then, a portion of the stage 1 PCR product is used in a second round PCR with primers specific to methylated or unmethylated alleles. This approach improves the sensitivity of MSP for examining biological fluids such as sputum, which is highly contaminated with normal cells, or formalin-fixed tissues, where usually the nucleic acids might be highly degraded.

5. GENES RELATED WITH CHEMOSENSITIVITY

14-3-3σ gene: 14.3.3σ (stratifin) is a member of the large family of 14.3.3 proteins. This protein interacts in cell cycle progression, apoptosis, and mitogenic signaling, by binding and sequestrating phosphorylated proteins [22,23]. Multiple effectors are involved in the checkpoints that permit the transitions between cell phases. 14.3.3σ promotes cell cycle arrest in G_2 following DNA damage. Its expression is induced by p53 [23–25]. This function has been analyzed in the human colorectal cell line, HCT 116, which has intact p53 and 14.3.3σ [22–25]. After irradiation of HCT 116 cells, 14.3.3σ sequestered Cdc2/cyclin B1 complexes in the cytoplasm, thus arresting cells in G2 and preventing them from initiating mitosis before repair to their damage DNA [25] (Figure 17.1A).

Colon carcinoma cells lacking 14.3.3σ treating with adriamycin can still initiate, but not maintain, G_2 arrest, leading to mitotic catastrophe and cell death [25] (Figure 17.1B). Based on the data provided, the expression of 14.3.3σ is reduced by *p53* gene inactivation and by silencing of *14.3.3σ* gene by methylation of CpG island [3,26].

Many studies support the idea that hypermethylation at the 14.3.3σ is a consistent molecular alteration in breast cancer that is involved in gene silencing [27]. Moreover by proteomic analysis, 14.3.3σ was undetectable in breast cancer samples [28]. Similar effects of the 14.3.3σ hypermethylation have been supported in many tumors, including lung [29], gastric [30], ovarian [31], prostate [32], and hepatocellular [33] carcinomas.

14.3.3σ shows to be methylated in 43% of 60 gastric cancers. In gastric cell lines, either with wild type or mutated p53, the sensibility of them to Adriamycin was conserved. However, 14.3.3σ methylation negative cell lines, either with wild type or mutated p53, were resistant [30]. The cell cycle checkpoint plays an important role in maintaining the integrity of the cells. So, defects in it may contribute to chemosensitivity. On the basis of these data, we reasoned that patients with 14.3.3σ methylation positive tumors would derive greater benefit from cisplatin-based chemotherapy.

CHFR: Another important gene that is affected by promoter hypermethylation is *CHFR*. CHFR is a checkpoint gene that coordinates an early mitotic phase by delaying chromosome condensation in response to microtubules poisons [34]. It contains a RING finger domain that is essential for the cell cycle checkpoint, playing a role in the ubiquitination of many proteins [35,36] and also contains an FHA domain that recognizes phosphorylated proteins [37].

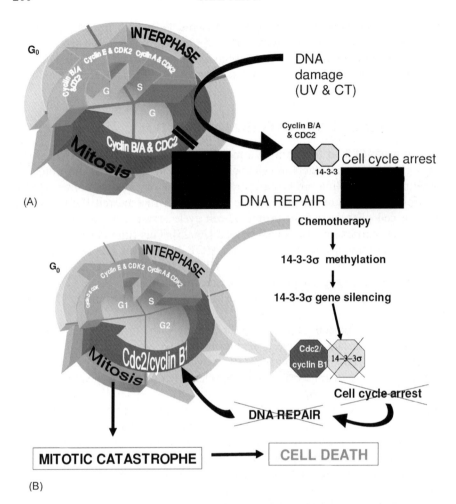

Figure 17.1. Transition through the cell cycle involves a series of precise events that ensure order (DNA replication, spindle assembly, nuclear division, and cytokinesis) and directionality. (A) illustrates the function of 14–3-3σ in G2 cell cycle checkpoint control in response to DNA damage. Following ionizing irradiation or adriamycin, 14–3-3σ sequesters Cdc2/cyclin B1 complexes in the cytoplasm, arresting the cells in G2. (B) illustrates that following chemotherapy, 14–3-3σ methylation-positive tumor cells cannot maintain G2 arrest, leading to mitotic catastrophe and cell death.

The deletion of this domain showed an attenuation in the CHFR function [35]. The inactivation of CHFR can occur by promoter hypermethylation or, less common, by missense mutation. The overall frequency of *CHFR* is quite high, between 15 and 50% in various tumor types [38].

In spite of the importance of this gene in human cancer, its molecular function is poorly understood. *CHFR*, as a mitotic stress checkpoint gene, regulates a prophase delay in cells exposed to agents that disrupt microtubules, such as nocodazole and taxol. A recent study shows that *CHFR* delays cell cycle at early prophase after exposure to these agents. These cells arrested in the early prophase do not show signs of chromosome condensation. This event involves the inhibition of the Cyclin B1 in the nucleus by *CHFR* [38].

FANCF: Another important event where the epigenetic silencing of the gene leads to sensitivity to drugs in ovarian cancer cells is FANCF. This gene is one of seven members of the Fanconi Anemia genes whose protein products were found to interact with proteins involved in DNA repair pathways. FANC interacts with BRCA1 through ubiquination of FANCD2. Defects in Fanconi Anemia BRCA (FA-BRCA) pathways are associated with increased sensitivity to DNA damaging agents, such as cisplatin [39]. The hypermethylation of FANCF in ovarian cancer cells lines contributes to selective sensitivity to platinum salts and has been observed in about 20% of primary ovarian cancers [40].

6. METHYLATION PATTERNS AND CHEMOSENSITIVITY IN NSCLC: CLINICAL APPLICATION

With this example, we want to illustrate the relationship between aberrant patterns of DNA methylation in circulating serum DNA and chemosensitivity in patients.

This study was a multicenter prospective trial where we investigated whether 14.3.3σ methylation in pretreatment serum DNA could predict survival in advanced NSCLC cancer patients treated with cisplatin plus gemcitabine.

From a total of 115 patients enrolled, the median follow up of patients still alive at the time of analysis was 17 months (range 1–30.7) [41].

For the analysis of circulating DNA, we used 10 mL of peripheral blood that were collected in clot activator tubes and the serum was separated from cells by centrifugation. DNA was extracted from serum by using commercial kit and resuspended in a final volume of 50 μl. Methylation-specific polymerase (MSP) chain reaction was performed with primers specific for either methyalted or the modified unmethylated DNA spanning the regions between CpG dinucleotides 3 and 9 within *14.3.3σ* gene. We obtained representative results of the MSP chain reaction analysis that are illustrated in the following Figure 17.2.

Table 17.1 shows the clinicopathologic characteristics of the 115 patients according to methylation status.

Paired tumor and serum DNA from an independent group of 28 surgical resected NSCLC patients was used in order to validate the results obtained in circulating serum DNA; seven of them were methylated positive in both tumor and serum and the remaining 21 were methylation-negative (Figure 17.2B).

In this study we have observed that overall time to progression for all 115 patients was 6.9 months (95% CI, 5.3–8.5). Median survival for all 115 patients was 10.9 months (Figure 17.3).

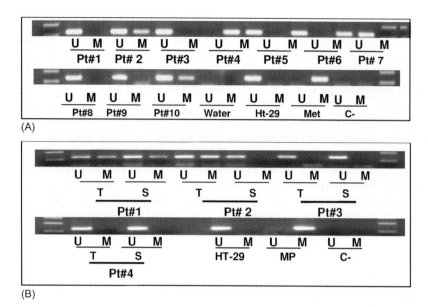

Figure 17.2. Methylation of 14-3-3σ in serum DNA, M: methylated; U, unmethylated; HT29; human colorectal cell line used as unmethylated control; MP, in vitro Sss1 bisulfite-modified placental DNA used as methylated control, C— no template control. (A) Serum DNA obtained from Gem/Cis Patients. (B) Tumoral and paired serum DNA used as validation tool. MP: placental DNA treated in vitro with Sss I methyltransferase (New England Biolabs, Beverly Hills, MA); HT-29: colorectal cell line HT-29 (ATCC, Rockville, MD); C– and H_2O: negative control.

Correlation between time to progression and 14.3.3σ methylation status was 6.3 months (95% CI, 4.5–8.2) for the methylation-negative group and 8 months for the methylation-negative groups ($P = 0.0027$ by the two-side log rack test) (Figure 17.4).

The correlation between survival and 14.3.3σ methylation status was 9.8 months (95% CI, 7.3–12.5) for the methylation-negative group, compared with 15.1 months (95% CI, 9.7–20.6) for the methylation-positive group ($P = 0.004$ by the two-side log rack test) (Figure 17.5).

Data show that only 14.3.3σ methylation status was identified to be an independent prognostic factor for time to progression.

Kaplan-Meier curves for survival of responders according to 14.3.3σ methylation status [42,43] showed that median survival for 22 to 14.3.3σ methylation-positive responders has not been reached, while for 29 14.3.3σ methylation-negative responders, it was 11.3 months (95% CI, 9.0–13.5) ($P = 0.001$ by the two-sided log rank test). The estimated survival rate at 18 months is 64% (95% CI, 44–94%) for the methylation-positive responders and 21% (95% CI, 9–47%) ($P = 0.017$ by the two-sided log rank test) for the methylation-negative responders (Figure 17.6).

Table 17.1. Patient characteristics for all patients by 14-3-3σ methylation status (N = 115)

	Total	14-3-3σ Methylation-negative	14-3-3σ Methylation-positive
No. Patients	115	76	39
Age	62	63	61
Median range	31–81	40–81	31–78
Sex			
Male	108 (93.95%)	70 (92.1%)	38 (97.4%)
Female	7 (6.1%)	6 (7.9%)	1 (2.6%)
Smoking status			
Smoker	99 (86.1%)	64 (84.2%)	35 (89.7%)
Non-smoker	16 (13.9%)	12 (15.8%)	4 (10.3%)
ECOG Performance Status			
0	32 (27.8%)	21 (27.6%)	11 (28.2%)
1	83 (72.2%)	55 (72.4%)	28 (71.8%)
Histology			
Adenocarcinoma	51 (44.7%)	38 (50.7%)	13 (33.3%)
Squamous cell carcinoma	42 (36.8%)	23 (30.7%)	19 (48.7%)
Large cell carcinoma	21 (18.4%)	14 (18.7%)	7 (17.9%)
Pleural effusion			
Yes	25 (21.7%)	16 (21.1%)	9 (23.1%)
No	90 (78.3%)	60 (78.9%)	30 (76.9%)
Prior surgery			
Yes	8 (7%)	5 (6.6%)	3 (7.7%)
No	107 (93%)	71 (93.4%)	36 (92.3%)
Response			
Complete response	2 (1.8%)	2 (2.8%)	0
Parcial response	49 (44.5%)	27 (38%)	22 (56.4%)
Stable disease	26 (23.6%)	21 (29.6%)	5 (12.8%)
Progressive disease	33 (30%)	21 (29.6%)	12 (30.8%)

Methylation-negative responders had four times greater risk of death than the methylation-positive responders (hazards ratio = 3.95 [95% CI, 1.57–9.94; P = 0.004] by the Cox model).

7. DISCUSSION

Chemotherapy resistance is an important impediment in the success of tumor treatment. In general, the mechanisms by which this process is carried out are difficult to understand as they involved many cellular pathways: (1) p53 and K-ras mutations; (2) aberrant expression of genes that prevents apoptosis [44]; (3) methylation-dependent gene silencing [45]; and (4) excessive activation of mitogenic pathways by mutations in the epidermal growth factor receptor [46,47].

In this context, molecular changes associated with gene silencing may serve as marker for risk assessment, diagnosis, and prognosis [3]. Studies of many genes methylated, involved in cell checkpoints, have reported the importance of them

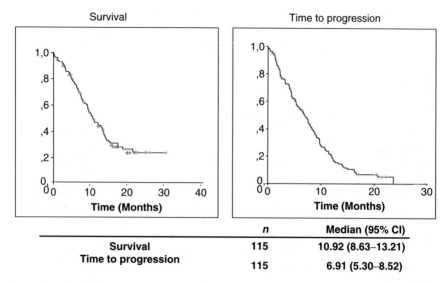

	n	Median (95% CI)
Survival	115	10.92 (8.63–13.21)
Time to progression	115	6.91 (5.30–8.52)

Figure 17.3. Survival and time to progression for 115 patients treated with Gem/Cisp.

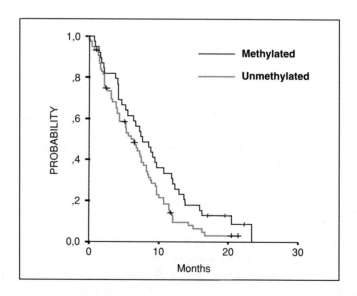

	N(%)	Media (95% CI)	*p*-value
Methylated	39(33,91)	8.03(5.3–10.7)	0.027
Unmethylated	76(66,09)	6.35(4.5–8.2)	

Figure 17.4. Kaplan-Meier probability of time to progression according to 143-3σ methylation status.

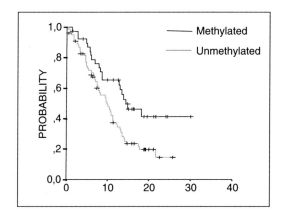

	N (%)	Media (95% CI)	p-value
Methylated	39(33,91)	15.13 (9.7–20.6)	0.0047
Unmethylated	76(66,09)	9.87 (7.3–12.5)	

Figure 17.5. Kaplan-Meier probability of survival according to 143-3σ methylation status.

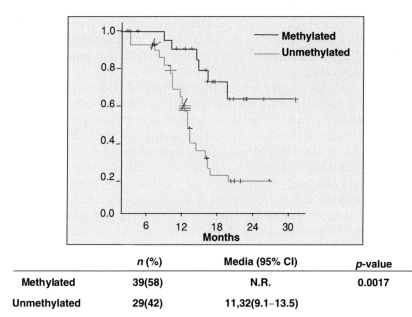

	n (%)	Media (95% CI)	p-value
Methylated	39(58)	N.R.	0.0017
Unmethylated	29(42)	11,32(9.1–13.5)	

Figure 17.6. Kaplan-Meier probability of survival for responders with the estimated survival rates and 95% CI.

as predicting factors of response to chemotherapeutic agents, e.g., FANCF, that confers sensitivity in ovarian cancer cell lines [40] and CHFR, that improves response to microtubule inhibitors [34]. However, none of the various cell cycle checkpoints markers has been tested in clinical studies. The deregulation of 14-3-3σ expression by methylation has been observed in a wide variety of human cancers [26,27,29–33]. Lack of 14-3-3σ in colorectal cancer cells was shown to sensitize them to chemotherapy-induced apoptosis [24] and 14-3-3σ expression was upregulated in chemoresistant pancreatic adenocarcinoma cells [48]. Besides its G_2/M checkpoint function, 14-3-3σ has an antiapoptotic role. 14-3-3σ sequesters phosphorylated BAD (proapoptotic protein) in the cytosol, preventing the union between phosphorylated BAD and BCL2 (antiapoptotic protein). Then, the complex 14-3-3σ/BAD is sequestered from the mitochondrial localized BCL-xL, inhibiting BAD-induced apoptosis. In addition, data suggest that BRCA1, which is a differential regulator of chemotherapy-induced apoptosis, seems to induce 14-3-3σ expression after DNA damage in mouse embryonic stem cells. BRCA1 seems to act synergistically with p53 to active 14-3-3σ expression [49]. Decreased BRCA1 mRNA expression in human breast cancer cell line led to greater sensitivity to cisplatin, but a greater resistance to the microtubule-interfering agents paclitaxel and vincristine [50]. We have observed that locally advanced NSCLC patients with the lowest BRCA1 mRNA expression obtain the maximum benefit from neoadjuvant gemcitabine plus cisplatin chemotherapy, whereas those with the highest levels had the poorest survival [51].

Since double DNA in the form of chromatin fragments is released in the blood serum of cancer patients [11] and DNA methylation of the serum is a biomarker [18,52], we have investigated 14-3-3σ methylation in the serum DNA of advance NSCLC patients treated with first-line cisplatin plus gemcitabine.

We have shown that:
- 14-3-3σ is methylated in the sera of the one third of NSCLC patients and is related to significantly better median survival for these patients overall.
- 14-3-3σ had even greater influence on survival in responder groups.
- The risk of death for the 14-3-3σ methylation-negative responders was almost five times that of the methylation-positive responders.
- This study sheds light on the influence of methylated 14-3-3σ in serum, which can identify a subgroup of patients with a significantly better survival and help to explain the striking differences in survival among NSCLC patients [8].

Translational research studies in advance NSCLC are limited by the scarcity of availably tumor biopsy tissue. The 14-3-3σ serum assay circumvents the need for the tumor tissue and offers a novel and accurate method to select patients for cisplatin-based chemotherapy.

REFERENCES

1. Das, P.M. and Singal, R. (2004) DNA methylation and cancer. *J Clin Oncol* 22: 4632–4642.
2. Baylin, S.B. and Herman, J.G. (2000) DNA hypermethylation in tumorigenesis: epigenetics joins genetics. *Trends Genet* 16: 168–174.

3. Herman, J.G. and Baylin, S.B. (2003) Gene silencing in cancer in association with promoter hypermethylation. *N Engl J Med* 349: 2042–2054.
4. Baylin, S.B., Esteller, M., Rountree, M.R., Bachman, K.E., Schuebel, K. and Herman, J.G. (2001) Aberrant patterns of DNA methylation, chromatin formation and gene expression in cancer. *Hum Mol Genet* 10: 687–692.
5. Esteller, M., Garcia-Foncillas, J., Andion, E., Goodman, S.N., Hidalgo, O.F., Vanaclocha, V., Baylin, S.B. and Herman, J.G. (2000) Inactivation of the DNA-repair gene MGMT and the clinical response of gliomas to alkylating agents. *N Engl J Med* 343: 1350–1354.
6. Balana, C., Ramirez, J.L., Taron, M., Roussos, Y., Ariza, A., Ballester, R., Sarries, C., Mendez, P., Sanchez, J.J. and Rosell, R. (2003) O6-methyl-guanine-DNA methyltransferase methylation in serum and tumor DNA predicts response to 1,3-bis(2-chloroethyl)-1-nitrosourea but not to temozolamide plus cisplatin in glioblastoma multiforme. *Clin Cancer Res* 9: 1461–1468.
7. Schiller, J.H., Harrington, D., Belani, C.P., Langer, C., Sandler, A., Krook, J., Zhu, J. and Johnson, D.H. (2002) Comparison of four chemotherapy regimens for advanced non-small-cell lung cancer. *N Engl J Med* 346: 92–98.
8. Hoang, T., Xu, R., Schiller, J.H., Bonomi, P. and Johnson, D.H. (2005) Clinical model to predict survival in chemonaive patients with advanced non-small-cell lung cancer treated with third-generation chemotherapy regimens based on eastern cooperative oncology group data. *J Clin Oncol* 23: 175–183.
9. Fossella, F., Pereira, J.R., von Pawel, J., Pluzanska, A., Gorbounova, V., Kaukel, E., Mattson, K.V., Ramlau, R., Szczesna, A., Fidias, P., Millward, M. and Belani, C.P. (2003) Randomized, multinational, phase III study of docetaxel plus platinum combinations versus vinorelbine plus cisplatin for advanced non-small-cell lung cancer: the TAX 326 study group. *J Clin Oncol* 21: 3016–3024.
10. Shapiro, B., Chakrabarty, M., Cohn, E.M. and Leon, S.A. (1983) Determination of circulating DNA levels in patients with benign or malignant gastrointestinal disease. *Cancer* 51: 2116–2120.
11. Leon, S.A., Shapiro, B., Sklaroff, D.M. and Yaros, M.J. (1977) Free DNA in the serum of cancer patients and the effect of therapy. *Cancer Res* 37: 646–650.
12. Stroun, M., Anker, P., Maurice, P., Lyautey, J., Lederrey, C. and Beljanski, M. (1989) Neoplastic characteristics of the DNA found in the plasma of cancer patients. *Oncology* 46: 318–322.
13. Sorenson, G.D., Pribish, D.M., Valone, F.H., Memoli, V.A., Bzik, D.J. and Yao, S.L. (1994) Soluble normal and mutated DNA sequences from single-copy genes in human blood. *Cancer Epidemiol Biomarkers Prev* 3: 67–71.
14. Nawroz, H., Koch, W., Anker, P., Stroun, M. and Sidransky, D. (1996) Microsatellite alterations in serum DNA of head and neck cancer patients. *Nat Med* 2: 1035–1037.
15. Chen, X.Q., Stroun, M., Magnenat, J.L., Nicod, L.P., Kurt, A.M., Lyautey, J., Lederrey, C. and Anker, P. (1996) Microsatellite alterations in plasma DNA of small cell lung cancer patients. *Nat Med* 2: 1033–1035.
16. Sozzi, G., Musso, K., Ratcliffe, C., Goldstraw, P., Pierotti, M.A. and Pastorino, U. (1999) Detection of microsatellite alterations in plasma DNA of non-small cell lung cancer patients: a prospect for early diagnosis. *Clin Cancer Res* 5: 2689–2692.
17. Sanchez-Cespedes, M., Monzo, M., Rosell, R., Pifarre, A., Calvo, R., Lopez-Cabrerizo, M.P. and Astudillo, J. (1998) Detection of chromosome 3p alterations in serum DNA of non-small-cell lung cancer patients. *Ann Oncol* 9: 113–116.
18. Esteller, M., Sanchez-Cespedes, M., Rosell, R., Sidransky, D., Baylin, S.B. and Herman, J.G. (1999) Detection of aberrant promoter hypermethylation of tumor suppressor genes in serum DNA from non-small cell lung cancer patients. *Cancer Res* 59: 67–70.
19. Herman, J.G., Graff, J.R., Myohanen, S., Nelkin, B.D. and Baylin, S.B. (1996) Methylation-specific PCR: a novel PCR assay for methylation status of CpG islands. *Proc Natl Acad Sci USA* 93: 9821–9826.
20. Nuovo, G.J., Plaia, T.W., Belinsky, S.A., Baylin, S.B. and Herman, J.G. (1999) In situ detection of the hypermethylation-induced inactivation of the p16 gene as an early event in oncogenesis. *Proc Natl Acad Sci USA* 96: 12754–12759.

21. Palmisano, W.A., Divine, K.K., Saccomanno, G., Gilliland, F.D., Baylin, S.B., Herman, J.G. and Belinsky, S.A. (2000) Predicting lung cancer by detecting aberrant promoter methylation in sputum. *Cancer Res* 60: 5954–5958.
22. Hermeking, H., Lengauer, C., Polyak, K., He, T.C., Zhang, L., Thiagalingam, S., Kinzler, K.W. and Vogelstein, B. (1997) 14-3-3 sigma is a p53-regulated inhibitor of G2/M progression. *Mol Cell* 1: 3–11.
23. Hermeking, H. (2003) The 14-3-3 cancer connection. *Nat Rev Cancer* 3: 931–943.
24. Chan, T.A., Hwang, P.M., Hermeking, H., Kinzler, K.W. and Vogelstein, B. (2000) Cooperative effects of genes controlling the G(2)/M checkpoint. *Genes Dev* 14: 1584–1588.
25. Chan, T.A., Hermeking, H., Lengauer, C., Kinzler, K.W. and Vogelstein, B. (1999) 14-3-3 sigma is required to prevent mitotic catastrophe after DNA damage. *Nature* 401: 616–620.
26. Ferguson, A.T., Evron, E., Umbricht, C.B., Pandita, T.K., Chan, T.A., Hermeking, H., Marks, J.R., Lambers, A.R., Futreal, P.A., Stampfer, M.R. and Sukumar, S. (2000) High frequency of hypermethylation at the 14-3-3 sigma locus leads to gene silencing in breast cancer. *Proc Natl Acad Sci USA* 97: 6049–6054.
27. Umbricht, C.B., Evron, E., Gabrielson, E., Ferguson, A., Marks, J. and Sukumar, S. (2001) Hypermethylation of 14-3-3 sigma (stratifin) is an early event in breast cancer. *Oncogene* 20: 3348–3353.
28. Vercoutter-Edouart, A.S., Lemoine, J., Le Bourhis, X., Louis, H., Boilly, B., Nurcombe, V., Revillion, F., Peyrat, J.P. and Hondermarck, H. (2001) Proteomic analysis reveals that 14-3-3 sigma is down-regulated in human breast cancer cells. *Cancer Res* 61: 76–80.
29. Osada, H., Tatematsu, Y., Yatabe, Y., Nakagawa, T., Konishi, H., Harano, T., Tezel, E., Takada, M. and Takahashi, T. (2002) Frequent and histological type-specific inactivation of 14-3-3 sigma in human lung cancers. *Oncogene* 21: 2418–2424.
30. Suzuki, H., Itoh, F., Toyota, M., Kikuchi, T., Kakiuchi, H. and Imai, K. (2000) Inactivation of the 14-3-3 sigma gene is associated with 5' CpG island hypermethylation in human cancers. *Cancer Res* 60: 4353–4357.
31. Akahira, J., Sugihashi, Y., Suzuki, T., Ito, K., Niikura, H., Moriya, T., Nitta, M., Okamura, H., Inoue, S., Sasano, H., Okamura, K. and Yaegashi, N. (2004) Decreased expression of 14-3-3 sigma is associated with advanced disease in human epithelial ovarian cancer: its correlation with aberrant DNA methylation. *Clin Cancer Res* 10: 2687–2693.
32. Urano, T., Takahashi, S., Suzuki, T., Fujimura, T., Fujita, M., Kumagai, J., Horie-Inoue, K., Sasano, H., Kitamura, T., Ouchi, Y. and Inoue, S. (2004) 14-3-3 sigma is down-regulated in human prostate cancer. *Biochem Biophys Res Commun* 319: 795–800.
33. Iwata, N., Yamamoto, H., Sasaki, S., Itoh, F., Suzuki, H., Kikuchi, T., Kaneto, H., Iku, S., Ozeki, I., Karino, Y., Satoh, T., Toyota, J., Satoh, M., Endo, T. and Imai, K. (2000) Frequent hypermethylation of CpG islands and loss of expression of the 14-3-3 sigma gene in human hepatocellular carcinoma. *Oncogene* 19: 5298–5302.
34. Satoh, A., Toyota, M., Itoh, F., Sasaki, Y., Suzuki, H., Ogi, K., Kikuchi, T., Mita, H., Yamashita, T., Kojima, T., Kusano, M., Fujita, M., Hosokawa, M., Endo, T., Tokino, T. and Imai, K. (2003) Epigenetic inactivation of CHFR and sensitivity to microtubule inhibitors in gastric cancer. *Cancer Res* 63: 8606–8613.
35. Scolnick, D.M. and Halazonetis, T.D. (2000) CHFR defines a mitotic stress checkpoint that delays entry into metaphase. *Nature* 406: 430–435.
36. Mizuno, K., Osada, H., Konishi, H., Tatematsu, Y., Yatabe, Y., Mitsudomi, T., Fujii, Y. and Takahashi, T. (2002) Aberrant hypermethylation of the CHFR prophase checkpoint gene in human lung cancers. *Oncogene* 21: 2328–2333.
37. Li, J., Lee, G.I., Van Doren, S.R. and Walker, J.C. (2000) The FHA domain mediates phosphoprotein interactions. *J Cell Sci* 113 (Pt 23): 4143–4149.
38. Summers, M.K., Bothos, J. and Halazonetis, T.D. (2005) The CHFR mitotic checkpoint protein delays cell cylce progression by excluding Cyclin B1 from the nucleus. *Oncogene* 24: 2589–2598.

39. Olopade, O.I. and Wei, M. (2003) FANCF methylation contributes to chemoselectivity in ovarian cancer. *Cancer Cell* 3: 417–420.
40. Taniguchi, T., Tischkowitz, M., Ameziane, N., Hodgson, S.V., Mathew, C.G., Joenje, H., Mok, S.C. and D'Andrea, A.D. (2003) Disruption of the Fanconi anemia-BRCA pathway in cisplatin-sensitive ovarian tumors. *Nat Med* 9: 568–574.
41. Parmar, M.K.B. and Machin, D. (eds) (1996) *Survival analysis: apratical approach.* Wiley, Chichester, UK.
42. Mantel, N. (1966) Evaluation of survival data and two new rank order statistics arising in its consideration. *Cancer Chemother Rep* 50: 163–170.
43. Kaplan, E.L., Meier, P. (1958) *Non parametric estimation from incomplete observation.*
44. Osada, H. and Takahashi, T. (2002) Genetic alterations of multiple tumor suppressors and oncogenes in the carcinogenesis and progression of lung cancer. *Oncogene* 21: 7421–7434.
45. Shiraishi, M., Sekiguchi, A., Terry, M.J., Oates, A.J., Miyamoto, Y., Chuu, Y.H., Munakata, M. and Sekiya, T. (2002) A comprehensive catalog of CpG islands methylated in human lung adenocarcinomas for the identification of tumor suppressor genes. *Oncogene* 21: 3804–3813.
46. Sordella, R., Bell, D.W., Haber, D.A. and Settleman, J. (2004) Gefitinib-sensitizing EGFR mutations in lung cancer activate anti-apoptotic pathways. *Science* 305: 1163–1167.
47. Lynch, T.J., Bell, D.W., Sordella, R., Gurubhagavatula, S., Okimoto, R.A., Brannigan, B.W., Harris, P.L., Haserlat, S.M., Supko, J.G., Haluska, F.G., Louis, D.N., Christiani, D.C., Settleman, J. and Haber, D.A. (2004) Activating mutations in the epidermal growth factor receptor underlying responsiveness of non-small-cell lung cancer to gefitinib. *N Engl J Med* 350: 2129–2139.
48. Logsdon, C.D., Simeone, D.M., Binkley, C., Arumugam, T., Greenson, J.K., Giordano, T.J., Misek, D.E., Kuick, R. and Hanash, S. (2003) Molecular profiling of pancreatic adenocarcinoma and chronic pancreatitis identifies multiple genes differentially regulated in pancreatic cancer. *Cancer Res* 63: 2649–2657.
49. Aprelikova, O., Pace, A.J., Fang, B., Koller, B.H. and Liu, E.T. (2001) BRCA1 is a selective co-activator of 14-3-3 sigma gene transcription in mouse embryonic stem cells. *J Biol Chem* 276: 25647–25650.
50. Quinn, J.E., Kennedy, R.D., Mullan, P.B., Gilmore, P.M., Carty, M., Johnston, P.G. and Harkin, D.P. (2003) BRCA1 functions as a differential modulator of chemotherapy-induced apoptosis. *Cancer Res* 63: 6221–6228.
51. Taron, M., Rosell, R., Felip, E., Mendez, P., Souglakos, J., Ronco, M.S., Queralt, C., Majo, J., Sanchez, J.M., Sanchez, J.J. and Maestre, J. (2004) BRCA1 mRNA expression levels as an indicator of chemoresistance in lung cancer. *Hum Mol Genet* 13: 2443–2449.
52. Herman, J.G. (2004) Circulating methylated DNA. *Ann NY Acad Sci* 1022: 33–39.

CHAPTER 18
PHARMACOGENOMICS AND COLORECTAL CANCER

HEINZ-JOSEF LENZ

Professor of Medicine and Preventive Medicine, Division of Medical Oncology, University of Southern California/Norris Comprehensive Cancer Center, Keck School of Medicine, Los Angeles, CA 90033, USA

Abstract: The current methods of treating cancer patients with chemotherapeutics do not account for interpatient variability in the expression of particular target genes. This variability leads to unpredictable tumor responses and host toxicity. The approach we have taken is to determine gene expression levels in the metabolic pathways of drugs used in the treatment of gastrointestinal tumors. One of the main obstacles in the evaluation and determination of these markers has been the limitations of available technology. Many advances have been made in the development of more sophisticated techniques and the ability to perform these techniques on paraffin-embedded tumor tissue. In fact, with the identification of genetic polymorphisms, these markers may be obtained from peripheral blood specimens, thus making access to tissues a moot issue. An immediate goal is the application of this nascent technology and incorporation of these data in prospective clinical trials that would stratify patients according to their molecular profile. The ability to predict with a high degree of accuracy which patients are likely to respond to treatment and identify those who are not likely to respond will significantly influence the design of new treatment regimens with fluoropyrimidines and platinum. Tumors with high TS, TP, and DPD expression levels should be treated with such non-TS-directed anticancer drugs as irinotecan or oxaliplatin, or in combination with 5-FU. Patients with high expression of ERCC1 should be treated with nonplatinum-based regimens, whereas patients with low levels would be good candidates for cisplatin or oxaliplatin. We now understand that molecular determinants play an important role in response to 5-FU. With the development of new effective anticancer drugs such as irinotecan and oxaliplatin, it is important to gain a better understanding about the metabolism of these new active agents and mechanisms of resistance. It is essential to understand why some patients develop life-threatening toxicity and why some tumors are resistant to irinotecan or oxaliplatin. With the integration of

Affiliation: Heinz-Josef Lenz, M.D., F.A.C.P., Professor of Medicine and Preventive Medicine. USC/Norris Comprehensive Cancer Center, USC Keck School of Medicine, 141 Eastlake Aveneu, Suite 3456, Los Angeles, CA 90033, Phone: 323-865-3955, Fax: 323-865-0061, E-mail: lenz@usc.edu

novel-targeted therapies such as Erbitux and Avastin, molecular characterization and profiling will become more important for patient selection. Preliminary data suggest that germ line polymorphisms of cyclin D and gene expression levels of VEGF are associated with efficacy of Erbitux therapy.

1. INTRODUCTION

Colorectal cancer remains one of the most common causes of death from cancer [1]. The success of chemotherapeutic strategies depends on various factors. These factors include gender, age, overall performance status, and histological subtype of the tumor [2–4]. A few promising polymorphisms that may predict chemotherapy success and toxicity in colorectal cancer have been identified. All of these potential predictive polymorphisms have been evaluated retrospectively in relatively small patient numbers. The differences in drug effects, however, between different genotypes can be dramatic (e.g., DPD and UGT1A1). Therefore, it should become a standard procedure in clinical studies to include analyses of known functional genetic polymorphisms that might impact the efficacy and toxicity of the drugs used. Haplotype analyses are critical in allowing us to identify marker patterns for certain treatment combinations. Large-scale evaluation and confirmation studies will generate data from a reliable number of patients to identify polymorphisms that could be included in prospective, randomized trials to develop individualized treatment for colorectal cancer patients.

Pharmacogenomic studies have demonstrated their effectiveness in developing such strategies. Genetically determined variability of function of certain key enzymes has been shown to influence toxicity, response, and survival to chemotherapy. These enzymes may be involved in metabolism, influx, and efflux of anticancer drugs and procedures influencing cell viability such as DNA-repair mechanisms. The identification of individual genetic characteristics in pharmacogenomic analyses is gaining more importance as a growing number of therapeutic substances and strategies are being developed at a rapid pace, making alternative treatment options available. Although we are continuously generating new data that elucidates relationships between the outcome of current therapeutic strategies in colorectal cancer and interindividual genetic markers, clinical efficacy, and toxicity of a given chemotherapy in most cases is still unpredictable for the individual patient. We are only beginning to use the knowledge of the genome and newest high-throughput technology to acquire more information to develop individualized therapies. The crucial step will be the rapid transfer of new data from research laboratory into clinic. Prospective clinical trials are needed to confirm preliminary results in pharmacogenomics in the chemotherapy of colorectal cancer. This review focuses on the implications of known genetic polymorphisms effecting success and side effects of current chemotherapy in the treatment of colorectal cancer (Table 18.1 and 18.2).

Table 18.1. Germline polymorphisms and clinical significance

Protein	Polymorphisms	Function therapy	Clinical significance	
Thymidylate synthase toxicity (TS)	28 bp tandem repeat (5-UTR)	↑ TS activity	5-FU	↓ Response, ↓
Dihydropyrimidine hydrogenase (DPD)	Exon 14 skipping mutation	↓↓↓ DPD activity	5-FU	↑↑↑ Toxicity
Thiopurinmethyl-transferase (TPMT)	G238C transversion, G460A and A719G transition	↑ TPMT catabolism	↑ Toxicity, ↑ response ↑ Second primary tumors	
5,10-Methylente-trahydrofolatreduktase (MTHFR)	C677T transition	Folate imbalance	MTX, CMF ↑↑ toxicity Raltitrexed	
Glutathion S-transferase P1 (GSTP1)	A313G substitution	↓ GSTP1 activity	Oxaliplatin ↑ survival	
Xeroderma pigmentosum Gruppe D (XPD) Protein	C751A substitution	↓ DNA-repair	Oxaliplatin ↑survival, ↑ response	
Glutathion S-transferase T1 and M1 (GSTT1 and GSTM1)	Deletionspoly morphisms	Decreased activity	Alkylating ↑ response, ↑ toxicity agents	

Table 18.2. Gene expression and clinical significance

Gene	Factor	Clinical implications
5-Fluorouracil chemotherapy		
TS	mRNA expression	Prediction of response and overall survival in first-line and second therapy
DPD	mRNA expression	Prediction of response and overall survival in disseminated CRC
TP	mRNA expression	Prediction of response and overall survival in disseminated CRC
Oxaliplatin chemotherapy		
ERCC1	mRNA expression	Prediction of overall survival in second-line therapy

2. PHARMACOGENETIC ASPECTS OF SURVIVAL, RESPONSE, AND TOXICITY TO FLUOROPYRIMIDINES

2.1. Thymidylate synthase

Thymidylate synthase (TS) catalyzes the intracellular conversion of deoxyuridylate to deoxythymidylate — the sole de novo source of thymidylate in the cell [5]. The active metabolite of 5-FU, 5-fluorodeoxyuridylate (5-FdUMP), binds to TS and inhibits it by forming a stable ternary complex [6]. Several studies focused on the predictive value of an earlier identified genomic polymorphism in the *TS* gene for clinical outcome to 5-FU-based chemotherapy. Horie et al. [7] described a 28 bp sequence of the 5′-region of the *TS* gene to be polymorphic with either two repeats (2R) or three (3R) repeats. Caucasians and African Americans almost exclusively possess double repeats (2R) or triple repeats (3R) for this polymorphism. Multiple repeats (4R, 5R, and 9R) have also been reported within certain Asian and African populations [8–10], but their functional significance has not yet been determined. The impact of the double repeats and triple repeats of this 28 base sequence of the gene on TS function has been intensively studied during the last few years. It was postulated that this 2R/3R polymorphism is implicated in modulating TS mRNA expression [11] and TS mRNA translational efficiency [9]. At this point it is not exactly clear whether the polymorphic site in the 5′-region alters mRNA expression, translational efficacy, or both.

First results by Horie et al. [7] from in vitro experiments revealed an association between the 3R variant of the *TS* gene and increased TS mRNA levels compared with individuals homozygous for the 2R allele. This finding was confirmed in human tissue samples by different investigators [12,13]. For example Pullarkat et al. [13] found an almost fourfold increased TS mRNA level in metastatic colorectal tumor tissue (9.42 [95% CI 5.51, 16.12]) for homozygous carriers of the 3R TS variant, when compared with individuals homozygous for the 2R variant (2.60 [95% CI 1.39, 4.87], $p = 0.004$). The heterozygotes demonstrated intermediate TS mRNA levels (5.53 [95% CI 3.68, 8.31], $p = 0.011$).

The rational to incorporate these findings into clinical analyses and protocols was provided by earlier studies by Leichman et al. [14] and others [15–17]. Leichman et al. were the first to demonstrate a significant inverse relationship between intratumoral *TS* gene expression and response to 5-FU-based chemotherapy. In a retrospective analysis of 42 colorectal tumor samples, pretreatment TS expression was determined using a RT-PCR-based protocol. The median TS expression level among the 12 patients (29%) who responded to 5-FU/LV was 1.7×10^{-3} (as TS/β-actin ratio). Nonresponding tumors showed a median TS expression level of 5.6×10^{-3} (as TS/β-actin ratio). Patients with low *TS* gene expression levels showed a superior median survival of 13.6 months, compared with 8.2 months in patients whose tumors had an increased TS-level (TS/β-actin ratio of 3.5×10^{-3} was determined as cutoff) ($p = 0.02$) [14]. Further clinical evaluation identified low TS expression as predictive for superior response to 5-FU/oxaliplatin [17], raltitrexed [15] or other fluoropyrimidine-based chemotherapy [16,18] in the treatment of colorectal cancer.

Clinical analysis of the 2R/3R TS polymorphism among chemonaive patients with advanced colorectal cancer who received 5-FU/LV chemotherapy showed that this genetic variation of the TS 5′-region segregates responders and nonresponders. Patients homozygous for the double repeats (2R) showed a significantly better response rate to 5-FU compared with those with a homozygous 3R TS genotype (50% vs. 9%, $p = 0.04$) [13]. Further retrospective analyses confirmed these results [19]. A study by Villafranca et al. demonstrated that this TS polymorphism is also associated with tumor downstaging in rectal cancers. T-downstaging ($T_{Pretreatment}$ vs. $T_{Posttreatment}$) was correlated with the TS variants among 65 rectal cancer patients who received preoperative 5-FU-based chemoradiation [20]. Patients with a homozygous 3R/3R genotype showed a lower probability of downstaging compared with the 2R/3R + 2R/2R group

(22% vs. 60%, $p = 0.036$) [20]. This information may be crucial for determination of the optimal treatment schedule in those patients. In addition, the TS 3R/3R genotype was more often associated with local recurrence of locally advanced rectal cancers [21]. In addition to its predictive value to fluoropyrimidine chemotherapy the 2R/3R 5′-TS polymorphism results from a retrospective analysis showed a possible association with toxicity of this treatment. Patients harboring the 3R/3R TS genotype showed less toxicity compared with patients possessing a 2R variant [13]. The authors stated that the high TS level (3R/3R) leads to less efficient TS inhibition and subsequently decreased cell death rates also in the normal tissue, resulting in less toxicity.

A prospective study by Etienne et al. [22] among 103 colorectal cancer patients who received 5-FU-based chemotherapy identified the 2R/2R TS genotype as most favorable for survival (median survival was 19 months for 2R/2R, 10 months for 2R/3R, and 14 months for 3R/3R, $p = 0.025$). Interestingly, the group with TS 2R/3R genotype demonstrated the shortest survival. Iacopetta et al. [23] demonstrated in an analysis of 221 Dukes' C colorectal cancer patients who received 5-FU that patients possessing two 3R alleles derive no survival benefit from chemotherapy in contrast to patients who harbor at least one 2R allele (RR = 0.52, 95% CI 0.33–0.82, $p = 0.005$). However, detailed analysis showed that a sizeable fraction of patients with a 3R/3R genotype gained some short-term benefit form 5-FU chemotherapy [23]. Conversely, a recent study among 135 Japanese patients with colorectal cancer, who received 5-FU-based oral adjuvant chemotherapy, failed to confirm a prognostic value of this 5′TS polymorphism [24]. Furthermore, up to 25% of patients homozygous for the 3R/3R TS genotype showed low TS mRNA expression in a different study [13]. These observations raise the question whether other modulators impact TS expression and hereby clinical outcome to 5-FU-based chemotherapy. Identification of further modifiers of TS expression may shed additional light into the potential role of TS polymorphisms as a prognosticator in the adjuvant setting. A recent analysis by Mandola et al. [25] furthered our understanding of the regulation of TS function. The authors identified a new G→C SNP within the 28 bp repeat polymorphism, that solely appears within the 3R variant of the *TS* gene and disrupts a binding site for the transcriptions factor USF (Figure 18.1).

Further analysis revealed that TS mRNA of a 3RC variant is significantly decreased when compared with a 3RG variant and is similar to 2R TS variant (Figure 18.1). Two groups of TS expression have been postulated based on these results: high TS expression (3RG/3RG, 3RG/3RC, and 2R/3RG) and low TS expression (3RC/3RC, 2R/3RC, and 2R/2R). The frequency of these polymorphisms is significantly different between various ethnic groups (Figure 18.2). These results may at least in part explain why some patients with 3R/3R TS genotype show low TS expression and good response to 5-FU chemotherapy. Although data linking these TS polymorphisms to differences in TS mRNA expression appear convincing, the possible impact of these genetic variations on additional mechanism has to be considered. In contrast to aforementioned

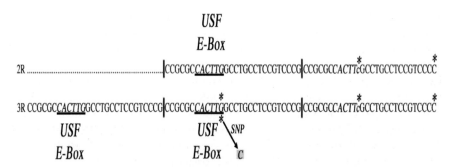

Figure 18.1. Scheme of repeat polymorphism in the 5-UTR region of TS and binding site of USF.

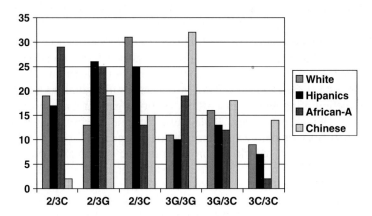

Figure 18.2. Frequency of TS polymorphisms in different ethnic groups.

results Kawakami et al. [9,26,27] demonstrated that the 2R/3R polymorphism as well as the new SNP within the 3R repeat maybe related to modifications of translational efficacy of TS.

Furthermore it is important to realize that reported studies analyzing associations between clinical outcome to 5-FU and TS polymorphisms are utilizing DNA from different cell types (tumor tissue versus white blood cells). The TS locus at the short arm of chromosome 18 has been reported to be altered in some tumors including colorectal cancer [28]. Due to loss of heterozygosity (LOH) at this locus individuals that are heterozygous 2R/3R in "normal tissue" (e.g., white blood cells) may acquire the 2R/loss or the 3R/loss genotype in the tumor. Recent analyses by Uchida et al. [29] comparing normal and tumor tissue revealed that the response rates of 2R/loss carriers to fluoropyrimidines is significant superior (8/10) when compared with individuals whose tumor harbors a 3R/loss genotype (1/7, $p = 0.029$). It has been concluded that analyses of LOH should be considered when studying the TS polymorphism [29,30].

At the moment, less information is available regarding a recently identified 6bp deletion polymorphism in the 3'-region of the *TS* gene (1494–99del). This polymorphism might alter mRNA stability, secondary structure, or expression as it has been demonstrated for alterations of the 3'-region in other genes [31]. Recent data by Mandola et al. demonstrated a significant association between the 6 bp deletion and low TS mRNA expression in colorectal tumor tissue. Patients harboring a +6bp/+6bp genotype demonstrated a more than fourfold expression of mRNA (11.4 [95% CI 6.4, 20.0]) when compared with a homozygous deletion genotype (2.7 [95% CI 1.2, 6.3], $p = 0.007$) [32]. The heterozygous genotype showed an intermediate expression level [32]. However, most recent data by McLeod et al. [33] showed, that the 6bp deletion in the 3'UTR of TS may be associated with less favorable outcome to 5-FU-based combination therapy in colorectal tumors. This is in contrast to the gene expression data by Mandola et al. [32], since most studies demonstrated an association between low TS levels and favorable clinical response [14–18]. Further analyses are needed identifying the regulators and/or mechanisms by which this deletion/insertion polymorphism alters TS expression. This may help to interpret these preliminary conflicting data.

Summarizing the recent knowledge regarding TS polymorphisms its association with clinical response to 5-FU-based chemotherapy, it appears that genetic variations of the *TS* gene may alter its expression in different ways (e.g., transcription and translation). In order to better understand the linkage and the importance of associations between TS polymorphisms, gene expression, protein expression, and LOH a comprehensive study is warranted analyzing these interactions. This may provide us with a more defined conclusion of which of these TS characteristics are finally important as predictive markers and which source tissue to use for their determination.

2.2. Dihydropyrimidine dehydrogenase and orotate phosphoribosyl transferase

Dihydropyrimidine dehydrogenase (DPD) represents the key enzyme of 5-FU metabolism. This rate-limiting enzyme of 5-FU catabolism inactivates >80% of the drug in the liver [34]. Retrospective analyses of gene expression revealed that low DPD levels might be important for superior response to fluoropyrimidine treatment [35]. It has been suggested that low levels of DPD may increase bioavailability of the drug, thereby improving response. Seventeen different mutations within the *DPD* gene have been identified as being associated with decreased activity of the enzyme [36]. Consequently, impaired 5-FU catabolism leads to accumulation of active 5-FU metabolites. This genetically determined decrease of DPD activity can be the cause of increased hematopoietic and gastrointestinal toxicity [37–39]. An allele frequency of 0.91% for a G→A substitution in the invariant GT splice donor site flanking exon 14 (IVS14 + 1G > A) was reported for Caucasians [39], and appears to be the most common known DPD variant [40]. This G→A mutation causes a lack of the exon in corresponding mRNA, thereby dramatically diminishing DPD activity [41]. Patients harboring one or two of these nonfunctional DPD-alleles have been shown to experience severe and even lethal toxicities to 5-FU-based regimens. The drastic consequences of the polymorphism for those patients require the necessity of its screening prior to 5-FU therapy.

Recently, Isshi et al. [42] demonstrated that high levels of orotate phosphorybosyl transferase (OPRT) may be associated with increased sensitivity to 5-FU-based chemotherapy. OPRT catalyzes the reduction of FUDP to the actively TS-inhibiting metabolite FdUMP. A recent study by Ichikawa et al. [43] indicated that a newly identified SNP of OPRT exon 3 may be critical in predicting toxicity to 5-FU-based chemotherapy. A 213G> substitution increases OPRT activity. In a small cohort of 52 colorectal cancer patients that received 5-FU chemotherapy those who possessed one or two OPRT A alleles experienced a higher frequency of adverse events, including diarrhea and leucopenia ($p < 0.0001$). If confirmed in a larger cohort this polymorphism of the OPRT may as well be a considerable candidate for prediction of toxicity to fluoropyrimidines.

3. PHARMACOGENETIC ASPECTS OF SURVIVAL, RESPONSE, AND TOXICITY TO IRINOTECAN

Irinotecan (CPT-11), a camptothecin derivate, acts as a topoisomerase I poison and has become a standard player in the chemotherapy of colorectal cancer, usually in combination with fluoropyrimidines or platinum compounds. Distinct differences of tolerability and response rates of CPT-11 have been reported. Genetically defined variations of drug metabolism may explain, in part, interindividual differences of drug effects. The metabolic pathways of CPT-11 are highly complex, involving numerous enzymes for activation, transport, and deactivation. Several potentially important genetic polymorphisms have been identified within these pathways.

The prodrug irinotecan has to be transformed into its active player SN-38. Tissue carboxylesterase (CE)-mediated hydrolysis appears to be essential for this activation step. Humerickhouse et al. identified the human carboxylesterase-2 (hCE-2), localized in the liver as an enzyme with high affinity to CPT-11 indicating its central role in CPT-11 activation [44]. Intestinal CE was demonstrated to be required for CPT-11 activation within the gastrointestinal tract [45]. Recently, Charasson et al. [46] showed the existence of nine allelic variation of hCE-2 constituting nine different haplotypes in addition to hCE-2 wild-type sequence. None of the variants alters hCE-2 protein structure, but one of these polymorphisms is located within the promoter of the gene, and is therefore potentially important as a modulator of *hCE-2* gene expression. The impact of these variants on hCE-2 function and/or expression is currently under investigation. The polymorphisms of hCE-2 enzymes may become critical for gastrointestinal toxicity of irinotecan, representing the main limiting factor of its use. A large study revealed high-grade gastrointestinal toxicity (grade 3 and grade 4) after CPT-11 in 31% of patients who received CPT-11 alone and in 23% of patients who received CPT-11/5-FU chemotherapy [47].

CYP3A4 and CYP3A5 generate the oxidized CPT-11 metabolites APC (7-ethyl-10[4-N-(5-amino-pentanoicacid0-1-piperidino]) carbonyloxycamptothecin) that is inactive and NPC (7-ethyl-10(4amino-1piperidino) carbonyloxycamptothecin), which can be hydrolyzed to the active SN-38 [48–50]. Several polymorphisms have been identified in the CYP3A4 and CYP3A5 [51] genes. Analyses in human liver tissues indicated that CYP3A4 was primarily responsible for APC formation, but an analyzed 392A>G polymorphism ($CYP3A4^*1B$) did not show any genotype-dependent differences of APC formation. Although the oxidization of CPT-11 by CYP enzymes represents only a minor metabolic path of fate of the drug, studies analyzing possible association between variants of these enzymes are important, because functional polymorphisms may cause imbalances between the different activation/inactivation pathways. These imbalances may contribute in a significant way to irinotecan loss or accumulation.

Most pharmacogenomic analyses of CPT-11 metabolism have been done on human UDP-glucuronyltransferase (UGT). Hepatic uridine diphosphate glucosyl transferase (UGT) 1A1 enzyme is the key deactivator of SN-38. UGT1A1 inactivates SN-38 by glucuronidation [52]. UGT activity appears to be significantly altered by genetic polymorphisms. Reduction of UGT1A1 activity in vitro and in vivo has been shown to be associated with a promoter polymorphism of UGT1A1, consisting of a two base pair insertion $(TA)_7TAA$ ($UGT1A1^*28$) [53–60]. A case–control study revealed that patients possessing at least one $UGT1A1^*28$ allele are seven times more likely to experience severe toxicity after CPT-11 administration if compared with individuals homozygous for the UGT1A1 wild type (OR 7.2, 95% CI 2.5, 22.3, $p < 0.001$) [61]. The same group studied a newly identified genetic variant (3263T > G) in the phenobarbital-responsive enhancer module of the UGT1A1 promoter, that had been shown to be associated with decreased transcriptional activity of UGT1A1, in 119 CPT-11-treated patients. The homozygous group for the 3263T > G polymorphism showed an association with severe toxicity, but this effect disappeared after including $UGT1A1^*28$ genotypes in the analysis. However, patients harboring both unfavorable UGT genotypes ($UGT1A1^*28$ and homozygous 3263T > G) were 6.2 times as likely to encounter severe toxicity if compared with the wild type (OR = 6.2, 95% CI 2.2–17.9) [62].

Besides hepatic UGT1A1, it has been shown that the additional hepatic enzyme UGT1A9 and the extra-hepatic form UGT1A7 are deeply involved in the glucuronidation-mediated detoxification of SN-38 [63]. First results of a functional variant of UGT1A9 have been presented. A G > A variation at position 766 in exon 1 of the $UGT1A9$ gene leads to the amino acid substitution D256N. In vitro studies demonstrated a dramatically decreased SN-38 glucuronidation efficacy of the D256N variant of less than 5% if compared with the wild-type UGT1A9 [64]. This highly functional polymorphism should be incorporated into analyses evaluating associations between UGT genotypes and toxicity and/or response to CPT-11.

Most recently Mathijssen et al. reported that a polymorphism in the MRD1 P-glycoprotein (*ABCB1*) gene might be associated with differences of exposure to irinotecan and its active metabolite SN-38. Among 65 patients, who received CPT-11 infusion, those homozygous for the T allele of ABCB1 1236C > T were at an increased exposure for irinotecan and SN-38 [65]. The variant is located at codon 411 of the P-glycoprotein. These data suggest a possible functionality of this polymorphism, although the exact mechanism has yet to be defined. If confirmed in a larger study this ABCB1 polymorphism might be of assistance in dose optimization of irinotecan chemotherapy.

3.1. Frequency of UGT1A1 polymorphism in Hispanics

We assayed the UGT1A1 genotype in 67 Hispanic patients diagnosed with colon polyps. An approximate 100 bp region of the *UGT1A1* gene containing the polymorphic site was PCR amplified using a P-33 end-labeled forward primer and an unlabeled reverse primer. PCR products were separated on sequencing gels and autoradiographed, and products were sized by comparison to a known 6/7 heterozygote. Among these 67 Hispanic subjects, the frequency of the Gilbert's allele (35%) was similar to the frequencies observed in Scottish and Dutch populations (36% and 40%, respectively). However, in addition to the (TA) 6 and (TA) 7 alleles, we observed shorter alleles in 5 (7%) of subjects. These short alleles were of two sizes: (1) 2 bp shorter than the (TA) 6 allele (3 subjects); and (2) 1 bp shorter (2 subjects). Direct sequencing of the forward and reverse strands of these PCR products was carried out to determine the nature of these newly identified alleles. Although heterozygosity made the sequence difficult to read, three subjects clearly carried (TA) 5 alleles. The sequence of the 2 other short alleles could not be confidently determined. We plan to clone these newly identified alleles to obtain a source of pure DNA for sequencing and for functional studies. Unpublished data presented at ASCO 2003 by Dr. McLeod indicate that UGT1A1 may predict efficacy of CPT-11-based chemotherapy. We propose to identify and correlate the various UGT1A1 polymorphisms with irinotecan efficacy and toxicity.

In summary, we state that most valid analyses of irinotecan toxicity and efficacy involve UGT genotypes. Results from recent studies suggest that UGT1A1 and UGT1A9 genotype analyses should be part of all studies evaluating pharmacogenetic aspects of CPT-11. The dramatic toxicity *UGT1A1*28* carriers experience after CPT-11 administration warrants determination of UGT1A1 genotypes prior to chemotherapy. This is especially true for combination treatments with fluoropyrimidines, since additional unfavorable genotypes of the 5-FU pathway (e.g., DPD IVS14 + 1G > A) may increase the toxic side effects to even life-threatening events.

In June 2005, the FDA recommended the screening for UGT1A1 polymorphism and suggested dose modification for CPT-11 in the presence of 7/7 genotype for UTG 1A1.

4. PHARMACOGENETIC ASPECTS OF SURVIVAL, RESPONSE, AND TOXICITY TO OXALIPLATIN

The platinum compound oxaliplatin has been proven to be effective in chemotherapy regimens of colorectal cancers. Large clinical trials including several hundreds of colorectal cancer patients with disseminated disease demonstrated the superiority of 5-FU/LV/oxaliplatin vs. 5-FU/LV in terms of response (51% vs. 22%, $p = 0.0001$) and progression-free survival (median, 9.0 vs. 6.2 months, $p = 0.0003$) [66] and identified 5-FU/LV/Oxaliplatin vs. 5-FU/LV/CPT-11 as the more favorable chemotherapy regimen in response rate ($p = 0.02$) and time to progression ($p = 0.004$) [67].

DNA adducts caused by the bulky 1,2-diamino-cyclohexane ring containing oxaliplatin are considered to mediate increased cytotoxicity and more effectively block DNA replication if compared with adducts formed by other platinum agents [68]. The excision repair cross-complementing (*ERCC*) gene family plays a central role in removing platinum-mediated DNA adducts. An inverse relationship between impaired DNA-repair capacity (e.g., low ERCC activity) and increased response rates has been demonstrated for platinum drugs. Analysis of gene expression in colorectal cancer patients treated with second line 5-FU/oxaliplatin revealed that patients whose tumor showed low levels of *ERCC1* gene amplification had superior overall survival. The relative risk (RRs) of dying was 4.2-fold (95% CI 1.4, 13.3, $p = 0.008$) in the ERCC1 high expresser group [17].

Functional polymorphisms have been reported in different nucleotide excision repair (*NER*) genes (Table 18.4 and Table 18.5). Preliminary data suggested that [69] an A > C variation at codon 751 of the *ERCC2* gene may be associated with clinical outcome to 5-FU/oxaliplatin chemotherapy. The amino acid exchange from lysine (Lys) to glutamine (Gln) by this base substitution was associated with a less favorable response rate and survival time. Lys/Lys genotypes showed 24% response compared with 10% in the Lys/Gln and 10% in the Gln/Gln group ($p = 0.015$). The median survival for those with the Lys/Lys genotype was 17 (95% C.I. 7.9, 27) vs. 13 (95% C.I. 8.5, 26) and 3.3 (95% C.I. 1.4, 6.5) months for patients with the Lys/Gln and Gln/Gln, respectively ($p = 0.002$) [69]. These results were confirmed in a larger cohort of patients using multivariate analysis [70]. However, recent data by McLeod et al. could not confirm an association between the ERCC2 K751Q polymorphism and clinical outcome to 5-FU/oxaliplatin treatment [71]. Further clinical studies are needed to determine the true association between this ERCC2 polymorphism and oxaliplatin metabolism. Additionally, biochemical evaluation of ERCC2 and oxaliplatin interactions are critical to assess the potential of this gene as a marker for clinical prediction to chemotherapy, since most experiments exploring the mechanism of interaction are done using cisplatin.

Less repair of oxaliplatin damage by proteins of the NER may result in superior response rates and survival. But the efficacy of oxaliplatin maybe impacted by additional mechanisms. Glutathione-S transferases (GST's) catalyze the conjugation of toxic and carcinogenic electrophilic molecules with glutathione. Through this reaction the GST's (Alpha, Pi, Mu, Theta, Zeta) protect cellular macromolecules from damage [72]. The GSTP1 enzyme appears to be predominantly expressed in colorectal tissue and tumors [73]. Drug resistant tumors contain very high levels of GSTP1 [74]. Goto et al. [75] showed a direct participation of GSTP1 in detoxification of cisplatin (CDDP) by formation of cisplatin-glutathione adducts (DDP-GSH). Furthermore, transfection analysis with GSTP1 antisense complementary DNA in colon cancer cell lines revealed GSTP1 to be responsible for both intrinsic and acquired resistance to platinum [76]. It has been demonstrated that a 313A>G variation of the gene leads to an amino acid exchange (isoleucine → valine) affecting its substrate-binding domain. Activity measurements in human tissues revealed that the GSTP1 ^{105}Val form correlates with impaired GSTP1 activity [77]. Based on these biochemical results an analysis was undertaken that evaluated a possible association between the GSTP1 polymorphism and clinical outcome to oxaliplatin chemotherapy. Patient possessing less active GSTP1 molecules were expected to accumulate the administrated drug and subsequently to show an increased response. (Table 18.1).

Among 107 colorectal cancer patients, who were treated with 5-FU/oxaliplatin combination as second line chemotherapy those whose tumor possessed two Val alleles benefited most from the therapy. Patients harboring two Val alleles experienced the most favorable survival (median, 25 months, 95% CI 9.4, 25+ months) while the Ile/Ile genotype was associated with the least favorable survival (median, 7.9 months, 95% CI 5.4, 9.6 months). Intermediate survival outcome was noted for the Ile /Val carriers (median, 13 months, 95% CI 8.4, 24 months; p<0.001) [78]. The results of this retrospective study have recently been contrasted by McLeod et al. [71] who could not confirm an association between this GSTP1 polymorphism and oxaliplatin-based chemotherapy. One possible explanation might be the fact that McLeod et al. were able to analyze samples from first line chemotherapy patients. The effect of pretreatment might have impacted the data of the smaller study. But again, we first need to identify the true biochemical interaction between GSTP1 and oxaliplatin and second to include this GSTP1 polymorphism in large clinical analyses in the future. This appears to be important since the GSTP1 polymorphism represents alterations of drug accumulation as an additional mechanism for genetically based interpatient variability of oxaliplatin effects.

4.1. Epidermal growth factor receptor: predictive and prognostic marker?

Epidermal growth factor receptor (EGFR) is a commonly expressed transmembrane glycoprotein of the tyrosine kinase growth factor family. EGFR are frequently overexpressed in many types of human cancers, including CRC (colon and rectal cancers) and their overexpression typically confers a more aggressive clinical behavior. The level of EGFR expression is primarily regulated by the abundance of its mRNA and the nature of the EGFR overexpression is believed to be due to an increase in the rate of EGFR transcription. A recent study showed that EGFR gene transcription activity declines with increasing numbers of a highly polymorphic dinucleotide repeat (CA repeat) in Intron 1 [79]. We hypothesized that patients with short CA repeat (16/16) may show poor clinical outcome compared with the patients with long CA repeat (16/20). We assessed the EGFR polymorphic dinucleotide repeat (CA repeat) of 100 patients with metastatic colorectal cancer who previously had failed 5-FU-based chemotherapy and determined their response and overall survival to 5-FU/oxaliplatin combination treatment.

An association between EGFR signaling pathway and response of cancer cells to ionizing radiation has been reported. Recently, a polymorphic variant in the *EGFR* gene that leads to an arginine-to-lysine substitution in the extracellular domain at codon 497 within subdomain IV of EGFR has been identified. The variant EGFR (HER-1 497K) may lead to attenuation in ligand binding, growth stimulation, tyrosine kinase activation, and induction of protooncogenes MYC, FOS, and JUN. A $(CA)_n$ repeat polymorphism in intron 1 of the *EGFR* gene that alters EGFR expression in vitro and in vivo has also been described. In the current pilot study, we assessed both polymorphisms in 59 patients with locally advanced rectal cancer treated with adjuvant or neoadjuvant chemoradiation therapy, using PCR-RFLP and a 5′-end ^{33}P-γATP-labeled PCR protocol. We tested whether either polymorphism alone or in combination can be associated with local recurrence in the setting of chemoradiation treatment. We found that patients with HER-1 497 *Arg/Arg* genotype or lower number of CA repeats (both alleles <20) tended to have a higher risk of local recurrence ($p = 0.24$ and 0.31, respectively). Combined analysis showed the highest risk for local recurrence was seen in patients who possessed both a HER-1 497 *Arg*-allele and <20 CA repeats ($p = 0.05$, log rank test). Our data suggest that the HER-1 R497K and EGFR intron 1 $(CA)_n$ repeat polymorphisms may be potential indicators of radiosensitivity in patients with rectal cancer treated with chemoradiation (Figure 18.3) [80].

4.2. Interleukin 8: a predictor of clinical outcome

Interleukin 8 (IL-8), a member of the CXC chemokine family has been shown to be involved in tumor cell growth and metastasis in colorectal cancer [81]. Its receptor CXCR1 and CXCR2 play an important role in tumor progression and angiogenesis [82]. Studies show that expression of IL-8 and its receptors CXCR1 and CXCR2 contribute to tumor progression and development of metastases in vitro and in vivo. Polymorphisms in the promoter regions of *IL-8* gene (T-251A) and a novel polymorphism in exon 2 of *CXCR1* gene (Ser + 2607Thr) may influence the expression of IL-8 and its receptor and therefore

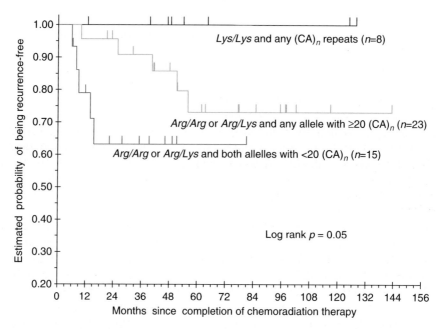

Figure 18.3. Recurrence-free survival of patients with rectal cancer by EGFR polymorphisms. The vertical hash marks denote the time of last follow-up for those patients who were still recurrence-free at the time of the analysis of data. All censored patients and those who were locally recurrent are accounted for.

may influence the clinical outcome of patients with metastatic colorectal cancer [83–85]. We tested the hypothesis that patients with genomic polymorphisms associated with higher expression or activity of IL would have a poorer prognosis. We analyzed these polymorphisms in a retrospective 5-FU/oxaliplatin chemotherapy study to correlate the association of aforementioned genetic polymorphism and the survival of patients with advanced colorectal cancer. *IL-8* (T-251A) and *CXCR1* (Ser+2607Thr) gene polymorphism in 106 patients with metastatic colorectal cancer who received 5-FU/oxaliplatin combination chemotherapy were examined using PCR-RFLP method and the association between polymorphism and overall survival was evaluated using RRs of dying and the log rank test. Patients with Ser/Thr genotype for the CXCR1 polymorphism had an RR = 1.65 (95% C.I = 0.87–3.13) compared with patients with Ser/Ser genotype. After stratification by performance status and histology and prior treatment, the RR was 2.50 (95% CI = 1.17–5.35) ($p = 0.025$). The median follow-up period was 11.9 months (95% C.I 3.2–30.9) and median survival was 9.3 months (95% C.I 3.5–12.8). Patients with Ser/Thr genotype survived a median of 5.9 months compared with those with Ser/Ser genotype, who survived a median of 11.4 months ($p < 0.05$). The IL-8 polymorphism was not associated with survival. Our present data suggest an association between CXCR1 exon2 (Ser + 2607Thr) polymorphism, which is associated with a higher/lower IL-8 activity/expression, and survival of patients with advanced colorectal cancer receiving 5-FU/oxaliplatin chemotherapy. This finding and its mechanism need to be confirmed in future prospective studies [86].

5. FUTURE DEVELOPMENTS

We have evidence of germ line polymorphisms as well as gene expression levels of enzymes involved in DNA repair, oxidative stress and metabolic pathway of 5-FU that appear to be of clinical significance however these data are from small retrospective studies. It is critical to include these functional analyses into large, prospective clinical studies. Most research laboratories are capable of PCR-based RFLP techniques, while extraction and storage of human DNA samples are routine in these laboratories. All samples from each clinical trial should be sent to one laboratory to ensure a high quality analysis. To include gene expression analysis into the routine of clinical trials might be more difficult since the techniques for extraction (e.g., RNA extraction from paraffin) are more complex and the technology is not broadly accessible. When gene expression analyses are part of clinical trials, specialized laboratories that have long experiences with quantitative mRNA measurements should be contacted in the beginning and analysis centralized. Meanwhile data from cDNA microarray analyses will supply information on other potential pathways that might have to be considered if patients are treated with fluoropyrimidines, irinotecan, and oxaliplatin. In addition, genes involved in metabolism, accumulation and action of newly developed drugs like antibodies against EGFR or VEGF have to be screened for polymorphisms that impact the function of the gene products. One example of a possible candidate in this new field is a CA repeat polymorphism in intron 1 of the *EGFR* gene that affects EGFR transcription [79,87]. The consequent analysis of pharmacogenetic markers in clinical settings is critical to guarantee the application and incorporation of these promising data into clinical care, allowing us to stratify patients according to their genetic makeup.

REFERENCES

1. Midgley, R. and Kerr, D. Colorectal cancer. *Lancet* 353: 391–399, 1999.
2. Hermanek, P., Wiebelt, H., Staimmer, D., and Riedl, S. Prognostic factors of rectum carcinoma — experience of the German Multicentre Study SGCRC. German Study Group Colo-Rectal Carcinoma. *Tumori* 81: 60–64, 1995.
3. Wolmark, N., Wieand, H. S., Rockette, H. E., Fisher, B., Glass, A., Lawrence, W., Lerner, H., Cruz, A. B., Volk, H., Shibata, H., et al. The prognostic significance of tumor location and bowel obstruction in Dukes B and C colorectal cancer. Findings from the NSABP clinical trials. *Ann Surg* 198: 743–752, 1983.
4. McArdle, C. S. and Hole, D. Impact of variability among surgeons on postoperative morbidity and mortality and ultimate survival. *BMJ* 302: 1501–1505, 1991.
5. Kundu, N. G. and Heidelberger, C. Cyclopenta(f)isoquinoline derivatives designed to bind specifically to native deoxyribonucleic acid. 3. Interaction of 6-carbamylmethyl-8-methyl-7H-

cyclopenta(f)isoquinolin-3(2H)-one with deoxyribonucleic acids and polydeoxyribonucleotides. *Biochem Biophys Res Commun* 60: 561–568, 1974.
6. Danenberg, P. V. Thymidylate synthetase — a target enzyme in cancer chemotherapy. *Biochim Biophys Acta* 473: 73–92, 1977.
7. Horie, N., Aiba, H., Oguro, K., Hojo, H., and Takeishi, K. Functional analysis and DNA polymorphism of the tandemly repeated sequences in the 5′-terminal regulatory region of the human gene for thymidylate synthase. *Cell Struct Funct* 20: 191–197, 1995.
8. Luo, H. R., Lu, X. M., Yao, Y. G., Horie, N., Takeishi, K., Jorde, L. B., and Zhang, Y. P. Length polymorphism of thymidylate synthase regulatory region in Chinese populations and evolution of the novel alleles. *Biochem Genet* 40: 41–51, 2002.
9. Kawakami, K., Salonga, D., Park, J. M., Danenberg, K. D., Uetake, H., Brabender, J., Omura, K., Watanabe, G., and Danenberg, P. V. Different lengths of a polymorphic repeat sequence in the thymidylate synthase gene affect translational efficiency but not its gene expression. *Clin Cancer Res* 7: 4096–4101, 2001.
10. Marsh, S., Collie-Duguid, E. S., Li, T., Liu, X., and McLeod, H. L. Ethnic variation in the thymidylate synthase enhancer region polymorphism among Caucasian and Asian populations. *Genomics* 58: 310–312, 1999.
11. Kaneda, S., Takeishi, K., Ayusawa, D., Shimizu, K., Seno, T., and Altman, S. Role in translation of a triple tandemly repeated sequence in the 5′-untranslated region of human thymidylate synthase mRNA. *Nucleic Acids Res* 15: 1259–1270, 1987.
12. Marsh, S., McKay, J. A., Cassidy, J., and McLeod, H. L. Polymorphism in the thymidylate synthase promoter enhancer region in colorectal cancer. *Int J Oncol* 19: 383–386, 2001.
13. Pullarkat, S. T., Stoehlmacher, J., Ghaderi, V., Xiong, Y. P., Ingles, S. A., Sherrod, A., Warren, R., Tsao-Wei, D., Groshen, S., and Lenz, H. J. Thymidylate synthase gene polymorphism determines response and toxicity of 5-FU chemotherapy. *Pharmacogenomics J* 1: 65–70, 2001.
14. Leichman, C. G., Lenz, H. J., Leichman, L., Danenberg, K., Baranda, J., Groshen, S., Boswell, W., Metzger, R., Tan, M., and Danenberg, P. V. Quantitation of intratumoral thymidylate synthase expression predicts for disseminated colorectal cancer response and resistance to protracted-infusion fluorouracil and weekly leucovorin. *J Clin Oncol* 15: 3223–3229, 1997.
15. Farrugia, D., Ford, H., Cunningham, D., Danenberg, K., Danenberg, P., Brabender, J., McVicar, A., Aherne, G., Hardcastle, A., McCarthy, K., and Jackman, A. Tymidylate synthase expression in advanced colorectal cancer predicts for response to ralitrexed. *Clin Cancer Res* 9: 792–801, 2003.
16. Ichikawa, W., Uetake, H., Shirota, Y., Yamada, H., Nishi, N., Nihei, Z., Sugihara, K., and Hirayama, R. Combination of dihydropyrimidin dehydrogenase and tymidylate synthase gene expression in primary tumors as predictive parameters for the efficacy of fluoropyrimidine-based chemotherapy for metastatic colorectal cancer. *Clin Cancer Res* 9: 786–791, 2003.
17. Shirota, Y., Stoehlmacher, J., Brabender, J., Xiong, Y. P., Uetake, H., Danenberg, K. D., Groshen, S., Tsao-Wei, D. D., Danenberg, P. V., and Lenz, H. J. ERCC1 and thymidylate synthase mRNA levels predict survival for colorectal cancer patients receiving combination oxaliplatin and fluorouracil chemotherapy. *J Clin Oncol* 19: 4298–4304, 2001.
18. Kubota, T., Watanabe, M., Otani, Y., Kitayima, M., and Fukushima, M. Different pathways of 5-fluorouracil metabolism after continuous venous or bolus injection in patients with colon carcinoma: possible predictive value of thymidylate synthase mRNA and ribonucleotide reductase for 5-fluorouracil. *Anticancer Res* 22: 3537–3540, 2002.
19. Kawakami, K., Omura, K., Kanehira, E., and Watanabe, Y. Polymorphic tandem repeats in the thymidylate synthase gene is associated with its protein expression in human gastrointestinal cancers. *Anticancer Res* 19: 3249–3252, 1999.
20. Villafranca, E., Okruzhnov, Y., Dominguez, M. A., Garcia-Foncillas, J., Azinovic, I., Martinez, E., Illarramendi, J. J., Arias, F., Martinez Monge, R., Salgado, E., Angeletti, S., and Brugarolas, A. Polymorphisms of the repeated sequences in the enhancer region of the thymidylate synthase gene promoter may predict downstaging after preoperative chemoradiation in rectal cancer. *J Clin Oncol* 19: 1779–1786, 2001.

21. Lu, B., Gil, B., Zhang, W., Stoehlmacher, J., Zahedy, S., Mallik, N., and Lenz, H. J. Thymidylate synthase polymorphism predicts pelvic recurrence in rectal cancer patients treated with combined modalities [abstract 2237]. *Proc Am Soc Clin Oncol* 21, 2002.
22. Etienne, M. C., Chazal, M., Laurent-Puig, P., Magne, N., Rosty, C., Formento, J. L., Francoual, M., Formento, P., Renee, N., Chamorey, E., Bourgeon, A., Seitz, J. F., Delpero, J. R., Letoublon, C., Pezet, D., and Milano, G. Prognostic value of tumoral thymidylate synthase and p53 in metastatic colorectal cancer patients receiving fluorouracil-based chemotherapy: phenotypic and genotypic analyses. *J Clin Oncol* 20: 2832–2843, 2002.
23. Iacopetta, B., Grieu, F., Joseph, D., and Elsaleh, H. A polymorphism in the enhancer region of the thymidylate synthase promoter influences the survival of colorectal cancer patients treated with 5-fluorouracil. *Br J Cancer* 85: 827–830, 2001.
24. Tsuji, T., Hidaka, S., Sawai, T., Nakagoe, T., Yano, H., Haseba, M., Komatsu, H., Shindou, H., Fukuoka, H., Yoshinaga, M., Shibasaki, S., Nanashima, A., Yamaguchi, H., Yasutake, T., and Tagawa, Y. Polymorphism in the thymidylate synthase promoter enhancer region is not an efficacious marker for tumor sensitivity to 5-fluorouracil-based oral adjuvant chemotherapy in colorectal cancer. *Clin Cancer Res* 9: 3700–3704, 2003.
25. Mandola, M., Stoehlmacher, J., Muller-Weeks, S., Cesarone, G., Yu, M. C., Lenz, H. J., and Ladner, R. D. A novel single nucleotide polymorphism within the 5' tandem repeat polymorphism of the *thymidylate synthase* gene abolishes USF-1 binding and alters transcriptional activity. *Cancer Res* 63: 2898–2904, 2003.
26. Kawakami, K. and Watanabe, G. Single nucleotide polymorphism in the tandem-repeat sequence of thymidylates synthase gene: A candidate for an additional predictor in 5-FU-based chemotherapy [abstract 509]. *Proc Am Soc Clin Oncol* 2003.
27. Kawakami, K. and Watanabe, G. Identification and functional analysis of single nucleotide polymorphism in the tandem repeat sequence of thymidylate synthase gene. *Cancer Res* 63: 6004–6007, 2003.
28. Zinzindohoue, F., Ferraz, J. M., and Laurent-Puig, P. Thymidylate synthase promoter polymorphism. *J Clin Oncol* 19: 3442, 2001.
29. Uchida, K., Danenberg, K., Kawakami, K., Schneider, S., Park, J., Kuramochi, H., Hayashi, K., Takasaki, K., and Danenberg, P. Loss of heterozygosity at the TS locus affects tumor response and survival in individuals heterozygous for a TS promoter polymorphism [abstract 492]. *Proc Am Soc Clin Oncol* 2003.
30. Kawakami, K., Ishidida, Y., Danenberg, K., Omura, K., Watanabe, G., and Danenberg, P. Functional polymorphism of the thymidylate synthase gene in colorectal cancer accompanied by frequent loss of heterozygosity. *Jpn J Cancer Res* 93(11): 1221–1229, 2002.
31. Schaaf, M. and Cidlowski, J. AUUUA motifs in the 3'UTR of human glucocorticoid receptor alpha and beta mRNA destabilize mRNA and decrease receptor protein expression. *Steroids* 67(7): 627–636, 2002.
32. Mandola, M. V., Stoehlmacher, J., Zhang, W., Groshen, S., Yu, M. C., Iqbal, S., Lenz, H. J., and Ladner, R. D. A 6 bp polymorphism in the thymidylate synthase gene causes message instability and is associated with decreased intratumoral TS mRNA levels. *Pharmacogenetics* 14: 319–327, 2004.
33. McLeod, H., Sargent, D., Marsh, S., Fuchs, C., Ramanathan, R., Williamson, S., Findlay, B., Thibodeau, S., Peterson, G., and Goldberg, R. Pharmacogenetic analysis of systemic toxicity and response after 5-fluorouracil (5-FU)/CPT-11, 5-FU/oxaliplatin (oxal), or CPT-11/oxal therapy for advanced colorectal cancer (CRC): Results from an intergroup trial. Abstract # 1013. In: *Proceedings of the American Society of Clinical Oncology* 2003.
34. Heggie, G. D., Sommadossi, J. P., Cross, D. S., Huster, W. J., and Diasio, R. B. Clinical pharmacokinetics of 5-fluorouracil and its metabolites in plasma, urine, and bile. *Cancer Res* 47: 2203–2206, 1987.
35. Salonga, D., Danenberg, K. D., Johnson, M., Metzger, R., Groshen, S., Tsao-Wei, D. D., Lenz, H. J., Leichman, C. G., Leichman, L., Diasio, R. B., and Danenberg, P. V. Colorectal tumors

responding to 5-fluorouracil have low gene expression levels of dihydropyrimidine dehydrogenase, thymidylate synthase, and thymidine phosphorylase. *Clin Cancer Res* 6: 1322–1327, 2000.
36. McLeod, H. L., Collie-Duguid, E. S., Vreken, P., Johnson, M. R., Wei, X., Sapone, A., Diasio, R. B., Fernandez-Salguero, P., van Kuilenberg, A. B., van Gennip, A. H., and Gonzalez, F. J. Nomenclature for human DPYD alleles. *Pharmacogenetics* 8: 455–459, 1998.
37. Diasio, R. B., Beavers, T. L., and Carpenter, J. T. Familial deficiency of dihydropyrimidine dehydrogenase. Biochemical basis for familial pyrimidinemia and severe 5-fluorouracil-induced toxicity. *J Clin Invest* 81: 47–51, 1988.
38. Johnson, M. R., Hageboutros, A., Wang, K., High, L., Smith, J. B., and Diasio, R. B. Life-threatening toxicity in a dihydropyrimidine dehydrogenase-deficient patient after treatment with topical 5-fluorouracil. *Clin Cancer Res* 5: 2006–2011, 1999.
39. van Kuilenburg, A. B., Muller, E. W., Haasjes, J., Meinsma, R., Zoetekouw, L., Waterham, H. R., Baas, F., Richel, D. J., and van Gennip, A. H. Lethal outcome of a patient with a complete dihydropyrimidine dehydrogenase (DPD) deficiency after administration of 5-fluorouracil: frequency of the common IVS14 + 1G > A mutation causing DPD deficiency. *Clin Cancer Res* 7: 1149–1153, 2001.
40. Van Kuilenburg, A. B., Vreken, P., Abeling, N. G., Bakker, H. D., Meinsma, R., Van Lenthe, H., De Abreu, R. A., Smeitink, J. A., Kayserili, H., Apak, M. Y., Christensen, E., Holopainen, I., Pulkki, K., Riva, D., Botteon, G., Holme, E., Tulinius, M., Kleijer, W. J., Beemer, F. A., Duran, M., Niezen-Koning, K. E., Smit, G. P., Jakobs, C., Smit, L. M., Van Gennip, A. H., et al. Genotype and phenotype in patients with dihydropyrimidine dehydrogenase deficiency. *Hum Genet* 104: 1–9, 1999.
41. Van Kuilenburg, A. B., Vreken, P., Beex, L. V., Meinsma, R., Van Lenthe, H., De Abreu, R. A., and van Gennip, A. H. Heterozygosity for a point mutation in an invariant splice donor site of dihydropyrimidine dehydrogenase and severe 5-fluorouracil related toxicity. *Eur J Cancer* 33: 2258–2264, 1997.
42. Isshi, K., Sakuyama, T., Gen, T., Nakamura, Y., Kuroda, T., Katayuma, T., and Maekawa, Y. Predicting 5-FU sensitivity using human colorectal cancer specimens: caomparison of tumor dihydropyrimidine dehydrogenase and orotate phosphoribosyl transferase activities with in vivo chemosensitivity to 5-FU. *Int J Clin Oncol* 7(6): 335–342, 2002.
43. Ichikawa, W. T. T., Nihei, Z., Shirota, Y., Suto, K., and Hirayama, R. Polymorphism of orotate phosphoribosyl transferase (OPRT) gene and tymidylate synthase tandem repeat (TSTR) predict adverse events (AE) in colorectal cancer (CRC) patients treated with 5-fluorouracil (FU) plus leucovorin (LV) [abstract 1063]. *Proc Am Soc Clin Oncol* 2003.
44. Humerickhouse, R., Lohrbach, K., Li, L., Bosron, W. F., and Dolan, M. E. Characterization of CPT-11 hydrolysis by human liver carboxylesterase isoforms hCE-1 and hCE-2. *Cancer Res* 60: 1189–1192, 2000.
45. Khanna, R., Morton, C. L., Danks, M. K., and Potter, P. M. Proficient metabolism of irinotecan by a human intestinal carboxylesterase. *Cancer Res* 60: 4725–4728, 2000.
46. Charasson, V., Bellott, R., Gorry, P., Longy, M., and Robert, J. Pharmacogenetics of human carboxylesterase 2, an enzyme involved in the activation of irinotecan into SN-38 [abstract 4668]. *Proc Am Assoc Cancer Res* 94, 2003.
47. Saltz, L. B., Cox, J. V., Blanke, C., Rosen, L. S., Fehrenbacher, L., Moore, M. J., Maroun, J. A., Ackland, S. P., Locker, P. K., Pirotta, N., Elfring, G. L., and Miller, L. L. Irinotecan plus fluorouracil and leucovorin for metastatic colorectal cancer. Irinotecan Study Group. *N Engl J Med* 343: 905–914, 2000.
48. Rivory, L. P., Riou, J. F., Haaz, M. C., Sable, S., Vuilhorgne, M., Commercon, A., Pond, S. M., and Robert, J. Identification and properties of a major plasma metabolite of irinotecan (CPT-11) isolated from the plasma of patients. *Cancer Res* 56: 3689–3694, 1996.
49. Haaz, M. C., Riche, C., Rivory, L. P., and Robert, J. Biosynthesis of an aminopiperidino metabolite of irinotecan [7-ethyl-10-[4-(1-piperidino)-1-piperidino]carbonyloxycamptothecine] by human hepatic microsomes. *Drug Metab Dispos* 26: 769–774, 1998.

50. Haaz, M. C., Rivory, L., Riche, C., Vernillet, L., and Robert, J. Metabolism of irinotecan (CPT-11) by human hepatic microsomes: participation of cytochrome P-450 3A and drug interactions. *Cancer Res* 58: 468–472, 1998.
51. Rebbeck, T. R., Jaffe, J. M., Walker, A. H., Wein, A. J., and Malkowicz, S. B. Modification of clinical presentation of prostate tumors by a novel genetic variant in CYP3A4. *J Natl Cancer Inst* 90: 1225–1229, 1998.
52. Iyer, L., King, C. D., Whitington, P. F., Green, M. D., Roy, S. K., Tephly, T. R., Coffman, B. L., and Ratain, M. J. Genetic predisposition to the metabolism of irinotecan (CPT-11). Role of uridine diphosphate glucuronosyltransferase isoform 1A1 in the glucuronidation of its active metabolite (SN-38) in human liver microsomes. *J Clin Invest* 101: 847–854, 1998.
53. Bosma, P. J., Chowdhury, J. R., Bakker, C., Gantla, S., de Boer, A., Oostra, B. A., Lindhout, D., Tytgat, G. N., Jansen, P. L., Oude Elferink, R. P., et al. The genetic basis of the reduced expression of bilirubin UDP-glucuronosyltransferase 1 in Gilbert's syndrome. *N Engl J Med* 333: 1171–1175, 1995.
54. Innocenti, F., Iyer, L., and Ratain, M. J. Pharmacogenetics of anticancer agents: lessons from amonafide and irinotecan. *Drug Metab Dispos* 29: 596–600, 2001.
55. Iyer, L., Das, S., Janisch, L., Wen, M., Ramirez, J., Karrison, T., Fleming, G. F., Vokes, E. E., Schilsky, R. L., and Ratain, M. J. UGT1A1*28 polymorphism as a determinant of irinotecan disposition and toxicity. *Pharmacogenomics J* 2: 43–47, 2002.
56. Iyer, L., Hall, D., Das, S., Mortell, M. A., Ramirez, J., Kim, S., Di Rienzo, A., and Ratain, M. J. Phenotype-genotype correlation of in vitro SN-38 (active metabolite of irinotecan) and bilirubin glucuronidation in human liver tissue with UGT1A1 promoter polymorphism. *Clin Pharmacol Ther* 65: 576–582, 1999.
57. Beutler, E., Gelbart, T., and Demina, A. Racial variability in the UDP-glucuronosyltransferase 1 (UGT1A1) promoter: a balanced polymorphism for regulation of bilirubin metabolism? *Proc Natl Acad Sci USA* 95: 8170–8174, 1998.
58. Yamamoto, K., Sato, H., Fujiyama, Y., Doida, Y., and Bamba, T. Contribution of two missense mutations (G71R and Y486D) of the bilirubin UDP glycosyltransferase (UGT1A1) gene to phenotypes of Gilbert's syndrome and Crigler-Najjar syndrome type II. *Biochim Biophys Acta* 1406: 267–273, 1998.
59. Monaghan, G., Ryan, M., Seddon, R., Hume, R., and Burchell, B. Genetic variation in bilirubin UPD-glucuronosyltransferase gene promoter and Gilbert's syndrome. *Lancet* 347: 578–581, 1996.
60. Iyer, L., Janisch, L., Das, S., Ramirez, J., Hurley-Buterman, C., DeMario, M., Vokes, E., Kindler, H., and Ratain, M. UGT1A1 Promoter Genotype Correlates with Pharmacokinetics of Irinotecan (CPT-11) [abstract 690]. *Proc Am Soc Clin Oncol* 2000.
61. Ando, Y., Saka, H., Ando, M., Sawa, T., Muro, K., Ueoka, H., Yokoyama, A., Saitoh, S., Shimokata, K., and Hasegawa, Y. Polymorphisms of UDP-glucuronosyltransferase gene and irinotecan toxicity: a pharmacogenetic analysis. *Cancer Res* 60: 6921–6926, 2000.
62. Ando, M., Kitagawa, H., Ando, Y., Sekido, Y., Wakai, K., Suzuki, R., Shimokata, K., and Hasegawa, Y. Genetic polymorphism in the phenobarbital-responsive enhancer module of the UDP-Glucuronosyltransferase (UGT) 1A1 gene and irinotecan toxicity in Japanese patients [abstract 496]. *Proc Am Soc Clin Oncol* 2003.
63. Gagne, J., Montminy, V., Belanger, P., Journault, K., Gaucher, G., and Guillemette, C. Common human UGT1A polymorphisms and the altered metabolism of irinotecan active metabolite 7-ethyl-10-hydroxycamptothecin (SN-38). *Mol Pharmacol* 62(3): 608–617, 2002.
64. Jinno, H., Saeki, M., Saito, Y., Tanaka-Kagawa, T., Hanioka, N., Sai, K., Kaniwa, N., Ando, M., Shirao, K., Minami, H., Ohtsu, A., Yoshida, T., Saijto, N., Ozawa, S., and Sawada, J. Functional characterization of human UDP-glucuronosyltransferase 1A9 variant, D256N, found in Japanese cancer patients. *J Phamacol Exp Ther* 306(2): 688–693, 2003.
65. Mathijssen, R. H., Marsh, S., Karlsson, M. O., Xie, R., Baker, S. D., Verweij, J., Sparreboom, A., and McLeod, H. L. Irinotecan pathway genotype analysis to predict pharmacokinetics. *Clin Cancer Res* 9: 3246–3253, 2003.

66. de Gramont, A., Figer, A., Seymour, M., Homerin, M., Hmissi, A., Cassidy, J., Boni, C., Cortes-Funes, H., Cervantes, A., Freyer, G., Papamichael, D., Le Bail, N., Louvet, C., Hendler, D., de Braud, F., Wilson, C., Morvan, F., and Bonetti, A. Leucovorin and fluorouracil with or without oxaliplatin as first-line treatment in advanced colorectal cancer. *J Clin Oncol* 18: 2938–2947, 2000.
67. Goldberg, R., Morton, R., Sargent, D., Fuchs, C., Ramanathan, R., Williamson, S., and Findlay, B. N9741: oxaliplatin (Oxal) or CPT-11 + 5-fluorouracil (5-FU)/leucovorin (LV) or oxal + CPT-11 in advanced colorectal cancer (CRC). Updated efficacy and quality of life (QOL) data from an intergroup study [abstract 1009]. *Proc Am Soc Clin Oncol* 2003.
68. Schmoll, H. J. and Cassidy, J. Integrating oxaliplatin into the management of colorectal cancer. *Oncologist* 6(Suppl. 4): 24–28, 2001.
69. Park, D. J., Stoehlmacher, J., Zhang, W., Tsao-Wei, D. D., Groshen, S., and Lenz, H. J. A Xeroderma pigmentosum group D gene polymorphism predicts clinical outcome to platinum-based chemotherapy in patients with advanced colorectal cancer. *Cancer Res* 61: 8654–8658, 2001.
70. Lenz, H., Park, D., Zhang, W., Tsao-Wei, D., Groshen, S., Zahedy, S., Mallik, N., Gil, B., and Stoehlmacher, J. A multivariate analysis of genetic markers for clinical response to 5-FU/oxaliplatin chemotherapy in advanced colorectal cancer [abstract 513]. *Proc Am Soc Clin Oncol* 2002.
71. McLeod, J., Sargent, D., Marsh, S., Fuchs, C., Ramanathan, R., Williamson, S., Findlay, B., Thibodeau, S., Petersen, G., and Goldberg, R. Pharmacogenetic analysis of systemic toxicity and response after 5-fluorouracil (5-FU)/CPT-11, 5-FU/oxaliplatin (oxal), or CPT-11/oxal therapy for advanced colorectal cancer (CRC): Results from an intergroup trial [abstract 1013]. *Proc Am Soc Clin Oncol* 2003.
72. Board, P., Baker, R., Chelvanayagam, G., and Jermiin, L. Zeta, a novel class of glutathion transferase in a range of species from plants to human. *Biochem J* 328: 929–935, 1997.
73. Moscow, J. A., Fairchild, C. R., Madden, M. J., Ransom, D. T., Wieand, H. S., O'Brien, E. E., Poplack, D. G., Cossman, J., Myers, C. E., and Cowan, K. H. Expression of anionic glutathione-S-transferase and P-glycoprotein genes in human tissues and tumors. *Cancer Res* 49: 1422–1428, 1989.
74. Tsuchida, S. and Sato, K. Glutathion transferases and cancer. *Crit Rev Biochem Mol Biol* 27: 337–384, 1992.
75. Goto, S., Iida, T., Cho, S., Oka, M., Kohno, S., and Kondo, T. Overexpression of glutathione S-transferase pi enhances the adduct formation of cisplatin with glutathione in human cancer cells. *Free Radic Res* 31: 549–558, 1999.
76. Ban, N., Takahashi, Y., Takayama, T., Kura, T., Katahira, T., Sakamaki, S., and Niitsu, Y. Transfection of glutathione S-transferase (GST)-pi antisense complementary DNA increases the sensitivity of a colon cancer cell line to adriamycin, cisplatin, melphalan, and etoposide. *Cancer Res* 56: 3577–3582, 1996.
77. Srivastava, S., Singhal, S., Hu, X., Awasthi, Y., Zimniak, P., and Singh, S. Differential catalytic efficiency of allelic variants of human glutathione S-transferase Pi in catalyzing the glutathione conjugation of thiotepa. *Arch Biochem Biophys* 366(1): 89–94, 1999.
78. Stoehlmacher, J., Park, D. J., Zhang, W., Groshen, S., Tsao-Wei, D. D., Yu, M. C., and Lenz, H. J. Association between glutathione S-transferase P1, T1, and M1 genetic polymorphism and survival of patients with metastatic colorectal cancer. *J Natl Cancer Inst* 94: 936–942, 2002.
79. Gebhardt, F., Zanker, K. S., and Brandt, B. Modulation of epidermal growth factor receptor gene transcription by a polymorphic dinucleotide repeat in intron 1. *J Biol Chem* 274: 13176–13180, 1999.
80. Zhang, W., Park, D. J., Lu, B., Yang, D. Y., Gordon, M., Groshen, S., Yun, J., Press, O. A., Vallbohmer, D., Rhodes, K., and Lenz, H. J. Epidermal growth factor receptor gene polymorphisms predict pelvic recurrence in patients with rectal cancer treated with chemoradiation. *Clin Cancer Res* 11: 600–605, 2005.
81. Xie, K. Interleukin-8 and human cancer biology. *Cytokine Growth Factor Rev* 12: 375–391, 2001.

82. Miller, L. J., Kurtzman, S. H., Wang, Y., Anderson, K. H., Lindquist, R. R., and Kreutzer, D. L. Expression of interleukin-8 receptors on tumor cells and vascular endothelial cells in human breast cancer tissue. *Anticancer Res* 18: 77–81, 1998.
83. Brew, R., Erikson, J. S., West, D. C., Kinsella, A. R., Slavin, J., and Christmas, S. E. Interleukin-8 as an autocrine growth factor for human colon carcinoma cells in vitro. *Cytokine* 12: 78–85, 2000.
84. Li, A., Varney, M. L., and Singh, R. K. Expression of interleukin 8 and its receptors in human colon carcinoma cells with different metastatic potentials. *Clin Cancer Res* 7: 3298–3304, 2001.
85. Renzoni, E., Lympany, P., Sestini, P., Pantelidis, P., Wells, A., Black, C., Welsh, K., Bunn, C., Knight, C., Foley, P., and du Bois, R. M. Distribution of novel polymorphisms of the interleukin-8 and CXC receptor 1 and 2 genes in systemic sclerosis and cryptogenic fibrosing alveolitis. *Arthritis Rheum* 43: 1633–1640, 2000.
86. Zhang, W., Stoehlmacher, J., Park, D. J., Yang, D., Borchard, E., Gil, J., Tsao-Wei, D. D., Yun, J., Gordon, M., Press, O. A., Rhodes, K., Groshen, S., and Lenz, H. J. Gene polymorphisms of epidermal growth factor receptor and its downstream effector, interleukin-8, predict oxaliplatin efficacy in patients with advanced colorectal cancer. *Clin Colorectal Cancer* 5: 124–131, 2005.
87. Liu, W., Innocenti, F., Chen, P., Das, S., Cook, E. J., and Ratain, M. Interethnic difference in the allelic distribution of human epidermal growth factor receptor intron 1 polymorphism. *Clin Cancer Res* 9(3): 1009–1012, 2003.

CHAPTER 19

DEVELOPMENT OF PHARMACOGENOMIC PREDICTORS FOR PREOPERATIVE CHEMOTHERAPY OF BREAST CANCER

LAJOS PUSZTAI

Associate Professor of Medicine, Department of Breast Medical Oncology, The University of Texas M.D. Anderson Cancer Center, Houston, TX, USA.

1. INTRODUCTION

"If it were not for the great variability among individuals, medicine might as well be a science and not an art". *Sir William Osler: The Principles and Practice of Medicine, 1892*

The selection of a particular chemotherapy or a combination of drugs for the treatment of breast cancer is currently more of an art than a science. There are many chemotherapy drugs that cause tumor shrinkage in metastatic breast cancer and several combination chemotherapy regimens have been shown to improve disease-free and overall survival when used as adjuvant therapy for early stage breast cancer. However, no chemotherapy regimen is universally effective for all patients. In the adjuvant (or neoadjuvant) treatment setting, choosing the best possible chemotherapy regimen for an individual is challenging and the right choice may make the difference between cure from breast cancer and relapse. Considering only some of the most effective regimens, physicians and patients could choose from sequential epirubicine/cyclophosphamide/5-fluorouracil and docetaxel, dose-dense every 2-week doxorubicin/ cyclophosphamide (AC) and paclitaxel, weekly paclitaxel followed by doxorubicin/cyclophosphamide/5-FU (FAC), or triple combination therapy with docetaxel/doxorubicin/cyclophosphamide [1–4].

Correspondence to: Lajos Pusztai, Department of Breast Medical Oncology of the University of Texas M.D. Anderson Cancer Center, Unit 424, 1515 Holcombe Boulevard, Houston, TX 77030-4009 Tel: 713-792-2817, Fax: 713-794-4385, e-mail: lpusztai@mdanderson.org

Currently, it is impossible to predict at the time of diagnosis which of these regimens will actually work for a particular individual [5].

Results from the traditionally designed clinical trials, which compare the activity of two or more adjuvant regimens in order to establish which is the "generally" most effective therapy offers little guidance for individualized treatment selection. This approach ignores the fact that many patients are cured with earlier generation adjuvant regimens and therefore are not well served by the administration of longer, potentially more toxic, and vastly more expensive second- and third-generation regimens. These studies also cannot reveal whether some cancers are particularly sensitive to a particular drug (or combination regimen) and therefore can only be cured by that treatment. The individually most effective treatment may or may not be the same regimen that is generally the most effective for the whole population. These theoretical concerns and the burgeoning choice of adjuvant chemotherapy treatment options have fueled a renewed interest in chemotherapy response prediction that may lead to more personalized treatment recommendations.

2. BREAST CANCER PHARMACOGENOMIC DISCOVERY PROGRAM AT THE NELLIE B. CONNALLY BREAST CENTER, UNIVERSITY OF TEXAS M. D. ANDERSON CANCER CENTER

Gene expression profiling represents a novel analytical tool that was developed in the early 1990s [6]. This technology enables investigators to measure, in a single experiment, the expression of all (or almost all) genes that are expressed in a small piece of cancer tissue. It is hypothesized that gene expression patterns measured by this technology will yield multigene markers that could predict clinical outcome more accurately than any single gene marker. Each new molecular analytical technique offers a hope to improve on existing diagnostic tests and fill a currently unmet medical need. However, finding the right diagnostic niche and determining if the technology is mature enough for routine clinical use is a complex and lengthy process. We prefer to think of genomic marker development as a process that is fundamentally similar to the clinical drug discovery process (Figure 19.1) [7]. During clinical drug development, Phase I studies establish the feasibility of administering a drug to humans and aim to identify the maximum tolerated dose. In Phase II studies, the anticancer activity of the drug is examined and the last phase includes randomized studies to demonstrate that the new drug provides clinical benefit, in a particular disease setting, over a standard treatment. We initiated a similar three-stage pharmacogenomic marker discovery and validation program to determine if regimen-specific multigene predictors of response to preoperative chemotherapy for breast cancer can be developed. During the Phase I of this program, we examined the feasibility of obtaining comprehensive gene expression profiling from needle biopsy specimens. We also studied the variables that can affect the results including the biopsy technique and profiling platform. We examined the intra- and inter-laboratory reproducibility of gene expression data and performed studies to estimate the poten-

**Pharmacogenomic predictive marker
discovery process**

<u>Clinical end-Point:</u>   Pathologic CR to preoperative chemo

<u>Technology:</u>   DNA microarray

<u>Trial design:</u>   Phase 1. Perform feasibility and assess potential accuracy

            Phase 2. Define optimum sample size, develop optimal predictor

            Phase 3. Validate independently, demonstrate clinical utility

            Phase 4. Built portfolio of predictors to several regimens

<u>Outcome:</u> A series of gene-signature-based, regimen-specific predictors of pathological CR

Figure 19.1. Schematic overview of the clinical development of pharmacogenomic predictors of response to therapy.

tial accuracy of multigene predictors of response to a complex chemotherapy regimen. During the second phase of the marker development program, our goals were to determine the necessary sample size to develop reliable multigene predictors, optimize the predictor for final validation and assess the robustness of the predictor in the face of experimental noise. A predictor of response to chemotherapy is the most useful if a portfolio of such tests exists. Each test could then be applied to the same comprehensive gene expression data to determine which regimen is the most effective for a particular tumor. To move towards this goal, we initiated several parallel studies to develop predictors for multiple different preoperative chemotherapy regimens. The third and most critical stage of the marker development process is to prove the clinical utility of the proposed new tests. A new test is clinically useful if it improves patient outcome. This usually can only be established through randomized studies. In the context of prediction of response to preoperative chemotherapy, an attractive two-arm clinical trial design that could prove clinical utility is randomization to receive either individualized chemotherapy based on test result or treatment with the currently "generally most effective" regimen. The study end point could be comparison of pathologic complete response (pCR) rates in the two arms. This design assumes that pCR after preoperative chemotherapy is a valid surrogate of the more important endpoint of prolonged overall survival [8,9].

3. PHASE I PHARMACOGENOMIC MARKER DISCOVERY STUDIES

3.1. Feasibility of gene expression profiling from needle biopsies of breast cancer

The principal assumption behind using preoperative chemotherapy as the clinical trial setting to discover molecular predictors of response is that tumor response can be monitored in the primary site and pCR is a reliable early surrogate of long-term survival. Thus, pCR can be used as an indicator of clinically important benefit from chemotherapy. To what extent patients who achieve less than pCR benefit from preoperative therapy with regards of improvement in survival is unknown. In order to develop molecular markers in this clinical setting, one needs to be able to analyze small amounts of tissue that are obtained with needle biopsy. There are two basic needle biopsy techniques that are commonly used to acquire tissue for diagnosis. These include fine needle aspiration (FNA) and core needle biopsy (CNB). Multiple investigators have shown that both FNA and CNB yield sufficient RNA for gene expression profiling with DNA microarrays [10–12]. On average, needle biopsies of breast cancer yield 1–3 μg of total RNA. More recently reagents and profiling platforms have been developed to retrieve and analyze RNA from formaldehyde-fixed paraffin-embedded tissue specimens [13,14]. These technological advances provide clinical investigators with a repertoire of choice for tissue sampling that suits their trial design best.

It is important to realize, that different sampling methods yield tissues with different cellular composition and distinct gene expression profiles [15]. On average, 40–50% of all cells in CNB are stromal cells including adipocytes, fibroblasts, blood vessels, and leukocytes (Figure 19.2). On the other hand, FNA of breast cancer contain 80–90% neoplastic cells; the remaining cells are mostly leukocytes. Red cell contamination is less of a problem for gene expression profiling studies because mature erythrocytes have no nucleus and contain virtually no RNA. However, excessive amounts of blood in a specimen can interfere with the RNA extraction process. Not surprisingly, the gene expression profiles of FNA and CNB obtained from the same tumor, at the same time and from the same location are slightly but systematically different. This difference is due to stromal gene expression that can be detected in CNB, but absent in FNA. This is important because stromal genes may contain important prognostic or predictive information. It was recently reported that the gene expression profiles of some breast cancers contain a gene signature that is similar to fibroblast in healing wounds. This "wound healing signature" that is most likely contributed by stromal cells was predictive of poor prognosis [16]. On the other hand, stromal genes may contribute little to other types of outcome predictions.

3.2. The impact of profiling platforms on gene expression data

There are multiple different gene expression profiling platforms that can be used to study cancer tissue. It is important to determine to what extent results observed on one platform hold up in another platform. This information is critical for the design of validation studies. It could be argued that if the technology is robust, measuring the expression of the same gene with different microarray platforms should not matter because similar measurements are expected. However, different platforms measure the expression of the same gene with different precision, on a different relative scale and with different dynamic

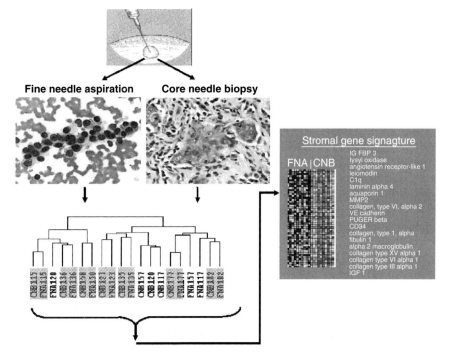

Figure 19.2. Different biopsy techniques yield tissues with different cellular composition from the same tumor. Fine needle aspirations (FNA) yield relatively pure neoplastic cells compared with core needle biopsies (CNB) that contain a mixture of stroma and neoplastic cells. The dendogram generated by unsupervised hierarchical clustering shows that seven of ten matching FNA/CNB pairs cluster together. Indicating similarity between the gene expression profiles. However, pairwise comparison of matching FNA and CNB from the same tumor also revealed important differences due to the abundance of stromal cells in CNB.

range [17,18]. Gene expression values are expressed as normalized relative units and different normalization procedures could result in different numerical values even from the same data. Furthermore, different microarray platforms invariably contain distinct sets of genes that only partially overlap. This makes cross platform testing of multigene predictors particularly challenging. Even when the same genes are represented on distinct platforms, the actual probes usually target different sequences of the target mRNA.

As part of the evaluation of the technology as a potential diagnostic tool, we examined how closely normalized gene expression measurements for the same gene correlate between cDNA and Affymetrix platforms and how well multigene predictors developed on one platform hold up on data generated by the other platform [19]. We profiled the same 33 RNA specimens on cDNA arrays proprietary to Millennium Pharmaceuticals and also on Affymetrix U133A GeneChips. This analysis was not designed to compare the accuracy of the two platforms. This would have required a third, external, gold standard measurement of mRNA expression. The U133A chip contained 22,215 probe sets representing 13,736 unique UniGene clusters, whereas the cDNA arrays contained 30,720 clones corresponding to 21,594 unique UniGene clusters. Only 9,402 clusters were common to both platforms. Some of the most informative genes associated with response to therapy on any particular platform were not present on the other platform. Only 30% of all corresponding gene expression measurements on the two platforms had Pearson correlation coefficient $r \geq 0.7$. There was substantial variation in correlation between different Affymetrix probe sets that matched to the same cDNA probe. In general, genes with higher than median expression values and probes that targeted the same or adjacent sequence of the target mRNA showed the highest correlation between the two platforms (Figure 19.3). This indicates that when probes on different platforms map to the same target sequence then expression measurements can be highly concordant. However, this condition is rarely met when probes from two distinct platforms are compared. Our experiments also showed that multigene predictors optimized for one platform loose discriminating accuracy when applied to data generated by another platform. These findings are consistent with several other reports that indicated modest overall correlation of gene expression measurements across different gene expression profiling platforms (Figure 19.4) [20–22]. Indeed, essentially all published results that attempted cross platform testing of informative genes, regardless of methodology, reported diminished classification accuracy on data generated by other than the original platform [23,24].

From a practical point of view, experimental noise could be considered acceptable if it does not alter the conclusion from an experiment. In the case of multigene predictors, if the same outcome prediction can be made on data from the original and from the replicate samples of the same RNA then the experimental noise is practically insignificant. To examine this, we developed response predictors using different numbers of genes in combination with three commonly used class prediction algorithms (kNN, DLDA, and SVM) corresponding to 111 different predictors that were trained on 62 samples. These multigene predictors were then tested on 31 independent cases that were profiled in replicates. The goal was to examine the prediction agreement on the replicate sets of data. The mean percentage agreement for the DLDA-based predictors was 92% (95% CI: 0.90–0.93, SD 0.009), for the kNN predictors it was 94% (95% CI: 0.92–0.95, SD 0.008) and for the SVM predictors it was 97% (95% CI: 0.96–0.98, SD 0.006). Each prediction method performed well with a broad number of informative genes included (Figure 19.5). These results indicate that the technical reproducibility of gene expression data is sufficiently high to yield essentially identical prediction results from duplicate samples. This degree of reproducibility is the same or better than reported for ER immunohistochemistry or HER-FISH performed on the same samples in different clinical laboratories [25,26]. A large collaborative project between industry and academia

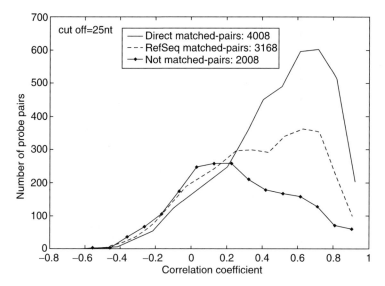

Figure 19.3. Distribution of Pearson correlation coefficients of matching gene expression measurements generated from the same RNA with Affymetrix GeneChips and cDNA microarrays. 33 RNA specimens were hybridized to Affymetrix U133A chips and to cDNA microarrays of Millennium Pharmaceuticals. cDNA and oligonucleotide probes that targeted overlapping sequences on the cRNA for at least 25 nucleotides (direct match, n = 4008 probes) produce highly correlated results. Probes that match to a common RefSeq sequence for at least 25 nucleotides, but at different location along the cRNA (n = 3168) show moderate correlation. Probes that only share a common UniGene annotation, but are unmatched when using RefSeq sequences show poor correlation across the two platforms.

also addressed the cross-platform as well as the intra- and inter-laboratory reproducibility of gene expression data in the context of toxicogenomics (The Hepatotoxicity Working Group of the International Life Science Institute, http:// www.ILSI.org). This research consortium also reported high levels of reproducibility of gene expression data between different commercial and academic laboratories when the same platform and similar experimental procedures were used [27,28].

3.4. Estimated accuracy of multigene predictors of response to preoperative chemotherapy

Several small studies have shown that the gene expression profiles of breast cancers that are sensitive to chemotherapy are different from those tumors that are resistant to treatment (Table 19.1). Numerous mathematical methods can be used to combine these differentially expressed genes into multigene pharmacogenomic predictors of response [29]. In one study, investigators analyzed gene expression data from core needle biopsies of 24 patients with locally advanced breast cancer

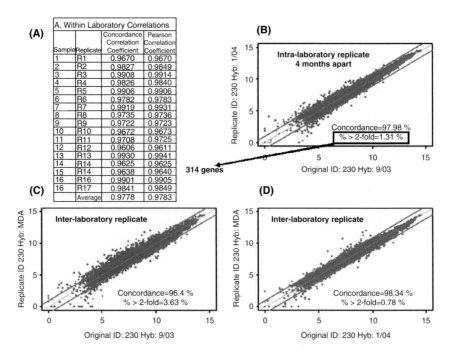

Figure 19.4. Reproducibility of gene expression data. (A) Concordance correlation coefficients of 16 replicate experiments within the same laboratory. (B) The same RNA was profiled on Affymetrix U133A chip in the same laboratory 4 months apart. Note that even if the concordance is 97.98%, 314 genes show greater than twofold difference. (C and D) The same RNA as shown in panel B profiled at a different laboratory using the same technical protocol.

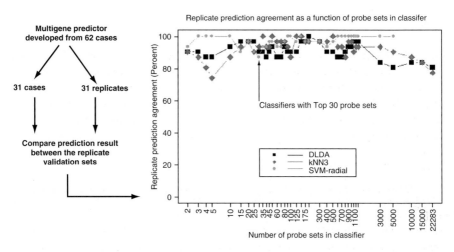

Figure 19.5. Prediction agreements between replicate experiments as function of number of probe sets included in the classifier. kNN, DLDA, and SVM class predictors were trained on 62 cases and tested on replicate data from 31 independent cases. The number of genes included in the prediction models is shown on the X-axis. The Y-axis shows how often the same outcome was predicted on matched pairs of replicate results.[42]

Table 19.1. Phase I pharmacogenomic marker discovery studies in breast cancer

Patients and sampling method	Treatment regimen	Informative genes and platform	Predictive accuracy	Reference
n = 24 + 6 CNB	Docetaxel	92 genes Affymetrix HgU95	88%*	[12]
n = 24 + 18 FNA	Weekly paclitaxel followed by 5-FU/doxorubicin/ cyclophosphamide	74 genes Millennium cDNA array	78%**	[30]
n = 89 FFPE/CNB	Doxorubicin/paclitaxel followed by weekly paclitaxel	86 genes Genomic health RT-PCR	Not reported	[14]
n = 75 CNB	Weekly paclitaxel	23 genes Agilent cDNA	100%*	[31]
n = 10 CNB	Doxorubicin/ cyclophosphamide	37 genes NCI cDNA array	Not reported	[10]
n = 49 CNB	Doxorubicin/cyclophosphamide or doxorubicin/docetaxel	30 genes NKI cDNA array	Not reported	[33]*
n = 21 CNB	Paclitaxel or doxorubicin/ docetaxel or 5-FU/ doxorubicin/ cyclophosphamide	241 genes Millennium cDNA array	Not reported	[32]

The table summarizes the currently available published results on pharmacogenomic predictors of response to preoperative chemotherapy for breast cancer. CNB, core needle biopsy; FNA, fine needle aspiration; FFPE, formaldehyde-fixed paraffin-embedded tissue.
*overall cross-validation accuracy; **overall accuracy in independent validation.

who received preoperative treatment with single agent docetaxel [12]. Sensitivity to therapy was defined as residual diseases volume that was 25% or less than the pretreatment clinical tumor volume. The investigators identified 92 genes that were significantly correlated with the volume of residual cancer, and a genomic predictor was developed. In cross-validation analysis of the same data, ten of 11 sensitive tumors (91% specificity) and 11 of 13 resistant tumors (85% sensitivity) were correctly classified yielding an overall response prediction accuracy of 88%. Our group used gene expression signatures from fine needle aspirates of 24 breast cancers to develop a multigene predictor of pCR to preoperative sequential weekly paclitaxel followed by 5-fluorouracil, doxorubicin, and cyclophosphamide chemotherapy [30]. An overall predictive accuracy of 78% in 21 independent cases was observed in that study. Other investigators have also identified gene expression signatures associated with response to preoperative chemotherapy. Paraffin-embedded biopsy materials from 89 patients were analyzed by real-time RT-PCR to identify which of the 384 candidate genes were associated with a complete pathologic response [14]. Only 86 genes identified in univariate analysis were associated with response to sequential doxorubicin and paclitaxel-based chemotherapy. Genes associated with response to single-agent weekly paclitaxel preoperative chemotherapy also have been reported [31].

An interesting approach to pharmacogenomic research is profiling serial tumor biopsies during chemotherapy [10,32–34]. As a practical diagnostic test this approach may have limitations because predicting response with an imperfect surrogate marker after treatment has begun is less appealing than selecting the right drug before therapy. However, the analysis of transcriptional response to therapy in serial biopsies could lead to new biological insights into mechanisms of response and resistance to chemotherapy. Some consistent observations already emerged from these individually small studies. It is clear that pre- and posttreatment samples from the same tumor cluster together. This indicates that the transcriptional changes induced by chemotherapy combined with the resampling differences are minor compared with the interindividual differences among breast tumors. The limited data also suggest that there is substantial interindividual difference in transcriptional response to the same treatment. In general, cancers that respond to treatment appear to have more transcriptional changes than resistant tumors [33,34].

4. PHASE II PHARMACOGENOMIC MARKER OPTIMIZATION

All of the pharmacogenomic marker discovery studies reviewed in the previous sections represent small, exploratory trials. They established the feasibility of this approach to diagnostics and suggest that multigene predictors may have accuracy that is within a clinically useful range. However, we may be reluctant to commit substantial resources to the validation of a marker that has been developed from 20–30 cases. The goal of the second phase of marker development, after technical feasibility is established, is to optimize the predictor for independent validation. Intuitively, predictors developed from the analysis of larger sample sets are likely to be more accurate. This may particularly hold true for multigene predictors that utilize machine-learning algorithms. These methods rely on supervised classification, in which statistical learning algorithms formulate classification rules that connect gene expression values to observed patient outcomes (e.g., response) [35]. Several different mathematical methods may be used for this purpose, including linear discriminant analysis (LDA), support vector machines (SVM), neural networks (NN), k-nearest neighbors (kNN), and recursive partitioning. In general, these methods are trained on cases with known outcome, the training set, and combine the expression values from multiple genes to predict the outcome in new cases (i.e., test set). The informative genes entered into these algorithms are most commonly picked from a list of differentially expressed genes that was generated from the training data by comparing the two outcome groups of interest. An important goal of the predictive marker optimization process is to estimate what particular algorithm yields the best performance and how many genes need to be included in the prediction model. This is frequently done by comparing the performance of various predictors using different numbers of genes in leave-n-out cross-validation. During this process, a subset of cases is omitted from the training set; informative genes are identified from the remaining subset and a class predictor is constructed which is then tested on the held out cases [36]. This process is repeated many times, with different sets of cases left out on each occasion, and the predictive accuracies are averaged to yield estimated error rates for each particular classifier method. Typically, the classifier that requires the least number of genes to achieve an acceptable rate of correct classification is considered the best. To determine the true predictive accuracy of this optimized classifier, it will need to be tested on an independent set of data.

A common feature of machine learning as well as human learning is that performance improves with training. This implies that the accuracy of learning algorithm-based predictors may improve as the predictor is trained on larger data sets until it reaches an inherent plateau [37]. This leads to a second important question, what the ideal training sample size is to develop a predictor that shows optimal performance. Several different methods have been suggested to calculate sample size for multigene marker discovery. A simple strategy is to base sample-size calculations on the number of cases needed to ensure adequate power for univariate screening of discriminating genes. Assuming that the array data are approximately normally distributed, then one can use standard two-sample testing methods to perform sample-size calculations [38]. Two important limitations of this relatively simple approach are that it requires specification of standard deviation, which can vary from gene to gene, and it can also lead to high false discovery rates (FDR = the proportion of genes identified as being significantly different when, in fact, they are not) due to many fold more variables (genes) than samples in each group. However, preliminary gene expression data from 15–20 cases in each outcome group can be used to estimate standard deviation for genes and methods to control FDR have also been developed. One such method treats the distribution of p values computed for all candidate genes as a mixture of two distributions, one from truly differentially expressed genes and one from the nondifferentially expressed genes [39]. Using this approach, one can specify an acceptable FDR threshold and find a p value threshold that will lead to that FDR. Another appealing strategy to estimate optimal sample size for machine learning-based marker discovery is based on using preliminary results to extrapolate classification results for larger sample sizes. Learning curves may be generated for any particular predictor by fitting inverse power-law models to the preliminary data to predict how fast the classification performance improves as the training set size increases.

We used gene expression data from 82 patients who received preoperative paclitaxel/FAC chemotherapy to develop and optimize a predictor of pCR to this regimen. We evaluated the performance of 780 different class predictors that represented 20 different classification methods (DLDA, SVM with lineal, polynomial and radial kernels, kNN with k = 3, 5, 7, 9, 11, 13, 15, 17, 19, 21, 23, 25, 27, 29, 31, and 33) in combination with 39 distinct gene sets spanning the range from 1 to 22,283 probes ranked by p values and spaced approximately equally on the log scale. Area above the ROC (AAC) curves calculated for each predictor from the means of 100 iterations of fivefold complete cross-validation indicated that classifier performance improved with increasing numbers of genes leveling off at about 80 genes. The majority of the predictors performed equally well compared with each other [40]. Among the classifiers with fewer than 80 genes, a DLDA 30 gene classifier showed the best performance. In fivefold cross-validation in the training set, this predictor showed mean AAC of 18% (95%CI: 0–37%), sensitivity 75% (35–100%), specificity 73% (48–97%), and positive (PPV) and negative predictive values (NPV) of 50% (20–79%) and 90% (75–100%), respectively. To determine if it performed significantly better than chance we did permutation testing in cross-validation. The permutation test p value was 0/1000, none of the 1000 permuted data sets had performance as high or higher than that calculated from the original class labels. This 30-probe set DLDA classifier (DLDA-30) was selected for independent validation. However, the mean AAC point estimates for most of the other classifiers fall within the confidence interval shown by DLDA-30. This indicates that picking the best classifier is a somewhat arbitrary process because many prediction methods and gene sets can perform equally well.

We also estimated how the performance of different predictors may improve with larger training set size by fitting learning curves to existing data and extrapolating the results to larger sample sizes using the method described by Mukherjee et el. [37]. For most classifiers learning curves started to plateau after around 50–60 training cases (including 1/3 of cases with pCR). Figure 19.6 shows an example of learning curves of kNN predictors including various numbers of genes and trained on increasing number of cases. Data beyond 80 cases represent projected results of computer simulation. The results suggest that around 80–100 cases may be sufficient to develop multigene predictors of pCR that operate close to maximal accuracy. However, this number is likely to vary substantially from prediction problem to prediction problem. The necessary sample size depends on the structure of the gene expression data (i.e., extent of difference in gene expression between outcome groups) and frequency of outcome to be predicted.

5. PHASE III VALIDATION OF CLINICAL UTILITY

Once a candidate predictor has been identified and its predictive accuracy estimated during the phase II marker optimization process, the goal of an independent validation study is to (1) define the true sensitivity, specificity and the PPV and NPV with greater precision and (2) to prove clinical utility of the test. No phase III randomized validation studies have been completed so far to demonstrate clinical utility and establish the true predictive values of any proposed pharmacogenomic response predictor. A two-arm phase III validation trial is currently under way at the Nellie B. Connally Breast Center of the University of Texas M. D. Anderson Cancer Center. This study tests the accuracy and regimen specificity of a 26-gene predictor of pCR to paclitaxel/

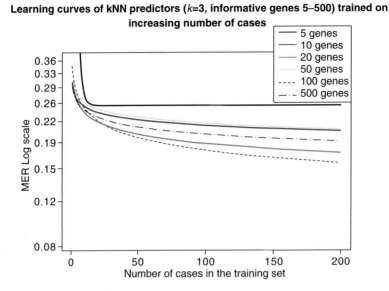

Figure 19.6. Learning curves of kNN predictors including various numbers of genes and trained on increasing number of cases.

FAC chemotherapy that was developed from the analysis of gene expression data from 82 patients. The design of this trial is illustrated in Figure 19.7. An FNA is obtained for transcriptional profiling before starting preoperative chemotherapy, subsequently patients are randomized to receive either six courses of FAC or 12 courses of weekly paclitaxel followed by four courses of preoperative FAC chemotherapy. It is expected that patients with paclitaxel/FAC-sensitive profile will have a 60% pCR rate (corresponding to 60% positive predictive value), whereas the rest of the patients will have a low pCR rate of 5%. Response data from the 110 patients who will receive this therapy will be used to test these assumptions. Data from the remaining patients who received FAC chemotherapy will be used to assess if the predictor is chemotherapy regimen-specific. Since both regimens include FAC, we expect that patients with a paclitaxel/FAC sensitive profile will also have a somewhat higher rate of pCR to FAC only therapy compared with the patients with other profiles. However, we hope that in the sensitive group of patients the pCR rates to FAC only therapy will be significantly lower (40%) compared with paclitaxel/FAC treatment (60%). Final results from this ongoing study are expected within 2 years. Similar studies are underway at other academic centers to determine if drug or treatment-regimen-specific multigene predictors of response can be developed. Preliminary results from a study that aims to develop pharmacogenomic predictors of response to single agent docetaxel and doxorubicin/cyclophosphamide preoperative chemotherapies suggest that

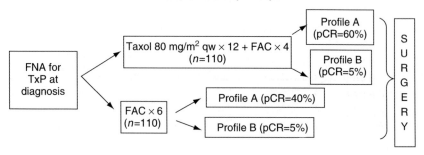

Figure 19.7. Trial to test the accuracy and regimen specificity of a 26-gene predictor of pCR to paclitaxel/FAC chemotherapy.

different sets of genes are associated with response to these two different types of treatments (J. C. Chang, personal communication).

6. CONCLUSIONS

Gene expression profiling of cancer is a promising new technology that may be used in the future to select the most cost-effective and least toxic chemotherapy for an individual based on the molecular profile of her cancer. However, before this technique can be adopted in the clinic, several challenges will need to be solved. The preliminary observations will have to be confirmed, and the true predictive accuracy of these tests will need to be determined in larger clinical trials. To demonstrate clinical utility, randomized, phase III marker validation studies will need to be completed. In order to be most useful, pharmacogenomic predictors will need to be regimen-specific, and a portfolio of such predictors will need to be developed. Multigene predictors utilizing machine-learning algorithms may have several unique features. The number of genes, as well as the individual sequences, included in the prediction model may change over time as more data becomes available for marker discovery to train the predictor. This may result in the introduction of a series of predictors for the same purpose with increasing accuracy. It is also increasingly clear that many different mathematical methods and distinct combinations of genes could yield several different predictors with reasonably similar performance from the same data. This could make the commercial development and regulatory approval of pharmacogenomic predictors challenging. It is likely that if adopted for clinical use, several different profiling platforms may coexist as optional diagnostic tests. Limited profiling using RT-PCR or small, specialized arrays that measure only the genes that contribute to a particular prognostic signature are already commercially available (OncotypeDX®, Genomic Health, Inc, Redwood City, CA; MammaPrint®, Agendia, Amsterdam, The Netherlands). Similar products for chemotherapy prediction may appear on the market within a few years. It is also important to recognize that mRNA-based assessment of the expression of estrogen receptor (ER) and HER-2 gene amplification corresponds very closely to routine immunohistochemistry and FISH results [41]. Microarray-based determination of clinical ER and HER-2 status is quite feasible. Therefore, an "all-in-one" diagnostic service may emerge in the future where traditional single gene markers are reported along with the new multigene markers that predict prognosis and response to various types of treatments, all from a single comprehensive gene expression profile data obtained from a needle biopsy at the time of diagnosis.

REFERENCES

1. Roche H, Fumoleau P, Spielmann M, Canon JL, Delozier T, Kerbrat P, Serin D, Lortholary A, de Ghislain C, Viens P, Bergerat JP, Geneve J, Martin AL, Asselain B. Five years analysis of the PACS 01 trial: 6 cycles of FEC100 vs 3 cycles FEC100 followed by 3 cycles of docetaxel (D) for the adjuvant treatment of node positive breast cancer. *Breast Cancer Res Treat* 88(1): S16, 2004 (Abstr 27).
2. Nabholtz JM, Vannetzel JM, Llory JF, Bouffette P.Advances in the use of taxanes in the adjuvant therapy of breast cancer. *Clinical Breast Cancer* 4: 187–192, 2003.
3. Green MC, Buzdar AU, Smith T, Ibrahim NK, Valero V, Rosales MF, Cristofanilli M, Booser DJ, Pusztai L, Rivera E, Theriault RL, Carter C, Frye D, Hunt KK, Symmans WF, Strom EA, Sahin AA, Sikov W, Hortobagyi GN. Weekly paclitaxel improves pathologic complete remission in operable breast cancer when compared with paclitaxel once every 3 weeks. *J Clin Oncol* 23(25): 5983–5992, 2005.
4. Citron ML, Berry DA, Cirrincione C, Hudis C, Winer EP, Gradishar WJ, et al. Randomized trial of dose-dense versus conventionally scheduled and sequential versus concurrent combination chemotherapy as postoperative adjuvant treatment of node-positive primary breast cancer: first report of Intergroup Trial C9741/Cancer and Leukemia Group B Trial 9741. *J Clin Oncol* 21: 1431–1439, 2003.
5. Pusztai L, Gianni L. Prediction of response to preoperative chemotherapy in operable breast cancer. *Nature Clin Practice Onc* 1: 44–50, 2004.
6. Fodor SP, Rava RP, Huang XC, Pease AC, Holmes CP, Adams CL. Multiplexed biochemical assays with biological chips. *Nature* 364(6437): 555–556, 1993.
7. Pusztai L. Perspectives and challenges of clinical pharmacogenomics in cancer. *Pharmacogenomics* 5(5): 451–454, 2004.
8. Fisher B, Bryant J, Wolmark N, et al. Effect of preoperative chemotherapy on the outcome of women with operable breast cancer. *J Clin Oncol* 16: 2672–2685, 1998.
9. Kuerer HM, Newman LA, Smith TL, et al. Clinical course of breast cancer patients with complete pathologic primary tumor and axillary lymph node response to doxorubicin-based neoadjuvant chemotherapy. *J Clin Oncol* 17: 460–469, 1999.
10. Sotiriou C, Powles TJ, Dowsett M, et al. Gene expression profiles derived from fine needle aspiration correlate with response to systemic chemotherapy in breast cancer. *Breast Cancer Res* 4: R3, 2002.
11. Assersohn L, Gangi L, Zhao Y, et al. The feasibility of using fine needle aspiration from primary breast cancers for cDNA microarray analyses. *Clin Cancer Res* 8: 794–801, 2002.

12. Chang JC, Wooten EC, Tsimelzon A, et al. Gene expression profiling for the prediction of therapeutic response to docetaxel in patients with breast cancer. *Lancet* 362: 362–369, 2003.
13. Baunoch D, Moore M, Reyes M, Cotter P, Bloom K, Erlander M, Ma X-J, Sgroi D. Microarray analysis of formalin fixed paraffin-embedded tissue: the development of a gene expression staging system for breast carcinoma. *Breast Cancer Res Treat* 82(Suppl. 1): S116, 2003 (Abstr 474).
14. Gianni L, Zambetti M, Clark K, Baker J, Cronin M, Wu J, Mariani G, Rodriguez J, Carcangiu M, Watson D, Valagussa P, Rouzier R, Symmans WF, Ross JS, Hortobagyi GN, Pusztai L, Shak S. Gene expression profiles in paraffin-embedded core biopsy tissue predict response to chemotherapy in women with locally advanced breast cancer. *J Clin Oncol* 23(29): 7265–7277, 2005.
15. Symmans WF, Ayers M, Clark E, et al. Fine needle aspiration and core needle biopsy samples of breast cancer provide similar total RNA yield, but different stromal gene expression profiles cancer. *Cancer* 97: 2960–2971, 2003.
16. Nuyten DSA, Chang HY, Sneddon JB, Hart GAM, van't Veer LJ, Peterse HL, Bartelink H, Brown PO, van de Vijver MJ. Reproducibility of several gene expression signatures in predicting outcome in breast cancer. *Breast Cancer Res Treat* 88(1): S22, 2004 (Abstr 36).
17. Ali TR, Li MS, Langford PR. Monitoring gene expression using DNA arrays. *Methods Mol Med* 71: 119–134, 2003.
18. de Bolle X, Bayliss CD. Gene expression technology. *Methods Mol Med* 71: 135–146, 2003.
19. Stec J, Wang J, Coombes K, Ayers M, Hoersch S, Gold D, Ross JS, Hess KR, Tirrell S, Linette J, Hortobagyi GN, Symmans WF, Pusztai L. Comparison of the predictive accuracy of DNA array-based multigene classifiers across cDNA arrays and Affymetrix GeneChips. *J Mol Diagn* 7(3): 357–367, 2005.
20. Kuo WP, Jenssen TK, Butte AJ, Ohno-Machado L, Kohane IS. Analysis of matched mRNA measurements from two different microarray technologies. *Bioinformatics* 18: 405–412, 2002.
21. Yuen T, Wurmbach E, Pfeffer RL, Ebersole BJ, Sealfon SC. Accuracy and calibration of commercial oligonucleotide and custom cDNA microarrays. *Nucleic Acids Res* 30(10 e48): 1–9, 2002.
22. Tan PK, Downey TJ, Spitznagel EL Jr., Xu P, Fu D, Dimitrov DS, Lempicki RA, Raaka BM, Cam MC. Evaluation of gene expression measurements from commercial microarray platforms. *Nucleic Acids Res* 31(19): 5676–5684, 2003.
23. Sorlie T, Tibshirani R, Parker J, Hastie T, Marron JS, Nobel A, Deng S, Johnsen H, Pesich R, Geisler S, Demeter J, Perou CM, Lonning PE, Brown PO, Borresen-Dale AL, Botstein D. Repeated observation of breast tumor subtypes in independent gene expression data sets. *Proc Natl Acad Sci USA* 100: 8418–8423, 2003.
24. Sotiriou C, Neo SY, McShane LM, Korn EL, Long PM, Jazaeri A, Martiat P, Fox SB, Harris AL, Liu ET. Breast cancer classification and prognosis based on gene expression profiles from a population-based study. *Proc Natl Acad Sci USA* 100: 10393–10398, 2003.
25. Roche PC, Suman VJ, Jenkins RB, Davidson NE, Martino S, Kaufman PA, Addo FK, Murphy B, Ingle JN, Perez EA. Concordance between local and central laboratory HER2 testing in the breast intergroup trial N9831. *J Natl Cancer Inst* 94(11): 855–7, 2002.
26. Rhodes B, Jasani DM, Barnes L, Bobrow G, Miller KD. Reliability of immunohistochemical demonstration of oestrogen receptors in routine practice: interlaboratory variance in the sensitivity of detection and evaluation of scoring systems. *J Clin Pathol* 53: 125–130, 2000.
27. Chu TM, Deng S, Wolfinger R, Paules RS, Hamadeh HK. Cross-site comparison of gene expression data reveals high similarity. *Environ Health Perspect* 112(4): 449–455, 2004.
28. Baker VA, Harries HM, Waring JF, Duggan CM, Ni HA, Jolly RA, Yoon LW, De Souza AT, Schmid JE, Brown RH, Ulrich RG, Rockett JC. Clofibrate-induced gene expression changes in rat liver: a cross-laboratory analysis using membrane cDNA arrays. *Environ Health Perspect* 112(4): 428–438, 2004.
29. Radmacher MD, McShane LM, Simon R. A paradigm for class prediction using gene expression profiles. *J Comput Biol* 9: 505–511, 2002.

30. Ayers M, Symmans WF, Stec J, Damokosh A, Clark E, Hess K, et al. Gene expression profiles predict complete pathologic response to neoadjuvant paclitaxel/FAC chemotherapy in breast cancer. *J Clin Oncol* 22: 2284–2293, 2004.
31. Yoshimoto M. Prediction of the therapeutic response to paclitaxel by gene expression profiling in neoadjuvant chemotherapy for breast cancer. *Proc Am Soc Clin Oncol* 22(14S), 2004 (Abstr 500).
32. Buchholz TA, Stivers D, Stec J, Ayers M, Brown J, Bolt A, Sahin AA, Symmans FW, Valero V, Hortobagyi GN, Pusztai L. Global gene expression changes during neoadjuvant chemotherapy of human breast cancer. *Cancer J* 8: 461–468, 2002.
33. Hannemann J, Oosterkamp HM, Bosch CAJ, et al. Changes in gene expression profiling due to primary chemotherapy in patients with locally advanced breast cancer. *Proc Am Soc Clin Oncol* 22(14S), 2004 (Abstr 502).
34. Modlich D, Priscak H-B, Munnes M, Audretsch W, Bojar H. Immediate gene expression changes after first course of neoadjuvant chemotherapy in patients with primary breast cancer disease. *Clin Cancer Res* 10: 6418–6431, 2004.
35. Pusztai L, Hess KR. Clinical trial design for microarray predictive marker discovery and assessment. *Ann Oncol* 15: 1731–1737, 2004.
36. Shoa J. Linear Model Selection by Cross-Validation, *J Am Stat Assoc* 88: 422, 1993.
37. Mukherjee S, Tamayo P, Rogers S, et al. Estimating dataset size requirements for classifying DNA microarray data. *J Comput Biol* 10: 119–142, 2003.
38. Simon R, Radmacher MD, Dobbin K. Design of studies using DNA microarrays. *Genet Epidemiol* 23: 21–36, 2002.
39. Pounds S, Morris SW. Estimating the occurrence of false positives and false negatives in microarray studies by approximating and partitioning the empirical distribution of p-values. *Bioinformatics* 19: 1236–1242, 2003.
40. Pusztai.L, Hess K, Anderson K, Hortobagyi GN, Symmans WF. Affymetrix gene chip profile predicts pathologic complete response to preoperative chemotherapy in breast cancer. San Antonio Breast Symposium 2005 (Abstr 306).
41. Pusztai L, Ayers M, Stec J, Clark E, Stivers D, Damokosh A, Sneige N, Buchholz TA, Esteva FJ, Arun B, Booser D, Rosales M, Valero V, Adams C, Hortobagyi GN, Symmans WF. Gene expression profiles obtained from single passage fine needle aspirations (FNA) of breast cancer reliably identify prognostic/predictive markers such as estrogen (ER) and HER-2 receptor status and reveal large-scale molecular differences between ER-negative and ER-positive tumors. *Clin Cancer Res* 9: 2406–2415, 2003.
42. Anderson K, Hess KR, Kapoor M, Tirrell S, Courtemanche J, Wang B, Wu Y, Gong Y, Hortobagyi GN, Symmans WF, Pusztai L. Reproducibility of gene expression signature based predictions in replicate experiments. *Clin Cancer Res* 12(6): 1721–1727, 2006.

CHAPTER 20

VASCULAR ENDOTHELIAL GROWTH FACTOR INHIBITORS IN COLON CANCER

EDUARDO DÍAZ-RUBIO

Hospital Clínico San Carlos, Madrid, Spain

Abstract: Angiogenesis, the process by which new blood vessels are formed from preexisting vessels, is essential for growth and progression of solid tumors. A number of growth factors regulate angiogenesis, but vascular endothelial growth factor (VEGF) is the most important. Activation of the VEGF pathway triggers a range of signaling processes, stimulating vascular endothelial cell growth, migration, survival, and permeability. VEGF is expressed by most human cancers and its expression often correlates with tumor progression and poor prognosis. In contrast, VEGF plays a limited role in adult physiological angiogenesis, making it an ideal target for therapeutic agents designed to inhibit tumor angiogenesis. Disruption of VEGF signaling can occur through inhibition of the VEGF ligand or receptors, and preclinical models have shown that VEGF and VEGF receptors are viable targets for antiangiogenic agents.

Many anti-VEGF/VEGF receptor agents are in development for the treatment of various cancers. Combining these agents with chemotherapy regimens results in enhanced antitumor activity, possibly due to normalization of the existing tumor vasculature or increased endothelial/tumor cell apoptosis. Only two such agents have been evaluated in phase III clinical trials. Bevacizumab (Avastin), an anti-VEGF monoclonal antibody, has been investigated in combination with the most commonly used chemotherapy regimens for the treatment of metastatic colorectal cancer (mCRC). Phase III trial data show that bevacizumab and concomitant chemotherapy significantly improve patient outcomes in the first- and second-line settings. Bevacizumab is the only antiangiogenic agent approved for the treatment of mCRC. PTK787 (vatalanib), a VEGF receptor tyrosine kinase inhibitor (TKI), is the only other agent with phase III data in mCRC. The CONFIRM-1 trial showed that PTK787 in combination with FOLFOX4 produced a small increase in progression-free survival that is neither clinically or statistically significant. Anti-VEGF/VEGF receptor agents are generally well tolerated and do not increase the incidence of chemotherapy-related toxicities. However, they are associated with side effects, including hypertension and thromboembolism. Additionally, drug-specific adverse events may be observed, such as gastrointestinal perforation with bevacizumab and neurological effects with PTK787.

The ongoing development of anti-VEGF/VEGF receptor agents is expected to improve the management of mCRC. Trial data from studies using bevacizumab in the adjuvant setting and first line in combination with FOLFOX and XELOX are keenly awaited. It will also be interesting to evaluate the overall survival data from CONFIRM-1, as well as the second-line CONFIRM-2 trial, to clarify the role of PTK787 in CRC.

1. INTRODUCTION

Angiogenesis, the process whereby new blood vessels are formed from preexisting vessels, is essential for both embryonic and infant growth and development. Angiogenesis continues throughout development, with new vessels forming by sprouting or splitting from existing vessels, followed by remodeling. However, in healthy adults, angiogenesis occurs at a greatly decreased rate and is associated primarily with wound healing and the menstrual cycle.[1]

Both normal and tumor cells rely on angiogenesis to ensure a constant supply of nutrients and growth factors; tumor cells also require this new blood supply to allow metastasis to distant sites and local increases in blood vessel density are known to precede rapid tumor growth.[2] Tumors <2 mm in size are dormant and receive their nutrients simply by diffusion. These tumors may remain dormant for many years before angiogenesis allows them to grow and metastasize. Growth of a tumor to a size larger than 1–2 mm in diameter requires access to a vascular supply. This involves stimulating the neighboring host mature vasculature to sprout new blood vessel capillaries, which grow towards, and subsequently infiltrate, the tumor mass.[3]

Angiogenesis is regulated by a balance of pro- and antiangiogenic factors. The process by which the tumor becomes able to stimulate the development of blood vessels is often referred to as the "angiogenic switch," which results from tumor mutation that produces an imbalance between pro- and antiangiogenic factors in favor of pro-angiogenic factors.[4] Research has shown that of the known proangiogenic factors, VEGF is the key mediator of both normal and tumor angiogenesis.[5,6]

The "angiogenic switch" itself leads to a complex series of events, including endothelial cell activation and the release of proteolytic enzymes, followed by endothelial cell migration, proliferation, and capillary tube formation.[7–12] Factors such as activation of oncogenes and loss of tumor suppressor genes are also capable of inducing the "angiogenic switch."[13] Once initiated, tumor angiogenesis allows growth of the primary tumor and the resulting tumor vasculature is also a route for metastatic spread of individual cancer cells.[14]

As the tumor vasculature results from abnormal angiogenic stimulation, it has an abnormal structure and function. Tumor blood vessels are immature and dependent on VEGF, irregularly shaped, abnormally branched, imperfectly lined with endothelial cells, and prone to hemorrhage, do not always connect properly to other vessels, and are not organized into definitive venules, arterioles, and capillaries. Tumor vessels do not possess the tight endothelial layer that is essential for normal barrier function, leading to leakage from the vessels[15] and consequently an increase in interstitial fluid pressure.[16] As a result of these irregularities, blood flow within tumor vessels is generally slow, which can lead to dysfunctional capillaries,[17] and normal pressure gradients are reversed compromising drug delivery.

Much research has been devoted to developing agents that will target tumor angiogenesis. The VEGF pathway has been identified as a key target for inhibition.

2. VASCULAR ENDOTHELIAL GROWTH FACTOR

VEGF is a member of a family of angiogenic and lymphangiogenic growth factors. The VEGF family comprises six glycoproteins: VEGF-A to E and placental growth factors (PlGF)-1 and -2.[1, 18, 19] VEGF (also known as VEGF-A) is a homodimeric glycoprotein of 34–46 kDa in size that exists in at least seven isoforms, resulting from alternative exon splicing of VEGF messenger ribonucleic acid (mRNA).[1, 20] The most common VEGF isoform is $VEGF_{165}$, so called because there are 165 constituent amino acids in the protein. VEGF has a number of biological functions that contribute to angiogenesis. These include growth and proliferation of vascular endothelial cells, migration of vascular endothelial cells, survival of immature endothelial cells through prevention of apoptosis, and increased vascular permeability in capillaries.[1, 20]

Expression of the *VEGF* gene is regulated by various factors, including hypoxia, pH, cellular transformation, growth factors, hormones, and oncogenes. Of these, hypoxia is the most important. Hypoxia induces VEGF expression by signaling through hypoxia-inducible transcription factors (HIFs), which upregulate many angiogenic genes.[11] HIF-1 is a heterodimer that binds to a specific region within the *VEGF* gene known as the hypoxia-response element (HRE), activating *VEGF* gene transcription. In the presence of oxygen, HIF-1 is bound to the tumor suppressor von Hippel-Lindau (VHL) protein and remains inactive. Under hypoxic conditions, HIF-1 is phosphorylated and interacts with transcription factors in the nucleus to activate genes, including *VEGF* and *VEGF receptor-1*.[21]

VEGF is expressed by colorectal cancers (CRCs) as well as a variety of other solid tumors, and expression is significantly correlated with the extent of tumor angiogenesis,[22] advanced lymph node status and distant metastasis.[23] VEGF is also closely correlated with other prognostic indicators,[24] and early preclinical data have implicated VEGF in tumor growth.[25]

2.1. VEGF receptors

VEGF binds primarily to two receptors situated on endothelial cells,[26] VEGF receptor-1 (also known as Flt-1) and VEGF receptor-2 (also known as Flk-1 or KDR). Each of these receptors has an extracellular domain, which binds VEGF, a single transmembrane region and a tyrosine kinase intracellular domain. VEGF receptor-1 is crucial for physiologic and developmental angiogenesis and was originally thought to act as a "decoy receptor" for VEGF,[27] reducing the number of VEGF molecules able to bind to VEGF receptor-2. However, recent studies have shown that VEGF receptor-1 is also able to induce a mitogenic signal.[1, 28] VEGF receptor-2 mediates the majority of downstream signaling events stimulated by VEGF in angiogenesis and there is evidence to suggest that activation of VEGF receptor-2 alone is enough to efficiently stimulate angiogenesis.[29] A third VEGF receptor has also been described (VEGF receptor-3). This receptor was originally cloned from a human leukemia cell line and human placenta,[30, 31] and does not bind VEGF, but preferentially binds VEGF-C and VEGF-D.

2.2. The role of VEGF in angiogenesis

VEGF is recognized as the key mediator of angiogenesis. Activation of the VEGF pathway triggers multiple signaling events that result in endothelial cell survival, mitogenesis, migration and differentiation, and mobilization of endothelial progenitor cells from the bone marrow into the peripheral circulation.[32] Thus, VEGF is a key factor in controlling many of the processes essential to tumor angiogenesis: it stimulates vascular endothelial cells to grow and migrate to form blood vessels; and it enables these vessels to survive, even if they are immature or abnormally formed.[33] Furthermore, the activity of VEGF has other effects that influence the characteristics of tumor blood vessels. Tumor blood vessels are hyperpermeable and VEGF is known to be a permeability factor; hyperpermeability causes leakage of fluids from vessels into the tumor[34, 35] and an increase in interstitial pressure, effectively reversing the normal pressure gradients within tissues and preventing the penetration of molecules, including chemotherapeutic agents.[16] VEGF may also be involved in lymphangiogenesis, and stimulation of the formation of giant lymphatic vessels in tumors may provide another possible route for metastatic spread.[36]

Based on this, targeting VEGF as a therapeutic strategy has the potential to prevent tumor angiogenesis. By targeting a factor that influences the growth of endothelial cells lining tumor blood vessels rather than tumor cells, the need for drugs to penetrate the tumor is removed. Furthermore, unlike tumor cells, vascular endothelial cells are genetically stable and therefore there is potentially less risk of acquired resistance to therapy.[37]

In addition to promoting tumor growth by inducing angiogenesis, VEGF may also support tumor cell growth and survival through alternative mechanisms, including the induction of antiapoptotic signals,[38] inducing resistance to radiotherapy, increasing the likelihood of metastasis and suppressing the antitumor immune response.[39]

2.3. VEGF as a target for anticancer therapy

As previously stated, the role of VEGF and angiogenesis in adulthood is limited. Thus, targeting VEGF would be expected to inhibit tumor angiogenesis and thus growth without compromising the function of normal vasculature in healthy tissues. VEGF signaling can be targeted by developing agents that bind to and inhibit the VEGF ligand or its receptor(s), preventing receptor activation and downstream signaling. Both are potentially viable targets for anticancer therapy and a number of different treatment options are currently in development. Differences exist between agents that inhibit VEGF and those that inhibit VEGF receptor (Table 20.1).

Examples of agents that inhibit VEGF or its receptors include monoclonal antibodies, soluble receptors and small molecule TKIs. Monoclonal antibodies (e.g., bevacizumab [Avastin]) bind to VEGF, effectively removing circulating free VEGF and preventing it from binding to its receptor.[39, 40] There are several benefits to be gained from targeting VEGF in this way. Monoclonal antibodies are custom designed and consequently are highly specific and bind only to the desired molecule. They are able to inhibit VEGF activity on all receptors with which the ligand interacts. Therefore, the proangiogenic effects of VEGF mediated through all of the receptors to which it binds can be inhibited. Soluble receptors, such as VEGF-Trap, also bind directly to the VEGF ligand to inhibit its binding to VEGF receptors.

Monoclonal antibodies can also be used to target the VEGF receptor, making the receptor unavailable for binding with VEGF. Again, the ability to custom design monoclonal antibodies means that a single receptor can be targeted. This means, however, that although VEGF signaling through a single receptor may be blocked, signaling through other VEGF receptors will not be affected

Table 20.1. Selected differences between VEGF and VEGF receptor inhibitors

Agents targeting VEGF	Agents targeting VEGF receptor
Inhibit activities of ligand on endothelial and nonendothelial cells	Small molecule inhibitors of VEGF receptor have variable specificity
Inhibit VEGF activity on all of its receptors	May inhibit VEGF signaling through a single receptor but not all receptors
May not affect the function of other VEGF family ligands	Inhibit the activity of other VEGF family members signaling through the same receptor

and the effects of VEGF may still be expressed. Alternatively, the effects of other ligands acting through the same receptor may be inhibited, causing unwanted effects.

Another method for targeting VEGF receptors is the use of small molecule TKIs, which directly inhibit VEGF receptor tyrosine kinase within the cell, effectively inhibiting VEGF receptor signaling. The potential drawbacks of utilizing TKIs are a lack of specificity and failure to bind and inhibit all VEGF receptors, allowing receptor stimulation and angiogenesis by circulating free VEGF.

Based on the known activity of VEGF in tumor angiogenesis, the likely effects of VEGF inhibition are: (1) inhibition of new tumor vessel development, inhibiting tumor growth and metastasis; (2) regression of immature vessels, causing tumor cell death and growth inhibition; and (3) normalization of the function of remaining vessels, improving the delivery of other therapies.

3. VEGF INHIBITORS IN CLINICAL DEVELOPMENT FOR USE IN COLORECTAL CANCER

To date, several VEGF inhibitors have been developed and are under clinical investigation for the treatment of CRC. Two main classes of inhibitor are currently being evaluated: (1) monoclonal antibodies and (2) TKIs.

3.1. Monoclonal antibodies

Several anti-VEGF or anti-VEGF receptor monoclonal antibodies are under investigation, with bevacizumab being the best developed. Others include IMC-1121B and 2C3. IMC-1121B (ImClone) is a fully humanized monoclonal antibody targeted against VEGF receptor-2. It is currently undergoing phase I evaluation in a dose-escalation study conducted at the Fox Chase Cancer Center and the University of Colorado. The primary endpoint of the study is to establish the safety profile and maximum tolerated dose of weekly IMC-1121B administered to patients who have previously failed standard treatment for advanced cancer or to those patients for whom no standard therapy is available.

2C3 is a monoclonal antibody that selectively blocks the binding of VEGF to VEGF receptor-2. In preclinical studies, 2C3 has been shown to be effective at controlling the growth of human tumor xenografts in mice.[41, 42]

The remainder of this section focuses on bevacizumab. A4.6.1 is an anti-human VEGF monoclonal antibody produced in mice that binds to human VEGF. It has no effect on mouse VEGF[43] and binds at a specific site within the human VEGF molecule with high affinity (dissociation constant = 8×10^{-10} M). A4.6.1 is unsuitable for use in humans because it is likely to provoke an immune response that could lead to anaphylactic shock. Consequently, a humanized form of A4.6.1, bevacizumab, has been developed. Bevacizumab inhibits the binding of all VEGF isoforms to their receptors,[43] with a binding affinity of 1.8 nM, close to that of its parent antibody, A4.6.1. Bevacizumab is currently the only antiangiogenic and anti-VEGF agent approved for use in CRC, or any other form of cancer, and it is also undergoing clinical evaluation in a variety of other cancers, including breast and non-small cell lung cancer.

3.1.1. Preclinical studies of bevacizumab

In preclinical studies, bevacizumab was shown to inhibit VEGF-induced proliferation of endothelial cells and to inhibit angiogenesis and the growth of human tumor xenografts in mice.[43] Bevacizumab has also been shown to have synergistic activity in combination with different chemotherapy regimens. For example, the addition of bevacizumab to capecitabine (Xeloda) resulted in improved inhibition of growth of colorectal tumor xenografts in mice.[44] This synergy has been attributed to the different mechanisms of action and targets of the therapies used. Preclinical toxicity data indicated that bevacizumab should be well tolerated both as a single dose and in repeat doses over 6 months.[45] Therefore, clinical studies were implemented to test the efficacy and safety of bevacizumab in human patients.

3.1.2. Clinical development of bevacizumab

Early phase I trials were carried out to investigate the activity and safety of bevacizumab as a single agent[46] and in combination with different chemotherapy regimens.[47] In the monotherapy study, 25 patients were accrued and all were eligible for safety assessment. Dose levels of 0.1, 0.3, 1.0, 3.0, and 10.0 mg/kg were investigated and bevacizumab was administered on days 0, 28, 35, and 42. There were no grade III or IV infusion-related adverse events; however, a small number of patients reported grade 1 or 2 asthenia, headache, fever, and rash as the most common adverse events. There were also reported cases of increased blood pressure and bleeding, which were thought to be related to bevacizumab. Bevacizumab demonstrated a favorable pharmacokinetic profile (demonstrating a half-life of approximately 21 days at doses ≥ 0.3 mg/kg) and preliminary efficacy results were encoraging, with 12 out of 23 patients achieving stable disease 70 days after commencing treatment. No patients achieved objective partial or complete responses.

Margolin et al.[47] investigated the combination of bevacizumab with doxorubicin (Adriamycin), carboplatin (Paraplatin)/paclitaxel (Taxol) or 5-fluorouracil (5-FU)/leucovorin (LV) in 12 patients with advanced solid tumors. Patients were excluded if they had evidence of central nervous system (CNS) disease, pregnancy, hypercalcemia or other major medical illness. Bevacizumab was administered initially as a 90-min intravenous infusion, followed by a second infusion over 60 min and all subsequent infusions over 30 min in the absence of any adverse events. Doxorubicin was administered as a 50 mg/m^2 dose every 28 days; carboplatin and paclitaxel were also administered every 28 days (area under the curve [AUC] = 6 and 175 mg/m^2, respectively); 5-FU (500 mg/m^2) was administered weekly for 6 weeks with LV (20 mg/m^2) every 8 weeks. The limited number of patients enrolled in this study prevents any definite conclusions regarding the antitumor effects of the combination therapies being made. Grade 3 toxicities were rare and were attributed to the chemotherapy regimen, not bevacizumab. Reported toxicities were diarrhea (one patient), thrombocytopenia (two patients), and leukopenia (one patient). The authors concluded that the regimens were well tolerated and appropriate for further study in randomized clinical trials.

Following this demonstration of the safety of bevacizumab in these early studies, bevacizumab entered phase II and III trials in CRC and other tumor types. In CRC, bevacizumab has been investigated with irinotecan (Camptosar)/5-FU/LV (IFL),[48] 5-FU/LV,[49] 5-FU/LV plus oxaliplatin (Eloxatin) (FOLFOX4),[50] capecitabine plus oxaliplatin (XELOX),[51] and cetuximab (Erbitux) with or without irinotecan.[52]

3.1.3. Investigation of bevacizumab first line in metastatic CRC

The initial clinical trial of bevacizumab in metastatic CRC was a phase II trial investigating the safety, efficacy, and pharmacokinetics of the combination of two different doses of bevacizumab (5 mg/kg and 10 mg/kg) with 5-FU/LV.[49] A total of 104 patients were randomly assigned to one of three treatment arms 5-FU/LV alone, bevacizumab 5 mg/kg every 2 weeks plus 5-FU/LV or bevacizumab 10 mg/kg every 2 weeks plus 5-FU/LV. All patients received 5-FU/LV weekly for the first 6 weeks of an 8-week cycle according to the Roswell Park regimen.[53] Bevacizumab was administered as a continuous infusion over 90 min every 2 weeks for up to 48 weeks or until disease progression.

Efficacy appeared best in the bevacizumab 5 mg/kg arm (progression-free survival of 9 months compared with 7.2 months for the 10 mg/kg arm and 5.2 months for the control arm) (Table 20.2), with a statistically significant 61% reduction in the hazard of progressing and a reduction in the hazard of death of 37%. The latter result was not statistically significant, possibly because 22 of the 36 patients in the 5-FU/LV alone arm crossed over to receive single-agent bevacizumab 10 mg/kg every 2 weeks at progression. In these 22 patients, median duration of therapy was 2 months, two patients achieved a partial response and seven patients had stable disease. These data suggested that bevacizumab might have efficacy in patients with pretreated metastatic CRC.

Grade 3/4 adverse events were comparable in both bevacizumab treatment arms, but were increased in incidence compared with the control arm (Table 20.3). This increase could be a result of the longer duration of treatment in the

Table 20.2. Efficacy results from a phase II trial of 5-FU/LV with or without bevacizumab[49]

	Control ($n = 36$)	Bevacizumab		
		5 mg/kg ($n = 35$)	10 mg/kg ($n = 33$)	Pooled ($n = 68$)
Median time to progression (months)	5.2	9.0	7.2	7.4
Response rate (%)	17	40	24	32
Median survival (months)	13.8	21.5	16.1	—

Table 20.3. Selected grade 3/4 adverse events in a phase II trial of 5-FU/LV with or without bevacizumab[49]

		Bevacizumab	
	Control ($n = 35$)	5 mg/kg ($n = 35$)	10 mg/kg ($n = 32$)
Any event (%)	54	74	78
Diarrhea (%)	37	28	31
Leukopenia (%)	3	6	3
Stomatitis (%)	0	0	0
Fever (%)	0	0	3
Headache (%)	0	0	3
Rash (%)	0	3	0
Chills (%)	0	0	0
Abdominal pain (%)	3	9	12
Weight loss (%)	0	3	0
Gastrointestinal hemorrhage (%)	0	0	9
Epistaxis (%)	0	0	0
Hypertension (%)	0	9	25
Infection (%)	0	0	3
Thrombotic events (%)	3	14	6

bevacizumab arms. Bevacizumab therapy was associated with hypertension, proteinuria, and epistaxis. The incidence of thrombosis was identified as a cause for concern as it was fatal in one patient and caused bevacizumab discontinuation in three other patients. Notwithstanding, bevacizumab was generally viewed as well tolerated.

These preliminary study results demonstrated that bevacizumab, in combination with 5-FU/LV, increases response rate and prolongs time to progression compared with 5-FU/LV alone in patients with metastatic CRC. The bevacizumab 5 mg/kg every 2 weeks dose was selected for use in further trials based on these data.

A large phase III trial of bevacizumab was initiated in 1999 when IFL was the standard therapy in the USA.[48] In this trial, patients with previously untreated metastatic CRC were randomized to IFL plus placebo, IFL with bevacizumab 5 mg/kg every 2 weeks or 5-FU/LV (Roswell Park) plus bevacizumab 5 mg/kg every 2 weeks. Eight hundred and thirteen patients were randomly assigned to IFL plus bevacizumab or IFL plus placebo. The third treatment arm (5-FU/LV plus bevacizumab) was discontinued after approximately 100 patients had been enrolled when a planned interim safety analysis concluded that the combination of IFL plus bevacizumab had an acceptable safety profile. The primary endpoint was overall survival; secondary endpoints were progression-free survival, response rate, duration of response, safety, and quality of life.

Patients were excluded if they had previously undergone chemotherapy or biological therapy for metastatic disease, any radiotherapy within 14 days of study commencement, major surgery within 28 days of study commencement, clinically significant cardiovascular disease or known CNS metastases. Bevacizumab (5 mg/kg) was administered every 2 weeks, concurrently with chemotherapy. All patients were followed until death, loss to follow-up or termination of the study.

This trial demonstrated that adding bevacizumab to IFL significantly increases overall survival from 15.6 to 20.3 months (Table 20.4). Median time to progression and response rate were also increased in the bevacizumab treatment arm. Adverse events are summarized in Table 20.5. The incidence of any grade 3/4 adverse event was approximately 10% higher in those patients treated with IFL plus bevacizumab compared with the control arm. Importantly, there were no significant differences in the incidence of events leading to hospitalization or in the 60-day any cause mortality rate. Bevacizumab-related events included hypertension, proteinuria, and minor bleeding, all of which were either asymptomatic or manageable. However, gastrointestinal perforation, which had not been observed in other trials of bevacizumab, also occurred, albeit uncommonly. This

Table 20.4. Efficacy results from the phase III trial of IFL with or without bevacizumab[48]

	IFL/placebo ($n = 411$)	IFL/bevacizumab ($n = 402$)	P
Median survival (months)	15.6	20.3	<0.001
1-year survival rate (%)	63.4	74.3	<0.001
Progression-free survival (months)	6.2	10.6	<0.001
Overall response rate (%)	34.8	44.8	0.004
Complete response (%)	2.2	3.7	–
Partial response (%)	32.6	41.0	–
Median duration of response (months)	7.1	10.4	0.001

Reproduced with permission. Copyright 2004 Massachusetts Medical Society. All rights reserved.

Table 20.5. Safety results from the phase III trial of IFL with or without bevacizumab[48]

	IFL/placebo ($n = 397$)	IFL/bevacizumab ($n = 393$)
Any grade 3/4 adverse event (%)	74.0	84.9
Adverse event leading to hospitalization (%)	39.6	44.9
Adverse event leading to treatment discontinuation (%)	7.1	8.4
Adverse event leading to death (%)	2.8	2.6
Death within 60 days (%)	4.9	3.0
Leukopenia (grade 3/4, %)	31.1	37.0
Diarrhea (grade 3/4, %)	24.7	32.4
Hypertension		
Any	8.3	22.4
Grade 3	2.3	11.0
Any thrombotic event (%)	16.2	19.4
Deep thrombophlebitis (%)	6.3	8.9
Pulmonary embolus (%)	5.1	3.6
Bleeding (grade 3/4, %)	2.5	3.1
Proteinuria		
Any (%)	21.7	26.5
Grade 3 (%)	0.8	0.8
Gastrointestinal perforation (%)	0.0	1.5

Reproduced with permission. Copyright 2004 Massachusetts Medical Society. All rights reserved.

event has subsequently been found to affect approximately 1.5% of bevacizumab-treated patients and can be fatal. Therefore, it is essential that physicians using bevacizumab are aware that this event can occur and are familiar with the symptoms to enable early diagnosis.

Subsequently, data from the third arm of this trial have been reported. In agreement with the results of the phase II trial conducted by Kabbinavar et al. (2003),[49] these also show that adding bevacizumab to 5-FU/LV improves patient outcomes.[54] Further support is added by the results of another phase II trial comparing the same regimens, but in patients not suitable for irinotecan therapy.[55] This trial enrolled 209 patients with previously untreated metastatic CRC; these patients had a higher median age than those treated in other bevacizumab trials in metastatic CRC. The primary endpoint of this study was overall survival and secondary endpoints included progression-free survival, objective response rate, response duration and change in quality of life as measured by the Functional Assessment of Cancer Therapy-Colorectal (FACT-C) quality of life questionnaire.

The median duration of therapy was 23 weeks in the placebo treatment arm and 31 weeks in the bevacizumab treatment arm. The addition of bevacizumab to FU/LV was associated with increases in overall survival, median progression-free survival, response rate and median duration of response (Table 20.6). Bevacizumab was again well tolerated, with side effects similar to those observed in other trials (Table 20.7).

To further evaluate the efficacy of adding bevacizumab to 5-FU/LV, a combined analysis was also carried out.[56] Efficacy data from the bevacizumab 5 mg/kg every 2 weeks plus 5-FU/LV arms of the three studies described were compared with data for 5-FU/LV alone or IFL, the control arm in the phase III trial. The inclusion of IFL-treated patients in the control arm biases against seeing a benefit of adding bevacizumab to 5-FU/LV because IFL is a more effective regimen than 5-FU/LV. This combined analysis allows a more robust assessment of the efficacy of bevacizumab plus 5-FU/LV.

In total, 241 patients were included in the combined control group (receiving either FU/LV or IFL) and 249 patients received 5-FU/LV/bevacizumab. The primary efficacy endpoint was duration of survival; progression-free survival, median time to disease progression and objective response during first-line therapy were also investigated. Efficacy results are summarized in Table 20.8.

Table 20.6. Efficacy results from a phase II trial of 5-FU/LV with or without bevacizumab in patients unsuitable for irinotecan[55]

	5-FU/LV/placebo (*n* = 105)	5-FU/LV/bevacizumab (*n* = 104)	*P*
Median survival (months)	12.9	16.6	0.160
Progression-free survival (months)	5.5	9.2	0.0002
Overall response rate (%)	15.2	26.0	0.055
Duration of response (months)	6.8	9.2	0.088

Table 20.7. Safety results from a phase II trial of 5-FU/LV with or without bevacizumab in patients unsuitable for irinotecan[55]

	5-FU/LV/placebo (n = 104)	5-FU/LV/bevacizumab (n = 100)
Any grade 3/4 adverse event (%)	71	87
Adverse event leading to study discontinuation (%)	12	10
Adverse event leading to death (%)	7	4
All-cause mortality at 60 days (%)	14	5
Diarrhea (grade 3/4, %)	40	39
Leukopenia (grade 3/4, %)	7	5
Hypertension		
Any (%)	5	32
Grade 3 (%)	3	16
Thrombotic events (any, %)	18	18
Deep thrombophlebitis (%)	9	6
Pulmonary embolus (%)	2	3
Arterial thrombotic event (any, %)	5	10
Bleeding (grade 3/4, %)	3	5
Proteinuria		
Any (%)	19	38
Grade 3 (%)	0	1
Gastrointestinal perforation (%)	0	2

Reproduced with permission from the American Society of Clinical Oncology.

Median duration of survival, response rate and the median duration of progression-free survival were all increased in the bevacizumab treatment arm, indicating that the addition of bevacizumab to 5-FU/LV provides significantly greater benefit to patients with previously untreated metastatic CRC than 5-FU/LV alone. The safety analysis was complicated by the inclusion of irinotecan in the combined control group but overall, tolerability was as expected (Table 20.9).

Data from these trials have led to the approval of bevacizumab for the first-line treatment of metastatic CRC worldwide. The precise approved use varies, but currently available data solidly support the use of bevacizumab in combination with 5-FU/LV with or without irinotecan.

3.1.4. Bevacizumab in combination with oxaliplatin-containing regimens

Bevacizumab has also been combined with oxaliplatin-containing regimens to great effect. The phase III E3200 trial investigated the combination of

Table 20.8. Efficacy results from the combined analysis of 5-FU/LV with bevacizumab[56]

	FU/LV or IFL (n = 241)	FU/LV/bevacizumab (n = 249)
Median duration of survival (months)	14.6	17.9
Median progression-free survival (months)	5.5	8.8
Total objective response (%)	24.5	34.1
Complete response (%)	0.8	2.4
Partial response (%)	23.7	31.7

Table 20.9. Safety results from the combined analysis of 5-FU/LV with bevacizumab[56]

	FU/LV or IFL ($n = 237$)	FU/LV/bevacizumab ($n = 244$)
Any grade 3/4 adverse event (%)	73	81
Adverse event leading to study discontinuation (%)	8	10
Diarrhea (grade 3/4, %)	34	37
Leukopenia (grade 3/4, %)	19	5
Hypertension		
Any (%)	3	16
Grade 3 (%)	3	16
Thrombotic events (any, %)	17	17
Deep thrombophlebitis (%)	6	7
Pulmonary embolus (%)	3	3
Arterial thrombotic event (any, %)	3	5
Bleeding (grade 3/4, %)	2	5
Proteinuria		
Any (%)	19	32
Grade 3 (%)	0	1
Gastrointestinal perforation	0	1

Reproduced with permission from the American Society of Clinical Oncology.

bevacizumab, at a higher than standard dose of 10 mg/kg every 2 weeks, with FOLFOX4 in patients with previously treated advanced CRC.[50] This phase III trial accrued 829 patients over 17 months who were randomized to receive either bevacizumab plus FOLFOX4, FOLFOX4 alone or bevacizumab alone. The majority of patients had received prior chemotherapy for advanced disease and a small number of patients had received adjuvant chemotherapy. The primary objective was overall survival and secondary endpoints included response rate, progression-free survival and safety. The combination of bevacizumab and FOLFOX4 was superior to either agent alone in terms of efficacy (Table 20.10).

Median overall survival, the primary endpoint, was increased to 12.9 months in the combination arm compared with 10.8 months in the FOLFOX4 alone arm and 10.2 months in the bevacizumab alone arm. Similarly, median progression-free survival was increased in the combination arm (7.2 months compared with 4.8 months in the FOLFOX4 alone arm, and 2.7 months in the

Table 20.10. Efficacy results from trial E3200 of FOLFOX with or without bevacizumab in patients with pretreated metastatic CRC

	FOLFOX4/bevacizumab ($n = 289$)	FOLFOX4 ($n = 290$)	Bevacizumab ($n = 243$)
Median overall survival (months)	12.9	10.8	10.2
Median progression-free survival (months)	7.2	4.8	207
	($n = 271$)	($n = 271$)	($n = 230$)
Overall response (%)	21.8	9.2	3.0
Complete response (%)	1.9	0.7	0.0
Partial response (%)	19.9	8.5	3.0
Stable disease (%)	51.7	45.0	29.1

bevacizumab alone arm). Response rate was also improved in the combination arm, at 21.8% compared with 9.2% for FOLFOX4 alone and 3% for bevacizumab alone. The most commonly reported grade 3/4 adverse events were hypertension, neuropathy, and vomiting, which were all increased in the combination arm with the exception of hypertension, which was increased in the bevacizumab alone treatment arm. Gastrointestinal perforation was infrequent and occurred only in patients treated with bevacizumab (Table 20.11).

The TREE-2 study is investigating the combination of bevacizumab with each of three oxaliplatin-based regimens: FOLFOX4, bFOL (a bolus 5-FU-containing regimen), and XELOX.[57] The primary endpoint is the overall incidence of grade 3/4 toxicities during the first 12 weeks of the study; secondary endpoints include efficacy and comparative safety. Two hundred and twenty three patients have been enrolled, of whom 213 were eligible for treatment. The addition of bevacizumab to all oxaliplatin-containing regimens was well tolerated and was also found to improve overall response rate (Table 20.12). Overall and progression-free survival data are awaited with interest.

Thus, available data to date suggest that the addition of bevacizumab to oxaliplatin-based regimens will improve patient outcomes in the first-line setting.

Table 20.11. Safety results from trial E3200 of FOLFOX with or without bevacizumab in patients with pretreated metastatic CRC

	FOLFOX4/ bevacizumab ($n = 287$)		FOLFOX4 ($n = 284$)		Bevacizumab ($n = 234$)		P^*
	Grade 3	Grade 4	Grade 3	Grade 4	Grade 3	Grade 4	
Hypertension (%)	5	1	2	<1	7	0	0.018
Bleeding (%)	3	<1	<1	0	2	0	0.011
Neuropathy (%)	16	<1	9	<1	<1	<1	0.016
Vomiting (%)	9	1	3	<1	5	0	0.01
Proteinuria (%)	1	0	0	0	<1	0	0.25

*FOLFOX4/bevacizumab versus FOLFOX4.

Table 20.12. Efficacy and safety in trial TREE 2 of first-line bevacizumab in combination with oxaliplatin-containing therapy

	FOLFOX6 plus bevacizumab ($n = 75$)	bFOL plus bevacizumab ($n = 74$)	XELOX plus bevacizumab ($n = 74$)
Response rate (%)	63.4 (46.9)	42.9 (32.0)	56.9 (37.5)
	$n = 71$	$n = 70$	$n = 72$
Number progressed (%)	49 (78)	63 (76)	40 (65)
Median time to treatment failure (months)	5.2 (6.4)	5.3 (4.9)	5.4 (4.4)
Adverse events (%)			
Gastrointestinal perforation	2.8	4.2	2.8
Wound healing complications	7.0	2.9	8.3
Sepsis	5.6	1.8	0.0
Treatment-related deaths	0.0	4.3	2.8

Numbers in brackets refer to the equivalent result without bevacizumab.

Ongoing phase III trials will provide data to support this, but bevacizumab is already widely used with oxaliplatin-based regimens in the USA and an observational study indicates that the combination is well tolerated.[58]

3.2. Tyrosine Kinase Inhibitors

The majority of the anti-VEGF/VEGF receptor agents in clinical development are TKIs. Tyrosine kinases are enzymes whose function is to phosphorylate the amino acid tyrosine, an important event in mediating signal transduction through cell surface receptors, and which triggers a cascade of intracellular signaling events that can result in uncontrolled cell proliferation. Tyrosine kinase inhibition represents an attractive therapeutic target as receptor signaling can be inhibited, preventing downstream signaling. TKIs compete with adenosine triphosphate for binding to the kinase region. If an inhibitor preferentially binds to the kinase region, the kinase is unable to initiate protein phosphorylation, and signaling ceases.

3.2.1. TKIs in clinical development

A number of VEGF receptor TKIs are in development, but the most, well characterized in CRC is PTK-787. PTK-787 (vatalanib) is an oral inhibitor of VEGF receptors 1, 2, and 3 that is administered once daily and is currently in phase III trials for CRC. Preliminary phase I studies demonstrated potential antitumor activity both as a single agent[59] and in combination with chemotherapy[60] in patients with liver metastases and glioblastoma multiforme, respectively. In CRC, PTK-787 has been evaluated in combination with 5-FU/LV plus irinotecan (FOLFIRI)[61] and FOLFOX.[62] A phase I/II study of PTK-787 plus FOLFIRI in 21 patients with metastatic CRC showed that the combination was well tolerated with moderate activity. Similar trials of PTK-787 with FOLFOX suggested that the combination was very effective, producing high response rates and encoraging progression-free survival.

The results of the CONFIRM-1 trial, a randomized, double-blind, placebo-controlled phase III study that was expected to demonstrate the first-line efficacy of PTK-787 plus FOLFOX, have recently been published[63] (Table 20.13). In this trial, patients with metastatic CRC received first-line chemotherapy with

Table 20.13. Efficacy results from a trial of first-line FOLFOX4 with or without PTK-787 (CONFIRM-1)

	FOLFOX4 plus PTK-787 ($n = 585$)	FOLFOX4 plus placebo ($n = 583$)	P
Complete response (%)	3	3	NS
Partial response (%)	39	43	NS
Stable disease (%)	37	33	NS
Progressive disease (%)	16	15	NS
Unknown (%)	6	6	NS

NS = not significant.

FOLFOX4 in combination with PTK-787 or placebo. The results showed a 12% reduction in the risk of progression in those patients receiving FOLFOX4 plus PTK-787, but this was not statistically significant by independent radiological review. However, PTK-787 combination therapy was found to benefit those patients with high levels of lactate dehydrogenase. PTK-787-associated toxicities include hypertension, thrombosis, and a higher incidence of dizziness; many patients receiving PTK-787 experienced PTK-associated neurological syndrome (Table 20.14). Novartis have also recently announced that the CONFIRM-2 trial, which is examining the same therapies but in patients with pretreated metastatic CRC, is unlikely to reveal any improvement in survival when PTK-787 is added to FOLFOX. Until final survival data for both trials are available, the future of PTK-787 is uncertain.

3.2.2. Combining VEGF inhibitors with chemotherapy: why does it work?

It was originally hypothesized by Teicher (1996)[64] that combining antiangiogenic and cytotoxic theapies should lead to increased benefit to patients as the antiangiogenic agent would starve the tumor cells of nutrients, leading to death, and the cytotoxic agent would kill tumor cells directly. This hypothesis is apparently supported by the results of several clinical trials demonstrating that the addition of anti-VEGF agents to chemotherapy increases survival.[48] The reasons for this improvement are not fully understood, but may be explained by the proposed mode of action for anti-VEGF therapy.

The proposed mode of action of anti-VEGF therapy has two stages. In the first stage, the vasculature undergoes a process of modification in which the tumor vasculature is pruned of immature and inefficient blood vessels.[65, 66] This leads to a more efficient or "normalized" vascular network, with improved blood flow and reduced intratumoral pressure, which would aid the delivery of chemotherapy, increasing cell kill. Modification of the tumor vasculature also improves oxygen flow to the tumor. Radiotherapy requires oxygen to be

Table 20.14. Safety results from a trial of first-line FOLFOX4 with or without PTK-787 (CONFIRM-1)

Adverse event (%)	FOLFOX4 plus PTK-787 (n = 579)		FOLFOX4 plus placebo (n = 574)	
	Grade 3	Grade 4	Grade 3	Grade 4
Neutropenia	17	14	21	11
Thrombocytopenia	6	<1	4	<1
Diarrhea	15	<1	10	<1
Nausea	9	–	5	–
Vomiting	7	<1	6	<1
Abdominal pain	4	<1	5	<1
Peripheral neuropathy	9	<1	7	<1
Hypertension	21	<1	6	<1
Dizziness	7	<1	2	0
Venous thrombosis	7	<1	4	<1
Pulmonary embolism	–	6	–	1

effective. Therefore, inhibiting VEGF reduces areas of tumor hypoxia, leaving the tumor more susceptible to radiotherapy.[67] Tumors in patients with rectal cancer treated with bevacizumab in combination with chemotherapy and radiation have been shown to have decreased blood flow, tumor density, vascular volume, and interstitial fluid pressure 2 weeks after a single injection of bevacizumab.[68] Similar reductions in blood flow have also been demonstrated in patients treated with PTK-787.[69]

Continuing anti-VEGF therapy is likely to cause direct and rapid changes to the vasculature by increasing endothelial cell apoptosis.[32] This would ultimately lead to vascular regression and is supported by preclinical and clinical studies. In a phase I study, a single infusion of anti-VEGF therapy resulted in a 29–59% reduction in microvascular density.[68] Regression of existing vasculature may result in tumor growth arrest and a reduction in both primary tumors and metastases.[70]

Another effect of anti-VEGF therapy is the blockade of new vessel formation. Revascularization is needed to sustain tumor growth and also provides a route for transport of tumor cells to distant sites where metastases can form.[14] Preclinical studies have shown that anti-VEGF therapy inhibits tumor growth[71] and reduces the number and size of liver metastases in nude mice.[72] Removal of anti-VEGF therapy results in revascularization, providing the tumor with the microenvironment required for further growth.[73]

3.2.3. Future directions for agents targeting the VEGF/VEGF receptor pathway

Clinical trials of agents targeting the VEGF/VEGF receptor pathway in metastatic CRC have produced contrasting results to date. These trials having demonstrated that the humanized anti-VEGF monoclonal antibody bevacizumab can be combined with the commonly used chemotherapy regimens for metastatic CRC, with outcomes including progression-free and overall survival being improved. However, the VEGF receptor TKI PTK-787, while producing promising data in phase I/II trials in combination with FOLFOX and FOLFIRI, has failed to demonstrate significant benefit in phase III trials in combination with first- and second-line FOLFOX. Thus, the future development of these two types of agent is likely to be significantly different.

In the case of PTK-787, overall survival data from the CONFIRM-1 and CONFIRM-2 trials are needed before a definitive conclusion regarding its potential efficacy in metastatic CRC can be made. However, even positive overall survival data could create questions. Grothey et al. have shown that survival is influenced more by the number of active agents that patients receive during therapy for metastatic CRC than the order in which these agents are given.[74] Therefore, results would have to be reviewed carefully to ensure that no imbalances between the treatment arms of CONFIRM-1 or CONFIRM-2 could have resulted in survival differences. In addition, if overall survival is improved but progression-free survival is not, the potential for broad clinical use of PTK-787 has to be questioned when an agent that can improve both outcomes in a clinically significant manner is available.

If the CONFIRM-1 and CONFIRM-2 trials prove to be negative, which currently appears to be the most likely outcome, the question is why. Theoretically VEGF receptor TKIs should be as effective as a VEGF inhibitor because many of the same pathways are inhibited. The affinity of PTK-787 for the VEGF receptors and its inhibitory activity for VEGF receptor-mediated signaling are well defined;[75–77] therefore, it appears unlikely that PTK-787 does not inhibit signaling. Other factors that could have affected the efficacy of PTK-787 in these trials include pharmacokinetics and dosing. The half-life of PTK-787 is 3–6 h;[69,78] however, it is administered once daily.[69] This creates the possibility that for periods of the 24-h dosing interval, VEGF receptor tyrosine kinase activity is not completely inhibited and that VEGF can stimulate intracellular signaling and angiogenic activity. Allowing tumor angiogenesis to occur would most likely result in continued tumor growth, which would translate into the failure to see a benefit from PTK-787 therapy. In this situation, administering PTK-787 twice or even thrice daily may result in an improvement in outcomes. However, whether this dose frequency would be acceptable to patients is difficult to assess. An alternative explanation could be that the PTK-787 dose administered is not high enough to effectively inhibit VEGF receptors. PTK-787 has dose-limiting toxicities that drove selection of the dose used in clinical trials.[78,79] Furthermore, PTK-787 has side effects, such as the commonly observed neurotoxicity syndrome, that could be related to its broad VEGF receptor inhibition or even penetration of the blood–brain barrier. It is therefore possible that higher doses may be required to achieve full clinical efficacy. However, this appears to be unlikely based on the results of phase I/II trials using the same dose as that used in phase III trials, which suggested that the addition of PTK-787 to chemotherapy had efficacy. Thus, the future development of PTK-787 will need to be guided by careful consideration of pharmacokinetics, dosing and side effects, and carefully designed trials will be needed to determine whether altering dosing is likely to affect outcomes. Whether the failure of the CONFIRM-1 and CONFIRM-2 trials will affect investigation of other TKIs in CRC also remains to be determined.

The future development of bevacizumab is clearer. With proven efficacy in combination with 5-FU/LV alone and in combination with irinotecan first line, and in combination with FOLFOX and cetuximab and irinotecan second line, this agent has a confirmed place in the treatment of metastatic CRC. The survival benefit when bevacizumab is combined with first-line therapy is greater than that seen with second-line therapy, which, together with its mechanism of action and the changing role of VEGF as tumors progress, indicates that it should be used first line for the treatment of metastatic CRC. An extensive program of ongoing trials is designed to demonstrate that bevacizumab can be effectively combined with first-line oxaliplatin-containing and capecitabine- containing regimens, and to confirm that combinations with FOLFIRI are at least as effective as those with IFL. These trials should provide data over the next 2 years.

With a mechanism of action that includes inhibition of the formation of new tumor vasculature as well as regression of existing vasculature, there is also a strong rationale to examine bevacizumab in the adjuvant setting, where it could inhibit the formation of new metastases and the growth of existing micrometastases. Large phase III trials are already underway, with a National Surgical Adjuvant Breast and Bowel Project (NSABP) trial in the USA and a Roche-sponsored trial in the rest of the world actively recruiting patients. These trials will reveal the efficacy of bevacizumab combined with oxaliplatin-containing regimens, the new gold standard in the adjuvant therapy of CRC. Further phase III trials are also investigating whether bevacizumab can be effectively combined with 5-FU-based regimens in this setting. In addition to adjuvant use, smaller phase II trials are also examining the potential neoadjuvant use of bevacizumab. This use is affected by the observation that bevacizumab causes wound healing complications, but appropriate patient management during the use of bevacizumab makes neoadjuvant treatment a possibility.

Questions remaining to be answered regarding bevacizumab are whether it has any role in second-line metastatic CRC and the appropriate dose for use in this setting. The E3200 trial showed that bevacizumab improves survival when added to second-line FOLFOX, but a 10 mg/kg dose was used. It is not known whether this dose is necessary and it appears unlikely that this trial will be repeated with the standard 5 mg/kg dose because the number of patients entering second-line therapy trials who have not received bevacizumab will decrease rapidly. This raises a further question, which is whether bevacizumab will provide benefit when used after disease progression. The mechanism of action suggests that this may be the case, but the design of trials to demonstrate this needs to be carefully considered due to the long half-life of bevacizumab and the potential continued activity of the drug after its withdrawal.

Finally in relation to the future development of bevacizumab, its potential in other tumors cannot be ignored. Survival benefit has already been demonstrated in metastatic non-small cell lung and breast cancer, as has improved time to progression in renal cell cancer.[80–82] Phase III trials are ongoing in these indications as well as pancreatic cancer, and are planned in ovarian cancer. In addition, phase II trials are ongoing or have already reported promising data in cancers as diverse as hematological malignancies, prostate cancer, and melanoma.[83–85] Thus, bevacizumab has the potential to be a true pan-tumor therapy.

4. CONCLUSIONS

A wealth of evidence exists to support the use of antiangiogenic agents for the treatment of solid tumors. The humanized anti-VEGF monoclonal antibody bevacizumab has demonstrated a survival benefit in combination with standard chemotherapies in patients with metastatic CRC and studies are underway to evaluate this agent in other solid tumors. Antiangiogenic agents are generally well tolerated, although specific adverse events have been observed, including an increased risk of hypertension, proteinuria, and minor bleeding. TKIs targeted against VEGF receptors represent a second antiangiogenic strategy and several agents are currently undergoing clinical evaluation in CRC. Whether these prove to be as effective as bevacizumab and how best to use them remains to be defined.

REFERENCES

1. N. Ferrara, H.P. Gerber, and J. LeCouter. The biology of VEGF and its receptors. *Nat Med* 9(6): 669–676 (2003).
2. G.H. Algire, H.W. Chalkley, F.Y. Legallais, and H.D. Park. Vascular reactions of normal and malignant tissues in vivo: I. Vascular reactions of mice to wounds and to normal and neoplastic transplants. *J Natl Cancer Inst* 6: 73–85 (1945).
3. R.S. Kerbel. Tumor angiogenesis: past, present and the near future. *Carcinogenesis* 21(3): 505–515 (2000).
4. D. Hanahan and J. Folkman. Patterns and emerging mechanisms of the angiogenic switch during tumorigenesis. *Cell* 86(3): 353–364 (1996).
5. R.K. Jain. Molecular regulation of vessel maturation. *Nat Med* 9(6): 685–693 (2003).
6. N.L. Lewis and N.J. Meropol. Development of new agents for the treatment of advanced colorectal cancer. *Clin Colorectal Cancer* 3(3): 154–164 (2003).
7. R.S. Herbst and I.J. Fidler. Angiogenesis and lung cancer: potential for therapy. *Clin Cancer Res* 6(12): 4604–4606 (2000).
8. S. Brem. Angiogenesis and cancer control: from concept to therapeutic trial. *Cancer Control* 6(5): 436–458 (1999).
9. S. Brem, H. Brem, J. Folkman, D. Finkelstein, and A. Patz. Prolonged tumor dormancy by prevention of neovascularization in the vitreous. *Cancer Res* 36(8): 2807–2812 (1976).
10. S. Liekens, E. De Clercq, and J. Neyts. Angiogenesis: regulators and clinical applications. *Biochem Pharmacol* 61(3): 253–270 (2001).
11. P. Carmeliet. Angiogenesis in health and disease. *Nat Med* 9(6): 653–660 (2003).
12. P. Carmeliet and R.K. Jain. Angiogenesis in cancer and other diseases. *Nature* 407(6801): 249–257 (2000).
13. J. Rak, J.L. Yu, G. Klement, and R.S. Kerbel. Oncogenes and angiogenesis: signaling three-dimensional tumor growth. *J Investig Dermatol Symp Proc* 5(1): 24–33 (2000).
14. G. Neufeld, O. Kessler, Z. Vadasz, and Z. Gluzman-Poltorak. The contribution of proangiogenic factors to the progression of malignant disease: role of vascular endothelial growth factor and its receptors. *Surg Oncol Clin N Am* 10(3): 339–356 (2001).
15. H. Hashizume, P. Baluk, S. Morikawa, J.W. McLean, G. Thurston, S. Roberge, R.K. Jain, and D.M. McDonald. Openings between defective endothelial cells explain tumor vessel leakiness. *Am J Pathol* 156(4): 1363–1380 (2000).
16. R.K. Jain, L.L. Munn, and D. Fukumura. Dissecting tumour pathophysiology using intravital microscopy. *Nat Rev Cancer* 2(4): 266–276 (2002).
17. G. Bergers, S. Song, N. Meyer-Morse, E. Bergsland, and D. Hanahan. Benefits of targeting both pericytes and endothelial cells in the tumor vasculature with kinase inhibitors. *J Clin Invest* 111(9): 1287–1295 (2003).
18. K.A. Houck, N. Ferrara, J. Winer, G. Cachianes, B. Li, and D.W. Leung. The vascular endothelial growth factor family: identification of a fourth molecular species and characterization of alternative splicing of RNA. *Mol Endocrinol* 5(12): 1806–1814 (1991).
19. E. Tischer, R. Mitchell, T. Hartman, M. Silva, D. Gospodarowicz, J.C. Fiddes, and J.A. Abraham. The human gene for vascular endothelial growth factor. Multiple protein forms are encoded through alternative exon splicing. *J Biol Chem* 266(18): 11947–11954 (1991).
20. N. Ferrara. Role of vascular endothelial growth factor in regulation of physiological angiogenesis. *Am J Physiol Cell Physiol* 280(6): C1358–1366 (2001).
21. E. Ikeda, M.G. Achen, G. Breier, and W. Risau. Hypoxia-induced transcriptional activation and increased mRNA stability of vascular endothelial growth factor in C6 glioma cells. *J Biol Chem* 270(34): 19761–19766 (1995).

22. S. Zheng, M.Y. Han, Z.X. Xiao, J.P. Peng, and Q. Dong. Clinical significance of vascular endothelial growth factor expression and neovascularization in colorectal carcinoma. *World J Gastroenterol* 9(6): 1227–1230 (2003).
23. M.J. Chung, S.H. Jung, B.J. Lee, M.J. Kang, and D.G. Lee. Inactivation of the PTEN gene protein product is associated with the invasiveness and metastasis, but not angiogenesis, of breast cancer. *Pathol Int* 54(1): 10–15 (2004).
24. J.C. Lee, N.H. Chow, S.T. Wang, and S.M. Huang. Prognostic value of vascular endothelial growth factor expression in colorectal cancer patients. *Eur J Cancer* 36(6): 748–753 (2000).
25. Y. Kondo, S. Arii, A. Mori, M. Furutani, T. Chiba, and M. Imamura. Enhancement of angiogenesis, tumor growth, and metastasis by transfection of vascular endothelial growth factor into LoVo human colon cancer cell line. *Clin Cancer Res* 6(2): 622–630 (2000).
26. M. Shibuya, N. Ito, and L. Claesson-Welsh. Structure and function of vascular endothelial growth factor receptor-1 and -2. *Curr Top Microbiol Immunol* 237: 59–83 (1999).
27. J.E. Park, H.H. Chen, J. Winer, K.A. Houck, and N. Ferrara. Placenta growth factor. Potentiation of vascular endothelial growth factor bioactivity, in vitro and in vivo, and high affinity binding to Flt-1 but not to Flk-1/KDR. *J Biol Chem* 269(41): 25646–25654 (1994).
28. D.I. Gabrilovich, H.L. Chen, K.R. Girgis, H.T. Cunningham, G.M. Meny, S. Nadaf, D. Kavanaugh, and D.P. Carbone. Production of vascular endothelial growth factor by human tumors inhibits the functional maturation of dendritic cells. *Nat Med* 2(10): 1096–1103 (1996).
29. S. Ogawa, A. Oku, A. Sawano, S. Yamaguchi, Y. Yazaki, and M. Shibuya. A novel type of vascular endothelial growth factor, VEGF-E (NZ-7 VEGF), preferentially utilizes KDR/Flk-1 receptor and carries a potent mitotic activity without heparin-binding domain. *J Biol Chem* 273(47): 31273–31282 (1998).
30. K. Pajusola, O. Aprelikova, E. Armstrong, S. Morris, and K. Alitalo. Two human FLT4 receptor tyrosine kinase isoforms with distinct carboxy terminal tails are produced by alternative processing of primary transcripts. *Oncogene* 8(11): 2931–2937 (1993).
31. F. Galland, A. Karamysheva, M.J. Pebusque, J.P. Borg, R. Rottapel, P. Dubreuil, O. Rosnet, and D. Birnbaum. The FLT4 gene encodes a transmembrane tyrosine kinase related to the vascular endothelial growth factor receptor. *Oncogene* 8(5): 1233–1240 (1993).
32. D.J. Hicklin and L.M. Ellis. Role of the vascular endothelial growth factor pathway in tumor growth and angiogenesis. *J Clin Oncol* 23(5): 1011–1027 (2005).
33. N. Ferrara. VEGF: an update on biological and therapeutic aspects. *Curr Opin Biotechnol* 11(6): 617–624 (2000).
34. B.K. Zebrowski, W. Liu, K. Ramirez, Y. Akagi, G.B. Mills, and L.M. Ellis. Markedly elevated levels of vascular endothelial growth factor in malignant ascites. *Ann Surg Oncol* 6(4): 373–378 (1999).
35. B.K. Zebrowski, S. Yano, W. Liu, R.M. Shaheen, D.J. Hicklin, J.B. Putnam Jr., and L.M. Ellis. Vascular endothelial growth factor levels and induction of permeability in malignant pleural effusions. *Clin Cancer Res* 5(11): 3364–3368 (1999).
36. J.A. Nagy, E. Vasile, D. Feng, C. Sundberg, L.F. Brown, M.J. Detmar, J.A. Lawitts, L.E. Benjamin, X. Tan, E.J. Manseau, A.M. Dvorak, and H.F. Dvorak. Vascular permeability factor/vascular endothelial growth factor induces lymphangiogenesis as well as angiogenesis. *J Exp Med* 196(11): 1497–1506 (2002).
37. R. Kerbel and J. Folkman. Clinical translation of angiogenesis inhibitors. *Nat Rev Cancer* 2(10): 727–739 (2002).
38. G.P. Pidgeon, M.P. Barr, J.H. Harmey, D.A. Foley, and D.J. Bouchier-Hayes. Vascular endothelial growth factor (VEGF) upregulates BCL-2 and inhibits apoptosis in human and murine mammary adenocarcinoma cells. *Br J Cancer* 85(2): 273–278 (2001).
39. D.I. Gabrilovich, T. Ishida, S. Nadaf, J.E. Ohm, and D.P. Carbone. Antibodies to vascular endothelial growth factor enhance the efficacy to cancer immunotherapy by improving endogenous dendritic cell function. *Clin Cancer Res* 5(10): 2963–2970 (1999).
40. A. Vitaliti, M. Wittmer, R. Steiner, L. Wyder, D. Neri, and R. Klemenz. Inhibition of tumor angiogenesis by a single-chain antibody directed against vascular endothelial growth factor. *Cancer Res* 60(16): 4311–4314 (2000).

41. W. Zhang, S. Ran, M. Sambade, X. Huang, and P.E. Thorpe. A monoclonal antibody that blocks VEGF binding to VEGFR2 (KDR/Flk-1) inhibits vascular expression of Flk-1 and tumor growth in an orthotopic human breast cancer model. *Angiogenesis* 5(1–2): 35–44 (2002).
42. R.A. Brekken, J.P. Overholser, V.A. Stastny, J. Waltenberger, J.D. Minna, and P.E. Thorpe. Selective inhibition of vascular endothelial growth factor (VEGF) receptor 2 (KDR/Flk-1) activity by a monoclonal anti-VEGF antibody blocks tumor growth in mice. *Cancer Res* 60(18): 5117–5124 (2000).
43. L.G. Presta, H. Chen, S.J. O'Connor, V. Chisholm, Y.G. Meng, L. Krummen, M. Winkler, and N. Ferrara. Humanization of an anti-vascular endothelial growth factor monoclonal antibody for the therapy of solid tumors and other disorders. *Cancer Res* 57(20): 4593–4599 (1997).
44. B.Q. Shen, S. Stainton, D. Li, N. Pelletier, and T.F. Zioncheck. Combination of Avastin™ and Xeloda® synergistically inhibits colorectal tumor growth in a COLO205 tumor xenograft model. *Proc Am Assoc Cancer Res* 45: 508 (Abstract 2203) (2004).
45. A.M. Ryan, D.B. Eppler, K.E. Hagler, R.H. Bruner, P.J. Thomford, R.L. Hall, G.M. Shopp, and C.A. O'Neill. Preclinical safety evaluation of rhuMABVEGF, an antiangiogenic humanized monoclonal antibody. *Toxicol Pathol* 27(1): 78–86 (1999).
46. M.S. Gordon, K. Margolin, M. Talpaz, G.W.J. Sledge, E. Holmgren, R. Benjamin, S. Stalter, S. Shak, and D. Adelman. Phase I safety and pharmacokinetic study of recombinant human anti-vascular endothelial growth factor in patients with advanced cancer. *J Clin Oncol* 19(3): 843–850 (2001).
47. K. Margolin, M.S. Gordon, E. Holmgren, J. Gaudreault, W. Novotny, G. Fyfe, D. Adelman, S. Stalter, and J. Breed. Phase Ib trial of intravenous recombinant humanized monoclonal antibody to vascular endothelial growth factor in combination with chemotherapy in patients with advanced cancer: pharmacologic and long-term safety data. *J Clin Oncol* 19(3): 851–856 (2001).
48. H. Hurwitz, L. Fehrenbacher, W. Novotny, T. Cartwright, J. Hainsworth, W. Heim, J. Berlin, A. Baron, S. Griffing, E. Holmgren, N. Ferrara, G. Fyfe, B. Rogers, R. Ross, and F. Kabbinavar. Bevacizumab plus irinotecan, fluorouracil, and leucovorin for metastatic colorectal cancer. *N Engl J Med* 350(23): 2335–2342 (2004).
49. F. Kabbinavar, H.I. Hurwitz, L. Fehrenbacher, N.J. Meropol, W.F. Novotny, G. Lieberman, S. Griffing, and E. Bergsland. Phase II, randomized trial comparing bevacizumab plus fluorouracil (FU)/leucovorin (LV) with FU/LV alone in patients with metastatic colorectal cancer. *J Clin Oncol* 21(1): 60–65 (2003).
50. B.J. Giantonio, P.J. Catalano, N.J. Meropol, P.J. O'Dwyer, E.P. Mitchell, S.R. Alberts, M.A. Schwartz, and A.B. Benson. High-dose bevacizumab improves survival when combined with FOLFOX4 in previously treated advanced colorectal cancer: results from the Eastern Cooperative Oncology Group (ECOG) study E3200. *J Clin Oncol* 23(June 1 Suppl.): 1s (Abstract 2) (2005).
51. N. Fernando, D. Yu, M. Morse, G. Blobe, L. Odogwu, J. Crews, A. Polito, W. Honeycutt, A. Franklin, and H. Hurwitz. A phase II study of oxaliplatin, capecitabine and bevacizumab in the treatment of metastatic colorectal cancer. *J Clin Oncol* 23(June 1 Suppl.): 260s (Abstract 3556) (2005).
52. L.B. Saltz, H. Lenz, H. Hochster, S. Wadler, P. Hoff, N. Kemeny, E. Hollywood, M. Gonen, S. Wetherbee, and H. Chen. Randomized phase II trial of cetuximab/bevacizumab/irinotecan (CBI) versus cetuximab/bevacizumab (CB) in irinotecan-refractory colorectal cancer. *J Clin Oncol* 23(June 1 Suppl.): 248s (Abstract 3508) (2005).
53. N. Petrelli, H.O. Douglass Jr., L. Herrera, D. Russell, D.M. Stablein, H.W. Bruckner, R.J. Mayer, R. Schinella, M.D. Green, F.M. Muggia, et al. The modulation of fluorouracil with leucovorin in metastatic colorectal carcinoma: a prospective randomized phase III trial. Gastrointestinal Tumor Study Group. *J Clin Oncol* 7(10): 1419–1426 (1989).
54. H.I. Hurwitz, L. Fehrenbacher, J.D. Hainsworth, W. Heim, J. Berlin, E. Holmgren, J. Hambleton, W.F. Novotny, and F. Kabbinavar. Bevacizumab in combination with fluorouracil and leucovorin: an active regimen for first-line metastatic colorectal cancer. *J Clin Oncol* 23(15): 3502–3508 (2005).

55. F.F. Kabbinavar, J. Schulz, M. McCleod, T. Patel, J.T. Hamm, J.R. Hecht, R. Mass, B. Perrou, B. Nelson, and W.F. Novotny. Addition of bevacizumab to bolus fluorouracil and leucovorin in first-line metastatic colorectal cancer: results of a randomized phase II trial. *J Clin Oncol* 23(16): 3697–3705 (2005).
56. F.F. Kabbinavar, J. Hambleton, R.D. Mass, H.I. Hurwitz, E. Bergsland, and S. Sarkar. Combined analysis of efficacy: the addition of bevacizumab to fluorouracil/leucovorin improves survival for patients with metastatic colorectal cancer. *J Clin Oncol* 23(16): 3706–3712 (2005).
57. H.S. Hochster, L. Welles, L. Hart, R.K. Ramanathan, J. Hainsworth, G. Jirau-Lucca, A. Shpilsky, S. Griffing, R. Mass, and D. Emanuel. Safety and efficacy of bevacizumab (Bev) when added to oxaliplatin/fluoropyrimidine (O/F) regimens as first-line treatment of metastatic colorectal cancer (mCRC): TREE 1 & 2 studies. *J Clin Oncol* 23(June 1 Suppl.): 249s (Abstract 3515) (2005).
58. M. Kozloff, A. Cohn, N. Christiansen, P. Flynn, F. Kabbinavar, R. Robles, M. Ulcickas-Yood, S. Sarkar, J. Hambleton, and A. Grothey. Safety of bevacizumab (BV) among patients (pts) receiving first-line chemotherapy (CT) for metastatic colorectal cancer (mCRC): preliminary results from a larger registry in the US. *J Clin Oncol* 23(June 1 Suppl.): 262s (Abstract 3566) (2005).
59. J. Drevs, K. Mross, M. Medinger, M. Muller, D. Laurent, D. Reitsma, A. Henry, J. Xia, D. Marme, and C. Unger. Phase I dose-escalation and pharmacokinetic (PK) study of the VEGF inhibitor PTK787/ZK 222584 (PTK/ZK) in patients with liver metastases. *Proc Am Soc Clin Oncol* 22: 284 (Abstract 1142) (2003).
60. D. Reardon, H.S. Friedman, W.K.A. Yung, M. Brada, C. Conrad, J. Provenzale, E.F. Jackson, H. Serajuddin, D. Laurent, and D. Reitsma. A phase I trial of PTK787/ZK 222584 (PTK/ZK), an oral VEGF tyrosine kinase inhibitor, in combination with either temozolomide or lomustine for patients with recurrent glioblastoma multiforme (GBM). *Proc Am Soc Clin Oncol* 22: 103 (Abstract 412) (2003).
61. N. Schleucher, T. Trarbach, U. Junker, M. Tewes, E. Masson, D. Lebwohl, S. Seeber, D. Laurent, and U. Vanhoefer. Phase I/II study of PTK787/ZK 222584 (PTK/ZK), a novel, oral angiogenesis inhibitor in combination with FOLFIRI as first-line treatment for patients with metastatic colorectal cancer. *J Clin Oncol* 22(July 15 Suppl.): (Abstract 3558) (2004).
62. W.P. Steward, A. Thomas, B. Morgan, B. Wiedenmann, C. Bartel, U. Vanhoefer, T. Trarbach, U. Junker, D. Laurent, and D. Lebwohl. Expanded phase I/II study of PTK787/ZK 222584 (PTK/ZK), a novel, oral angiogenesis inhibitor, in combination with FOLFOX-4 as first-line treatment for patients with metastatic colorectal cancer. *J Clin Oncol* 22(July 15 Suppl.): (Abstract 3556) (2004).
63. J.R. Hecht, T. Trarbach, E. Jaeger, J. Hainsworth, R. Wolff, K. Lloyd, G. Bodoky, M. Borner, D. Laurent, and C. Jacques. A randomized, double-blind, placebo-controlled, phase III study in patients (Pts) with metastatic adenocarcinoma of the colon or rectum receiving first-line chemotherapy with oxaliplatin/5-fluorouracil/leucovorin and PTK787/ZK 222584 or placebo (CONFIRM-1). *J Clin Oncol* 23(June 1 Suppl.): 2s (Abstract LBA3) (2005).
64. B.A. Teicher. A systems approach to cancer therapy. (Antioncogenics + standard cytotoxics — mechanism(s) of interaction). *Cancer Metastasis Rev* 15(2): 247–272 (1996).
65. R.K. Jain. Normalizing tumor vasculature with anti-angiogenic therapy: a new paradigm for combination therapy. *Nat Med* 7(9): 987–989 (2001).
66. R.K. Jain. Normalization of tumor vasculature: an emerging concept in antiangiogenic therapy. *Science*. 307(5706): 58–62 (2005).
67. C.-G. Lee, M. Heijin, E. di Tomaso, G. Griffon-Etienne, M. Ancukiewicz, C. Koike, K.R. Park, N. Ferrara, R.K. Jain, H.D. Suit, and Y. Boucher. Anti-vascular endothelial growth factor treatment augments tumor radiation response under nomoxic or hypoxic conditions. *Cancer Res* 60(19): 5565–5570 (2000).
68. C.G. Willett, Y. Boucher, E. di Tomaso, D.G. Duda, L.L. Munn, R.T. Tong, D.C. Chung, D.V. Sahani, S.P. Kalva, S.V. Kozin, M. Mino, K.S. Cohen, D.T. Scadden, A.C. Hartford, A.J.

Fischman, J.W. Clark, D.P. Ryan, A.X. Zhu, L.S. Blaszkowsky, H.X. Chen, P.C. Shellito, G.Y. Lauwers, and R.K. Jain. Direct evidence that the VEGF-specific antibody bevacizumab has antivascular effects in human rectal cancer. *Nat Med* 10(2): 145–147 (2004).
69. B. Morgan, A.L. Thomas, J. Drevs, J. Hennig, M. Buchert, A. Jivan, M.A. Horsfield, K. Mross, H.A. Ball, L. Lee, W. Mietlowski, S. Fuxuis, C. Unger, K. O'Byrne, A. Henry, G.R. Cherryman, D. Laurent, M. Dugan, D. Marme, and W.P. Steward. Dynamic contrast-enhanced magnetic resonance imaging as a biomarker for the pharmacological response of PTK787/ZK 222584, an inhibitor of the vascular endothelial growth factor receptor tyrosine kinases, in patients with advanced colorectal cancer and liver metastases: results from two phase I studies. *J Clin Oncol* 21(21): 3955–3964 (2003).
70. J. Huang, J.S. Frischer, A. Serur, A. Kadenhe, A. Yokoi, K.W. McCrudden, T. New, K. O'Toole, S. Zabski, J.S. Rudge, J. Holash, G.D. Yancopoulos, D.J. Yamashiro, and J.J. Kandel. Regression of established tumors and metastases by potent vascular endothelial growth factor blockade. *Proc Natl Acad Sci USA* 100(13): 7785–7790 (2003).
71. K.J. Kim, B. Li, J. Winer, M. Armanini, N. Gillett, H.S. Phillips, and N. Ferrara. Inhibition of vascular endothelial growth factor-induced angiogenesis suppresses tumour growth in vivo. *Nature* 362(6423): 841–844 (1993).
72. R.S. Warren, H. Yuan, M.R. Matli, N.A. Gillett, and N. Ferrara. Regulation by vascular endothelial growth factor of human colon cancer tumorigenesis in a mouse model of experimental liver metastasis. *J Clin Invest* 95(4): 1789–1797 (1995).
73. P. Baluk, H. Hashizume, and D.M. McDonald. Cellular abnormalities of blood vessels as targets in cancer. *Curr Opin Genet Dev* 15(1): 102–111 (2005).
74. A. Grothey, D. Sargent, R.M. Goldberg, and H.J. Schmoll. Survival of patients with advanced colorectal cancer improves with the availability of fluorouracil-leucovorin, irinotecan, and oxaliplatin in the course of treatment. *J Clin Oncol* 22(7): 1209–1214 (2004).
75. J.M. Wood, G. Bold, E. Buchdunger, R. Cozens, S. Ferrari, J. Frei, F. Hofmann, J. Mestan, H. Mett, T. O'Reilly, E. Persohn, J. Rosel, C. Schnell, D. Stover, A. Theuer, H. Towbin, F. Wenger, K. Woods-Cook, A. Menrad, G. Siemeister, M. Schirner, K.-H. Thierauch, M.R. Schneider, J. Drevs, G. Martiny-Baron, F. Totzke, and D. Marme. PTK787/ZK 222584, a novel and potent inhibitor of vascular endothelial growth factor receptor tyrosine kinases, impairs vascular endothelial growth factor-induced responses and tumor growth after oral administration. *Cancer Res* 60(8): 2178–2189 (2000).
76. B. Lin, K. Podar, D. Gupta, Y.T. Tai, S. Li, E. Weller, T. Hideshima, S. Lentzsch, F. Davies, C. Li, E. Weisberg, R.L. Schlossman, P.G. Richardson, J.D. Griffin, J. Wood, N.C. Munshi, and K.C. Anderson. The vascular endothelial growth factor receptor tyrosine kinase inhibitor PTK787/ZK222584 inhibits growth and migration of multiple myeloma cells in the bone marrow microenvironment. *Cancer Res* 62(17): 5019–5026 (2002).
77. H. Hess-Stumpp, M. Haberey, and K.H. Thierauch. PTK 787/ZK 222584, a tyrosine kinase inhibitor of all known VEGF receptors, represses tumor growth with high efficacy. *Chembiochem* 6(3): 550–557 (2005).
78. A.L. Thomas, B. Morgan, M.A. Horsfield, A. Higginson, A. Kay, L. Lee, E. Masson, D. Laurent, and W.P. Steward. Phase I study of the safety, tolerability, pharmacokinetics, and pharmacodynamics of PTK787/ZK 222584 administered twice daily in patients with advanced cancer. *J Clin Oncol* 23(18): 4162–4171 (2005).
79. C. Conrad, H. Friedman, D. Reardon, J. Provenzale, E. Jackson, H. Serajuddin, D. Laurent, B. Chen, and W.K.A. Yung. A phase I/II trial of single-agent PTK 787/ZK 222584 (PTK/ZK), a novel, oral angiogenesis inhibitor, in patients with recurrent glioblastoma multiforme (GBM). *J Clin Oncol* 22(July 15 Suppl.): (Abstract 1512) (2004).
80. A.B. Sandler, R. Gray, J. Brahmer, A. Dowlati, J.H. Schiller, M.C. Perry, and D.H. Johnson. Randomized phase II/III trial of paclitaxel (P) plus carboplatin (C) with or without bevacizumab (NSC # 704865) in patients with advanced non-squamous non-small cell lung cancer (NSCLC): an Eastern Cooperative Oncology Group (ECOG) trial — E4599. *J Clin Oncol* 23(June 1 Suppl.): 2s (Abstract LBA4) (2005).

81. K.D. Miller, M. Wang, J. Gralow, M. Dickler, M.A. Cobleigh, E.A.S. Perez, T.N., and N.E. Davidson. E2100. A randomised phase III trial of paclitaxel versus paclitaxel plus bevacizumab as first-line therapy for locally recurrent or metastatic breast cancer. *Presented at the 41st Annual Meeting of the American Society of Clinical Oncology, Orlando, FL, USA, May 13–17* (2005).
82. J.C. Yang, L. Haworth, R.M. Sherry, P. Hwu, D.J. Schwartzentruber, S.L. Topalian, S.M. Steinberg, H.X. Chen, and S.A. Rosenberg. A randomized trial of bevacizumab, an anti-vascular endothelial growth factor antibody, for metastatic renal cancer. *N Engl J Med* 349(5): 427–434 (2003).
83. A.T. Stopeck, W. Bellamy, J. Unger, L. Rimsza, M. Iannone, R.I. Fisher, and T.P. Miller. Phase II trial of single agent bevacizumab (Avastin) in patients with relapsed, aggressive non-Hodgkin's lymphoma (NHL): Southwest Oncology Group Study S0108. *J Clin Oncol* 23(June 1 Suppl.): 583s (Abstract 6592) (2005).
84. B. Rini, V. Weinberg, L. Fong, and E. Small. A phase 2 study of prostatic acid phosphatase-pulsed dendritic cells (APC8015; Provenge) in combination with bevacizumab in patients with serologic progression of prostate cancer after local therapy. *Presented at the Prostate Cancer Symposium, Orlando, FL, USA, Feb 17–19* (Abstract 251) (2005).
85. W.E. Carson, J. Biber, N. Shah, K. Reddy, C. Kefauver, P.D. Leming, K. Kendra, and M. Walker. A phase 2 trial of a recombinant humanized monoclonal anti-vascular endothelial growth factor (VEGF) antibody in patients with malignant melanoma. *Proc Am Soc Clin Oncol* 22: 715 (Abstract 2873) (2003).

CHAPTER 21

MOLECULAR IMAGING OF CANCER USING PET AND SPECT

ANDREAS KJÆR

Department of Clinical Physiology, Nuclear Medicine and PET, Rigshospitalet and Cluster for molecular imaging, Faculty of Health Sciences, University of Copenhagen, Denmark

Abstract: Molecular imaging allows for the study of molecular and cellular events in the living intact organism. The nuclear medicine methodologies of positron emission tomography (PET) and single photon emission computer tomography (SPECT) posses several advantages, which make them particularly suited for molecular imaging of cancer. Especially the possibility of a quick transfer of methods developed in animals to patients (translational research) is an important strength. This article will briefly discuss the newest applications and their importance and perspective in relation to the shift in paradigm in medicine towards more individualized treatment.

1. INTRODUCTION

Molecular imaging allows for noninvasive in vivo studies of physiological and pathophysiological processes at the cellular and molecular levels. Molecular imaging may be achieved using nuclear medicine techniques, optical imaging as well as magnetic resonance imaging. However, several circumstances favor the use of nuclear medicine techniques for molecular imaging. Below the current status and future perspectives regarding the use of nuclear medicine imaging modalities for molecular imaging of cancer will be reviewed.

2. PET AND SPECT

Imaging with nuclear medicine techniques are divided into those using positron emitters and those using gamma emitters.

Currently used positron emitters are ^{11}C, ^{15}O, ^{13}N, and ^{124}I. These radionuclides may be coupled to biomolecules and their destiny and distribution imaged with a PET scanner. When a positron is emitted it will almost immediately be

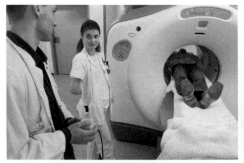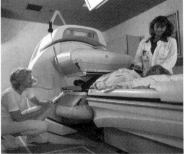

Figure 21.1. PET/CT and SPECT/CT scanners at the Department of Clinical Physiology, Nuclear Medicine and PET, Rigshospitalet, Copenhagen, Denmark.

neutralized by an electron (annihilated) and two photons will be emitted traveling in opposite direction (180°). Actually it is these two simultaneously emitted photons that are detected by a PET scanner (coincidence detection). It is not possible to discriminate between different positron emitters since the energy of the radiation is always the same (511 KeV). The most commonly used isotopes have half-lives in the range of few minutes to hours, which often demands that they are produced locally, making access to a cyclotron necessary.

Commonly used gamma-emitting isotopes suited for imaging are ^{99m}Tc, ^{111}In og, and ^{123}I. These isotopes may also be coupled to biomolecules allowing for imaging of their distribution using a gamma camera. When three-dimensional images are obtained using tomographic technique it is referred to as SPECT. Different isotopes emit gamma radiation of different energies allowing for discrimination. Thus, it is possible to label different biomolecules with different isotopes and simultaneously image the distribution of each of them. The most used gamma emitters for imaging have half-lives in the range of hours to days. They are either generator produced locally or delivered by external vendors (Figure 21.1).

In theory, all biomolecules may be labeled with radioactive isotopes suited for imaging. Thus the typical workflow in development of new tracers will be: (1) identification of process of pathophysiological relevance; (2) identification of relevant biomolecules; and (3) radioactive labeling of molecule. The amount of biomolecule that has to be labeled is very small and accordingly there is no real risk of side effects of the biomolecule.

3. THE ROLE OF MOLECULAR IMAGING IN CHANGING THE PARADIGM OF MEDICINE

The change of medicine in direction of individualized, tailored therapy has led to an increasing need for diagnosing at the molecular level. Several currently used molecular biology methods are limited by the need for tissue samples, i.e.,

Table 21.1. Some currently used PET or SPECT tracers for cancer

Radiotracer	PET/SPECT	Process visualized	Use (example)
[^{18}F]FDG	PET	Glycolytic activity	Several cancer forms
[^{11}C]Cholin	PET	Membrane synthesis	Prostate cancer
[^{18}F]FLT	PET	DNA synthesis/ cell proliferation	Response to therapy, grading
[^{11}C]methionin	PET	Amino acid transport	Certain cancer forms
[^{X}Cu]ATSM	PET	Hypoxia	Radiation planning
[^{18}F]fluoroestradiol	PET	Estrogen receptors	Breast cancer
[^{111}In]Octreotid	SPECT	Somatostatin receptors	Neuroendocrine tumors
[^{18}F]galacto-RDG	PET	Angiogenesis	Before antiangiogenesis therapy
[^{99m}Tc]Annexin V	SPECT	Apoptosis	Radiation planning

FDG: Fluorodeoxyglucose; FLT: Fluorodeoxythymidine; ATSM: Diacetylmethylthiosemicarbazon; ^{X}Cu: several Cu isotopes, e.g., ^{64}Cu.

invasive procedures. In contrast, molecular imaging allows for noninvasive use in living, intact animals and humans. At present, molecular imaging may be used for demonstration of a variety of processes, e.g., metabolism, apoptosis, cellular trafficking, receptor expression, and gene expression. Examples of some currently used tracers for molecular imaging in cancer are shown in Table 21.1.

4. GLYCOLYTIC ACTIVITY

The major nuclear medicine imaging agent for cancer still is ^{18}F-fluorodeoxyglucose (FDG) for PET. FDG, which is a glucose analogue, is taken up mainly by GLUT-1, which is upregulated in cancer cells. Once taken up by the cells FDG is phosphorylated into FDG-6-phosphate, but does not undergo glycolysis and accumulates in the cells. The value of FDG-PET in several cancer forms has been extensively documented [1].

5. RECEPTOR IMAGING

Overexpression of certain receptors on tumors may be used as imaging targets. An example of such imaging strategy is the noninvasive imaging of estrogen receptor status in breast cancer using ^{18}F-fluoroestradiol and PET [2]. This has been used to predict outcome of tamoxifen treatment. Neuroendocrine tumors are imaged on a routine basis using the somatostatin receptor ligand ^{111}In-pentreotide and SPECT [3] (Figure 21.2). Also more generally expressed surface characteristics common for different cancer types may be targeted. Finally, in the future also noninvasive imaging of endogenous gene expression may be used for imaging of cancer.

6. CELL PROLIFERATION

Tumor cell proliferation may be used to grade tumors. However, in most cases limited value is added compared with traditional FDG-PET based on glycolytic activity. However, in selected tumors imaging of cell proliferation may add

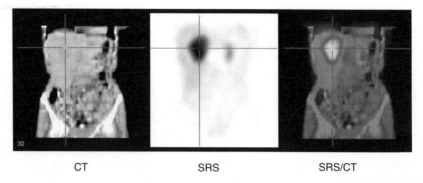

Figure 21.2. Scintigraphy of neuroendocrine tumor (VIPoma) in the liver using ^{111}In-octreotide scintigraphy. Scintigraphy has been image-fused with CT.

valuable information. More interesting might be that cell proliferation is a sensitive marker of response to therapy. Currently ^{18}F-fluorothymidine (^{18}F-FLT) is the most promising PET tracer for imaging of tumor cell proliferation [4]. Amino acid radioligands are indirect indicators of tumor cell proliferation since they reflect protein synthesis. For PET, especially ^{11}C-methionine (^{11}C-MET) has been used with some success in selected tumors.

7. HYPOXIA

Imaging of hypoxia is of great relevance in oncology since hypoxia is known to lead to radioresistance. Accordingly, imaging of hypoxia may influence treatment plans. Several compounds have been used for imaging of hypoxia, e.g., ^{18}F-FMISO, ^{18}F-FETA, ^{18}F-FAZA, and Cu-ATSM (several Cu-isotopes). In general ^{18}F-FMISO has been somewhat disappointing whereas ^{18}F-FETA, ^{18}F-FAZA, and Cu-ATSM may be more promising [5–7]. The compounds vary with respect to uptake and clearance and the final choice may depend on the specific tumors studied.

8. ANGIOGENESIS

Since angiogenesis is an important part of tumor progression and metastasizing, angiogenesis imaging would be of great value in oncology. Furthermore, several antiangiogenesis drugs are currently under evaluation for cancer treatment [8]. Imaging of angiogenesis could therefore potentially allow for selection and prediction of tumors suited for antiangiogenesis treatment as well as for therapy monitoring. Two major approaches are currently under investigation. Firstly ligands for $\alpha_v \beta_3$ integrins that are expressed in activated endothelial cells and are specific for angiogenesis, e.g., ^{18}F-galacto-RDG [9]. Secondly, MMP inhibitors.

9. APOPTOSIS

Apoptosis has been suggested as a marker for cellular radio sensitivity and as a prognostic marker. Accordingly, rapid increase in apoptosis following radiation could indicate a favorable response. Specific markers for apoptosis are available for radionuclide imaging with both PET and SPECT. So far, the most used compound has been Annexin V labeled with ^{99m}Tc for SPECT [10], but also PET imaging has been achieved with either ^{124}I or ^{18}F labeling.

10. MONITORING OF GENE THERAPY

The use of gene therapy has led to a need for monitoring the effectiveness of gene expression. By use of a reporter gene that is controlled by the same promoter as the therapeutic gene the two genes are expressed simultaneously. Accordingly, expression of the reporter gene indicates that the therapeutic gene is also expressed. An example of a reporter gene is the one coding for herpes simplex virus type 1 thymidine kinase (HSV1-tk) [11]. In order to visualize where HSV1-tk is expressed a reporter probe is used. In the case of HSV1-tk the reporter probe is a radioactively labeled compound that easily passes the cell membrane and is phosphorylated by HSV1-tk where after it cannot leave the cell (Figure 21.3). In this way, radioactivity is accumulated in cells where

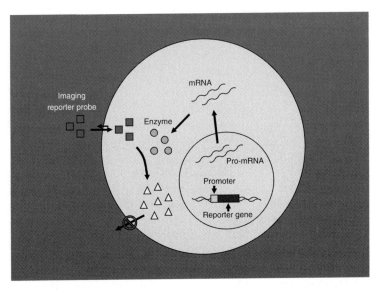

Figure 21.3. Imaging of gene expression using the reporter gene technique. The reporter gene encodes for herpes simplex virus 1 thymidine kinase (HSV1-tk). For further details, please see text.

HSV1-tk and thereby also the therapeutic gene are expressed. The imaging may be repeated by new administration of the reporter probe [12].

Another promising reporter gene for use with PET and SPECT is the gene encoding the sodium–iodine symporter (NIS) [13]. Expression of this gene leads to uptake of iodine, and the reporter probe in this case is simply iodine. Radioactive iodine for imaging is available both for PET (e.g., ^{124}I) and SPECT (e.g., ^{123}I).

11. RADIONUCLIDE THERAPY

When a tumor has been imaged using a specific ligand, this may be used as basis for localized radionuclide therapy. The ligand, that was labeled for imaging purposes may instead be labeled with α- or β-emitters whereby the tumor and metastases may be irradiated locally with a minimum of radiation to the non-cancer tissue (Figure 21.4). This principle is already in use for selected types of cancer, e.g., neuroendocrine tumors: imaging based on somatostatin receptor overexpression is obtained with ^{111}In-Octreotidscintigrafi and therapy is based on somatostatin receptor analogues labeled with ^{90}Y and/or ^{177}Lu that are β-emitters [14]. α- or β-emitters may also be coupled to monoclonal antibodies for tumor treatment, e.g., in the treatment of ovarian cancer (^{211}At) [15] or certain types of lymphoma (^{131}I) [16].

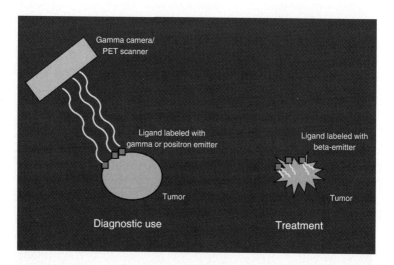

Figure 21.4. Use of the same ligand for imaging and radionuclide treatment. The ligand is labeled with different radioactive isotopes suited for imaging and treatment, respectively. For further details, please see text.

12. INTENSITY-MODULATED RADIATION THERAPY

Intensity-modulated radiation therapy (IMRT) is used for optimizing the radiation dose delivered to the different regions of a tumor. For this purpose, molecular profiling using molecular imaging will become increasingly important. Whereas the information considering FDG uptake already is routinely integrated in IMRT planning for certain tumor types, integration of further information from molecular imaging like hypoxia will refine this technique further.

13. IMAGE FUSION

Today most molecular imaging of cancer using PET and SPECT is image fused with anatomical information from CT or MRI.

14. CONCLUSION

Molecular imaging allows for the noninvasive study of pathophysiological processes in humans and animals. With availability of these techniques the advances for molecular biology may be used for early diagnosis and better characterization of cancer, which allows for more tailored and individualized treatment. Among techniques for molecular imaging PET and SPECT have the biggest translational potential, since techniques developed in animal models easily are transferred to humans. In order to exploit fully the potential of these techniques it is important to have access to both animal and clinical imaging facilities.

REFERENCES

1. Gutte H, Hojgaard L, Kjaer A. Early clinical experience and impact of FDG-PET. *Nucl Med Commun* 2005, 26: 989–94.
2. Jonson SD, Welch MJ. PET imaging of breast cancer with fluorine-18 radiolabeled estrogens and progestins. *Q J Nucl Med* 1998; 42(1): 8–17.
3. Krenning EP, Kwekkeboom DJ, Bakker WH, Breeman WA, Kooij PP, Oei HY, et al. Somatostatin receptor scintigraphy with [111In-DTPA-D-Phe1]- and [123I-Tyr3]-octreotide: the Rotterdam experience with more than 1000 patients. *Eur J Nucl Med* 1993; 20(8): 716–731.
4. Shields AF, Grierson JR, Dohmen BM, Machulla HJ, Stayanoff JC, Lawhorn-Crews JM, et al. Imaging proliferation in vivo with [F-18]FLT and positron emission tomography. *Nat Med* 1998; 4(11): 1334–1336.
5. Barthel H, Wilson H, Collingridge DR, Brown G, Osman S, Luthra SK, et al. In vivo evaluation of [18F]fluoroetanidazole as a new marker for imaging tumour hypoxia with positron emission tomography. *Br J Cancer* 2004; 90(11): 2232–2242.
6. Dehdashti F, Mintun MA, Lewis JS, Bradley J, Govindan R, Laforest R, et al. In vivo assessment of tumor hypoxia in lung cancer with 60Cu-ATSM. *Eur J Nucl Med Mol Imaging* 2003; 30(6): 844–850.
7. Sorger D, Patt M, Kumar P, Wiebe LI, Barthel H, Seese A, et al. [18F]Fluoroazomycinarabinofuranoside (18FAZA) and [18F]Fluoromisonidazole (18FMISO): a comparative study of their selective uptake in hypoxic cells and PET imaging in experimental rat tumors. *Nucl Med Biol* 2003; 30(3): 317–326.

8. Rehman S, Jayson CG. Molecular imaging of antiangiogenic agents. *Oncologist* 2005; 10: 92–103.
9. Haubner R, Kuhnast B, Mang C, Weber WA, Kessler H, Wester HJ, et al. [18F]Galacto-RGD: synthesis, radiolabeling, metabolic stability, and radiation dose estimates. *Bioconjug Chem* 2004; 15(1): 61–69.
10. Green AM, Steinmetz ND. Monitoring apoptosis in real time. *Cancer J* 2002; 8(2): 82–92.
11. Gambhir SS, Barrio JR, Phelps ME, Iyer M, Namavari M, Satyamurthy N, et al. Imaging adenoviral-directed reporter gene expression in living animals with positron emission tomography. *Proc Natl Acad Sci USA* 1999; 96(5): 2333–2338.
12. Avril N, Bengel FM. Defining the success of cardiac gene therapy: how can nuclear imaging contribute? *Eur J Nucl Med Mol Imaging* 2003; 30(5): 757–771.
13. Dadachova E, Carrasco N. The Na/I symporter (NIS): imaging and therapeutic applications. *Semin Nucl Med* 2004; 34(1): 23–31.
14. Krenning EP, Kwekkeboom DJ, Valkema R, Pauwels S, Kvols LK, de JM. Peptide receptor radionuclide therapy. *Ann NY Acad Sci* 2004; 1014: 234–245.
15. Andersson H, Elgqvist J, Horvath G, Hultborn R, Jacobsson L, Jensen H, et al. Astatine-211-labeled antibodies for treatment of disseminated ovarian cancer: an overview of results in an ovarian tumor model. *Clin Cancer Res* 2003; 9(10 Pt 2): 3914S–3921S.
16. Vose JM. Bexxar: novel radioimmunotherapy for the treatment of low-grade and transformed low-grade non-Hodgkin's lymphoma. *Oncologist* 2004; 9(2): 160–172.

CHAPTER 22

MRS AS ENDOGENOUS MOLECULAR IMAGING FOR BRAIN AND PROSTATE TUMORS: FP6 PROJECT "eTUMOR"

B. CELDA[1], D. MONLEÓN[1], M.C. MARTÍNEZ-BISBAL[1], V. ESTEVE[1],
B. MARTÍNEZ-GRANADOS[1], E. PIÑERO[1], R. FERRER[1], J. PIQUER[2],
L. MARTÍ-BONMATÍ[3], AND J. CERVERA[4]

[1]*Aplicaciones Biofísicas y Biomédicas de la RMN, Depto. Química Física, Universitat de Valencia;*
[2]*Hospital La Ribera, Alzira;* [3]*Clínica Quirón, Valencia;* [4]*IVO, Valencia*

Abstract: Molecular imaging has become during the last years in an important tool for supporting cancer diagnosis and prognosis. PET and SPECT are the most common molecular imaging techniques, although very promising and specific biological molecular agent contrast for CT and MRI are being recently developed. However, the above imaging techniques require exogenous contrast agents and usually a sole molecular image can be obtained at once. On the contrary, in vivo magnetic resonance spectroscopy (MRS), in particular ^{1}H MRS can simultaneously provide several molecular images using endogenous metabolites. In addition to biochemical spatial information from molecular imaging spectroscopy, MRS can also provide average metabolite profile of the selected affected tissue region. Initially MRS, especially ^{1}H MRS, was extensively applied to complete and improve the diagnosis and prognosis of central nervous system (CNS) pathologies, in particular brain tumors. However, during the last years the MRS applications have been extent to the diagnosis of different very common cancer types such as breast, prostate, colon carcinoma, and ovarian, among others. Likewise, MRS has been also used for lymph node assessment. In this contribution, the added value of MRS for the diagnosis, prognosis, and treatment selection of two different, important types of cancer: (1) brain tumors and (2) prostate, will be presented and discussed. Brain tumors are the leading cause of death in children under 15, and although in adults, brain cancers are proportionally less common than other cancers, it is a devastating disease with high mortality. There is a great need to increase our understanding of brain tumor biology to improve diagnosis and to develop new treatments. ^{1}H MRS is currently the only noninvasive method that can be used to investigate molecular profile of brain tumors and also provide molecular images, more than six in one acquisition, of the distribution of chemicals in a tumor, which are also generally heterogeneous. A summary of the applications of ^{1}H MRS to the in vivo diagnosis and prognosis of brain tumors will be presented. In addition, examples of metabolite limits, infiltration and high cellularity location for neurosurgery applications by MRS molecular images will be shown. Likewise, new ex vivo methods of studying the detailed biochemistry of tumor biopsies as metabolomic (high resolution magic angle spinning [HR-MAS]) and transcriptomic (DNA microarrays) will be discussed as complementary to in vivo MRS (FP6 European project eTUMOR). A preliminary comparison between molecular images from PET and ^{1}H MRS will be also presented. Finally, the application of ^{1}H MRS to the improvement of prostate diagnosis and prognosis, the second leading cause of cancer death, will also discussed, with particular attention to the location cancer contribution from MRS molecular images.

1. INTRODUCTION OF MRS AS A MOLECULAR IMAGING FOR CHARACTERIZATION OF BRAIN TUMORS AND PROSTATE CANCER

The lifespan of the European population is increasing and accordingly, diseases that become prevalent in adult and old age, such as brain tumors and prostate cancer, will afflict a larger percentage of this population.

Brain tumors do not have a lifestyle-associated aetiology; hence prevention is not yet possible. Likewise, in children over 1 year of age, brain tumors are the most common solid malignancies that cause death from disease. The current gold standard classification of a brain tumor by histopathological analysis of biopsy is an invasive surgical procedure and incurs a risk of 1–2% morbidity, in addition to healthcare costs and stress to patients. For tumors that evolve slowly in malignancy (e.g., pilocytic astrocytoma in children) repeated biopsies might not be advisable or practical. There is a need to improve brain tumor classification, and to provide noninvasive methods for brain tumor diagnosis, prognosis, to aid patient management and treatment. Diagnosis by magnetic resonance imaging (MRI) is noninvasive, but only achieves 60–90% accuracy depending on the tumor type and grade, hence can only replace biopsy for particular cases. MRS provides a noninvasive method to obtain a profile of the biochemical constituents of the tumor improving the accuracy of diagnosis in some significant instances [1–3]. There are a large number of different tumor types and grades, and usually the most malignant are heterogeneous. Thus, to develop automated classification methods that are comprehensive, data from several hospitals must be combined to fully characterize the variability of tumor spectra. Furthermore, the robustness of the classification method must then be validated in a real clinical setting. In addition, the possibility of phenotyping tumors with DNA microarrays may reveal new subtypes of tumors on molecular grounds. Moreover, the extensive and more precise metabolic analysis of tumors by MRS at high fields (>11 T) in nonmanipulated tissues (ex vivo) can allow a better understanding of the tumor biochemistry and may also refine the classification of brain tumors. Finally, it is important to look for correlations of patient survival with MRS characteristics, to assess whether there are better prognostic indicators than the current grading system. However, MRS spectra are complex and require skilled interpretation, hence clinical routine use of MRS is still low. The aim of eTUMOR project [4] is to coordinate European scientific expertise in MRS and genetic analysis to improve tumor classification and provide health care professionals with a validated decision support system (DSS) for noninvasive diagnosis of brain tumors, and the monitoring of tumor progression, and response to future new therapies.

On the other hand, prostate cancer is the second leading cause of cancer death in man and now represents the 15% of all cancers in developed countries. However, standard clinical test diagnosis as prostate-specific antigen (PSA) test, biopsy Gleason score, and digital rectal examinations are not accurate enough. Likewise, transrectal ultrasound using 2D or 3D techniques has poor sensitivity and specificity for disease detection. The specificity remains low when T2-w MRI is used. On the contrary, ^{1}H MRS and different techniques of DCE-MRI and diffusion are increasing the specificity and sensitivity for detecting, localizing and grading prostate cancer. Similarly to brain tumors, a multicenter study (International Multicenter Assessment of Prostate MR Spectroscopy, IMAPS) is being developed for prostate cancer [5]. Its main objective is to prove that ^{1}H MRS can detect and localize prostate carcinoma in the two major areas of prostate: (1) peripheral zone and (2) central gland.

MRS can be performed along with conventional MRI to provide metabolite profiles of a single voxel (SV) of tumor tissue in less than 5 min (Figure 22.1) or to produce several molecular images of different tumor metabolites (Figure 22.2) in ca. 10 min using multivoxel (MV) techniques, MRSI.

As well as characterizing the tumor type and grade, molecular images may provide spatial information on tumor boundaries and infiltration, and distinguish necrosis and active tumor within the lesion.

MRSI can provide, as it can be seen in Figure 22.2 [3,6], several different molecular images from endogenous metabolites with distinct biochemical and physiological significance.

2. BRAIN TUMORS

The prevalence of brain tumors is reduced with respect to other cancer types, as for instance prostate, breast, and colon. However, primary brain tumors are among the most lethal of all cancers. In fact, brain tumors are the leading cause of death from cancer in children under 15 and the second leading cause of death from 15–34. In addition, a lack of responsiveness to current therapy is often detected in this type of cancer. Therefore, the selection of individual treatment is still a challenge. On the other hand, brain tumors prevention is yet not possible because there is not a direct lifestyle associated. The current "gold standard" for brain tumor diagnosis is histopathologic assessment. Unfortunately, histopathology has some limitations: (1) it is invasive and carries risks (3% mortality); (2) their prognosis capacity is limited by the fact of repeated biopsies; (3) it can not be properly used for therapy selection; and (4) it is not useful for differentiation between different treatments.

Therefore, there is a great need to increase our understanding of brain tumor biology to improve diagnosis and to develop new treatments. During the last

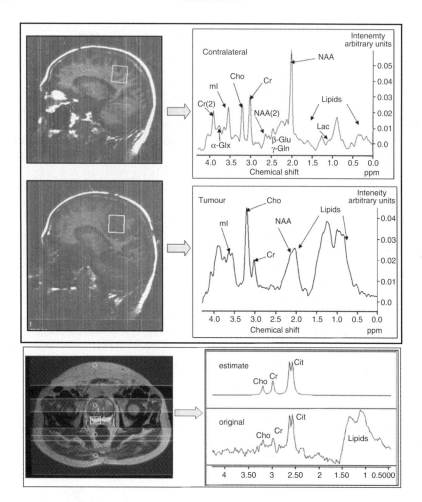

Figure 22.1. Metabolite profiles example of SV for: (A) *left*: brain tumour high-grade glioma: (i) *top*: tissue not affected; (ii) *down*: tumour tissue; (B) *right*: prostate tissue.

years it has become evident that there is a direct correlation between biochemical modifications in tumor tissues and histological modifications, as cellular proliferation, death, and density [7]. ^{1}H MRS is currently the only noninvasive method that can be used to investigate the molecular profile of brain tumors and also provide molecular images of the distribution of chemicals in a tumor, which are also generally heterogeneous. New methods of studying the detailed biochemistry of tumor biopsies are ^{1}H HR-MAS and DNA microarrays.

In this section the applications of different techniques (molecular imaging by MRS, ex vivo metabolomic, and ex vivo transcriptomic) for localizing, diagnostic, grade, tumor border definition, prognostic, and follow-up of brain tumors will be summarized.

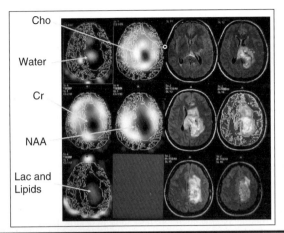
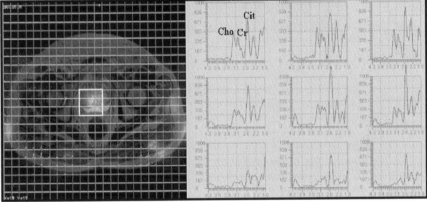

Figure 22.2. Top: example of five molecular images from MV for a GBM; *Bottom*: citrate image of prostate tissue together with six metabolite profiles from the selected volume.

2.1. Neurochemistry and in vivo metabolic diagnosis

The number of metabolites detectable by in vivo ^{1}H MRS at clinical MRI equipments (1.5 and 3 T) is mainly limited by two factors: (1) concentration in the selected volume, higher than 0.5 mM and (2) molecular correlation mobility directly related to T2 (transversal relaxation time). Hence, some interesting neurometabolites cannot be properly detected by the current MR technology, as for instance Dopamine by its low concentration. However, there are a large number of metabolites directly related with important anatomical and functional brain features. A summary of relevant metabolites detectable by ^{1}H MRS are gathered in Table 22.1.

Noteworthy, an excellent correlation between anatomic histopathological and metabolism alterations, detected by biochemical profiles (SV) and molecular images (MRSI) from ^{1}H MRS, in tumor tissues has been found [7]. As example, a

Table 22.1. Summary of brain metabolites observed by ¹H MRS

	CNS Neuro-metabolites
	NAA → neuronal marker. Localized in neurons and axons. Its disappearance is considered as neuronal dead and axonal damage.
	Cho → membrane turnover marker. Cho intensity changes are related to PCho and GPCho variations. Phosphatidilcholine no detectable by MRS when is in myelin, in membranes...
	Cr → energetic metabolism indicator. Cr + PCr is observed, approximately constant by: $H^+ + PCr + ADP^+ \Leftrightarrow Cr + ATP^+$
	Lac → final product of the glycolisis. Product of fail on oxidative mechanism. Lac is intra and extra-cellular. No detectable in healthy tissue.
	Mio-Inositol → Localized almost exclusively in astrocits, playing an osmolite role, regulating cellular volume.
	Lipids and polypeptides → Localized in myelin sheats and cellular and intracellular membranes. Detectable in MRS when are liberated from these structures.

summary of the correlation among the alterations in the concentration of different metabolites and distinct physiological modifications is enclosed in Table 22.2.

In addition to the common metabolites used for brain tumor classification and grade gathered in Table 22.1 and Table 22.2 there are other small biochemical compounds that can be used as biomarkers for increasing the sensitivity and specificity of ¹H MRS in brain diagnostic. Alanine (Ala), one of the simplest amino acids, is usually considered as a specific biomarker for benign or grade

Table 22.2. Metabolism by MRS vs. histopathology

Metabolism (MRS) vs. Histopathology		
*Metabolic changes	→	Precede cellular transformations Histologically detectables
• [NAA] ↓	⇔	**Neuronal Destruction** ↑
• [Cho] ↑	⇔	**Cellular Proliferation** ↑
[PCho]/[Cho] ↑ (not resolved In Vivo)	⇔	**Aggressiveness** ↑
• [Macromolecules] ↑		
(movil lipids, fatty acids & aminoacids)	⇔	**Necrosis** ↑
• [mI] ↓	⇔	**Aggressiveness**↑
		mI=mionositol in astrocits controls: brain hydrate equilibrium+protein C Kinasa activation proteolitic enzymes in malignant tumours
[mI] ↑ (regulation fail in mI production)	⇔	**Low grade Tumours** ↑

I meningiomas [8]. Likewise, Taurine (Tau) has been proposed as a particular biomarker to differentiate medulloblastomas from all other pediatric tumors [9].

2.2. Brain tumor detection and location

Precise tumor detection and location is critical for a proper diagnosis, prognosis and treatment selection. This fact is particularly important for brain tumors, for which the therapy selection can largely depend on the specific location of the most active part of the tumor.

During the last years it is becoming evident that MRSI has a larger number of advantages vs. SV methodology as:
(a) MRSI provides spatial information: more accurate spatial metabolite alteration distribution (better location of tumor affected regions not properly detected by MRI and SV) useful for biopsy retrieval guide and SV proper location
(b) MRSI can also determine infiltration process (primary vs. secondary tumor)
(c) MRSI enables tumor border definition (treatment selection and guide as radiotherapy or local chemotherapy)
(d) MRSI and SV allow evaluation of tumor progression (recurrence vs. radiation injury)

Choline compounds are directly related to cellular proliferation, dead and density [7] and therefore Cho MRSI images can provide spatial information about the more active and aggressive part of a brain tumor. This information is complementary to the standard imaging techniques as dynamic contrast enhancement, commonly used in clinical routine. An example of molecular Cho image for the detection and location of the most active part of the brain tumor is given in Figure 22.3. This information can be very useful for a more precise detection, location and different region definition of brain tumor.

2.3. Brain tumor classification and grade

Nevertheless, SV is still being a standard technique in many hospitals and can provide a better spectra resolution because of the higher field homogeneity achieved in a localized tissue region. There are two critical factors for an adequate and accurate diagnostic by SV MRS:
- SV location (i.e., avoiding necrotic and edema areas)
- parameters selection (mainly TR and TE)

For those brain tumors in which the contrast enhancement is low (gliomas grade II and III) or with a large heterogeneity (GBM) the accuracy of the diagnosis by SV strongly depends on adequate SV position on the most aggressive region of the tumor avoiding necrotic areas.

This methodology for SV location by MRSI combined with the acquisition of 2 TE (31 and 136 ms) has allowed a statistical classification of glial tumors (grade II, III, and IV) [1,3]. The combined used of short and long TE improves

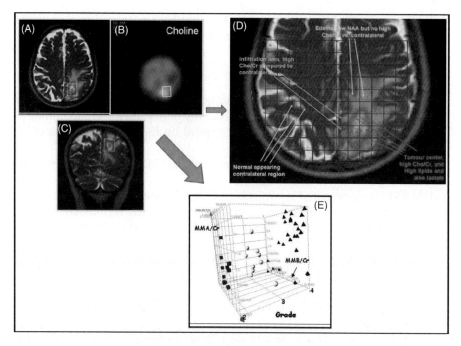

Figure 22.3. MRI anatomical and MRS molecular images for a glioma of grade II: (A) and (C) axial and coronal images after contrast enhancement. (B) Cho molecular image. (D) Clustering of gliomas of grade II, III, and IV classified by SV at two different TE (short 31 and long 136 ms echo times) previous volume interest selection by Cho image from MRSI.

the accuracy of brain tumors diagnosis (90–94%) as it has also being described in and previously reported [1,3]. Accurate diagnosis of tumors, in general and in brain ones, in particular, is crucial for a better management, treatment selection, and therapy response. As example, particularly important is the precise diagnostic of grades II and III gliomas, one of most common type of brain tumors, for the best treatment selection. In addition, both tumors may appear as nonenhancing or minimally enhancing lesions on T1-weighted MRI, hence their detection and location is difficult. The combination of MRSI images for an adequate location of more aggressive tumor regions and metabolic profiles from SV at two TE (31 and 136 ms) can allow a statistical differentiation among gliomas of grade II, III, and IV. As an example, the separation of grade II, III, and IV gliomas by two particular biomarkers (MMA and MMB) is shown in Figure 22.3E. Noteworthy, statistical significant differences among gliomas of grade II, III, and IV have been found for an extensive set of biomarkers as MMA/Cr, MMB/Cr, NAA/Cr, Cho/Cr, NAA/Cho, Cr/H_2O, mI/ H_2O, and MMA/ H_2O.

Probably one of the most interesting questions in brain tumor diagnosis is the differentiation between primary and secondary (metastases) lesions. Again, the metabolic spatial information provided by MRSI can be very useful for a proper differentiated diagnosis between high-grade gliomas, particularly GBM, and metastases. GBM are heterogeneous and infiltrating tumors.

Therefore, in principle, metabolic alteration can be detected in regions outside of the central core of the tumor, although not evident anatomical modifications can be observed. However, in metastases lesions infiltration is not expected, with the biochemical modifications localized just inside of the lesion. Examples of MRSI application for the molecular imaging differential diagnosis between GBM and metastases are enclosed in Figure 22.4. In the metastases lesion (Figure 22.4A), there is not evidence of infiltration either by anatomical images or from metabolic information by MRSI spectra, because Cho increment is not observed in the adjacent regions to the lesion. However, in the example of a GBM (Figure 22.4B), a clear increase of Cho concentration respect to normal

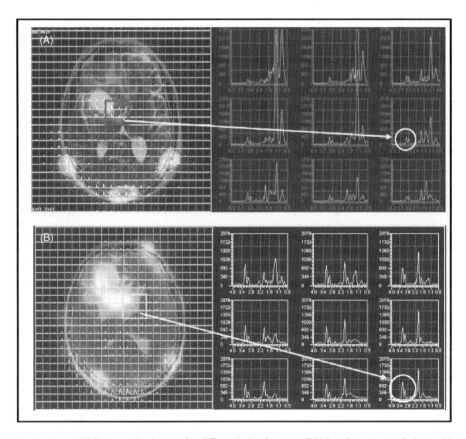

Figure 22.4. MRSI molecular images for differentiation between GBM and metastases lesions. (A) MRSI lipid image for a metastases lesion superimposed to T2 MRI image; not Cho concentration increment observed in the contiguous regions to lesion. (B) MRSI Cho image for a GBM, Cho concentration increase detected in infiltrated areas close to the tumor lesion.

tissue can be detected in regions contiguous to the tumor indicating an infiltration process related to aggressive gliomas.

2.4. MRSI molecular imaging for treatment planning

Biochemical spatial information enclosed in MRSI metabolic images can be very useful for treatment planning, as neurosurgery, chemotherapy, and radiation. In fact, currently, the standard approach to define radiation target volume for brain glioma treatment is to deliver a uniform dose to areas of imaging abnormalities determined by MRI images (FLAIR, postcontrast T1-weighted) or CT scans, with an additional uniform margin of 1.5–2 cm. This often causes the target volume to either cover too much noninvolved brain tissue or leave small areas of tumor infiltration untreated. Thus, it is critical that tumor heterogeneity and margins identified for different dose delivery are defined accurately. MRSI, contrarily to MRI, can provide better delineation of tumor heterogeneity, such as regions of higher grade elements in low grade tumor and region of microscopic disease, thus allowing more accurate prescription of radiation dose painting to treat gliomas more effectively while preserving nonaffected brain tissue [10]. In Figure 22.5 enclosed is an example of metabolic distribution by nosologic images from MRSI spectra biochemical spatial data.

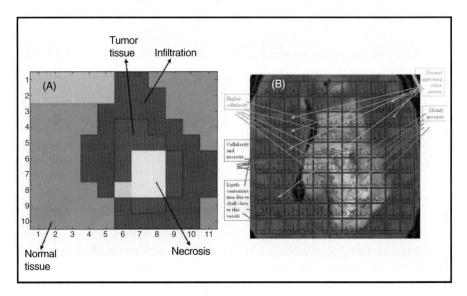

Figure 22.5. Nosological images from MRSI biochemical spatial information. (A) Nosological image of GBM with delineation of tumor tissue, necrotic areas, and region of metabolic infiltration. (B) MRSI spectra with the spatial metabolic distribution used for generating the nosological image.

2.5. Discrimination between recurrence and radionecrosis in brain tumors by MRSI spatial molecular information

Radiotherapy is a paramount therapeutic adjunct for patients with brain tumor. After radiation treatment, a considerable number of patients develop new lesions at or near the original tumor site. These lesions can be treatment-induced damage, tumor recurrence, or progressive tumor growth. The radiological differentiation of treatment effects from recurrent tumor is very difficult. Molecular information through the metabolic spatial distribution provided by MRSI images can be used for the diagnosis of radiation-induced injury respect to recurrence. Again, the Cho is a critical biomarker to distinguish areas of high cellularity, tumor recurrence, respect to necrotic regions, with large concentration of lactate and/or macromolecules/lipids, in radionecrosis. This is clearly illustrated in Figure 22.6 in which an example of recurrence and radiation-induced injury are shown.

An increase of Cho concentration indicating a high cellularity level and then tumor activity should be observed in those cases of recurrence. In the Figure 22.6A a clear decrease of NAA and slight increase of Cho concentrations indicating a neuronal/axonal damage together with cellular activity is detected. The clinical diagnosis was tumor recurrence. On the other hand, not NAA decrease and Cho increase concentrations are observed in the example of radionecrosis (Figure 22.6B).

2.6. ^{18}FDG PET and MRSI molecular imaging comparison for brain tumor location and classification

PET is an important molecular imaging technique for cancer detection and location that uses radioactive exogenous molecules. Similarly, the late development of MRI for whole body can also be used for screening the possible spread of cancer in one session using endogenous molecules. Both techniques are complementary and should be used individually or combined depending on the difficulty of diagnosis. However, there are few examples of combined use of PET and MRSI showing the potential applicability and limitations of each methodology in cancer diagnosis. In Figure 22.7 there are enclosed two combined studies of ^{18}FDG PET and MRSI for the diagnosis of brain lesions.

In both cases an excellent correlation between ^{18}FDG PET and MRSI molecular information was reached. Likewise, the diagnosis of both molecular techniques agreed with histopathology results. Particularly interesting is also the good spatial correlation between ^{18}FDG PET and MRSI as can be verified in Figure 22.7. A clear decrease in metabolic intensity in ^{18}FDG PET image is also observed almost in the same location that a high concentration of lipids and *N*-acetylated compounds is found in MRSI spectra for the cystic metastases of Figure 22.7 (*left part*). Contrarily, a high metabolic activity is detected in the ^{18}FDG PET image of the GBM of Figure 22.7 (*right part*) with a simultaneous increase of Cho and decrease of NAA concentrations in MRSI biochemical spectra.

296   CHAPTER 22

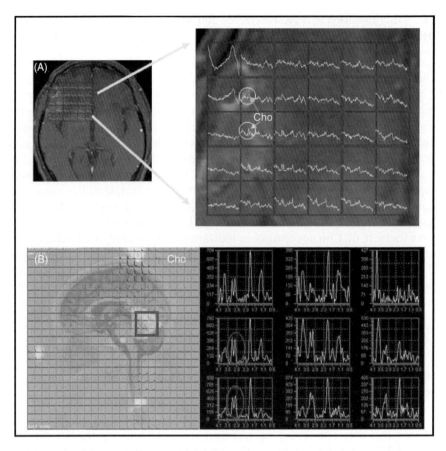

Figure 22.6. Examples of MRSI molecular images application for the differentiation between tumor recurrence and radionecrosis. (A) Overlay of transverse MRSI spectra and T2 image. In the image expansion can clearly be observed an increase of Cho concentration. (B) Sagittal MRSI molecular Cho image of radiation-induced injury.

2.7. Ex vivo metabolomics (HR-MAS)

One of the most critical limitations of MRS, either SV or MV, is the reduced spectral resolution (Figures 22.1 and 22.2) mainly due to distribution of magnetic susceptibilities in human tissue and, to a lesser extent, from residual dipolar couplings [11]. This limitation in part can be overcome by using high NMR field as 3 T. Nevertheless, during the last years it has been shown the possibility of acquiring high resolution NMR spectra (>7 T) in different intact type of tissues, which can be considered as ex vivo metabolomics [12], and their direct utility for increasing the specificity and sensitivity of different types of cancer [12,13]. The resolution of tissue spectra can be increased by combining techniques of high resolution NMR for liquids with solid state NMR, as using

MRS AS ENDOGENOUS MOLECULAR IMAGING FOR BRAIN   297

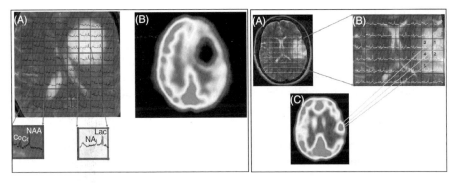

Figure 22.7. Molecular imaging comparison of MRSI vs. PET for brain tumors. *Left part* cystic metastases: (A) MRSI spectra superimposed to T2 image, enclosed are two expansions: normal tissue (*left*) and cystic spectrum (*right*); (B) PET image with a clear absence of metabolism activity in the cystic area. *Right part* GBM: (A) MRSI spectra superimposed to T2 image; (B) zoom of the tumor and contralateral regions. 1, 2, 3, and 4 voxels corresponding to the highest Cho concentration in high cellularity areas; voxels 5, 6, and 7 correspond to infiltration regions. (C) PET image, high metabolism region in good agreement with high Cho concentration area in MRSI (B).

magic angle together with high spin rotation of the sample (HR-MAS) [12]. Ex vivo HR-MAS spectra of brain, prostate, and breast tissues have allowed the identification of a great number of metabolites [11–13]. In principle, this large number of metabolites assigned at high resolution can be used for a more detailed differentiation among distinct types of brain tumors or even to distinguish between different possible tumor subtypes, as between GBM primary and secondary [11]. Examples of HR-MAS spectra are enclosed in Figure 22.8.

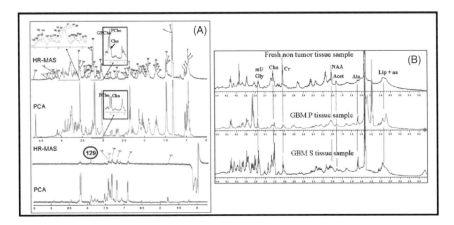

Figure 22.8. High NMR resolution spectra of brain tumor tissue. (A) Comparison between perchloric acid extracts (PCA) of brain tumor tissues with ex vivo tissue (HR-MAS). (B) HR-MAS ex vivo spectra of normal tissue (*top*); primary GBM (*middle*) and secondary GBM (*bottom*).

The resolution of ex vivo HR-MAS spectra (Figure 22.8A) is comparable with the one obtained from PCA extracts, but without the limitations of reproducibility of PCA methodology and also containing all metabolite information, including nonsoluble water compounds, as lipids and fatty acids, critical for a better brain tumor differentiation and avoiding the chemical pretreatment [11]. This large resolution enables the assignment of almost 130 resonances and the identification of more than 30 metabolites [11].

2.8. Location, classification and grade of brain tumors by in vivo MRSI, ex vivo metabolomics and transcriptomic: eTUMOR FP6 IP project

The recent developed technique of DNA microarray analysis can now provide a detailed genomic profile of tumor tissue. Genes that are associated with cellular proliferation, cellular mobility, and migration, are highly expressed in tumors, and particularly so in high-grade gliomas [14–16]. Differences in gene expression between grades of glioma [14–16] have been observed, and also between tumor types that are not easily distinguished by morphological appearance alone [17], and with prognosis [18]. Thus, gene profiling has potential to be an alternative to current histopathology and may provide subtler tumor classification and determination of an individual patient prognosis with respect to treatment [15].

Therefore, the combined use of in vivo and ex vivo MRS metabolic data with phenotype information from DNA microarrays can improve the diagnosis, prognosis, and therapy selection in brain tumors. This is one of the main objectives of the Integrated Project from the FP6 called eTUMOR (FP6-2002-LIFESCI-HEALTH 503094). The preliminary results for the first group of brain tumors studied by the whole set of described techniques is gathered in Figure 22.9. High resolved HR-MAS spectra can be used either for a better metabolite identification for in vivo MRS spectra and for a correct classification of different types of brain tumors. In addition, similar differentiation among the distinct types of brain tumors considered (GBM, astrocytomas, oligoastrocitomas, metastases, and meningiomas) has also been obtained by the pangenomic study carried out by DNA microarrays including more than 30,000 different human genes.

3. PROSTATE CANCER

Prostate carcinoma is the most common cancer affecting men in western developed countries [12]. The specificity and sensitivity of standard diagnostic techniques, as PSA, biopsy Gleason score, digital rectal, and transrectal ultrasound, examinations is low. Even the use of MRI techniques, as T2-weigthed images, does not increase the specificity of prostate cancer diagnosis. However, the metabolic information from ^{1}H MRS has provided additional data for a higher specificity and sensitivity for detection and location of prostate cancer [5].

Similarly to brain tumors, there are specific biomarkers from ^{1}H MRS for differentiating among normal tissue, benign prostatic hyperplasia (BPH) and

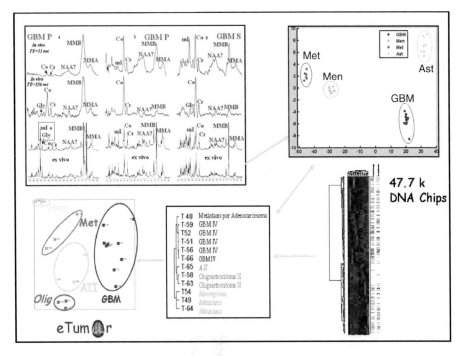

Figure 22.9. Brain tumor classification by combined use of: *left top*: in vivo MRS (two higher row spectra) and HR-MAS (two *bottom* spectra) with statistical canonical clustering (*right top*) and DNA microarrays (*bottom*) example of DNA chip + PCA clustering.

tumoral tissue. Cho is again a typical biomarker for high cellularity regions indicating tumor tissue, the higher the Cho concentration the higher the prostate cancer grade. In addition, Citrate (Cit) is a standard molecular marker for normal prostate tissue. An example of different metabolic profile between normal tissue and BPH is shown in Figure 22.10.

The combination of high Cho and a low Cit concentration is related with tumoral prostate tissue. Whereas, high Cit and low Cho concentrations correspond to normal prostate tissue. Finally, BPH is characterized by high Cho and medium Cit concentrations. These metabolic changes have been extensively correlated to histopathology data [5].

The metabolic profiles in Figures 22.10 and 22.11 have been obtained by using surface coils for detection [6]. This methodology is quite singular because usually endorectal coils are the standard methodology for prostate MRSI molecular imaging acquisition. Although this technique, provides a good signal/noise ratio and an adequate resolution, is invasive and cannot be tolerated by all the patients. Likewise, the endorectal coil implies additional expenses since it must be regularly despaired. However, the methodology proposed using surface coils can be easily incorporated to routine clinical situations. A good agreement

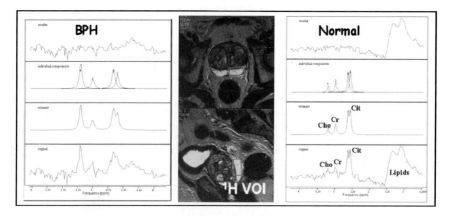

Figure 22.10. Metabolic profiles of prostate tissue by SV 1H MRS. *Left part*: Benign prostatic hyperplasia (BPH); *right part*: Normal tissue.

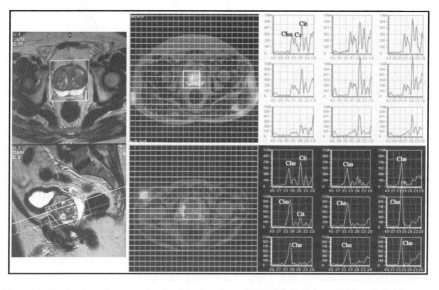

Figure 22.11. Prostate tissue molecular imaging from 1H MRS. *Left part*: T2 image for location of MV for MRSI. *Right part*: *Top*, example of BPH tissue; and *bottom*, example of prostate tumor tissue.

between metabolic profiles by a combination of SV and MRSI data with histopathology results has been obtained [6] demonstrating the potential use of surface coils as an alternative for a proper "virtual" in vivo biopsy for prostate location and detection.

ACKNOWLEDGMENTS

We thank the Radiology Services of Hospital Quirón (Valencia), Hospital Peset (Valencia) Hospital La Ribera-Alzira, IVO (Valencia) for their enthusiastic collaboration in the MRS acquisition data. We are grateful to the Neurosurgeons and Pathologist Services of Hospital La Ribera-Alzira and IVO (Valencia). Thanks are due to the SCSIE of the University of Valencia for providing access to the NMR facility. We also thank Bruker España S.A., Bruker Biospin France and Philips Ibérica España for financial support. The Association for the Development of Research in NMR (ADIRM), the Excellence Program of the Generalitat Valenciana (GV GRUPOS 03/072), the Spanish network "Imagen Médica Molecular y Multimodalidad. Análisis y Tratamiento de Imagen Médica" IM3(Institituo Salud Carlos III, G03/185), the Spanish National Project (SAF2004-06297) and the European Integrated Project eTUMUR (FP6-2002-LIFESCIHEALTH 503094) are acknowledged for financial support.

REFERENCES

1. M.C. Martínez-Bisbal, B. Celda, L. Martí-Bonmatí, P. Ferrer, A.J. Revert, J. Piquer, E. Mollá, E. Arana, and R. Dosdá. Contribución de la espectroscopía de resonancia magnética a la clasificación del glioma de alto grado. Valor predictivo de las macromoléculas. *Rev Neurol* 34(4): 309 (2002).
2. C. Majós, M. Juliá-Sapé, J. Alonso, M. Serrallonga, C. Aguilera, J.J. Acebes, C. Arús, and J. Gili. Brain tumour classification by proton MR spectroscopy: Comparison of diagnostic accuracy at short and long TE. *AJNR* 25: 1696–1704 (2004).
3. M.C. Martínez-Bisbal, R. Ferrer-Luna, B. Martinez-Granados, D. Monleón, V. Esteve, J. Piquer, A.J. Revert, E. Mollá, L. Martí-Bonmatí, and B. Celda. Glial tumours grading by a combination of 1H MRS short and medium echo time single voxel located by spectroscopic imaging. *MAGMA Magnetic Resonance Materials in Physics, Biology and Medicine* 2005; 18(S1), ESMRMB S68: 114 (2005).
4. D. Monleón, M.C. Martinez-Bisbal, B. Martinez-Granados, V. Esteve, R. Ferrer-Luna, E. Piñero-Sagredo, M. Matas, F.V. Pallardo, J. Piquer, L. Marti-Bonmati, and B. Celda. In vivo and ex vivo metabolomic and genomic data for brain tumour diagnosis and prognosis. A prospective study as part of eTUMOUR (FP6-2002-LSH503094). *MAGMA Magnetic Resonance Materials in Physics, Biology and Medicine* 2005; 18(S1), S194 ESMRMB S68: 325(2005).
5. T. Scheenen, E. Weiland, J. Futterer, P. van Hecke, P. Bachert, G. Villeirs, J. Lu, M. Lichy, B. Holshouser, S. Roell, J. Barentsz, and A. Heerschap. *Proc Intl Soc Mag Reson Med* 13: 260 (2005).
6. M.C. Martínez-Bisbal, B. Celda, B. Martínez-Granados, C. San Juan, and L. Martí-Bonmatí. Feasibility of quadrature body coil in prostate spectroscopic MR imaging European Congress of Radiology ECR 2005, Viena 4–8 Marzo 2005, *European Radiology* 15(S1): 449 (2005).
7. T.R. McKnight, T. Love, A. Bollen, M. Berger, S. Nelson, S. Chang, and K. Lamborn. Intratumoral correlation of MR spectroscopic and growth characteristics of grades II and III gliomas. *Proc Intl Soc Mag Reson Med* 13: 667 (2005).
8. M.C. Martínez-Bisbal, V. Esteve, D. Monleón, B. Martínez-Granados, J. Piquer, A.J. Revert, E. Mollá, L. Martí-Bonmatí, and B. Celda. 1H HR-MAS on brain tumour biopsies for the metabolical study of GBM, meningiomas and metastases. *MAGMA Magnetic Resonance Materials in Physics, Biology and Medicine* 2005; 18(S1), S194 ESMRMB S68: 278 (2005).

9. X. Liu, A. Panigraphy, I. González-Gómez, F.H. Pilles, G.J. McComb, M.D. Krieger, M.D. Nelson, and S. Bluml. Quantitative short echo time spectroscopy of untreated paediatric brain tumours. *Proc Intl Soc Mag Reson Med* 13: 665 (2005).
10. S. Takur, A. Narayana, J. Chang, S. Karimi, G. Perera, J. Koutcher, and W. Huang. Evidence of need of incorporation of MRSI data in radiation treatment planning gliomas. *Intl Soc Mag Reson Med* 13: 2089 (2005).
11. M.C. Martínez-Bisbal, L. Martí-Bonmatí, J. Piquer, A. Revert, P. Ferrer, J.L. Llácer, M. Piotto, O. Assemat, and B. Celda. 1H and 13C HR-MAS spectroscopy of intact biopsy samples ex vivo and in vivo 1H MRS study of human high grade gliomas. *NMR in Biomedicine* 17: 191–205 (2004).
12. C.E. Mountford, S. Doran, C.L. Lean, and P. Russell. Proton MRS can determine the pathology of human cancers with high level of accuracy. *Chem Rev* 104: 3677–3704 (2004).
13. B. Sitter, U. Sonnewald, M. Spraul, H.E. Fjösne, I.S. Gribbestad. High-resolution magic angle spinning MRS of breast cancer tissue. *NMR in Biomedicine* 15: 327–337 (2002).
14. D.S. Rickman, M.P. Bobek, D.E. Misek, R. Kuick, M. Blaivas, D.M. Kurnit, J. Taylor, and S.M. Hanash. Distinctive molecular profiles of high-grade and low-grade gliomas based on oligonucleotide microarray analysis. *Cancer Res* 61: 6885–6891 (2001).
15. B. Celda, D. Monleón, M.C. Martínez-Bisbal, and L. Martí-Bonmatí. Future developments expected from MR: status of the experimental work regarding tumour evolution prediction, lymph assessment. *MAGMA Magnetic Resonance Materials in Physics, Biology and Medicine* 2005; 18(S1), ESMRMB S68: 79 (2005).
16. P.S. Mischel, T.F. Cloughesy, and S.F. Nelson. DNA-microarray analysis of brain cancer: molecular classification for therapy. *Nat Rev Neurosci* 5: 782–792 (2004).
17. S.L. Pomeroy, P. Tamayo, M. Gaasenbeek, L.M. Sturla, M.E. Angelo McLaughlin, J.Y.H. Kim, L.C. Goumnerova, P.M. Black, C. Lau, J.C. Allen, D. Zagzag, J.M. Olson, T.Y. Curran, C. Wetmore, J.A. Blegel, T. Poggio, S. Mukherjee, R. Rifkin, A. Califano, G. Stoitzky, D. Louis, J.P. Mesirov, E.S. Lander, and T. Golub. Prediction of central nervous system embryonal tumour outcome based on gene expression. *Nature* 415: 436–442 (2002).
18. S.-L. Sallinen, P.K. Sallinen, H.K. Haapasalo, H.J. Helin, P.T. Helen, P. Schraml, O.-P. Kallioniemi, and J. Kononen. Identification of differentially expressed genes in human gliomas by DNA microarray and tissue chip techniques. *Cancer Res* 60: 6617–6622 (2001).

CHAPTER 23

FROM LINAC TO TOMOTHERAPY: NEW POSSIBILITIES FOR CURE?

G. STORME, D. VERELLEN, G. SOETE, N. LINTHOUT, J. VAN DE STEENE, M. VOORDECKERS, V. VINH-HUNG, K. TOURNEL, AND D. VAN DEN BERGE

Department of Radiation Oncology, Oncologic Center AZ-VUB, Laarbeeklaan, 101, 1090-Brussels, Belgium

Technical development in the last years has permitted, using the last imaging techniques, to deliver a "curative" dose without increasing toxicity to the adjacent organs.

Imaging techniques such as CT scan, MRI, and PET-scan are complementary and organ-specific (e.g., PET–CT for lung and MRI–CT for brain). Imaging allows us to define more accurately the macroscopic lesions (GTV) and delineate much better in our "daily" planning system.

On the other hand, modern radiotherapy machines allow for better targeting of the GTV by improving the positioning, and delivering an adequate conformal dose. Intensity modulated radiotherapy (IMRT) can in some cases, better envelop the planned target volume (PTV) without delivering a toxic dose in the surrounding sensitive healthy tissues.

With this perspective all imaging equipment (radiology department) is interconnected and all data are archived (PACS).

In accordance with the development of the electronic medical files, those images and protocols are available throughout the entire hospital. The system is linked with the hospital information network where also the departments of nuclear medicine and radiotherapy are connected. The exchange of all data even for consulting is complete.

In addition a direct link between imaging (radiology and nuclear medicine) and radiotherapy is a necessity in relation to the planning. In our department, this permitted the development of high precision radiotherapy using the Novalis "shaped beam surgery" or conformal radiotherapy, initially developed for

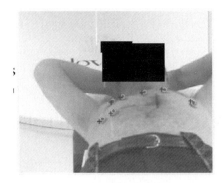 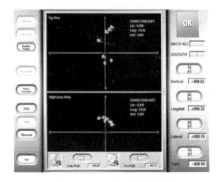

Figure 23.1. Patient set up with IR markers and if superimposed on crosses (*left*) the positioning is perfect expressed by blue lines in all directions (*right*).

radiosurgery to the brain. Here also fusion between CT and MRI is complementary as in the arteriovenous malformations the angiographies and CT.

The ExacTrac system permits using infrared reflectors (IR) for automated, computer-controlled patient setup yielding a positioning accuracy (Figure 23.1) exceeding those observed with traditional skin tattoos or fushin marks [1].

For now more than 5 years the Novalis Body system (1st prototype: 2 RX-tubes and one flat panel; currently 2 RX-tubes and two flat panels) has been used clinically, which allows us to evaluate with radiologic quality the bony structures or radio-opaque implanted markers (Figure 23.2). On the control screen, the measured 3D error is indicated and the correction is performed automatically using the IR system. An improvement is observed compared with the IR procedure [2]. Using both tools we have treated actually more than 450 patients with prostate

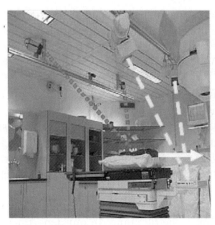 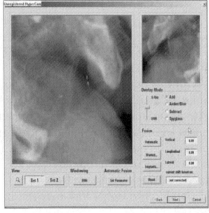

Figure 23.2. ExacTrac (red) and Novalis Body (blue) (*left*) help to position the patient at the millimeter level (*right*).

cancer without any major complication and millimeter accuracy has been realized when applying implanted markers [De Cock, M., et al., personal communication, 3]. Preliminary data shows that the actual 5-year expected survival for cT1-2 is 96.9% and for those who receive TAB is 83.5% [4].

In lung cancer where respiration influences tumor positioning the following sequence is used in clinic: (1) fusion between CT and PET (the last examination takes in to account the respiration correcting for the fast image acquisition obtained with last generation CT scanners which provide us snapshot images of the anatomy of the thorax); and (2) If possible radio-opaque markers are implanted transthoracic at the tumor level (Figure 23.3a) [5].

At the planimetry, after fusion, the lesion of the PET, which is always larger due to breathing, is used to define the internal margin (accounting for internal organ motion) for the PTV [6]. This precise localization, permits to deliver doses in the range of 10×4–5 Gy, 8×7.5 Gy and even 4×10 Gy and 2×20 Gy on the GTV taken in account of the proportion of the target volume on the total lung volume (Figure 23.3b).

Preliminary data on lung function (6 months follow-up for 54 patients) show no major toxicity, with the exception of one patient with G3 toxicity [7].

Liver metastasis is another location where conformal radiotherapy can replace surgery and or chemotherapy (Figure 23.4). The DKFZ has shown that in unifocal lesions, a dose between 20 and 25 Gy can obtain 80% nonevolution of the irradiated lesion at 2 years [8,9]. Our experience also shows that when multiple chemotherapy was already given, this approach could even prolong survival without toxicity for about 1.7 years [Vinh-Hung, V., et al., personal communication] and not those lesions do not to be treated.

Brain metastases (maximum 3 locations on MRI) with maximum diameter of 3 cm, taken in account the quality of survival in that category of patients, delivering 20 Gy in one fraction is a good indication (Figure 23.5). When these patients relapse within the brain the new location(s) can easily be retreated.

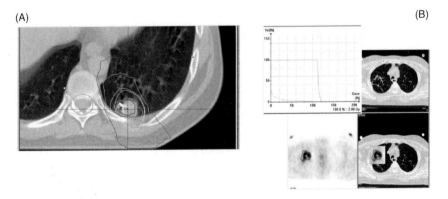

Figure 23.3. (A) Marker in tumor superimposed with isodose lines. (B) Green: volume of lung irradiated.

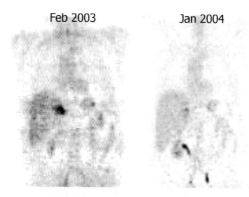

Figure 23.4. (A) Liver metastases from a colorectal cancer. (B) PET scan 11 months after treatment.

Obviously, primary brain tumors, neurinoma, and pituitary tumors, indications the Novalis was originally developed for are typical indications treated at the AZ-VUB.

If larger fields than 10×10 cm^2 are required, the system permits to navigate towards a second isocenter and deliver homogeneous dose to the entire GTV and CTV using IMRT. The planimetry takes in account the overlapping zone and homogenize in the overlapping fields in the range of ±5–8% [10].

The last development since this summer is the implementation of helical tomotherapy in the department [11]. This machine is in fact the result of integrating IMRT and volume imaging in 3D due to the use of megavoltage CT imaging. The main goal is to spare the major side effect of radiation in head and neck tumors: dryness of the mouth due to the functional abolition of the salivary glands (Figure 23.6). It was shown to reduce the dose to the parotic glands by + = 30% in comparison with the best other IMRT application [12]. The daily

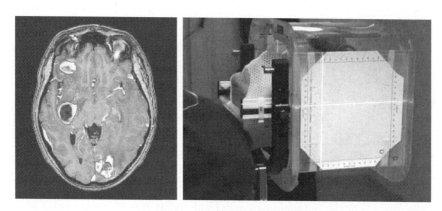

Figure 23.5. Brain metastases treated with a reallocable mask.

FROM LINAC TO TOMOTHERAPY: NEW POSSIBILITIES FOR CURE? 307

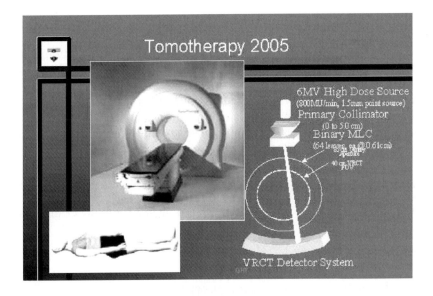

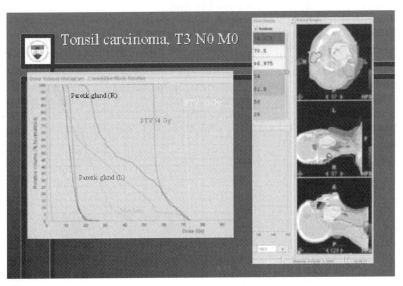

Figure 23.6. Tomotherapy facility and dose painting of a head and neck tumor (oropharynx) demonstrating that the heterolateral parotic gland (dark green) is under the 20 Gy level.

images are used for positioning the patient at millimeter level. Also the isodoses can be used to guide the positioning.

Here also we will integrate the approach launched with Novalis to irradiate metastatic locations within one session and improve the quality of life of the patients avoiding side effects.

Integrating the imaging modalities of the academic hospital in the department of radiation oncology permits us to precisely delineate the primary tumor as well as the critical organs. It permits us to deliver conformal radiotherapy, IMRT, navigating within the patient and eliminating toxic side effects to the adjacent normal surrounding tissues.

Actually we are implementing hypofractionation hoping we can obtain more local cure using the known radiobiological data and as such decrease the overall treatment time. The first data in prostate and lung cancer are encouraging not increasing toxicity and so far having excellent outcome.

ACKNOWLEDGMENTS

Part of this work is done by a grant of Brain-Lab and the Willy Gepts Fonds.

REFERENCES

1. Soete G et al. *Int J Radiat Oncol Biol Phys* 52: 694–698, 2002.
2. Soete G et al. *Int J Radiat Oncol Biol Phys* 54: 948–952, 2002.
3. Verellen D et al. *Radiother Oncol* 67: 129–141, 2003.
4. Soete G et al. *Int J Radiat Oncol Biol Phys* 2006 (in Press).
5. De Mey J et al. *J Vasc Interv Radiol* 16: 51–56, 2005.
6. Caldwell et al. *Int J Radiat Oncol Biol Phys* 55: 1381–1393, 2003.
7. Van De Steene J et al. *Radiother Oncol* 73(Suppl.): S428, 2004.
8. Herfarth KK et al. *J Clin oncol* 19: 164–170, 2001.
9. Herfarth KK et al. *Int J Radiat Oncol Biol Phys* 57: 444–451, 2003.
10. Linthout N et al. *Med Phys* 30: 846–855, 2003.
11. Hong TS et al. *Brit J Cancer* 92: 1819–1824 2005.
12. Van Vulpen M et al. *Int J Radiat Oncol Biol Phys* 62: 1535–1539, 2005.

CHAPTER 24

TARGETING mTOR FOR CANCER TREATMENT

BELEN RUBIO-VIQUEIRA AND MANUEL HIDALGO

The Sidney Kimmel Comprehensive Cancer Center at Johns Hopkins, The Johns Hopkins University, School of Medicine, Baltimore, MD.

Abstract: The mammalian target of rapamycin (mTOR) is involved in the control of cellular growth and proliferation. Abnormal activation of signaling pathways both proximal and distal to this kinase occurs frequently in human cancer suggesting that mTOR is an attractive target for antineoplasm therapies. Rapamycin and its analogs inhibit mTOR, and have showed potent antitumor activity in vitro and in xenograft models. Several phase I and phase II studies with rapamycin-like drug have been performed demonstrating antitumor activity in different types of refractory neoplasms. The clinical development of mTOR inhibitors exemplifies the challenges in developing targeted agents. mTOR inhibitors have been well tolerated at a wide range of doses, making the selection of phase II doses based solely on toxicity criteria difficult. Assessment of pharmacodynamic effects in surrogate tumor tissues has been used to determine pharmacodynamically active doses. Lack of parallel assessment of tumor tissue effects as well as the intrinsically high interpatient variability has limited the value of these studies. A better understanding of determinants of response to mTOR inhibitors could be used for patient selection in clinical trials. In conclusion, mTOR inhibitors are promising anticancer agents. Future studies are needed to properly develop these drugs as current cancer treatment.

Key words: mTOR, rapamycin, targeted therapy, CCI-779, RAD001, AP23573

1. INTRODUCTION

The development of cancer is controlled by a complex network of aberrant genes and signaling pathways [1]. Several of these genes and molecules have been identified, providing an abundant source of potential therapeutic targets. This research has been complemented by an impetus in drug discovery resulting

Corresponding authors: Manuel Hidalgo, M.D. The Sidney Kimmel Comprehensive Cancer Center at Johns Hopkins. The Baunting and Blaustein Cancer Research Building. 1650 Orleans St. Room 1M88. Baltimore, MD 21231. Phone: 410 502 9746; Fax: 410 614 9006; e-mail: mhidalg1@jhmi.edu

in a large variety of targeted drugs available for preclinical and clinical testing. The identification of different agents with demonstrated antineoplasic properties have also facilitated a better understanding of the targeted pathway. This is the case of the mTOR signaling pathway and the drug rapamycin. Originally discovered as an antifungal agent, the bacterial macrolide rapamycin (Sirolimus; Rapamune; Wyeth-Ayerst, PA, USA) was initially developed and received regulatory approval for the prevention of allograft rejection following organ transplantation due to its profound immunosuppressive actions [2]. After forming a complex with its cellular receptor, the FK506-binding protein (FKBP12), rapamycin binds and inhibits the function of the mTOR [3,4]. TOR proteins are "sensors" that control cell growth and proliferation [5]. Abnormal activation of signaling pathways both proximal and distal to mTOR occurs frequently in human cancer. This observation led to the interest in evaluating the antiproliferative effects of rapamycin in malignant neoplasm. Rapamycin demonstrated impressive preclinical activity against a broad range of human cancers. However, its poor aqueous solubility and chemical stability, along with lack of interest to promote the drug as an anticancer agent, precluded its clinical development. Recently, a series of rapamycin analogs with improved aqueous solubility and stability have been synthesized and evaluated. CCI-779 (Tensirolimus, Wyeth-Ayerst, PA, USA), RAD001 (Everolimus, Novartis Pharma AG, Basel, Switzerland), and AP23573 (Ariad, Pharmaceutics; Cambridge, MA) have been selected for development as anticancer agents based on their prominent antitumor profile and favorable pharmaceutical and toxicological characteristics in preclinical studies [6–9]. Several phase I and phase II studies with these analogs have been completed [10–20].

In this article we first review the mTOR signaling pathway to illustrate the potential role of this kinase as an attractive antineoplasic molecular target. We will discuss the preliminary results of experimental and clinical studies with this novel class of anticancer agents and the potential for individualized developmental strategies and clinical applications.

2. mTOR PATHWAY

The mammalian homologue of TOR protein (mTOR) [21–23] acts as a nutrient or growth factor sensor. Its best-known function is the regulation of translation initiation, mediated by the activation of the 40S ribosomal protein S6 kinase ($p70^{s6k}$) and the inactivation of 4E-binding protein (4E-BP1), increasing the translation of a subset of mRNAs, which protein products are required for traverse through the G_1 phase of the cell cycle.

3. UPSTREAM SIGNALING REGULATORS

The principal upstream regulator of mTOR is the phosphatidylinositol 3-kinase (PI3K)/protein kinase B (Akt) (PI3K/Akt) (Figure 24.1) [24]. PI3K is activated

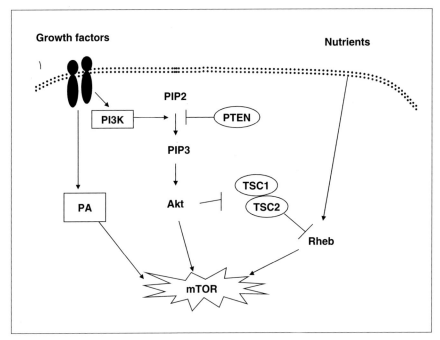

Figure 24.1. Upstream regulators de mTOR.

as a result of ligand-dependent or constitutive activation of tyrosine kinase receptors, G-protein-coupled receptors, or integrins. Receptor-independent activation can also occur, for example, in cells expressing constitutively active Ras proteins. Activated PI3K catalyzes the conversion of phosphatidylinositol (4, 5)-biphosphate (PIP2) to phosphatifylinositol-3, 4, 5-trisphospahte (PIP3), whereas PTEN dephosphorylates PIP_3 acting as a negative regulator of PI3K signaling. PI3K phosphorylates Akt, which in turn phosphorylates a host of other proteins that affect cell growth, cell cycle entry, and cell survival. While Akt can phosphorylate mTOR (Ser2448) directly, the tuberous sclerosis complex (TSC), comprised by harmartin (TSC1) and tuberin (TSC2) has emerged as an important intermediate in the regulation of mTOR by Akt. TSC binds to and inhibits mTOR [25–27]. Akt activation can directly phosphorylate TSC2 [25,28], inhibiting the formation of TSC1/2 complex, leading to derepression of mTOR, and consequently the kinase activity of mTOR increases. Signaling through protein kinase C or MEK-MAPK also phosphorylates and inactivates TSC2 leading to similar effects [29].

Proximal activators of mTOR regulated by nutrients and ATP are less well characterized. Ras homolog enriched in brain (Rheb) is a small G protein that functions upstream of mTOR in the nutrient signaling pathway. As in growth factor signaling, tuberin–hamartin heterodimer play also a role as an inhibitor of nutrient signaling through mTOR by inhibiting Rheb [26,27,30]. However, the identity of the sensor of nutrient signal remains controversial with studies showing that the phosphorylation state of TSC2 is nutrient-sensitive as opposed to others [31,32]. In addition to nutrients and mitogens, second messengers (such as phosphatidic acid), ATP levels and polyphosphates might also regulate TOR signaling. Moreover, it is known that mTOR forms a scaffold complex with other proteins, such as raptor and mLST8 [33], which act as potential modulators of mTOR function in response to nutrients.

4. DOWNSTREAM EFFECTS

The main downstream function of mTOR activation is the control of translational machinery, through the eukaryotic initiation factor 4E-BP1, and the 40S ribosomal protein S6 kinase ($p70^{s6k}$) (Figure 24.2) [34].

4E-BP1: The cap function in translation is mediated by a protein complex termed as eukaryotic translation initiation factor 4F (eIF4F) which consist of three subunits: (1) eIF4E; (2) eIF4A; and (3) eIF4G. In the unphosphorylated state, 4E-BP1 binds to eIF4E, blocking the assembly of the eIF4F complex at the 5′-cap structure of the mRNA template, thereby decreasing the efficiency of translation initiation [35]. In response to proliferative stimuli, 4E-BP1 becomes phosphorylated through the action of mTOR and other kinases that leads to the

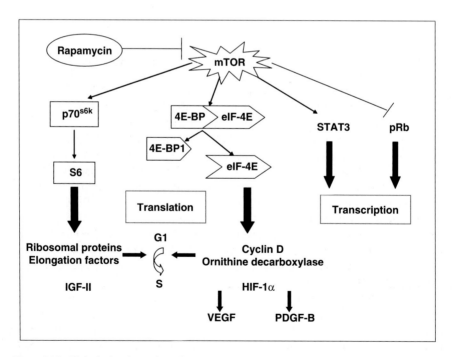

Figure 24.2. Biological actions of mTOR

release of eIF4E, which can form the multisubunit eIF4F complex. These interactions lead to an increase in the translation of mRNAs with regulatory elements such as those encoding c-myc, cyclin D1, ornithine decarboxylase, and hypoxia-inducible factor? 1-α, and consequently vascular endothelial growth factor (VEGF). It has been speculated that mTOR acts indirectly as an inhibitor of phosphatases, which function to dephosphorylate 4E-BP1 [36].

p70^{s6k}: Upon activation, mTOR activates p70^{s6k}, which, in turn, phosphorylates the S6 protein of the 40S ribosomal complex. The phosphorylation of S6 leads to the recruitment of the 40S ribosomal subunit into actively translating polysomes, enhancing the translation of mRNAs bearing a 5' terminal oligopolypyrimidine, that encode components of the translational machinery. These include ribosomal proteins and elongation factors, which are required to sustain the increased biosynthesis needed for cell growth. Other targets of p70^{s6k} are the antiapoptotic protein BAD [37], the transcription factor CREM [38], and the translational regulator eEF2 [39]. It is important to note that p70^{s6k} may also be activated by TOR-insensitive signaling pathways involving PDK1, MAPK, and SAPK.

At the transcriptional level, mTOR positively controls RNA polymerase activity either by inactivation of the retinoblastoma protein [40], or by stimulating the transcriptional activator STAT3 [41]. It has also been proposed the mTOR association with p53 and the apoptotic pathway.

5. MECHANISM OF ACTION AND EFFECTS OF RAPAMYCIN

Rapamycin binds intracellularly to FKBP12, a member of the immunophilin family of FK506-binding proteins, and the resultant complex interacts with and inhibits the mTOR kinase activity [3,4]. This results in inhibition of p70^{s6k} kinase, which leads to a reduction in the translation of 5'-terminal oligopyrimidine mRNAs involved in the synthesis of proteins essential for the cell cycle progression [42,43]. Rapamycin also blocks the phosphorylation of 4E-BPI. 4E-BPI, in its dephosphorylated state, binds to eIF4E, preventing the formation of the eIF4F complex and, consequently, inhibiting the cap-dependent translation initiation of mRNAs that encode for critical regulatory proteins [34,44]. These combine effects lead to cell cycle arrest at the G_1 phase of the cell cycle. Moreover, rapamycin interferes with other intracellular processes involved in cell cycle progression including increasing the turnover of cyclin D1 at the mRNA and protein level [45], upregulation of p27 and inhibition of cyclin-A-dependent kinase activity [46].

In addition to their cytostatic actions, rapamycin can induce apoptosis in some cell systems, like B-cells, rhabdomyosarcomas cells, renal tubular cells, and dendritic cells [47,48]. Huang et al. [47] suggest that cellular response to rapamycin is converted from G_1 cytostatic to apoptosis when p53 or p21 are absent. Preclinical studies have showed a potent rapamycin antiangiogenic effect. Rapamycin reduces the production of VEGF as well as blocks

VEGF-mediated stimulation of endothelial cells and tube formation [49]. In addition, because VEGF induction of the PI3k/Akt signaling pathway is important for endothelial cell survival [50,51], rapamycin treatment can induce apoptosis of VEGF-stimulated endothelial cells, potentially leading to tumor vessel thrombosis [52]. The rapamycin-effect of VEGF production could be explained by the correlation between mTOR activity and HIF-1α. The data shown by Majumder et al. suggest that a major component of the in vivo transcriptional response to activation of AKT is mTOR-dependent regulation of HIF-1α. They showed that mTOR inhibition by rapamycin analog RAD001 leads to inactivation of these *HIF-1α* genes [53].

6. PREDICTOR MARKERS OF RESPONSE TO MTOR INHIBITORS

It is becoming clear that there is a relationship between biological alterations that drive tumors development and progression and the effects of targeted therapies against such alterations. The identification of predictive markers of sensitivity or resistance will allow the selection of patients who are more likely to derive benefit from such a treatment. This concept has been demonstrated with several new drugs, e.g., trastuzumab (Herceptin, Genentech) in breast cancer patients overexpressing Her2neu; imatinib (Gleevec, Novartis) for patients with Philadelphia chromosome-positive CML or c-kit expression and mutations in gastrointestinal stromal tumors; and mutations in EGFR and the activity of gefitinib (Iressa, AstraZeneca) [54–57].

To date, there are no reports suggesting that activating mutations or overexpression of mTOR occur as primary event in malignant transformation. However, the PI3K/Akt/mTOR signaling pathway is frequently activated in human tumors secondary by either receptor activation (like EGFR, Her2neu, estrogen receptors, or IGFR, which are often overexpressed in several tumors) or activating mutations in downstream genes.

Loss of PTEN by deletion, silencing, or mutation leads to constitutive activation of Akt, and upregulation of mTOR-dependant pathways [58,59]. Studies in PTEN knockout mice, which are prone to developing tumors, have demonstrated that PTEN-deficient tumor cells have elevated levels of phosphorylated Akt and are extraordinarily sensitive to the growth inhibitory effects of rapamycin [60,61]. This correlation between PTEN-defective phenotype and increased sensitivity to mTOR inhibition has been also reported in studies with breast cancer [62,63] and myeloma cell lines [64]. In addition, several preclinical studies have shown that cell lines expressing a constitutively active Akt are more susceptible to mTOR inhibitors. High levels of Akt activity have been associated with resistance to standard therapy in many cancer types. mTOR inhibition potentates tumor response or restores the susceptibility of different Akt overexpressing cancer cell lines to many conventional anticancer agents [65–68].

However, a recent phase II study of CCI-779 in patients with glioblastoma, radiographic improvement was only significantly correlated with high levels of phosphorylated p70s6 kinase in baseline tumor samples, but not with other tissue biomarkers like Akt, phosphorylated Akt, or p70s6 kinase [69].

Mutations in *PI3K* gene have been discovered in up to 30% of colorectal and breast cancer patients [70]. Kang et al. [71] have reported that rapamycin inhibits the transformation induced by PI3K mutants in chicken embryonic fibroblast, suggesting that TOR is an essential component of the transformation process, and, therefore an important target in tumors carrying these mutations. However, these results differs from the selective involvement of FKHT and FKHRL1 observed in colorectal cancer cells, where, surprisingly, analysis of other AKT downstream targets, as mTOR, 4E-BP1, and p70^{S6K} showed no consistent differences in phosphorylation among the wild type and mutant clones [72]. Future work will be needed to determine if the targets of AKT activation, and therefore the rapamycin susceptibility, depend on the cell type, species or experimental system analyzed.

Disruptions of the *TSC1* or *TSC2* genes cause tuberous sclerosis [73], a human syndrome characterized by the widespread development of benign tumors [74]. Preclinical studies showed that inhibition of mTOR might have therapeutic benefit in TSC [75–77]. Although cancer development is rare in TSC, recently it has been reported the potential role of this mutation into the initiation or progression of some bladder tumors [73]. It would be interesting to determine the rate of these mutations in different tumors.

In addition to the factors mentioned above which are the best characterized and currently being tested in the clinic, other factors may also be implicated. Because rapamycin induces apoptosis in cells lacking functional p53, the assessment of this dysfunction in tumors may also be useful in predicting drug efficacy. Abnormalities of regulators of the G_1 checkpoint such as retinoblastoma protein, p16, p27, and cyclin D1 may also increase the sensitivity of tumors to rapamycin.

7. CLINICAL DEVELOPMENT

Currently, available inhibitors of mTOR are limited to rapamycin and the structurally related compounds CCI-779, RAD001, and AP23573. These agents share many biochemical and physiologic properties and are being developed for the treatment of different tumor types.

8. CCI-779

In the NCI human tumor cell line screen, CCI-779 and rapamycin demonstrated similar antitumor profiles and potencies, with IC_{50} values frequently below than 10^{-8} M. In vitro, human prostate, breast, small cell lung carcinomas, glioblastoma, melanoma, and T-cell leukemia cell lines were among the most sensitive to CCI-779. In in vivo studies, treatment of a variety of human tumor xenografts with CCI-779 caused significant tumor growth inhibition rather than tumor regression, suggesting that subsequent disease-directed trials should be designed to assess this potential outcome [6,78].

The results of published clinical trials with CCI-779 are displayed in Table 24.1. Two phase I studies where CCI-779 was administered as a 30-min iv infusion weekly and as a 30-min iv infusion daily for 5 days in every 2 weeks have been reported [10,11]. These studies were designed to determine the maximum tolerated dose (MTD) based on dose-limiting toxicities. The MTD of CCI-779 on the daily-for-5-day-every-2-week schedule were projected to be 15 and 24 mg/m^2/day in patients with minimal or extensive prior myelotoxic therapy, respectively, whereas the MTD was not determined for CCI-779 administered on a weekly schedule. The principal toxicities of CCI-779 included dermatologic toxicity (such as aseptic folliculitis, erythematous maculopapular rashes, dry skin, and nail disorders), myelosuppression, reversible elevations in liver function tests, and asymptomatic hypocalcemia. The principal hematologic toxicity was thrombocytopenia. Other toxicities, which were generally mild to moderate in severity, reversible, and noted over wide dosing ranges included mucositis, hypertriglyceridemia, hypercholesterolemia, and reversible decrements in serum testosterone. The pharmacokinetic studies indicate dose-dependent pharmacokinetics, elimination half-life values of approximately 15–17 h, and preferential partitioning of CCI-779 in red blood cells. Unexpectedly, based on of preclinical studies, in which the predominant therapeutic effect of CCI-779 was delayed tumor growth, major tumor responses (partial responses, PR: >50% reduction in the sum of the bidimensional product of all measurable lesions) and minor tumor responses (MR: <50% reduction in the sum of the bidimensional product of all measurable lesions) were noted. The fact that CCI-779 consistently induced tumor regressions at relatively nontoxic doses in the phase I studies suggests that its optimal therapeutic dose may be lower than the MTD.

Due to the high incidence of PTEN deletions in malignant gliomas, CCI-779 has been studied in patients with this disease. Since CCI-779 is principally metabolized by cytochrome P450 that are induced by many types of anticonvulsant agents commonly coadministered to these patients, the toxicities, pharmacokinetics, and optimal dose schedule of CCI-779 have been separately evaluated in this group. The MTD was established at 250 mg iv weekly for patients on enzyme-inducing antiepileptic agents (EIAEDs) [17], which is higher than for those not in EIAEDs in the phase II study, where the initial dose of 250 mg had to be decreased to 170 mg because of excessive toxicity. CCI-779 was well tolerated in recurrent glioma in a phase II study, with radiographic improvement in 36% of treated patients and significantly longer time to progression in the responder patients with respect to the nonresponders [69].

Phase I studies evaluating the feasibility of administering CCI-779 in combination with cytotoxic chemotherapeutics are ongoing. To date, enrollment is continuing in a phase I study of combination therapy of CCI-779 and interferon alpha (IFN), which has been generally well tolerated in patients with advanced RCC [18]. A phase I study of CCI-779 at 15 mg/m^2 in combination with

Table 24.1. Phase I clinical trials with CCI-779

Type of trial	Tumor type	No. of patients	Dose, mg/day	Regimen	Maximum tolerated dose, mg/day	Toxicities	Activity	References
Phase I	Renal, colorectal, soft tissue sarcoma, mesothelioma, NSCLC, breast, head and neck, melanoma, pancreatic, prostate, neuroendocrine carcinoma, and adreno-cortical carcinoma	24	7.5–220 mg/m²	Weekly IV	NA	Skin rash, stomatitis, thrombocytopenia, and bipolar disorder	8.3% PR 8% MR	[10]
Phase I	Renal cell carcinoma, NSCLC, soft tissue sarcomas, cervical and uterine carcinomas	51	0.75–19.1 mg/m²	Days 1–5 every 2 weeks IV	NA*	Hypocalcemia, skin rash, stomatitis, and thrombocytopenia	2% PR 15% MR or SD	[11]
Phase I	Renal cell carcinoma, NSCLC, myxoid chondrosarcoma, mesothelioma, leymiosarcoma	24	25–75 mg	Days 1–5 every 2 weeks orally	100	Stomatitis, rash, and hypertransaminemia	33% SD	[20]
Phase I	Glioblastoma	12	250–300 mg	CCI-779 weekly in combination with enzyme-inducing anti-epileptic drugs	250	Stomatitis, and hypertriglicerydemia		[17]
Phase I	Colorectal, gastric carcinoma, esophageal c, head and neck cancer	28	15–75 mg/m²	CCI-779 in combination with 5-FU/LV	250	Stomatitis/mucositis,	11% PR	[19]

PR = partial response; MR = minor response; SD = stable disease; OR = objective response; 5-FU = 5-fluorouracil; LV = leucovorin; DLT = dose-limiting toxicity; NSCLC: non-small cell lung cancer.
*15 mg/m² in heavily pretreated patients.

5-fluorouracil (2,600 mg/m^2) and leucovorin concluded that the combination used is not feasible, with mucositis being the dose-limiting toxicity [19].

An oral formulation of CCI-779 is also in study. The recommended oral dosage for phase II studies has been defined as 75 mg/day administered daily for 5 days every 2 weeks after the first phase I study [20].

Disease-directed efficacy studies of CCI-779 in a broad range of tumor types have been initiated (Table 24.2). In a phase II study, 111 patients with advanced, previously treated, renal cell carcinoma were randomized to treatment with CCI-779 at doses of 25, 75, or 250 mg/m^2 IV weekly. CCI-779 produced an objective response rate of 7% and minor responses in 26% of these patients. The median time to tumor progression (TTP) (5.8 months) and median survival (15 months) were also notable, particularly for heavily pretreated patients. Interestingly, there were no substantial differences in antitumor activity among the three dose levels tested [12]. While the response rate was low in these heavily pretreated patients, the high rate of PR/MR/SD and extended median TTP and survival are promising. Assessment of the pharmacodynamic effects in a subset of patients by measuring p70^{s6K} inhibition in peripheral blood mononuclear cells (PBMC) demonstrated a dose-independent inhibition of the target that was correlated with TTP [79]. In March 2002, the FDA designated CCI-779 for fast-track development for the treatment of RCC after failure of initial therapy; at this time phase III trial for RCC is ongoing comparing CCI-779 with interfero&nbdot; or the combination of both agents and using time to progression and survival as end point [80].

Another phase II study of CCI-779, in this case in women with previously treated with either a taxane or an anthracycline or both locally advanced or metastatic breast cancer, has been also performed [13]. One hundred nine patients were randomized to receive 75 or 250 mg/m^2 iv of CCI-79 on a weekly schedule. Clinical benefit was observed in 37% of patients, including ten partial responses and 26 patients with stable disease lasting 8 weeks. Efficacy was similar for both dose levels but toxicity was more common with the higher dose level. The most common reason for study discontinuation was lethargy and/or depression. Interestingly, none of the HER-2 negative tumors showed any significant response to CCI-779. Because of the association of hormone resistance and activation of the mTOR pathway, a randomized, double-blind, phase III study of oral CCI-779 in combination with letrozole versus letrozole alone in patients with estrogen-dependent breast cancer is ongoing [80].

Recently, a phase II of CCI-779 for relapsed mantle cell lymphoma has been reported [81]. Thirty-five patients were treated with 250 mg of CCI-779 iv every week. The overall response rate was 38% with one complete response (3%) and 12 partial responses (35%). The median time –to progression in all patients was 6.5 months, and the duration of response for the 13 responders was 6.9 months. Hematologic toxicities were the most common, being thrombocytopenia the most frequent cause of dose reductions.

Table 24.2. Phase II clinical trials with CCI-779

Type of trial	Tumor type	No. of patients	Dose, mg/day	Regimen	Toxicities	Activity	References
Phase II	Renal cell carcinoma	111	25, 75, 250 mg/m^2	Weekly	Rash, mucositis, asthenia, nausea, hyperglycemia, and hypophosphatemia	1% CR, 6,3% PR, 26% MR	[13]
Phase II	Metastatic or locally advanced breast carcinoma	109	75 or 250 mg/m^2	Weekly	Leucopenia, stomatitis, depression, and somnolence	10% PR, 24% SD	[14]
Phase II	Glioblastoma multiforme	65	250 mg	Weekly	Hematologic, hypercholesterolemia, hypertriglyceridemia, and hyperglycemia	36% evidence of improvement in neuroimaging	[69]
Phase II	Mantle cell non-Hodgkin's lymphoma	35	250 mg	Weekly	hrombocytopenia, hyperglycemia, Thypertriglyceridemia anemia, neutropenia, and mucositis	3% CR, 35% PR	[81]

PR = partial response; MR = minor response; SD = stable disease; OR = objective response.

9. RAD001

RAD001, a rapamycin ester analog, also has shown in preclinical studies dose-dependent inhibition of tumor growth and reduced tumor vascularity, as well as the ability to potentate the activity of a number of cytotoxics [7,82].

Two phase I studies of orally administered RAD001, were performed to identify the optimal biologically effective dose based on toxicity, pharmacokinetics, and pharmacodynamic assessments using the biomarker $p70^{s6k}$ activity in PBMCs [14] or p-S6, p-4E-BP1, and p-Akt in tumors [83]. RAD001 was well tolerated with only mild degrees of anorexia, fatigue, rash, mucositis, and gastrointestinal disturbance. Partial responses and stabilizations of >4 months were observed. Tabernero et al. [83] recommended a dosage of 10 mg daily for further phase II and III trials.

Phase I/II trials evaluating RAD001 in relapsed or refractory hematologic malignances or the combination of RAD001 and imatinib mesylate in patients with gastrointestinal stromal tumors that are refractory to imatinib mesylate, [84] with gefitinib in patients with NSCLC [85] or with different cytotoxic drugs are on going.

10. AP23573

AP23573, a rapamycin prodrug, has demonstrated prominent antiproliferative preclinical activity against several cancers [9].

Preliminary results of two phase I dose escalation studies of AP23573 in patients with refractory or advanced cancers have been reported, in which AP23573 was administered as a 30-min iv infusion daily × 5 days every 2 weeks for 4 weeks, with doses ranging from 3 to 28 mg in 5 dose level cohorts [15], and as a 30 min iv infusion weekly for 4 weeks cycles, with doses ranging from 6.25 to 25 mg [16]. No dose-limiting toxicity or related serious adverse events have been observed in any study. Side effects included grade 1 rash, grade 2 anemia and mucositis, and a transient grade 3 transaminase elevation. Inhibition of mTOR activity was reflected by decreased in phosphorylated 4E-BP1 in both studies in skin biopsies.

A series of phase II trials in refractory hematologic malignances, sarcomas, breast, brain, endometrial, and prostate cancer are underway, as well as a series of phase Ib trials evaluating this agent in combination with chemotherapeutic agents and other molecularly targeted drugs.

11. RAPAMYCIN

For reasons that are not totally clear, the development of rapamycin itself as an anticancer agent has not been attempted until recently. Based on its pharmacological and antitumor properties in preclinical models, there is no reason to believe that rapamycin will not be effective as an anticancer agent. In fact,

CCI-779 is essentially converted to parent rapamycin in plasma. Rapamycin alone and in combination with other immunosupressants has been safely administered to thousands of solid organ transplant patients [86–88]. There is an increasing effort to determine the optimal way of administration and to evaluate the efficacy of rapamycin in a clinical setting. Phase I and phase I/II studies are currently underway in pediatric refractory hematologic malignancies, glioblastoma multiforme, and pancreatic cancer patients.

12. CHALLENGES IN THE DEVELOPMENT OF mTOR INHIBITORS

The clinical development of the molecularly targeted drugs will require fundamental changes to the traditional clinical trial design and end points used for cytotoxic drugs. In this section of the review, we summarize some of the most relevant challenges to develop mTOR inhibitors.

Determination of the optimal dose remain the primary objective in phase I studies of molecularly targeted agents. The determination of the MTD as an solely end point may not be the most appropriate strategy because cytostatic agents may have a saturable target, and dose escalation beyond a certain threshold may not result in increased efficacy but only in increased toxicity. Furthermore the traditional definition of "intolerable toxicity" may be insufficient for chronically and/or continuously administered targeted agent that produces noticeable, moderate toxicities, which can be unacceptable toxicities if they become persistent. The selection of an appropriate dose for mTOR-targeted drugs is especially challenging since objective antitumor activity has been observed in patients treated with a wide range of doses, whereas toxicity has been more directly related to dose.

Ideally, phase I trials for target-based drugs should be designed to determine whether the target can be inhibited at a tolerable dose and schedule, and to estimate the dose required to achieve and maintain maximum inhibition of the target [89]. Drug effect on the target can be measured in tumor biopsy specimens obtained prior to and after the initiation of the therapy or in adequately validated surrogate tissues. The pharmacodynamic effects of mTOR inhibitors have been measured in PBMCs using a fully quantitative $p70^{s6k}$ assay [79]. Peralba et al. showed that the activity of $p70^{s6k}$ decreases in a linear fashion after exposure to an mTOR inhibitor, and there is a good correlation in magnitude of $p70^{s6k}$ inhibition between PBMCs and tumor tissues in mice. Additional clinical studies are currently in development to explore if this correlation also exists in patients in trials in which tumor tissue and PBMCs will be simultaneously collected. It is interesting that no relationship was observed between the administered dose of CCI-779 and inhibition of $p70^{s6k}$ in PBMCs in a small cohort of patients. Similar studies conducted with RAD001 and AP23573 have used pharmacodynamic effects in PBMC and skin biopsies to determine biologically active doses of the agents [14,16,83]. However these approaches have important limitations. First, there may not be a correlation between drug effects in normal tissues and in malignant tissues and, therefore, tumor biopsies may still need to be performed to document adequate tumor tissue effects. Second, inhibition of the target does not necessarily imply antitumor effects because tumors may still progress if there are others constitutively activated downstream mediators that are not regulated by upstream targets.

A second important question is the selection of appropriate end points for phase II clinical trials. Although regressions of well-established tumors have been noted in preclinical studies of rapamycin and its analogs, tumor growth delay is their principal therapeutic effect. Then, the sole use of tumor response as a marker of efficacy may be inadequate. In addition to the traditional end point, tumor regression, possible clinical end points that have been proposed for the evaluation of antiproliferative agents include time to progression, the proportion of patients without progressive disease as their best response, progression rate, symptomatic benefit, measures of target inhibition, PET scanning and reduction in tumor markers. None of these end points have, however, been well validated.

Another important issue in phase II design of target drug is that they should be tested in the patient group that is most likely to show a favorable effect. This approach should avoid spuriously negative or overall weak signals of clinical activity of otherwise very active drugs when used in the right group of patients, prevent unnecessary large costly trials, and limit the exposure of patients to drugs unlikely to produce any clinical benefit. Therefore, an important challenge during clinical development of mTOR inhibitors includes the ability to predict which tumors will be sensitive or resistant in base to their specific molecular abnormalities and select the patients for treatment based on predictor markers of sensitivity, as these we mentioned above. Several studies support the hypothesis that tumors became "addict" to their constitutively active pathways and that inhibition of these pathways result in important tumor growth arrest. Since cancer patients with, for example, PTEN-deficient tumor cells, amplification of Akt, or mutations in the *PI3K* gene can be identified using readily available tests, it may be possible to select those who may benefit most from rapamycin and rapamycin analogs. Thus far, Galanis et al. [69] are the first group who has tested several markers in the clinic and they have proposed that $p70^{s6k}$ activation is the only marker that appears related to outcome.

13. CONCLUSION

TOR proteins are "sensors" that appear to control cell growth and proliferation based on nutrient availability or growth factor stimulation through regulation of translation and transcription. Activation of the mTOR signaling pathway is frequent in human cancer and has become an attractive target for antineoplasm therapies. Rapamycin, CCI-779, RAD001, and AP23573 are mTOR inhibitors with potent antitumor activity in preclinical studies that have completed several phase I clinical evaluations and are currently undergoing broad disease-directed efficacy studies. These agents appear to be well tolerated at doses that have resulted in impressive antitumor activity in several types of refractory neoplasm. The clinical development of mTOR inhibitors exemplified the challenges and limitations regarding the development of novel anticancer agents and are fertile grounds to implement innovative approaches to clinical trials including innovative strategies for patient selection as well as rigorous incorporation of pharmacodynamic end points.

REFERENCES

1. Hanahan D, Weinberg RA. The hallmarks of cancer. *Cell* 2000, 100: 57–70.
2. Sehgal SN. Rapamune (Sirolimus, rapamycin): an overview and mechanism of action. *Ther Drug Monit* 1995, 17: 660–665.
3. Heitman J, Movva NR, Hall MN. Targets for cell cycle arrest by the immunosuppressant rapamycin in yeast. *Science* 1991, 253: 905–909.
4. Fruman DA, Wood MA, Gjertson CK, Katz HR, Burakoff SJ, Bierer BE. FK506 binding protein 12 mediates sensitivity to both FK506 and rapamycin in murine mast cells. *Eur J Immunol* 1995, 25: 563–571.
5. Schmelzle T, Hall MN. TOR, a central controller of cell growth. *Cell* 2000, 103: 253–262.
6. Gibbons JJ, Discafani C, Peterson R, Hernandez R, Skotnicki J, Frost J. The effect of CCI-779, a novel macrolide anti-tumor agent, on the growth of human tumor cells in vitro and in nude mouse xenograft in vivo. In: *Proc Am Assoc Cancer Res* 2000, 301.
7. Boulay A, Zumstein-Mecker S, Stephan C, Beuvink I, Zilbermann F, Haller R, Tobler S, Heusser C, O'Reilly T, Stolz B, et al. Antitumor efficacy of intermittent treatment schedules with the rapamycin derivative RAD001 correlates with prolonged inactivation of ribosomal protein S6 kinase 1 in peripheral blood mononuclear cells. *Cancer Res* 2004, 64: 252–261.
8. Lane HA, Preclinical and clinical pharmakinetic/pharmacodynamic (PK/PD) modeling to help to define an optimal biological dose for the oral mTOR inhibitor, RAD001, in oncology. *Proc Am Soc Clin Oncol* 2003, 22.
9. Clakson T, Metcalf CA III, Rozamus LW, et al. Regression of tumor xenografts in mice after oral administration of AP23573, a novel mTOR inhibitor that induces tumor starvation. In: *Proc Am Assoc Cancer Res* 2002, 43.
10. Raymond E, Alexandre J, Faivre S, Vera K, Materman E, Boni J, Leister C, Korth-Bradley J, Hanauske A, Armand JP. Safety and pharmacokinetics of escalated doses of weekly intravenous infusion of CCI-779, a novel mTOR inhibitor, in patients with cancer. *J Clin Oncol* 2004, 22: 2336–2347.
11. Hidalgo M, Rowinsky E, Erlichman C, Drengler R, Mashall B, Adjei A, et al. CCI-779, a rapamycin analog and multifaceted inhibitor of signal transduction: a phase I study. In *Proc Am Soc Clin Oncol* 2000, 187.
12. Atkins MB, Hidalgo M, Stadler WM, Logan TF, Dutcher JP, Hudes GR, Park Y, Liou SH, Marshall B, Boni JP, et al. Randomized phase II study of multiple dose levels of CCI-779, a novel mammalian target of rapamycin kinase inhibitor, in patients with advanced refractory renal cell carcinoma. *J Clin Oncol* 2004, 22: 909–918.
13. Chan S, Scheulen ME, Johnston S, Mross K, Cardoso F, Dittrich C, Eiermann W, Hess D, Morant R, Semiglazov V, et al. Phase II study of temsirolimus (CCI-779), a novel inhibitor of mTOR, in heavily pretreated patients with locally advanced or metastatic breast cancer. *J Clin Oncol* 2005, 23: 5314–5322.
14. O'Donnell A, Faivre S, Judson I, et al. A phase I study of the oral mTOR-inhibitor RAD001 as monotherapy to identify the optimal biologically effective dose using toxicity, pharmacokinetic (PK), and pharmacodynamic (PD) endpoints in patients with solid tumors. *Proc Am Soc Clin Oncol* 2003, 22.

15. Mita MM, Rowinsky EK, Goldton ML, et al. Phase I, pharmacokinetic and pharmacodynamic study of AP23573, an mTOR inhibitor, administered IV daily × 5 every other week in patients with refractoy or advanced malignancies. *Proc Am Soc Clin Oncol* 2004, 23.
16. Desai AA, Janish L, Berk LR, et al. A phase I trial of a novel mTOR inhibitor AP23573 administered weekly in patients with refractory or advanced malifnacies: A pharmacokinetic and pharmacodynamic analysis. *Proc Am Soc Clin Oncol* 2004, 23.
17. Chang SM, Kuhn J, Wen P, Greenberg H, Schiff D, Conrad C, Fink K, Robins HI, Cloughesy T, De Angelis L, et al. Phase I/pharmacokinetic study of CCI-779 in patients with recurrent malignant glioma on enzyme-inducing antiepileptic drugs. *Invest New Drugs* 2004, 22: 427–435.
18. Dutcher JP, Hudes G, Motzer R, Ko Y, Smith JW, Zonno K, et al. Preliminary report of a phase 1 study of intravenous (IV) CCI-779 given in combination with interferon-alfa (IFN) to patients with advanced renal cell carcinoma (RCC). In: *Pro Am Soc Clin Oncol* 2003, 213.
19. Punt CJ, Boni J, Bruntsch U, Peters M, Thielert C. Phase I and pharmacokinetic study of CCI-779, a novel cytostatic cell-cycle inhibitor, in combination with 5-fluorouracil and leucovorin in patients with advanced solid tumors. *Ann Oncol* 2003, 14: 931–937.
20. Farouzesh B, Buckner J, Adjei A, et al. Phase I bioavailability and pharmacokinetic study of oral dosage of CCI-779 administered to patients with advanced solid malignancies. *Eur J Cancer* 2002, 38 (Suppl. 7).
21. Brown EJ, Albers MW, Shin TB, Ichikawa K, Keith CT, Lane WS, Schreiber SL. A mammalian protein targeted by G1-arresting rapamycin-receptor complex. *Nature* 1994, 369: 756–758.
22. Sabatini DM, Erdjument-Bromage H, Lui M, Tempst P, Snyder SH. RAFT1: a mammalian protein that binds to FKBP12 in a rapamycin-dependent fashion and is homologous to yeast TORs. *Cell* 1994, 78: 35–43.
23. Bjornsti MA, Houghton PJ. The TOR pathway: a target for cancer therapy. *Nat Rev Cancer* 2004, 4: 335–348.
24. Downward J. Mechanisms and consequences of activation of protein kinase B/Akt. *Curr Opin Cell Biol* 1998, 10: 262–267.
25. Inoki K, Li Y, Zhu T, Wu J, Guan KL. TSC2 is phosphorylated and inhibited by Akt and suppresses mTOR signalling. *Nat Cell Biol* 2002, 4: 648–657.
26. Gao X, Zhang Y, Arrazola P, Hino O, Kobayashi T, Yeung RS, Ru B, Pan D. Tsc tumour suppressor proteins antagonize amino-acid-TOR signalling. *Nat Cell Biol* 2002, 4: 699–704.
27. Tee AR, Fingar DC, Manning BD, Kwiatkowski DJ, Cantley LC, Blenis J. Tuberous sclerosis complex-1 and -2 gene products function together to inhibit mammalian target of rapamycin (mTOR)-mediated downstream signaling. *Proc Natl Acad Sci USA* 2002, 99: 13571–13576.
28. Manning BD, Tee AR, Logsdon MN, Blenis J, Cantley LC. Identification of the tuberous sclerosis complex-2 tumor suppressor gene product tuberin as a target of the phosphoinositide 3-kinase/akt pathway. *Mol Cell* 2002, 10: 151–162.
29. Tee AR, Anjum R, Blenis J. Inactivation of the tuberous sclerosis complex-1 and -2 gene products occurs by phosphoinositide 3-kinase/Akt-dependent and -independent phosphorylation of tuberin. *J Biol Chem* 2003, 278: 37288–37296.
30. Tee AR, Manning BD, Roux PP, Cantley LC, Blenis J. Tuberous sclerosis complex gene products, Tuberin and Hamartin, control mTOR signaling by acting as a GTPase-activating protein complex toward Rheb. *Curr Biol* 2003, 13: 1259–1268.
31. Xia Y, Wen HY, Young ME, Guthrie PH, Taegtmeyer H, Kellems RE. Mammalian target of rapamycin and protein kinase A signaling mediate the cardiac transcriptional response to glutamine. *J Biol Chem* 2003, 278: 13143–13150.
32. Zhang Y, Gao X, Saucedo LJ, Ru B, Edgar BA, Pan D. Rheb is a direct target of the tuberous sclerosis tumour suppressor proteins. *Nat Cell Biol* 2003, 5: 578–581.
33. Hara K, Maruki Y, Long X, Yoshino K, Oshiro N, Hidayat S, Tokunaga C, Avruch J, Yonezawa K. Raptor, a binding partner of target of rapamycin (TOR), mediates TOR action. *Cell* 2002, 110: 177–189.
34. Gingras AC, Raught B, Sonenberg N. Regulation of translation initiation by FRAP/mTOR. *Genes Dev* 2001, 15: 807–826.

35. Sonenberg N, Gingras AC. The mRNA 5' cap-binding protein eIF4E and control of cell growth. *Curr Opin Cell Biol* 1998, 10: 268–275.
36. Di Como CJ, Arndt KT. Nutrients, via the Tor proteins, stimulate the association of Tap42 with type 2A phosphatases. *Genes Dev* 1996, 10: 1904–1916.
37. Harada H, Andersen JS, Mann M, Terada N, Korsmeyer SJ. p70S6 kinase signals cell survival as well as growth, inactivating the pro-apoptotic molecule BAD. *Proc Natl Acad Sci USA* 2001, 98: 9666–9670.
38. de Groot RP, Ballou LM, Sassone-Corsi P. Positive regulation of the cAMP-responsive activator CREM by the p70 S6 kinase: an alternative route to mitogen-induced gene expression. *Cell* 1994, 79: 81–91.
39. Wang X, Li W, Williams M, Terada N, Alessi DR, Proud CG. Regulation of elongation factor 2 kinase by p90(RSK1) and p70 S6 kinase. *EMBO J* 2001, 20: 4370–4379.
40. White RJ. Regulation of RNA polymerases I and III by the retinoblastoma protein: a mechanism for growth control? *Trends Biochem Sci* 1997, 22: 77–80.
41. Bromberg JF, Wrzeszczynska MH, Devgan G, Zhao Y, Pestell RG, Albanese C, Darnell JE Jr. Stat3 as an oncogene. *Cell* 1999, 98: 295–303.
42. Seufferlein T, Rozengurt E. Rapamycin inhibits constitutive p70s6k phosphorylation, cell proliferation, and colony formation in small cell lung cancer cells. *Cancer Res* 1996, 56: 3895–3897.
43. Grewe M, Gansauge F, Schmid RM, Adler G, Seufferlein T. Regulation of cell growth and cyclin D1 expression by the constitutively active FRAP-p70s6K pathway in human pancreatic cancer cells. *Cancer Res* 1999, 59: 3581–3587.
44. Brunn GJ, Hudson CC, Sekulic A, Williams JM, Hosoi H, Houghton PJ, Lawrence JC Jr., Abraham RT. Phosphorylation of the translational repressor PHAS-I by the mammalian target of rapamycin. *Science* 1997, 277: 99–101.
45. Hashemolhosseini S, Nagamine Y, Morley SJ, Desrivieres S, Mercep L, Ferrari S. Rapamycin inhibition of the G1 to S transition is mediated by effects on cyclin D1 mRNA and protein stability. *J Biol Chem* 1998, 273: 14424–14429.
46. Kawamata S, Sakaida H, Hori T, Maeda M, Uchiyama T. The upregulation of p27Kip1 by rapamycin results in G1 arrest in exponentially growing T-cell lines. *Blood* 1998, 91: 561–569.
47. Huang S, Liu LN, Hosoi H, Dilling MB, Shikata T, Houghton PJ. p53/p21(CIP1) cooperate in enforcing rapamycin-induced G(1) arrest and determine the cellular response to rapamycin. *Cancer Res* 2001, 61: 3373–3381.
48. Lieberthal W, Fuhro R, Andry CC, Rennke H, Abernathy VE, Koh JS, Valeri R, Levine JS. Rapamycin impairs recovery from acute renal failure: role of cell-cycle arrest and apoptosis of tubular cells. *Am J Physiol Renal Physiol* 2001, 281: F693–706.
49. Guba M, von Breitenbuch P, Steinbauer M, Koehl G, Flegel S, Hornung M, Bruns CJ, Zuelke C, Farkas S, Anthuber M, et al. Rapamycin inhibits primary and metastatic tumor growth by antiangiogenesis: involvement of vascular endothelial growth factor. *Nat Med* 2002, 8: 128–135.
50. Yu Y, Sato JD. MAP kinases, phosphatidylinositol 3-kinase, and p70 S6 kinase mediate the mitogenic response of human endothelial cells to vascular endothelial growth factor. *J Cell Physiol* 1999, 178: 235–246.
51. Suhara T, Mano T, Oliveira BE, Walsh K. Phosphatidylinositol 3-kinase/Akt signaling controls endothelial cell sensitivity to Fas-mediated apoptosis via regulation of FLICE-inhibitory protein (FLIP). *Circ Res* 2001, 89: 13–19.
52. Bruns CJ, Koehl GE, Guba M, Yezhelyev M, Steinbauer M, Seeliger H, Schwend A, Hoehn A, Jauch KW, Geissler EK. Rapamycin-induced endothelial cell death and tumor vessel thrombosis potentiate cytotoxic therapy against pancreatic cancer. *Clin Cancer Res* 2004, 10: 2109–2119.
53. Majumder PK, Febbo PG, Bikoff R, Berger R, Xue Q, McMahon LM, Manola J, Brugarolas J, McDonnell TJ, Golub TR, et al. mTOR inhibition reverses Akt-dependent prostate intraepithelial neoplasia through regulation of apoptotic and HIF-1-dependent pathways. *Nat Med* 2004, 10: 594–601.

54. Slamon DJ, Leyland-Jones B, Shak S, Fuchs H, Paton V, Bajamonde A, Fleming T, Eiermann W, Wolter J, Pegram M, et al. Use of chemotherapy plus a monoclonal antibody against HER2 for metastatic breast cancer that overexpresses HER2. *N Engl J Med* 2001, 344: 783–792.
55. Heinrich MC, Corless CL, Demetri GD, Blanke CD, von Mehren M, Joensuu H, McGreevey LS, Chen CJ, Van den Abbeele AD, Druker BJ, et al. Kinase mutations and imatinib response in patients with metastatic gastrointestinal stromal tumor. *J Clin Oncol* 2003, 21: 4342–4349.
56. Paez JG, Janne PA, Lee JC, Tracy S, Greulich H, Gabriel S, Herman P, Kaye FJ, Lindeman N, Boggon TJ, et al. EGFR mutations in lung cancer: correlation with clinical response to gefitinib therapy. *Science* 2004, 304: 1497–1500.
57. Lynch TJ, Bell DW, Sordella R, Gurubhagavatula S, Okimoto RA, Brannigan BW, Harris PL, Haserlat SM, Supko JG, Haluska FG, et al. Activating mutations in the epidermal growth factor receptor underlying responsiveness of non-small-cell lung cancer to gefitinib. *N Engl J Med* 2004, 350: 2129–2139.
58. Ali IU, Schriml LM, Dean M. Mutational spectra of PTEN/MMAC1 gene: a tumor suppressor with lipid phosphatase activity. *J Natl Cancer Inst* 1999, 91: 1922–1932.
59. Cantley LC, Neel BG. New insights into tumor suppression: PTEN suppresses tumor formation by restraining the phosphoinositide 3-kinase/AKT pathway. *Proc Natl Acad Sci USA* 1999, 96: 4240–4245.
60. Neshat MS, Mellinghoff IK, Tran C, Stiles B, Thomas G, Petersen R, Frost P, Gibbons JJ, Wu H, Sawyers CL. Enhanced sensitivity of PTEN-deficient tumors to inhibition of FRAP/mTOR. *Proc Natl Acad Sci USA* 2001, 98: 10314–10319.
61. Podsypanina K, Lee RT, Politis C, Hennessy I, Crane A, Puc J, Neshat M, Wang H, Yang L, Gibbons J, et al. An inhibitor of mTOR reduces neoplasia and normalizes p70/S6 kinase activity in Pten+/– mice. *Proc Natl Acad Sci USA* 2001, 98: 10320–10325.
62. DeGraffenried LA, Fulcher L, Friedrichs WE, Grunwald V, Ray RB, Hidalgo M. Reduced PTEN expression in breast cancer cells confers susceptibility to inhibitors of the PI3 kinase/Akt pathway. *Ann Oncol* 2004, 15: 1510–1516.
63. Yu K, Toral-Barza L, Discafani C, Zhang WG, Skotnicki J, Frost P, Gibbons JJ. mTOR, a novel target in breast cancer: the effect of CCI-779, an mTOR inhibitor, in preclinical models of breast cancer. *Endocr Relat Cancer* 2001, 8: 249–258.
64. Shi Y, Gera J, Hu L, Hsu JH, Bookstein R, Li W, Lichtenstein A. Enhanced sensitivity of multiple myeloma cells containing PTEN mutations to CCI-779. *Cancer Res* 2002, 62: 5027–5034.
65. Stromberg T, Dimberg A, Hammarberg A, Carlson K, Osterborg A, Nilsson K, Jernberg-Wiklund H. Rapamycin sensitizes multiple myeloma cells to apoptosis induced by dexamethasone. *Blood* 2004, 103: 3138–3147.
66. Mondesire WH, Jian W, Zhang H, Ensor J, Hung MC, Mills GB, Meric-Bernstam F. Targeting mammalian target of rapamycin synergistically enhances chemotherapy-induced cytotoxicity in breast cancer cells. *Clin Cancer Res* 2004, 10: 7031–7042.
67. deGraffenried LA, Friedrichs WE, Russell DH, Donzis EJ, Middleton AK, Silva JM, Roth RA, Hidalgo M. Inhibition of mTOR activity restores tamoxifen response in breast cancer cells with aberrant Akt Activity. *Clin Cancer Res* 2004, 10: 8059–8067.
68. Grunwald V, DeGraffenried L, Russel D, Friedrichs WE, Ray RB, Hidalgo M. Inhibitors of mTOR reverse doxorubicin resistance conferred by PTEN status in prostate cancer cells. *Cancer Res* 2002, 62: 6141–6145.
69. Galanis E, Buckner JC, Maurer MJ, Kreisberg JI, Ballman K, Boni J, Peralba JM, Jenkins RB, Dakhil SR, Morton RF, et al. Phase II trial of temsirolimus (CCI-779) in recurrent glioblastoma multiforme: a North Central Cancer Treatment Group Study. *J Clin Oncol* 2005, 23: 5294–5304.
70. Samuels Y, Wang Z, Bardelli A, Silliman N, Ptak J, Szabo S, Yan H, Gazdar A, Powell SM, Riggins GJ, et al. High frequency of mutations of the PIK3CA gene in human cancers. *Science* 2004, 304: 554.
71. Kang S, Bader AG, Vogt PK. Phosphatidylinositol 3-kinase mutations identified in human cancer are oncogenic. *Proc Natl Acad Sci USA* 2005, 102: 802–807.

72. Samuels Y, Diaz LA Jr., Schmidt-Kittler O, Cummins JM, Delong L, Cheong I, Rago C, Huso DL, Lengauer C, Kinzler KW, et al. Mutant PIK3CA promotes cell growth and invasion of human cancer cells. *Cancer Cell* 2005, 7: 561–573.
73. Adachi H, Igawa M, Shiina H, Urakami S, Shigeno K, Hino O. Human bladder tumors with 2-hit mutations of tumor suppressor gene TSC1 and decreased expression of p27. *J Urol* 2003, 170: 601–604.
74. van Slegtenhorst M, de Hoogt R, Hermans C, Nellist M, Janssen B, Verhoef S, Lindhout D, van den Ouweland A, Halley D, Young J, et al. Identification of the tuberous sclerosis gene TSC1 on chromosome 9q34. *Science* 1997, 277: 805–808.
75. Kenerson HL, Aicher LD, True LD, Yeung RS. Activated mammalian target of rapamycin pathway in the pathogenesis of tuberous sclerosis complex renal tumors. *Cancer Res* 2002, 62: 5645–5650.
76. Brugarolas JB, Vazquez F, Reddy A, Sellers WR, Kaelin WG Jr. TSC2 regulates VEGF through mTOR-dependent and -independent pathways. *Cancer Cell* 2003, 4: 147–158.
77. El-Hashemite N, Walker V, Zhang H, Kwiatkowski DJ. Loss of Tsc1 or Tsc2 induces vascular endothelial growth factor production through mammalian target of rapamycin. *Cancer Res* 2003, 63: 5173–5177.
78. Dudkin L, Dilling MB, Cheshire PJ, Harwood FC, Hollingshead M, Arbuck SG, Travis R, Sausville EA, Houghton PJ. Biochemical correlates of mTOR inhibition by the rapamycin ester CCI-779 and tumor growth inhibition. *Clin Cancer Res* 2001, 7: 1758–1764.
79. Peralba JM, DeGraffenried L, Friedrichs W, Fulcher L, Grunwald V, Weiss G, Hidalgo M. Pharmacodynamic Evaluation of CCI-779, an Inhibitor of mTOR, in Cancer Patients. *Clin Cancer Res* 2003, 9: 2887–2892.
80. www.clinicaltrials.gov Accesed June 23.
81. Witzig TE, Geyer SM, Ghobrial I, Inwards DJ, Fonseca R, Kurtin P, Ansell SM, Luyun R, Flynn PJ, Morton RF, et al. Phase II trial of single-agent temsirolimus (CCI-779) for relapsed mantle cell lymphoma. *J Clin Oncol* 2005, 23: 5347–5356.
82. Lane HA Tanka J, Kovaril T, et al. Preclinical and clinical pharmacokinetic/pharmacodynamic (PK/PD) modeling to help to define an optimal biological dose for the oral mTOR inhibitor, RAD001, in oncology. *Proc Am Soc Clin Oncol* 2003, 22.
83. Tabernero J, Ramos F, Burris H, Casado E, Macarulla T, Jones S, Dimitrijevic S, Hazell K, Shand N, Baselga J. A phase I study with tumor molecular pharmacodynamic (MPD) evaluation of dose and schedule of the oral mTOR-inhibitor Everolimus (RAD001) in patients (pts) with advanced solid tumors. *Proc Am Soc Clin Oncol abstract 3007*, 2005.
84. Van Oosterom AT, Dumez H, Desai S, et al. Combination signal transduction inhibition: A phase I/II trial of the oral mTOR-inhibitor everolimus (RAD001) and imatinib mesylate (IM) in patients with gastrointestinal stromal tumor (GISDT) refractory to IM. *Proc Am Soc Clin Oncol* 2004, 23.
85. Milton DT, Kris MG, Azzoli CG, Gomez JE, Heelan R, Krug LM, Pao W, Pizzo B, Rizvi NA, Miller VA. Phase I/II trial of Gefitinib and RAD001 (Everolimus) in Patients with Advanced Non-Small Cell Lung Cancer. *Proc Am Soc Clin Oncol abstract 7104*, 2005.
86. MacDonald AS: A worldwide, phase III, randomized, controlled, safety and efficacy study of a sirolimus/cyclosporine regimen for prevention of acute rejection in recipients of primary mismatched renal allografts. *Transplantation* 2001, 71: 271–280.
87. Kahan BD. Two-year results of multicenter phase III trials on the effect of the addition of sirolimus to cyclosporine-based immunosuppressive regimens in renal transplantation. *Transplant Proc* 2003, 35: 37S–51S.
88. Zimmerman JJ, Kahan BD. Pharmacokinetics of sirolimus in stable renal transplant patients after multiple oral dose administration. *J Clin Pharmacol* 1997, 37: 405–415.
89. Saijo N, Tamura T, Nishio K. Problems in the development of target-based drugs. *Cancer Chemother Pharmacol* 2000, 46 (Suppl.): S43–45.

CHAPTER 25

DUAL/PAN-HER TYROSINE KINASE INHIBITORS: FOCUS IN BREAST CANCER

JOAN ALBANELL

Medical Oncology Department, Hospital del Mar de Barcelona, Spain

Many human cancers, such as breast cancer, express multiple members of the HER receptor signaling family, a fact that points to the interest of developing inhibitors of both EGFR and HER2 (dual inhibitors), or inhibitors of the four receptors (pan-ErbB inhibitors). Among these agents, lapatinib (GW572016) is the one in a more advanced stage of clinical development. Lapatinib is a dual reversible TKI that potently inhibits both EGFR and HER2 and has a very slow off-rate for both EGFR and HER2 (t1/2 > 2 h). Lapatinib induces growth arrest and/or apoptosis in EGFR and HER2-dependent tumor cell lines and is a potent inhibitor of growth of tumor xenografts in mice. The marked reduction of tyrosine phosphorylation of EGFR and HER2 induced by lapatinib results in inhibition of MAPK and Akt. In fact, experimental evidence suggests that lapatinib may be better than EGFR-specific TKI's in inhibiting PI3K/Akt activation in vitro and in vivo. Complete inhibition of activated Akt in HER2 overexpressing cells correlated with a strong increase in apoptosis. Inhibition of activated Akt may be of therapeutic interest for the use of lapatinib as a monotherapy, or may enhance the antitumor activity of chemotherapeutics for which Akt may mediate chemoresistance. A potentially important advantage of lapatinib compared with monoclonal antibodies (MAbs) directed at the extracellular domain of HER2 (trastuzumab) is the inhibition of the phosphorylation of a truncated form of HER2, termed HER2p95 that lacks the extracellular domain but has TK activity. It is tempting to speculate that trastuzumab resistance may be mediated in part by the selection of HER2p95-expressing breast cancer cells capable of exerting potent growth and prosurvival signals through HER2p95-HER3 heterodimers. Whether this would explain the activity of lapatinib seen in patients with trastuzumab refractory breast cancers is as yet unknown. In addition to lapatinib activity seen in trastuzumab refractory patients, a study in first-line treatment of advanced breast cancer patients with FISH + tumors resulted in a promising response rate of 37.5%, further suggesting that this agent may have a role in the treatment of breast cancer. Newer agents inhibiting multiple ErbB receptors, such as CI-1033 or BMS-599626, have recently entered into the clinic.

1. THE EPIDERMAL GROWTH FACTOR RECEPTOR FAMILY

The effects of small molecules that inhibit the epidermal growth factor receptor (EGFR) tyrosine kinase (TK), also known as EGFR-TK inhibitors (EGFR-TKI), vary among different cells/tumors and this is due to the fact that the role of the EGFR in a given cell depends on the repertoire of coexpressed molecules that drive cell growth and survival.

1.1. Mechanisms of activation and signaling Pathways

The human EGFR family (also termed ErbB receptor family) consists of four homologue transmembrane receptor TKs: (1) the EGFR itself (ErbB1/EGFR/HER1); (2) HER2 (ErbB2); (3) HER3 (ErbB3); and (4) HER4 (ErbB4). These receptors have an extracellular ligand-binding domain, a transmembrane lipophilic segment, and an intracellular TK domain [1]. EGFR, HER2, and HER4 have intrinsic TK activity while HER3 is a kinase-defective receptor. Ligands that bind to the EGFR include epidermal growth factor (EGF), transforming growth factor alpha (TGF-a), amphiregulin, heparin-binding EGF, betacellulin, and epiregulin. A second class of ligands, known as neuregulins, bind directly to HER3 and/or HER4. The activation of these receptors involves dimerization, which can take place with an identical ErbB receptor as dimerization partner (homodimerization) or another member of the ErbB receptor family (heterodimerization).

Dimerized/oligomerized receptors undergo phosphorylation of specific tyrosines, which in turn, results in the recruitment of downstream signaling proteins. No ligand has yet been identified for HER2. Despite being a ligand-orphan receptor, HER2 is the preferred coreceptor for the EGFR, HER3, and HER4. Conversely, HER3 binds a number of ligands but since it has a defective TK, HER3 requires heterodimerization with another ErbB member to be activated [1–4].

The dimerization of the receptors results in activation of the intrinsic TK activity, and in autophosphorylation of the cytoplasmic domain. Each receptor complex may activate different signaling pathways that elicit specific cellular responses, resulting in an enormous signaling diversity.

Ultimately, the ErbB signaling family results in activation of a cascade of biochemical and physiological responses that regulate cell proliferation, survival, angiogenesis, and invasion [1–3].

There are many signaling pathways activated by ErbB-based dimers, such as Ras-Raf-MAPK, PI3K-Akt, PLC-g1, Src, STATs, and others [1] (Figure 25.1).

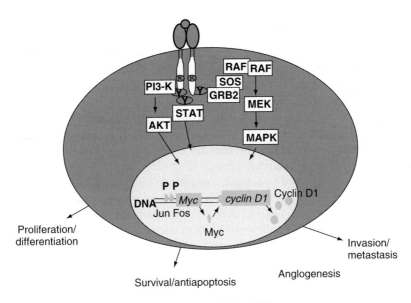

Figure 25.1. Signaling transduction pathways activated by ErbB receptors.

Activation of the Ras-Raf-MAPK pathway eventually regulates cell transcription and has been linked to cell proliferation, survival and transformation in laboratory studies [5] and more recently in studies in human tumors [6–8]. Another Signaling transduction pathway activated by the ErbB network is the PI3K/Akt pathway that plays an important role in cell survival.

Phosphatydilinositol-3 kinase (PI3K) is lipid kinase activated by HER2/HER3 dimers and other receptor complexes that eventually results in the phosphorylation and activation of Akt [9], the activation or inhibition of these pathways may play important roles in the resistance or response to EGFR-TKIs.

2. EARLY DEVELOPMENT OF EGFR TKIS AND PHARMACODYNAMIC ENDPOINTS

The development of EGFR TKIs was based on data that revealed that the TK activity of the EGFR was required for many of the biochemical responses induced by this receptor. A potential advantage of EGFR TKIs over MAbs is the ability to enter into the cell and to act on intracellular active forms of the receptor, such as EGFRvIII, or truncated HER2 receptor (HER2p95) [11–15,18–22].

Quinazoline compounds, such as PD153035 [16] and AG1478, represent the first of a class of competitive ATP inhibitors that were highly selective for the EGFR TK. Since then, a series of agents with different degrees of specificity towards single or multiple ErbB receptors has been produced. The inhibition of the receptor TK can be also reversible or irreversible. Based on the increasing complexity of the compounds available, an elegant classification of these agents has been recently proposed [14,17] (Table 25.1). The potential differences in activity or tolerability of dual EGFR/HER2 or pan-ErbB receptor inhibitors over highly specific EGFR TKIs are discussed later for individual agent that are under active clinical development [18].

The clinical development of EGFR receptor targeted agents has involved the possibility of selecting optimal biological doses (OBD) instead of using the maximally tolerated doses (MTD) that are used for the traditional development of nontargeted chemotherapeutic agents [7,18–23]. An OBD may be defined as the minimal dose required to achieve complete receptor inhibition. The methods used to help to select an optimal dose are based on pharmacokinetic, or preferably, pharmacodynamic assays (that is, demonstrating the desired biological effect on the target in vivo in posttreatment biopsies). Data on clinical activity and toxicity in phase I trials are however essential drivers to select of doses for further development. The potential advantages or disadvantages of each approach (OBD vs. MTD) are yet to be determined, although a priori the use of OBDs should avoid unnecessary side effects that would appear at doses above the ones that inhibit the target.

Pharmacodynamic studies may be also useful to define the biological response to the treatment, to discover biomarkers that predict response and to define potential mechanisms of therapeutic resistance. Several preclinical studies have been aimed to identify good surrogate markers of biological activity of EGFR TKIs. For instance, cell culture experiments using the TKI gefitinib and the MAb cetuximab, two EGFR-targeted agents, have shown that drug concentrations that caused a maximal decline in cell proliferation coincided with concentrations that reduced or abolished EGFR and MAPK activation [7]. These experiments supported the value of these markers for pharmacodynamic studies aimed at defining OBD. The promising value of EGFR and MAPK activation is strengthened by in vivo studies performed in mice bearing human breast cancer xenografts and treated with gefitinib [24]. Interestingly, Akt activation was also reduced, but to a lesser extent than EGFR or MAPK. This observation suggested the possibility that Akt might play a role in the in vivo response or resistance to EGFR-targeted agents. Many additional studies with a variety of EGFR targeted agents have supported the use of other downstream surrogates such as Akt, p27, cyclin D, and c-fos. However, the use of receptor downstream markers requires a careful interpretation of the results since our understanding

Table 25.1. ErbB receptor TK Inhibitors in clinical development

Selective EGFR TK inhibitors	Dual or pan-ErbB receptor TK inhibitors
Gefitinib (reversible)	Lapatinib (EGFR/HER2; reversible)
Erlotinib (irreversible)	CI-1033 (pan-ErbB; irreversible)
EKB-569 (irreversible)	BMS-599626

of EGFR dependent pathways in vivo in human tumors is yet limited. For instance, a significant relationship between expression of EGFR and downstream molecules (i.e., activated MAPK) has been reported in various tumor types, such as head and neck, breast, and gastric carcinomas [6–8,20]. Nevertheless, some specimens have high levels of activated MAPK without high levels of EGFR, thus suggesting that other receptors, or Ras mutations, are activating MAPK. The reverse is also true. The rapid knowledge gained since the discovery of EGFR TK mutations will be almost certainly of importance to better understand the basis of EGFR tumor dependency [10].

Hopefully, the use of transcriptional and proteomic profiling will also aid in the identification of tumors and pathways that are truly EGFR dependent in clinical samples.

3. DUAL ERBB RECEPTOR INHIBITORS

Many human cancer express multiple members of the ErbB receptor Signaling family, a fact that points to the interest of the clinical testing of inhibitors of both EGFR and HER2 (dual inhibitors), or inhibitors of the four receptors (pan-ErbB inhibitors) of the family, compared with inhibitors of single receptors (Table 25.2) [15,17,25,26].

Lapatinib (GW572016) is a dual reversible TKI that potently inhibits both EGFR and HER2 and has a very slow off-rate for both EGFR and HER2 (t1/2 > 2 h) [27]. Lapatinib induces growth arrest and/or apoptosis in EGFR and HER2-dependent tumor cell lines. Lapatinib also is a potent inhibitor of tumor growth xenografts in mice [28,29].

The marked reduction of tyrosine phosphorylation of EGFR and HER2 induced by lapatinib results in inhibition of MAPK and Akt. In fact, experimental evidence suggest that lapatinib may be better than EGFR-specific TKI's in inhibiting PI3K/Akt activation in vitro and in vivo [28,29]. Complete inhibition of activated Akt in HER2 overexpressing cells correlated with a strong increase in apoptosis. These observations were reproduced in vivo in human tumor xenografts. Inhibition of activated Akt may be of therapeutic interest for the use of lapatinib as a monotherapy, or may enhance the antitumor activity of chemotherapeutics for which Akt may mediate chemoresistance. The drug exhibited a favorable toxicity profile in rodents and dogs and no evidence of cardiac toxicity during high exposure over 6 and 9 months, respectively. A potentially important advantage of lapatinib compared with MAbs directed at the extracellular domain of HER2 (trastuzumab) is the inhibition of the

Table 25.2. Rationale to investigate dual/pan ErbB receptor tyrosine kinase inhibitors

ErbB receptor heterodimers are more potent than ErbB receptor homodimers
Coexpression of different ErbB receptors in human cancer specimens is common
Combined treatments of specific anti-EGFR agents and anti-HER2 agents result in greater antitumor activity than when each agent is used alone

phosphorylation of a truncated form of HER2, termed HER2p95, that lacks the extracellular domain but has TK activity. In fact, the expression of the NH2 terminally truncated HER2 receptor (HER2p95) in breast cancer correlates with metastatic disease progression compared with the expression of full-length HER2p185 [13,30]. It has been confirmed experimentally that HER2p85 forms heterodimers with HER3, but not with EGFR, thus suggesting that selective EGFR TKIs would not be able to inhibit this association. The dual EGFR/HER2 inhibitor lapatinib, however, inhibits baseline HER2p95 phosphorylation in BT474 cells and tumor xenografts, resulting in the inhibition of downstream phospho-MAPK and phospho-Akt [27]. It is tempting to speculate that trastuzumab resistance may be mediated in part by the selection of HER2p95-expressing breast cancer cells capable of exerting potent growth and prosurvival signals through HER2p95-HER3 heterodimers. Whether this would explain the activity of lapatinib seen in patients with trastuzumab refractory breast cancers is as yet unknown.

Phase I trials of oral lapatinib have been conducted [31]. Rash, diarrhea, nausea, and fatigue were the most common adverse events. Some evidence of clinical activity has been observed. Patients with a variety of tumors have had stable disease for up to 13 months; one patient exhibited a minor response, and one patient with head and neck tumor had a complete response and remained on study after 19 months.

Preliminary pharmacokinetic data indicate that the lapatinib serum concentrations were above the in vitro 90% inhibitory concentration at the 1,200 mg once-daily dose and pharmacokinetics appear to be linear over the tested dose range (up to 1,800 mg). A phase IB study of lapatinib in heavily pretreated metastatic cancer was conducted in patients with biopsable disease, and EGFR or HER-2 overexpression on immunohistochemistry, HER-2 gene amplification, or evidence of activated EGFR and HER-2 receptors on immunohistochemistry. Patients were randomized to receive lapatinib at doses of 500, 650, 900, 1,200, or 1,600 mg once daily [32]. In this study, 67 patients with metastatic solid tumors were evaluated. Tumor types in these patients consisted of breast cancer (33%), and others carcinomas (66%).

Treatment has been well tolerated, with no grade 4 and five grade 3 toxicity (gastrointestinal events and rash). The most common adverse events (all grade 1 or 2) were been diarrhea (42%), rash (31%), and nausea (13%). No treatment-related cardiac or pulmonary toxicity was observed.

Partial responses were observed in four patients at the 1,200-mg ($n = 2$), 900-mg ($n = 1$), and 650-mg ($n = 1$) doses, with those patients having received therapy for a median of 5.5 months (3–8 months). Each of the four patients had breast cancer, overexpressed ErbB2, and with one exception, coexpressed ErbB1.

Stable disease was observed in 24 patients with various other carcinomas, of whom ten received lapatinib for $> = 6$ months. The relationships between lapatinib dose or serum concentration and clinical response could not be adequately characterized due to the limited response data.

Overall, the tumor types responding to treatment (partial response or stable disease) have consisted of trastuzumab-refractory breast cancer ($n = 10$), colorectal cancer ($n = 3$), ovarian cancer ($n = 1$), lung cancer ($n = 5$), adenocarcinoma of unknown primary site ($n = 1$), granular cell carcinoma ($n = 1$), and head and neck cancer ($n = 3$). An on-day 28 biopsy in a patient with trastuzumab-refractory inflammatory breast cancer who had a rather dramatic partial response to lapatinib showed decrease in phospho-EGFR and phospho-HER2, phospho-MAPK index, cyclin D, and TGF were observed, with a dramatic increase in tumor cell apoptosis. Another patient that experienced a partial response to lapatinib had a marked increase in apoptosis. In contrast, correlates in a patient with progressive disease on lapatinib had no induction of apoptosis.

In on-study tumor biopsies, an inhibition of p-EGFR, p-HER2, p-MAPK, or p-Akt expression was reliably achieved at lapatinib doses of 650 mg and greater. Responders exhibited variable levels of inhibition of p-ErbB1, p-ErbB2, p-Erk1/2, p-Akt, cyclin D1, and TGF-a. Increased tumor cell apoptosis (TUNEL) occurred in patients with evidence of tumor regression but not in nonresponders. No significant cardiac toxicity has been reported [33,34].

The dose selected for further development a single agent was a daily dose of 1,500 mg, which is associated to desired pharmacodynamic effects and is yet below the MTD. Responses in phase I trials have been also seen in squamous cell carcinoma of the lung [35]. A phase II safety and efficacy study of oral lapatinib in 41 patients with metastatic breast cancer has been recently reported [36]. The investigators-assessed responses were one complete response and three partial responses (two of them confirmed by independent review). The proportion of progression-free patients at 16 weeks was 24%, as assessed by an independent committee. The activity of single agent lapatinib against breast cancer has been also observed in another phase II study in patients with advanced or metastatic breast cancer who progressed while receiving trastuzumab-containing regimens and had progressed to at least three prior chemotherapeutic regimens. In addition to its activity in trastuzumab-refractory patients, the combination with trastuzumab is also attractive.

Recent studies in cell lines showed that combining lapatinib, with anti-ErbB2 antibodies, either pAb or trastuzumab, in ErbB2-overexpressing breast cancer cells BT474 markedly downregulated survivin protein and enhanced tumor cell apoptosis. In addition to inhibit p-Erk1/2 and p-Akt, the combination of lapatinib with trastuzumab completely inhibited p-erbB2 steady state protein levels and marked inhibition of p-ErbB3 [37].

A recent phase I/II study of lapatinib and trastuzumab in 48 patients with HER2 overexpressing breast cancer has shown that the recommended dose of the combination was 1,000 mg/day of lapatinib plus the standard dose of trastuzumab [38]. Dose limiting toxicities consisted of fatigue, nausea, and anorexia. One complete response and four partial responses were observed in patients with heavily pretreated breast cancer, all of them progressing to prior trastuzumab. Based on these promising results, additional trials of trastuzumab and lapatinib are planned. Most of the responses have been observed in tumors with inflammatory features, leading to specific studies in patients with inflammatory breast cancer.

On the other hand, no objective responses have been seen in heavily pretreated (median of more than four prior chemotherapy regimens) non-HER2 overexpressing breast cancer patients that received lapatinib. However, studies addressing the role of lapatinib in less pretreated patients are warranted. Furthermore, a series of phase III trials of lapatinib in breast cancer are currently accruing patients [34]. In first line, lapatinib plus paclitaxel versus paclitaxel alone is investigated in a randomized, double-blind, placebo-controlled, two-arm, multicenter phase III trial. This study is preceded by a phase I trial of lapatinib in combination with paclitaxel. The optimal tolerated regimen was determined to be paclitaxel 175 mg/m^2 q 3 weeks and lapatinib 1,500 mg/day [39]. Importantly, three objective responses were observed in patients with taxane-resistant breast cancer. A phase III randomized, open-label, multicenter trial is comparing lapatinib plus capecitabine with capecitabine alone in patients with refractory advanced or metastatic breast cancer.

Recent evidence show that lapatinib was able to restore tamoxifen sensitivity in EGFR- or HER-2-expressing cells, while, in vivo, combined laptaninib and tamoxifen caused maximal regression of HER-2 overexpressing, tamoxifen-resistant MCF-7 xenografts, and to cooperate with tamoxifen to provide more rapid and profound cell-cycle arrest in hormone-resistant cells [40]. Based on this preclinical data, a number of small phase I/II trials have been initiated with TKIs in combination with tamoxifen, fulvestrant, or an aromatase inhibitor.

Finally, lapatinib plus letrozole versus letrozole alone is being compared in a randomized, double-blind, placebo-controlled, multicenter phase III trial in patients with estrogen/progesterone-receptor-positive advanced metastatic breast cancer. These trials should help to clarify the potential roles of lapatinib in advanced breast cancer. In addition to breast cancer, a wide program of development of lapatinib is in bladder [41] and kidney cancers and studies are ongoing in many other tumor types as well.

The results of this and further clinical studies will help to establish whether the above described properties and theoretical benefits of a potent, dual EGFR/HER2 TKI such as lapatinib will translate into benefits over inhibitors of either EGFR or HER2 alone.

4. PAN-ERBB RECEPTOR INHIBITORS

CI-1033 (Canertinib) is an orally available pan-erbB receptor TKI that, unlike the majority of receptor inhibitors, effectively blocks signal transduction through all four members of the erbB family [42]. CI-1033 is also unique in that it is an irreversible inhibitor, thereby providing prolonged suppression of erbB receptor-mediated signaling. Preclinical data have shown CI-1033 to be active against a variety of human tumors in mouse xenograft models, including breast carcinomas.

In a phase I study, CI-1033 has been shown to have an acceptable side effect profile at potentially therapeutic dose levels and demonstrates evidence of target biomarker modulation. As seen in similar studies with other agents such as gefitinib, there was no relationship between dose level and degree of modulation of biomarkers in tumor and skin [43].

Antitumor activity has also been observed, including one partial clinical response and stable disease in over 30% of patients, including one patient with heavily pretreated breast cancer. After examining different schedules of administration, a recommended phase II dose of 150 mg of CI-1033 given as continuous daily dosing has been shown to have predictable and tolerable toxicities [43–45]. The most common grade 1–2 toxicities observed in the continuous daily dosing regimen included rash, diarrhea, and stomatitis. It will be of interest to better characterize the activity and tolerability profile of this agent when compared with specific EGFR inhibitors or dual inhibitors.

BMS-599626 is an orally inhibitor of the HER1, HER2, and HER4 TKs, that was evaluated in phase I trials in patients with advanced solid tumors. The most important adverse events were grade 1–2 diarrhea, nausea/vomiting, rash, fatigue, cramp, and cough. To date, BMS-599626 has been well tolerated and no dose-limiting toxicities were observed. Dose escalation study is ongoing, in patients with metastatic solid tumors, refractory to standard therapies that expressed HER1 or HER2 [46,47].

5. PERSPECTIVES

Lapatinib, a dual EGFR/HER2 TK inhibitors, is particularly promising in breast cancer. Newer agents inhibiting multiple ErbB receptors, such as CI-1033 have recently entered into the clinic and BMS-599626 is in phase I trials. In addition to the use of these agents as single agents; many clinical studies are addressing the role of combining them with hormonal agents, biological agents, or chemotherapy.

An additional strategy is the use of TKIs that affect ErbB receptors and also the TK activity of other transmembrane receptors such as VEGFR or PDGFR. Examples of these agents are ZD6474 [48,49] and AEE788 [50].

REFERENCES

1. Rubin, I., Yarden, Y. The basic biology of HER2. *Ann Oncol* 12 (Suppl. 1): S3–8 (2001).
2. Olayioye, M.A., Neve, R.M., Lane, H.A., Hynes, N.E. The ErbB signaling network: receptor heterodimerization in development and cancer. *EMBO J* 19 (13): 3159–3167 (2000).
3. Lemmon, M.A., Schlessinger, J. Regulation of signal transduction and signal diversity by receptor oligomerization. *Trends Biochem Sci* 19: 459–463 (1994).

4. Normanno, N., Ciardiello, F., Brandt, R., Salomon, D.S. Epidermal growth factor-related peptides in the pathogenesis of human breast cancer. *Breast Cancer Res Treat* 29 (1): 11–27 (1994).
5. Lewis, T.S., Shapiro, P.S., Ahn, N.G. Signal transduction through MAP kinase cascades. *Adv Cancer Res* 74: 49–139 (1998).
6. Kenny, F.S., Willsher, P.C., Gee, J.M., Nicholson, R., Pinder, S.E., Ellis, I.O., Robertson, J.F. Change in expression of ER, bcl-2 and MIB1 on primary tamoxifen and relation to response in ER positive breast cancer. *Breast Cancer Res Treat* 65 (2): 135–144 (2001).
7. Albanell, J., Codony-Servat, J., Rojo, F., Del Campo, J.M., Sauleda, S., Anido, J., Raspall, G., Giralt, J., Rosello, J., Nicholson, R.I., Mendelsohn, J., Baselga, J. Activated extracellular signal-regulated kinases: association with epidermal growth factor receptor/transforming growth factor alpha expression in head and neck squamous carcinoma and inhibition by anti-epidermal growth factor receptor treatments. *Cancer Res* 61 (17): 6500–6510 (2001).
8. Rojo, F., Tabernero, J., Van Cutsem, E., Ohtsu, A., Albanell, J., Koizumi, W., Peeters, M., Averbuch, S., Gallagher, N., Baselga, J. Pharmacodynamic studies of tumor biopsy specimens from patients with advanced gastric carcinoma undergoing treatment with gefinitib (ZD1839). *Proc Am Soc Clin Oncol* 22, 191, 2003 (abstract 764).
9. Voges, D., Zwickl, P., Baumeister, W. The 26S proteasome: a molecular machine designed for controlled proteolysis. *Annu Rev Biochem* 68: 1015–1068 (1999).
10. Minna, J.D., Gazdar, A.F., Sprang, S.R., Herz, J. Cancer. A bull's eye for targeted lung cancer therapy. *Science* 304 (5676): 1458–1461 (2004).
11. Codony-Servat, J., Albanell, J., Lopez-Talavera, J.C., Arribas, J., Baselga, J. Cleavage of the HER2 ectodomain is a pervanadate-activable process that is inhibited by the tissue inhibitor of metalloproteases-1 in breast cancer cells. *Cancer Res* 59: 1196–1201 (1999).
12. Rojo, F., Albanell, J., Del Campo, J., Sauleda, S., Raspall, G., Giralt, J., Baselga, J. Activation of Mitogen-Activated Protein Kinase (MAPK) Is Associated with Epidermal Growth Factor Receptor (EGFR), Transforming Growth Factor (TGF)? and HER2 Receptor Overexpression in Head and Neck Tumors. *Proc Am Soc Clin Oncol* 19, 2000 (abstract 2417).
13. Molina, M.A., Saez, R., Ramsey, E.E., Garcia-Barchino, M.J., Rojo, F., Evans, A.J., Albanell, J., Keenan, E.J., Lluch, A., Garcia-Conde, J., Baselga, J., Clinton, G.M. NH(2)-terminal truncated HER-2 protein but not full-length receptor is associated with nodal metastasis in human breast cancer. *Clin Cancer Res* 8 (2): 347–353 (2002).
14. Gianni, L., Albanell, J., Eiermann, W., Bianchi, G., Bourquez, D., Viganò, L., Molina, R., Raab, G., Locatelli, A., Vanhauwere, B., and Baselga, J. Feasibility, Pharmacology, and Antitumor Activity of Herceptin® (H) with Doxorubicin and Taxol Followed by Weekly Taxol (AT&T) in Women with HER2-Positive Advanced Breast Cancer (ABC). *Proc Am Soc Clin Oncol* 20, 2001 (abstract 174).
15. Mendelsohn, J., Baselga, J. Status of epidermal growth factor receptor antagonists in the biology and treatment of cancer. *J Clin Oncol* 21 (14): 2787–2799 (2003).
16. Tan, P., Cady, B., Wanner, M., Worland, P., Cukor, B., Magi-Galluzzi, C., Lavin, P., Draetta, G., Pagano, M., Loda, M. The cell cycle inhibitor p27 is an independent prognostic marker in small (T1a,b) invasive breast carcinomas. *Cancer Res* 57 (7): 1259–1263 (1997).
17. Normanno, N., Bianco, C., De Luca, A., Maiello, M.R., Salomon, D.S. Target-based agents against ErbB receptors and their ligands: a novel approach to cancer treatment. *Endocr Relat Cancer* 10 (1): 1–21 (2003).
18. Baselga, J., Ibanell, J. Targeting epidermal growth factor receptor in lung cancer. *Curr Oncol Rep* 4 (4): 317–324 (2002).
19. Park, J.W., Hong, K., Kirpotin, D.B., Colbern, G., Shalaby, R., Baselga, J., Shao, Y., Nielsen, U.B., Marks, J.D., Moore, D., Papahadjopoulos, D., Benz, C.C. Anti-HER2 immunoliposomes: enhanced efficacy attributable to targeted delivery. *Clin Cancer Res* 8 (4): 1172–1181 (2002).
20. Arteaga, C.L., Baselga, J. Clinical trial design and end points for epidermal growth factor receptor-targeted therapies: implications for drug development and practice. *Clin Cancer Res* 9 (5): 1579–1589 (2003).

21. Arteaga, C.L., Baselga, J. Tyrosine kinase inhibitors: why does the current process of clinical development not apply to them? *Cancer Cell* 5 (6): 525–31 (2004).
22. Herbst, R.S., Fukuoka, M., Baselga, J. Gefitinib—a novel targeted approach to treating cancer. *Nat Rev Cancer* 4 (12): 956–965 (2004).
23. Normanno, N., Maiello, M.R., De Luca, A. Epidermal growth factor receptor tyrosine kinase inhibitors (EGFR-TKIs): simple drugs with a complex mechanism of action? *J Cell Physiol* 194 (1): 13–19 (2003).
24. Rojo, F., Albanell, J., Anido, J., Rodriguez-Viltro, I., Guzman, M., Averbuch, S., Baselga, J. *Proc Am Soc Clin Oncol* 21, 2002 (abstract 3893).
25. Albanell, J., Ross, J. *Molecular Oncology of Breast Cancer*, Ross, J. and Hortobagyi, G. (eds), Jones and Barlett Publishers, 2004, pp. 232–255.
26. Albanell, J., Muñoz, M., Gascon, P. *Rev Oncol* 6 (1): 1–10 (2004).
27. Xia, W., Liu, L.H., Ho, P., Spector, N.L. Truncated ErbB2 receptor (p95ErbB2) is regulated by heregulin through heterodimer formation with ErbB3 yet remains sensitive to the dual EGFR/ErbB2 kinase inhibitor GW572016. *Oncogene* 23 (3): 646–653 (2004).
28. Xia, W., Mullin, R.J., Keith, B.R., Liu, L.H., Ma, H., Rusnak, D.W., Owens, G., Alligood, K.J., Spector, N.L. Anti-tumor activity of GW572016: a dual tyrosine kinase inhibitor blocks EGF activation of EGFR/erbB2 and downstream Erk1/2 and AKT pathways. *Oncogene* 21 (41): 6255–6263 (2002).
29. Mullin, R.J., Keith, B.R., Murray, D.M., Onori, J. Combination therapy with the dual EGFR-ERBB-2 tyrosine kinase inhibitor GW572016. *Proc Am Soc Clin Oncol* 22, 2003 (abstract 970).
30. Albanell, J., Molina, M.A., Codony-Servat, J., Rojo, F., Lluch, A., Arribas, J., Baselga, *J Biol Ther Breast Cancer* 2 (4): 10–13 (2001).
31. Bence, A.K., Anderson, E.B., Halepota, M.A., Doukas, M.A., Desimone, P.A., Davis, G.A., Smith, D.A., Koch, K.M., Stead, A.G., Mangum, S., Bowen, C.J., Spector, N.L., Hsieh, S., Adams, V.R. Phase I pharmacokinetic studies evaluating single and multiple doses of oral GW572016, a dual EGFR-ErbB2 inhibitor, in healthy subjects. *Invest New Drugs* 23 (1): 39–49 (2005).
32. Burris, H.A., Hurwitz, H., Dees, C., Dowlati, A., Blackwell, K.L., O'Neil, B., Marcom, P.K., Ellis, M.J., Overmoyer, B., Jones, S.F., Harris, J.L., Smith, D.A., Koch, K.M., Stead, A., Mangum, S., Spector, N. Phase I safety, pharmacokinetics, and clinical activity study of lapatinib (GW572016), a reversible dual inhibitor of epidermal growth factor receptor tyrosine kinases, in heavily pretreated patients with metastatic carcinomas. *J Clin Oncol* 10; 23 (23): 5305–13 (2005).
33. Spector, N.L., Xia, W., Burris, H. III, Hurwitz, H., Dees, E.C., Dowlati, A., O'Neil, B., Overmoyer, B., Marcom, P.K., Blackwell, K.L., Smith, D.A., Koch, K.M., Stead, A., Mangum, S., Ellis, M.J., Liu, L., Man, A.K., Bremer, T.M., Harris, J., Bacus, S. Study of the biologic effects of lapatinib, a reversible inhibitor of ErbB1 and ErbB2 tyrosine kinases, on tumor growth and survival pathways in patients with advanced malignancies. *J Clin Oncol* 10; 23 (11): 2502–12 (2005).
34. Burris, H.A. III. Dual kinase inhibition in the treatment of breast cancer: initial experience with the EGFR/ErbB-2 inhibitor lapatinib. *Oncologist* 9 (Suppl. 3): 10–15 (2004).
35. Minami, H., Nakagawa, K., Kawada, K., Mukai, H., Tahara, M., Kurata, T., Uejima, H., Nogami, T., Sasaki, Y., Fukuoka, M. A phase I study of GW572016 in patients with solid tumors. *J Clin Oncol* 2004 ASCO Annual Meeting Proceedings (Post-Meeting Edition) 14S (July 15 Suppl.): 3048.
36. Blackwell, K.L., Kaplan, E.H., Franco, S.X., Marcom, P.K., Maleski, J.E., Sorensen, M.J., Berger, M.S. A phase II, open-label, multicenter study of GW572016 in patients with trastuzumab-refractory metastatic breast cancer. *J Clin Oncol* 2004 ASCO Annual Meeting Proceedings (Post-Meeting Edition) 22 (14S (July 15 Suppl.)): 3006.
37. Xia, W., Gerard, C.M., Liu, L., Baudson, N.M., Ory, T.L., Spector, N.L. Combining lapatinib (GW572016), a small molecule inhibitor of ErbB1 and ErbB2 tyrosine kinases, with therapeutic anti-ErbB2 antibodies enhances apoptosis of ErbB2-overexpressing breast cancer cells. *Oncogene* 15; 24 (41): 6213–6221 (2005).
38. Burris H.A. III, Storniolo, A., Overmoyer, B., Pegram, M., Jones, S., Peaccok, N., Loftiss, J., Ho, P., Koch, K.M., Paul, E., Beelen, A.P., Pandite, L. A phase I, open-label study of the safety,

tolerability and pharmacokinetics of lapatinib (GW572016) in combination with trastuzumab. *Proc. Breast Cancer San Antonio Meeting*, 2004 (abstract 3043).
39. Jones, S.F., Hainsworth, J.D., Spigel, D.R., Peacock, N.W., Willcutt, N.T., Pandite, L.N., Versola, M.J., Koch, K.M., Greco, F., Burris, H. A phase I study of the dual kinase inhibitor GW572016 in combination with paclitaxel (EGF10009). *J Clin Oncol* 2004 ASCO Annual Meeting Proceedings (Post-Meeting Edition) 22 (14S (July 15 Suppl.)): 2083.
40. Chu, I., Blackwell, K., Chen, S., Slingerland, J., The dual ErbB1/ErbB2 inhibitor, lapatinib (GW572016), cooperates with tamoxifen to inhibit both cell proliferation- and estrogen-dependent gene expression in antiestrogen-resistant breast cancer. *Cancer Res* 1; 65 (1): 18–25 (2005).
41. Machiels, J.-P., Wülfing, C., Richel, D.J., Beuzeboc, P., Garcia Del Muro, X., Grimm, M.-O., Farrell, J., Colman, J.R., El-Hariry, I.A. A single arm, multicenter, open-label phase II study of orally administered GW572016 as single-agent, second-line treatment of patients with locally advanced or metastatic transitional cell carcinoma of the urothelial tract. Interim analysis. *J Clin Oncol* 2004 ASCO Annual Meeting Proceedings (Post-Meeting Edition) 22 (14S (July 15 Suppl.)): 4615.
42. Slichenmyer, W.J., Fry, D.W. Anticancer therapy targeting the erbB family of receptor tyrosine kinases. *Semin Oncol* 28 (5 Suppl. 16): 67–79 (2001).
43. Zinner, R.G., Donato, N.J., Nemunaitis, J.J., Cunningham, C.C., Shin, H.J., Zentgraf, R.E., Ayers, G.D., Glisson, B.S., Khuri, F.R., Kies, M.S., Eiseman, I., Lenehan, P.F., Hong, W.K., Shin, D.M. Biomarker modulation in tumor and skin biopsy samples from patients with solid tumors following treatment with the pan-erbB tyrosine kinase inhibitor, CI-1033. *Proc Am Soc Clin Oncol* 21, 2002 (abstract 58).
44. Rinehart, J.J., Wilding, G., Willson, J., Krishnamurthi, S., Natale, R., Dasse, K.D., Olson, S., Marcy, P., Lenehan, P., Chakrabarti, D. A phase I clinical and pharmacokinetic (PK)/food effect study of oral CI-1033, a pan-erb B tyrosine kinase inhibitor, in patients with advanced solid tumors. *Proc Am Soc Clin Oncol* 22, 205, 2003 (abstract 821).
45. Rowinsky, E.K., Garrison, M., Lorusso, P., Patnaik, A., Hammond, L., De Bono, J., McCreery, H., Eiseman, I., Lenehan, P., Tolcher, A., Administration of CI-1033, an irreversible pan-erbB tyrsosine kinase (TK) inhibitor is feasible on a 7-day-on/7-day-off schedule: A phase I, pharmacokinetic (PK), and food effect study. *Proc Am Soc Clin Oncol* 22, 201, 2003 (abstract 807).
46. Garland, L., Pegram, M., Song, S., Mendelson, D., Parker, K., Martell, E.R.E., Gordon, M.S. Phase I study of BMS-599626, an oral pan-HER tyrosine kinase inhibitor, in patients with advanced solid tumors. *J Clin Oncol* 2005 ASCO Annual Meeting Proceedings. Vol 23, No 16S (June 1 Suppl.), 2005: 3152.
47. Soria, J.C., Cortes, J., Armand, J.P., Taleb, A., van Bree, L., Lopez, E., Song, S., Zeradib, K., Vazquez, F., Martell, R. E., Baselga, J. Phase I pharmacokinetic profile and early clinical evaluation of the PAN-HER inhibitor BMS-599626. *J Clin Oncol*, 2005 ASCO Annual Meeting Proceedings. Vol 23, No 16S (June 1 Suppl.), 2005: 3109.
48. Ciardiello, F., Caputo, R., Damiano, V., Troiani, T., Vitagliano, D., Carlomagno, F., Veneziani, B.M., Fontanini, G., Bianco, A.R., Tortora, G. Antitumor effects of ZD6474, a small molecule vascular endothelial growth factor receptor tyrosine kinase inhibitor, with additional activity against epidermal growth factor receptor tyrosine kinase. *Clin Cancer Res* 9 (4): 1546–1556 (2003).
49. Ciardiello, F., Bianco, R., Caputo, R., Damiano, V., Troiani, T., Melisi, D., De Vita, F., De Placido, S., Bianco, A.R., Tortora, G. Antitumor activity of ZD6474, a vascular endothelial growth factor receptor tyrosine kinase inhibitor, in human cancer cells with acquired resistance to antiepidermal growth factor receptor therapy. *Clin Cancer Res* 10 (2): 784–93 (2004).
50. Traxler, P., Allegrini, P.R., Brandt, R., Brueggen, J., Cozens, R., Fabbro, D., Grosios, K., Lane, H.A., McSheehy, P., Mestan, J., Meyer, T., Tang, C., Wartmann, M., Wood, J., Caravatti, G., AEE788: a dual family epidermal growth factor receptor/ErbB2 and vascular endothelial growth factor receptor tyrosine kinase inhibitor with antitumor and antiangiogenic activity. *Cancer Res* 15; 64 (14): 4931–4941 (2004).

CHAPTER 26

ANTITUMOR-ASSOCIATED ANTIGENS IGGS: DUAL POSITIVE AND NEGATIVE POTENTIAL EFFECTS FOR CANCER THERAPY

EMILIO BARBERÁ GUILLEM AND JAMES W. SAMPSEL

Celartia Research Laboratories, Powell, Ohio, USA

Abstract: Antitumor antigen antibodies are promising tools for cancer therapy, under the judgment of achieving targeted cell destruction. However, antibodies can not only kill tumor cells, but also trigger inflammation in the core of the tumor. Inflammation and cancer have been firmly associated for the last 10 years. Even if this connection was known by intuition since the late 1800s, solid demonstrations of molecular mechanisms behind it have been reported only recently. Nevertheless, basic antiinflammatory factors such as aspirin, and other COX inhibitors, all act somehow as good preventive drugs, but not as therapeutic agents. We have studied the inflammatory pathways associated with tumor cell invasion and metastasis, by analyzing triggers and brittle links in the chain of inflammatory events that promote cancer recurrence and metastasis. In our experiments we observed that signals through TNFα and lymphotoxin-alpha (LTα) constitute weak links in the tumor-promoting inflammatory scenario. Using gene-targeted mutations, we demonstrated that p55TNF-R blockade could reduce metastasis outcome in mice up to 50%. Likewise, LTα blockade reduces mortality in tumor-challenged and untreated mice by 10%, and 54% in mice treated with simple surgical tumor ablation. Conversely, p75TNF-R blockade increases metastasis outcome up to 200%. All taken together these results demonstrate that protumor inflammatory signals transmitted through TNF receptors are not complementary, but opposed: p55TNF-R mediates promalignancy inflammation and p75TNF-R quenches that pathway.

Among the triggers of promalignancy inflammatory mechanisms, we demonstrated, that IgGs developed against soluble and shed tumor associated antigens (sTTA) are a major trigger of protumor inflammation. We also demonstrated that by knocking out the B cell receptor (BCR), mice do not develop anti-sTTA IgGs, 90% of mice reject the tumor challenge entirely, and from the 10% that develop tumor, only 20% recur after tumor ablation.

Cloning and investigating the IgG-VH sequences, transcribed in lymphocytes and plasma cells, from bone marrow, spleen, and tumor stroma, we also observed that tumor infiltrating plasma cells produce a distinctive family of IgGs. The induction of random expression of these VH peptide sequences in mice, by in vivo transfection into muscle cells, with VH expressing vectors, reduced tumor progression in a significant manner. All these studies indicate that: (1) The use of TNF blockers (such as infliximab and adalimumab) and p55TNF-R blockers (such as lenercept) may have therapeutic benefit in oncology. (2) p75TNF-R blockers (such as etanercept) could be detrimental in oncology. (3) Active or passive immunization against sTAA, such as sTn and others, could be absolutely detrimental in cancer immune therapy. (4) Active or passive humoral immunization against membrane integrated tumor cell antigens should be carefully tested. (5) Investigation of IgG expressed in tumor infiltrating lymphoid cells, could convey important knowledge about the immune responsibility in tumor progression.

1. INTRODUCTION

We have previously insisted on the essential role of the humoral immune response to the most important traits of malignant growth as invasiveness and metastasis. Actually, antibodies raised against tumor antigens can promote tumor cell invasion. The mechanisms where by this works are multiple, but can be summarized as promoting local inflammation and tissue remodeling, including extensive angiogenesis.

Experimentally we collected evidences that G immunoglobulin (IgG) bound to specific tumor antigens, forming immune complexes can activate macrophages, which facilitate tumor cell progression through barriers of connective matrix [1]. In vivo experiments allowed us to collect consistent evidences that the presence of antitumor IgG favors tumor growth [2], as corroborated by others [3], and most important tumor cell invasion [2]. Moreover, combined in vitro and in vivo experiments showed that tumor angiogenesis is mainly supported by stroma cell-derived vascular endothelial cell growth factor (VEGF), not by tumor cell-derived VEGF, displacing the tumor cell expression of VEGF to a secondary, anecdotic responsibility in tumor growth biology [4]. Additionally, antitumor antigens IgGs promote macrophage infiltration of the tumor growth front, and consequent local VEGF paracrine release.

These results, and many reported by other researchers, are changing the conceptual core of tumor progression [5], which assumes that tumor malignant behavior is auto-fostered by the wrong behavior of tumor cells. Tumor cells carry genetic alterations that compromise the cell cycle control, changing patterns of growth and death. However, malignant tumors do not gain the qualification of "malignant" only from their vicious growth or extended cell life. They are malignant because the neoplastic tissue invades the surrounding structures and grows in wrong sites, distorting other organ functions.

Many published data show that the genetic changes, taking place in the tumor cell, concur with constellations of gene miss expressions, conductive to a variety of new cell phenotypes [6]. It is understandable that analyses of those alterations suggest that certain adequate phenotypes could have characteristic invasive and metastatic features.

Classic investigations pointed out the existence of tumor cells better prepared to invade and develop metastases (metastatic phenotype) [7]. These cell sorts can be selected and can maintain their metastatic ability in culture [8]. This concept was completed with experiments supporting the concept that specific environments are prone to host certain metastatic phenotypes [9], combining the "seed and soil" hypothesis, empirically initiated by Paget in late 1800s [10], which implies that cancer cells must find a suitable "soil" in a target organ for metastasis to occur. However, the definition of this soil and the differences between singular soils has not yet been described [11].

Typically, invasion and metastasis have been linked as two steps of the metastatic process or the metastasis cascade [12]. It is evident, that both share some common mechanism, and that metastasis development is dependent on prior successful invasion. Nonetheless, there is no vice versa. Tumors could be invasive, but emigrant tumor cells could fail in achieving metastatic growth [13], although, all metastasis are invasive.

In the classic vision of the metastasis cascade, invasion ends with tumor cell intravasation, and metastasis starts with tumor cell arrest, followed by tumor cell extravasation, which actually is an act of invasion, probably based on the same mechanisms by which tumor cells detach from the primary neoplasm and progress through the connective matrix. After that invasive act, neoangiogenesis and growth follow the same rules. Therefore, tumor cell arrest is the only singular act of the metastasis cascade that defines the metastatic phase.

Tumor cell arrest has been largely studied [14], and the most common finding is that it is governed by the same mechanisms of the inflammation-related cell arrest and tissue infiltration [15,16,17].

The fact that many released tumor cells do not form metastasis was described under the concept of "metastatic inefficiency" [18], a phenomenon attributed to diverse causes [19], including immune surveillance dependent elimination of tumor cells, before or after their ectopic arrest and colony formation [20].

Although the lack of effective tumor immune rejection in common circumstances suggests that the immune system actually ignores the tumor [21], there is old and broadly supported evidence showing that the immune system is aware of the existence of neoplastic tissue in the body. The immune system develops a humoral antitumor response characterized by the presence of antitumor antibodies in the patient's serum. Tumor infiltrating B-lymphocytes release Ig (predominantly IgG) reacting with autologous tumor targets and allogenic tumor targets of the same histology, but not against tumor targets of different histology [22]. Collectively it also develops a T cell response against the same Ags or others [23] that can even delete the tumor in experimental and infrequent circumstances [24,25]. The summarized conclusion of all these observations is that the immune system in natural scenarios reacts, stimulated by tumor antigens, but that reaction is not tumor-destructive. Moreover, classic experiments demonstrate that animals sensitized against an oncogenic virus or against certain tumors, are more susceptible to develop tumors when they are infected with the specific virus [26], or receive implants of the specific tumor [27].

This paradox can be explained by some of the natural costs of the humoral immune response: All immune responses induce local inflammation in the antigen environment, and there is accumulated experience to show that inflammatory cytokines and factors, carried by myeloid cells and lymphocytes into the tumor environment [28,29] are welcomed by the neoplastic cells and actually promote tumor growth [4], invasion and metastasis [30,31].

This immune-inflammatory tumor promoting result is not only a sustaining factor for well-established tumors; it is in fact, when chronic, a major oncogenic mechanism [32,33]. Remarkably, chronic inflammatory degenerative human pathologies, considered autoimmune diseases, such as rheumatoid arthritis and Crohn's disease, have in common bizarre humoral immune activity against tissue antigens, and myeloid cell infiltration of the damaged tissues. Actually, it is well accepted that long-term antiinflammatory treatments, similar to the ones used for autoimmune disorders, reduce the incidence of certain tumor types [34].

We have studied the influence of the antitumor humoral immune response, and the effect of the myeloid cell infiltration of the tumor environment, in the progression of the strictly malignant characteristics of neoplastic tissues: invasiveness, recurrent growth, and metastasis. These are all firmly associated with matrix degradation, tumor cell migration and angiogenesis.

Our in vitro models of tumor cell invasion showed that tumor cells alone were disappointingly poorly invasive, and truly significant invasion of synthetic matrix by those tumor cells was fully achieved only when concurred: (1) tumor cells, shedding a characteristic sTAA; (2) the specific anti-sTAA IgG; (3) macrophages; and (4) granulocytes [1]. Moreover, in vivo models demonstrate that exogenous anti-sTAA IgG supply exacerbates the progression of tumors shedding the precise TAA, promoting tumor invasion, recurrence, and metastasis, with concomitant monocyte infiltration [6].

2. RESULTS AND DISCUSSION

2.1. Tumor infiltrating macrophages contain TAA and IgGs in phagosomes

Tumor-associated macrophages (TAMs) are present in areas of the tumor tissue close to the tumor progression front and normally associated with areas of ongoing neoangiogenesis [35], in principle corresponding to hypoxic areas. TAMs are apparently attracted by low oxygen environments and respond to the levels of hypoxia found in tumors by upregulating transcription factors such as hypoxia-inducible factors 1 and 2 (HIF-1, 2), which can activate a broad array of genes, including mitogenic, proteolytic [36], proinvasive [37], and proangiogenic factors [38]. This could explain the empiric association of high numbers of TAMs and poor prognosis observed in various forms of cancer [39]. However, in human tumors, with a defined polarity such as colon cancer, TAMs accumulated close to the tumor surface were also eminent. That infiltration was present in connective matrix between dysplastic glands with apparently good blood supply. The monocyte-attracting factors of these tumor areas are not yet known [40].

We studied biopsies of human colon cancer and experimental adenocarcinomas, grown in mice, and the cytoplasm content of TAMs associated with both the tumor surface and the progressive front. In all cases the macrophage content of mucins or mucin-related antigens was prominent, as disclosed by peroxidase-conjugated B72.3-MAb, an ati-sTn monoclonal antibody (Figure 26.1). This shows that dysplastic cells are leaking mucin (i.e., sTTA) to the interstitial spaces, and that TAMs capture them. Therefore it is likely that TAMs are attracted to the tumor microenvironment, not only by hypoxia, but also by TAA concentration in the interstitial space.

Mucin phagocytosis by TAMs has a special meaning in the case of adenocarcinomas, because macrophages in contact with mucin produce cyclooxygenase-2 (COX-2), which favors local inflammation and recruitment of inflammatory cells [41], enhancing the probabilities of tumor cell invasion and growth. Therefore, it is predictable that these macrophages additionally contribute promoting tumor cell growth and invasion, through the same mechanisms observed in hypoxia-recruited monocytes.

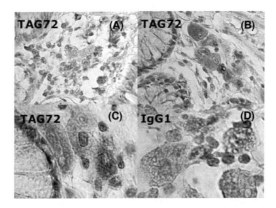

Figure 26.1. Tumor-associated antigen (TAG72, sTn antigen) and IgG phagocytosis by tumor infiltrating macrophages (TIM) in human colon cancer: (A) most frequent site of macrophage infiltration with TAG72 phagocytosis, in connective tissue, in close contact with dysplasia and early transformation. Immune staining with peroxidase-conjugated B72.3 MAb. (B) Detail of proximity between TIM and dysplasia, which cells do not contain TAG72. (C) Scavenger morphology of TIM. (D) Same group of TIM stained with peroxidase-conjugated antihuman IgG. TIM showing IgG phagocytosis.

Some in vitro experiments suggest a cytocidal role of TAMs as antitumor effectors; however, we never observed signs of tumor cell death in the proximities of TAM, but we frequently observed signs of basal membrane fading and dysplastic cell displacement towards the interstitial space.

Mucin and derivates have been shown to be potent TAA [42,43]; therefore, we tested the possibility of tumor bearing mice having spontaneous anti-sTn antibodies. All studied mice, developing an experimental isogenic adenocarcinoma, showed significant serum titers of anti-sTn antibodies, mainly IgG (not shown). Likewise, in cancer patients, the existence of serum titers antitumor antibodies is frequent [44,45].

We studied histology sections of human and experimental tumors stained with peroxidase-conjugated anti-IgG antibodies, and observed that many TAM, located in the areas with positive cytoplasm for sTn antigen, also showed positivity for IgG (Figure 26.1). This indicates that TAM could scavenge immune complexes formed with leaked TAA and anti-TAA IgG delivered in the environment, thus triggering local inflammation, modulated by the coordinate expression of activating and inhibitory Fc gamma receptors [46] (FcγR)

2.2. TAA are massively presented on follicular dendritic cells of the tumor regional lymph nodes

After the development of certain anti-TAA mouse monoclonal antibodies, multiple tumor cell targeting tests were performed. Using radio-labeled monoclonal antibodies many studies focused on pharmacokinetic analysis of antibody distribution after intravenous delivery [47]. Most of those studies concluded that tumor cell targeting was acceptable in experimental models, but in many cancer patients, tumor cells were not the main concentrators of those antibodies [48]. Most assays demonstrated that spleen and lymph nodes (LNs) concentrate heavy amounts of anti-TAA antibody [48].

That diversion of specific anti-TAA antibodies to lymphatic structures was initially interpreted as result of either the presence of tumor cells in those lymphatic structures or a sign of development of human anti-mouse antibodies (HAMA) [49]. However, histochemical staining of tumor cell markers in radio-labeled LNs failed to demonstrate the presence of tumor cells. Moreover, similar radio-labeled LNs (germinal centers) were also detected in patients with benign gastrointestinal diseases [50]. Those disappointing results opened a debate about the nature of the important concentration of radio-labeled anti-TAA in LNs lacking tumor cells [51,52].

Histochemical staining of the targeted TAA showed that the highest concentration of anti-TAA was in the germinal centers of those radio-labeled node follicles. One of us (Sampsel, J.W.), investigating human colon cancer samples, by autoradiography and histochemistry, showed that TAA coated follicular dendritic cells (FDCs) of spleen, regional LNs, and follicles of the intestinal mucosa, and was embedded in the tumor matrix. These results lead to the most acceptable interpretation: germinal centers were presenting TAA to B cells in a process of immune response and clonal selection (Figure 26.2).

To assess the influence of that massive TAA presentation, we retrieved clinical data from 15-year-old clinical trials of radio immune guided surgery [53] performed on stage 2 and 3 colon cancer patients. All these patients were classified in two groups: those in which all radio-labeled LNs were completely removed, and those in which the radio-labeled LNs were detected and recorded but not completely removed. Those data showed that radical selective lymphadenectomy of TAA-containing LNs, significantly reduced recurrence rate and improved 10-year survival of those cancer patients [54].

These conclusions indicated that the TAA presentation in germinal centers could be associated with IgG affinity maturation by somatic hypermutation of anti-TAA antibodies. These affinity-selected antibodies could be implicated in shed TAA recognition, immune complexes formation in the tumor environment, inflammatory cells recruitment, and tumor cell invasion promotion (Figure 26.3).

2.3. LTα– knockout mice show augmented tumor cell rejection, and reduction of tumor recurrence

LTα-deficient (LT-alpha–/– or LTα –/–) mice have congenital absence of LNs and Peyer's patches and defective spleen follicle structure. The splenic white pulp lack discrete T and B lymphocyte zones, and there are no FDC clusters or germinal centers (GCs) [55].

Experimentally immunized LT-alpha–/– mice with sheep red blood cells (SRBC) produced high levels of antigen-specific IgM, but no IgG in either primary or secondary responses, demonstrating failure of Ig class switching.

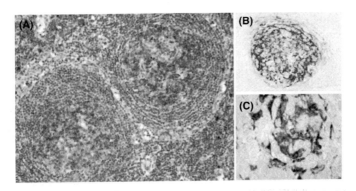

Figure 26.2. TAA (TAG72) presentation in germinal centers of tumor regional lymph nodes: (A) hyperplasic follicles with germinal centers showing intense peroxidase immune staining of TAG72 (with peroxidase-conjugated B72.3 MAb). (B and C) Details of TAG72-positive staining on follicular dendritic cells.

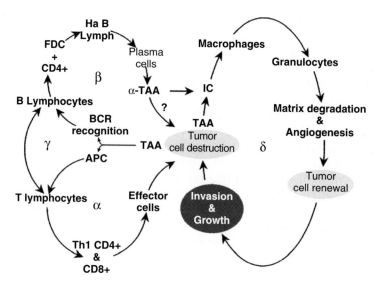

Figure 26.3. Diagram of the hypothetical tumor invasion agonistic and antagonistic effects of the antitumor immune response. T cells respond in a cyclic manner to exposed tumor-associated antigens (TAA), being activated and destroying tumor cells (tumor antagonistic alpha cycle). B cells respond in a cyclic manner to exposed tumor-associated antigens (TAA), being activated, producing antibodies that could destroy tumor cells (tumor antagonistic beta cycle). B and T interactions (gamma cycle) could either enhance tumor rejection or induce tolerization and quenching the alpha and beta cycles, becoming a tumor agonistic cycle. Anti-TAA antibodies (α-TAA), with TAA can form immune complexes (IC) in the intercellular space initiating and maintaining an inflammatory cycle (delta cycle) with evident tumor agonistic consequences.

The altered LN microenvironment characteristics of these mice results in impaired ability to switch to a productive IgG response [56].

Furthermore, these mice have dramatically impaired ability to generate high-affinity IgG, which reinforces the concept that germinal centers and FDCs antigen presentation are essential for affinity maturation [57].

To investigate the influence of the germinal center TAA presentation in promoting tumor progression we performed tumor growth and recurrence evaluation in LTα –/– mice, in which lymphocytes are normal but the germinal centers are absent.

A group of 6 LTα –/– mice were injected with 105 B16F10 melanoma cells into the spleen and sacrificed 7 days later. Seven days after tumor cell injection, livers from LTα –/– mice showed no metastases. One hundred percent of the liver sections did show only important lymphocyte clusters accumulated in the subendothelium of the portal venules (Figure 26.4). The same lymphocyte clusters and in the same frequency and size were observed in control LTα–/– mice, not challenged with tumor cells.

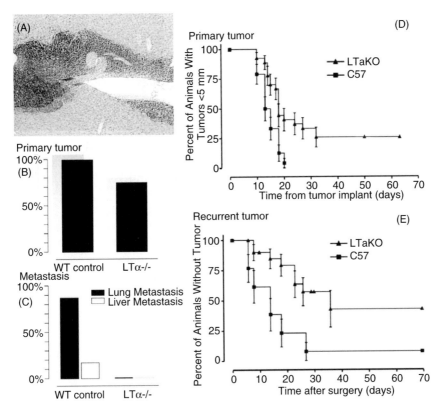

Figure 26.4. Tumor and metastasis development and in lymphotoxin alpha genetically deficient mice (LTα–/–): (A) Common lymphocyte granulomata developed in tumor challenge LTα–/– mice liver, without evidences of tumor cells. (B) Compared with normal mice, only 75–80% of LTα–/– mice developed tumor. (C) Compared with normal mice, LTα–/– mice did not develop either lung or liver metastasis in identical experimental circumstances. (D) Compared with normal mice, primary tumor development was slower in LTα–/– mice. (E) Compared with normal mice, recurrent tumor development was slower in LTα–/– mice, and in near 50% did not recur.

In order to increase the chance of metastasis, an additional group of 6 WT mice and one group of LTα –/– mice received 5× 107 B16F10 melanoma cells into the spleen and were sacrificed after 7 days. All WT mice showed visible macroscopic metastasis on the liver surface, but not the LTα –/– mice. Under the microscope, 100% of the WT mice livers did show increased number of metastases compared with the controls with low dose of tumor cells. However none of the LTα –/– livers showed metastases.

As an additional experiment, two groups of six mice each (WT and LTα –/– mice) were challenged with 106 Lewis lung carcinoma (3LL) cells injected subcutaneously in the flank. We selected this tumor because it is a lung-derived metastatic carcinoma, and because it expresses sTn antigens. After 2 weeks the subcutaneous tumors formed were surgically excised and the wound was properly treated. All mice were sacrificed and lungs and livers were explored for macroscopic and microscopic metastases after 15 days. Lungs of 80% of the WT mice showed metastases, and only 20% of them had liver metastases. However, no metastases were found in livers and only one metastasis was found in lungs of LTα −/− mice (Figure 26.5).

In specific experiments for tumor recurrence, groups of control mice (WT) and LTα −/− mice were injected with 5×10^5 Lewis lung tumor cells subcutaneously. Tumor development was followed, measuring tumor diameters daily. One hundred percent of control mice developed tumors of 5 mm diameter in less than 21 days. However, tumors developed slower in LTα −/− mice, and 25% of them never developed tumor (Figure 26.4A).

To test the rate of recurrence of those tumors in both host types, groups of both WT and LTα −/− mice were challenged with 10^6 tumor cells, injected subcutaneously in the left flank at 8 mm distance from the spinal cord and the hip. When the tumors were 5 mm in diameter they were carefully excised with wide margins, removing a skin patch of 10 mm diameter from the tumor center.

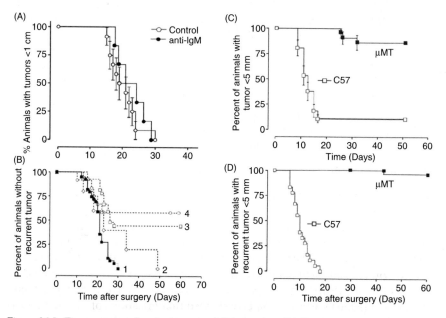

Figure 26.5. Tumor progression in immune deficient mice. (A) Compared with normal mice, B-lymphocyte depleted mice (repeated treatment with anti-IgM antibodies) were equally susceptible to primary tumor development. (B) Tumor recurrence frequency and progression in control immune competent mice (1), IgM-induced B cell-depleted mice (2), RAG−/− mice (3), SCID-beige mice (4). (C) Tumor development kinetics in μMT/μMT mice (μMT), compared with normal immune competent mice (C57). (D) Tumor recurrence kinetics in μMT/μMT mice (μMT), compared with normal immune competent mice (C57).

Subjacent aponeurosis was carefully cauterized with an electric cautery and, finally the skin sutured.

The evolution of the wound was followed in all mice. In the cases where tumors recurred locally, mice were sacrificed when tumors reached 10 mm in diameter, and autopsied. In other cases, 63 days after surgery all mice were sacrificed and autopsied.

Close to 50% of the WT mice developed local recurrence of more than 1 cm in less than 30 days, and 100% in 50 days. However, 50% of the LTα $-/-$ mice developed local recurrence in 80 days, and 50% never developed either local recurrent cancer or metastases (Figure 26.4B). Moreover, the recurrent tumors showed slower growth in LTα $-/-$ mice compared with controls (Figure 26.4C).

These results indicated that in hosts without LTα production, tumor cell invasion and tumor recurrence is less successful. The defect in tumor metastasis or local recurrence shown in LTα $-/-$ mice can be attributed to an IgG partial deficiency, caused by the absence of germinal centers and by B lymphocyte functional failure, because it has been demonstrated that LTα + lymphocytes are needed to support the development of strong secondary and memory IgG responses [58]. However, immunoglobulin affinity maturation is not totally suppressed in these LTα $-/-$ mice [59], and this could be the explanation for the partial protection against tumor growth and recurrence observed in these mice.

Other potential explanation to the tumor progression impairment shown in LTα $-/-$ mice, excluding the LNs defect, is the fact that LTα is an inflammatory cytokine [60] that shares receptors and is close in function to TNFα [61]. Therefore, the observed decreased tumor progression in LTα $-/-$ mice could be either secondary to incomplete B cell immune response or, alternatively, due to deficiencies of the particular LTα–dependent inflammatory response [62], or could be both.

2.4. B cell-deficient mice show reduced tumor growth and recurrence

Taking into consideration that LTα $-/-$ mice do not develop germinal centers and the B cell response against the tumor could be altered, and that tumor infiltrating macrophages contribute to the tumor progression in association with antitumor humoral immune response, we decide to further investigate the B cell involvement in tumor progression. We tried to reduce in normal mice the B cell response ability to the tumor challenge by treating them with high doses of anti-IgM to induce considerable B cell apoptosis [63], beginning the day after tumor challenge. Flow cytometry showed that injections of anti-IgM antibodies reduced the circulating B220+ cell population in all treated mice up to 70%. However, no significant effect was observed on tumor evolution (Figure 26.5A). Only a significant delay in recurrence progression was evident (Figure 26.5B).

Taking into account our LTα $-/-$ mouse model of tumor progression suggested, but did not clarify, the involvement of the B cell response to TAA, and clonal selection, in promoting tumor progression, we decided to investigate the tumor recurrence dynamics in µMT/µMT mice [64], which lack BCR.

In the absence of BCR, B cells do not respond to antigens and, therefore, ignore and do not present antigens, including TAA in tumor cell implantation experiments. Although this mutation does not involve either T cell function and or tumor cell rejection associated with antigen presentation in MHC context, it collaterally impairs T cell functions [65]. In these circumstances (total absence of humoral immune response, and feeble cellular immunity), antitumor immune surveillance should be weakened [66], and tumor progression should actually be enhanced.

As expected, tumor tissues of those BCR-KO mice did not show immunoglobulin embedding the matrix. However, against the logic of weakened immunosurveillance, experiments of tumor implant, tumor ablation, and tumor recurrence, showed that subcutaneous tumor only grew in 10% of the animals, compared with 100% in animal controls, injected with identical quantities and class of tumor cells (Figure 26.5C), and from that 10%, recurrence after tumor ablation was only 20% (Figure 26.5D).

Similar results were obtained by others experimenting with SCID mice with reconstituted T cell function by T cell transplantation. Those experiments also showed that tumors grew more slowly in and were rejected more frequently by the mice lacking B cells [67].

Recently, using a model of genetic elimination of mature B lymphocytes in a transgenic mouse model of inflammation-associated de novo epithelial carcinogenesis, it has been shown that neoplastic progression to development of epithelial hyperplasias fail to recruit innate immune cells in the absence of B lymphocytes. Moreover, adoptive transfer of B lymphocytes or serum from normal tumor bearing mice into B cell-deficient mice restores innate immune cell infiltration into premalignant tissue and reinstates chronic inflammation, neoangiogenesis, hyperproliferative epidermis, and finally, full malignancy [68].

This model, like we described before, links tumor progression to inflammatory cell infiltration, and implicate B cells and antibodies as the inflammatory cell recruiters. Therefore, supports the concept that B-lymphocytes are required for establishing chronic inflammatory states that promote tumor progression, including de novo carcinogenesis.

We did additional experiments to demonstrate the direct implication of tumor-primed B cells in promoting tumor progression. Met-129 mammary tumor cells were subcutaneously injected in isogenic immune competent mice. Using immunomagnetic procedures, CD19+ lymphocytes, or B220+ Lymphocytes 95% pure concentrates were obtained from either normal mice spleens, or from Met-129 growing tumors. One hundred thousand cells of either a mixture of 80% tumor cells and 20% spleen-derived B cells, or 80% tumor cells and 20% tumor-derived B cells were tail vein injected in three groups of full immune competent mice. Mice, which received tumor cells with normal lymphocytes, developed the same amount of lung metastases that mice injected with the same dose of tumor cells alone. However, all mice receiving tumor cells with tumor-derived B-lymphocytes developed significant more lung metastases (Figure 26.6).

The tumor growth timing of those last experiments (only 14 days evolution), indicated that tumor-primed B cells act directly as promoters, and not through a systemic immune system tolerization, caused by B cell TAA presentation in the priming phase, that could result in disabled CD4+ T cell help for CTL-mediated tumor immunity, as claimed by others [69]. Therefore, independently of other side effects [70] on the antitumor combined immune response, TAA reacting B-lymphocytes show the ability of promoting tumor growth as direct triggers of focal inflammation in the tumor core.

Actually, absence of CD4+ cells has been associated with decreased tumor growth [71] and metastasis [20]. However, this observation has been mainly explained as a consequence of T regulatory cells (Treg) [72,73,74] reduction, while those are CD4+ cells. To this concept contributed the fact that although cancer patients can show significant numbers of CD8 and CD4 T cells with specificities to tumor antigens, in most cases, such T cells fail to eradicate the tumor in vivo. Studies on T cell populations supported the explanation that the Ag-specific CD4(+)CD25(+) regulatory T cells interfere with the tumor-specific CD8 T cells, involving TGF-beta [75].

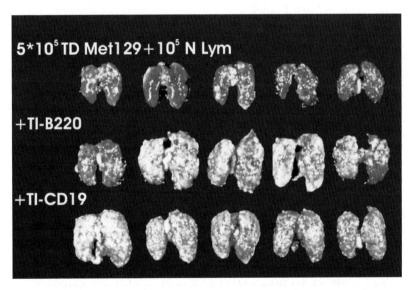

Figure 26.6. Transmitted predisposition to tumor metastasis by B cell transplant. Lung metastases developed in mice transplanted with normal B-lymphocytes obtained from normal donor mice (*upper* row), compared with lung metastases developed in mice transplanted with B-lymphocytes obtained from identical tumors grown in isogenic mice. B-lymphocytes were selected either by anti-CD19 bead-conjugated MAb (TI-CD19), or by anti-B220 bead-conjugated MAb (TI-B220). B-lymphocytes were administered by intravenous injection (tail vein) mixed with tumor cells in proportion of 1:5 lymphocyte/ tumor cells.

That explanation depends on the T cell function pondered in the tumor rejection context, but does not consider the tumor promotion effects observed in multiple experiments. Only if promotion is considered equal to lack of rejection could that explanation be completely acceptable. However, the inflammatory events in the tumor can either destroy tumor cells or actually promote tumor growth through neoangiogenesis, proteolysis, and many other inflammation-associated mechanisms. Without discharging the CTL possible impairment due to CD4+ Treg cells, we consider foremost that CD4+ T cells promote B cell proliferation [76], are the major contributor to B cell selection and isotype switching, and that CD4+ reduction inhibits B cell differentiation [77]. In this context our experiments of tumor recurrence in CD4-KO mice and SCID models can be well explained.

Using the same model of tumor recurrence evaluation, we tested the tumor behavior in different T cell deficient mice. Tumor growth was slower in CD4-KO mice, with unaltered CD8+ population, than in controls, and more than 50% of those mice never developed recurrence in 60 days experiments, compared with controls, which showed a 100% recurrence in less than 20 days (Figure 26.7). These results could be interpreted as a noninhibited CD8-mediated CTL function against tumor cells. However, practically identical results were obtained in experiments of tumor recurrence that we executed on RAG−/− and SCID-beige mice, in which all CD4 T, CD8 T, and B cell functions are absent (Figure 26.5). The common factor of all these experiments was an impaired B cell function, independently of the CTL activity, which is present in CD4−/− mice, absent except for NK activity in RAG−/− mice, and completely absent SCID-beige mice, and all of them showed 50% probability of recurrence versus 100% in full immune competent mice. However, experiments in which only B cell population is affected, tumor recurrence and B cell potential correlated (Figure 26.4B). In consequence we cannot discharge either the hypothesis that reduced tumor growth in CD4-KO mice is due, at least in part, to a handicapped B cell selection in the germinal center or in other lymphatic structures, a proposition also sustained by others [78].

All these observations reiterate the indication that a B cell system based autoimmune response against antigens presents in the tumor environment causes an inflammation cascade that promotes tumor growth, invasion, and metastasis. Like in most autoimmune diseases the nature of the antigens and antibodies involved in cancer pathology is elusive.

2.5. Tumor infiltrating lymphoplasmacytic cells produce a distinctive family of VH-IgGs

Some studies have focus on understanding the nature of TAAs that stimulate the production of specific antitumor antibodies. Analysis of human systemic immune responses against cancers allowed the identification of a number of TAAs [79]. The existence of these antigens indicates that cancers can be immunogenic.

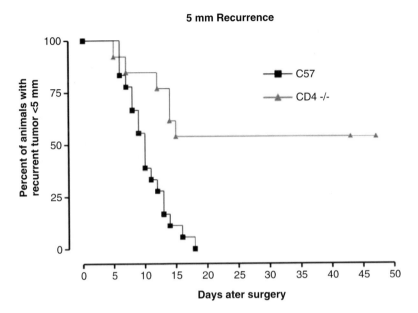

Figure 26.7. Tumor recurrence progression in CD4+ KO immune-deficient mice compared with normal mice.

Furthermore, tumor-associated antibody responses and circulating autoantibodies against tumor antigens have also been described in cancer patients [80,81,82]. The nature of these autoantibody responses is not fully understood, and a major concern exists about whether these autoantibodies directly relate to an immune response against the cancer process. Close molecular and clonal analysis of antibodies, produced by tumor infiltrating lymphoplasmacytic, have shown that there is a truly antitumor humoral immune response, and that this response is oligoclonal; therefore, indicating a response generated by specific antigens [83]. Curiously, some of the identified TAAs inducing autoantibody production are normal proteins, such as actin released and exposed during cell apoptosis [84] and huntingtin-interacting protein 1 (HIP1) [85], a nuclear protein, androgen receptor regulator [86]. However, whether these antigens are responsible for such damaging inflammatory responses remains unknown.

With the aim of identifying antigens and autoantibodies involved in this tumor-promoting immune response, we followed a new approach. We attempted to generate in the hosts an indirect anti-idiotypic response against antitumor autoantibodies produced in tumor bearing animals.

In a preliminary, highly hypothetical, experimental scheme we forced in hosts the expression of VH sequences (idiotypes) from a library obtained from tumor infiltrating lymphoplasmacytic cells. Hypothetically, the hosts will generate antibodies against these peptides (anti-idiotypes) as observed in humans after injection of antitumor antibodies [87], and these could downregulate the autoimmune response [88]. The rates of tumor growth and recurrence will be the preliminary indicators of a possible autoimmune control through these specific VH idiotypes.

Groups of mice received 3LL carcinoma cells subcutaneously, and developed tumors. When tumors grew up to 1 g they were dissected, the necrotic areas discarded, and the RNA extracted. Spleens of the same mice were also removed, the RNA extracted, and treated separately. PoliA RNAs were purified, and the hyper variable IgH V(D)J region was amplified by reverse transcriptase-PCR, using a primer that hybridizes to the third framework region of the immunoglobulins V segment and a second primer that hybridizes to the consensus sequence of J segments.

The VH cDNA segments were cloned and a library of VH sequences was made. This library was divided in four segments: From mice group #1, two segments corresponding to tumor-extracted RNA (T#1), and spleen-extracted RNA (S#1). From mice group #2, the same criteria were followed, and the sequences were classified according to the RNA source as T#2 and S#2 (Figure 26.8).

From that library a total of 25 clones were randomly selected from each segment. The total 100 cDNA clones were sequenced and clustered by phenetic distances [89], based on length and sequence of the reading frames (indirect aminoacid sequence).

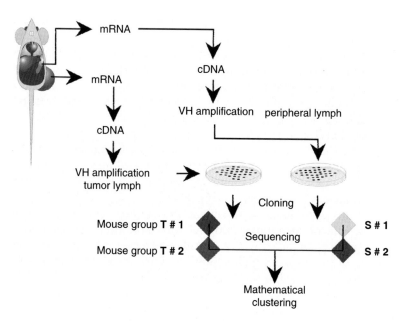

Figure 26.8. Diagram of the procedure followed to obtain and sort cloned VH sequences from tumor infiltrating lymphoplasmacytic cells, and from spleen lymphoplasmacytic cells.

Four clear clusters appeared (Figure 26.9) and from them, the most intriguing was one that contained a major proportion (>92.8%) of tumor derived sequences (Figure 26.9, C1, red and blue arrows). From these tumor-derived sequences 3 and 2 were hyper variations of two clones, but the others appeared to be originated by separated clones. The phenetic distance between these and the other clusters were high enough to consider that they constituted particular clones associated with the tumor infiltrating lymphoplasmacytic cells.

These sequences were selected and individually installed in pTracer vectors (Invitrogen), which are mammalian expression vectors engineered to constitutively express Cycle 3 GFP, an optimized version of green fluorescent protein, which allow identification of transfected cells 24–48 h posttransfection.

Vectors containing those sequences were combined in equal proportions and injected into the Tibialis anterior muscles of a group of mice, following a procedure designed to achieve a productive transfection of muscle cells [90]. As

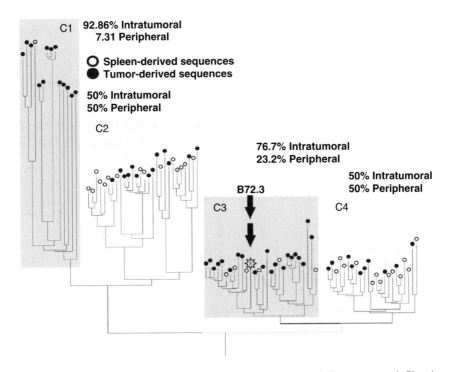

Figure 26.9. Phenetic dendrogram of VH sequences extracted from tumor infiltrating lymphoplasmacytic cells and spleen lymphoplasmacytic cells from tumor bearing mice. Solid circles represent tumor-derived clones and empty circles spleen-derived clones. The 100 sequences tested were clustered in four major groups. C1 and C3 have the majority of tumor-derived sequences. C2 and C4 have the same proportion of both.

controls, an additional group of mice were identically treated, but with pTracer vector without VH sequence. The injections were repeated every 4 days for better immunization [91] and after 7 days, two randomly selected mice were sacrificed, and sections of muscle observed under fluorescence microscopy to assess the GFP expression level.

After 20 days of the first injection, mice from a control and test groups, from which the random sample showed high expression of GFP, were challenged with 106 3LL cells, injected subcutaneously. Tumor growth was monitored by external measurement in control and test groups.

Two subgroups of control and test mice were separated when the tumor reached 5 mm diameter. In these mice tumors were removed, and the tumor recurrence process monitored.

The tumor progression study showed significant tumor growth retardation in animals transfected with tumor-derived VH sequences compared with control mice (Figure 26.10A). In addition, recurrence experiments showed a drastic reduction of tumor reappearance in mice transfected with tumor-derived VH sequences (Figure 26.10B).

These studies, although preliminary, indicate that the hypothesis of existing antitumor autoantibodies involved in promoting tumor progression, like in other autoimmune processes, could be real.

2.6. Tumor-infiltrating plasma cells may be specific and tumor auto supply of immune globulins

As described above the tumor regional LNs contain plasma cells and germinal centers, where TAA are massively presented on FDCs, suggesting that these

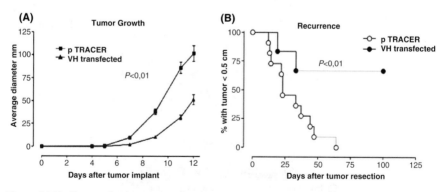

Figure 26.10. Tumor development and recurrence progression in mice transfected with VH sequences isolated from tumor infiltrating lymphoplasmacytic cells, compared with normal mice. (A) Kinetics of tumor growth. (B) Progression of tumor recurrence development. Controls (P-Tracer) were mice-transfected with the vector alone or with an irrelevant sequence. Test (VH-tranfected) was mice-transfected with the vector loaded with a set of VH sequences isolated from tumor infiltrating lymphoplasmacytic cells.

LNs could be supplying IgGs, sustaining a chronic inflammatory induction in the tumor core, where TAA is leaking into the connective matrix [92].

Many tumors show significant quantities of infiltrating plasma cells containing IgG that react with TAA [93], particular IgG-RNA sequences [94], and IgG that reacts with both TAA [95] and normal cell antigens [96].

In previous experiments we injected subcutaneously breast carcinoma cells expressing a specific TAA (sTn), together with 10% myeloma cells secreting monoclonal anti-sTn IgGs. Results of those experiments showed a spectacular increase of tumor growth and metastasis, and a mixture of carcinoma cells and myeloma cells in primary tumor and metastasis (not shown). These results reinforced the hypothesis of specific IgG-mediated tumor promotion, and highlighted the possibility that tumor-infiltrating plasma cells could be involved in such tumor-promoting effect in a paracrine manner.

In consequence, we studied statistical significant numbers of human colon carcinoma biopsies, H&E stained and immune-stained, in which infiltrating plasma cells were counted and topographically localized with coordinates.

Plasma cells were found in nonrandom distribution. Concentrated in limited areas of the tumor, showing spots of very high incidence (Figure 26.11) in association with small capillaries, lymphatic and hematic, most of them with red blood cells inside, suggesting that they were permeable.

A bird's eye view analysis of the plasma cell concentration areas, showed two sorts of them: one in the vicinity of dysplastic to neoplastic transformation (Figure 26.11 stars), and another located in the submucosa, opposite to the tumor invasion fronts (Figure 26.11 circles).

A close analysis of those plasma cells showed a trend of association: plasma cells are surrounded by lymphocytes and eosinophils, combinations compatible

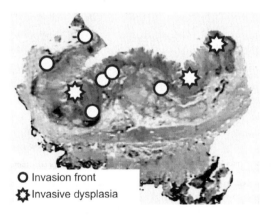

Figure 26.11. Map showing locations of highly concentrated plasma cell infiltration in a section of human colon cancer. Most of these clusters are located in the tumor invasion front (circle). Other typical location is the areas of dysplasia, in transformation to invasive tumor (stars).

with Th2 microenvironments [97] (Figure 26.12A). Completing that scenario was fibroblastic activity, and all in the near vicinity of capillaries and tumor cells, which were protruding through the basal membrane into the connective matrix (Figure 26.12A). That cellular association in neoplasia was especially evident in rare sclerosing tumors [98,99] and has not commonly reported in the most frequent tumor types except for Hodgkin lymphomas [100]. However, it was broadly repeated in all colon carcinomas studied.

Immune histochemistry staining studies showed the existence of TAA embedding the matrix in those spots, and frequently coating plasma cells (Figure 26.12B), together with IgGs.

The reiteration of this portrait in all tumor sections studied, pointed out the existence of a cellular structure defining foci of invasive dysplasia. These focal invasive dysplastic points could represent sites of antitumor invasion immune response, stimulated by the local presence of concentrated TAAs. The unanswered question remains as to which is the result of that focal confrontation.

Some experimental investigations insinuated that plasma cell infiltration could lead to tumor growth promotion [101]. A more general confirmation of

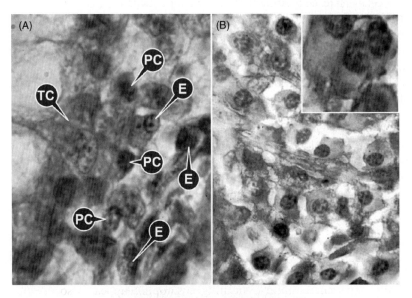

Figure 26.12. Aspects of tumor plasma cell infiltration. (A) Hematoxylin and eosin attaining. Image of focal invasive dysplasia, with the triad of components: tumor cells (TC) protruding into the connective matrix underneath the basal limit, plasma cells (PC), and eosinophils (E). (B) Peroxidase-conjugated antitumor-associated antigen (TAG72) MAb (B72.3) staining. Matrix embedded with tumor-associated antigen and plasma cells coated with tumor-associated antigen. Detail of plasma cell coated with tumor-associated antigen in *top-right* close caption.

this hypothesis will indicate that the above-mentioned confrontation terminates in favor of the tumor cell progression. However, some opinions, based on clinical data, sustain that the better clinical outcome of the medullary carcinoma of the breast is due to the presence of a prominent lymphocytic and plasma cellular infiltrates in the tumor stroma. These scientist suggest that in these tumors certain B cell clones were expanded, and transformed in plasma cells, possibly in response to specific tumor-associated stimuli, and that controlled the tumor progression, based on specific production of antitumor IgG [102,103]. On the other hand, broader studies on clinical cases of a variety of breast pathology, associate heavy plasma cell infiltration with malignancy [104]. Other studies on normal breasts, and in patients with hereditary high risk of breast cancer, have associated lymphocyte and plasma cell infiltration with premalignant alterations, which are prone to malignant transformation [105].

Our observations on a significant number of these points of focal invasive dysplasia showed no signs of cell death in any case, and a study of evolutive images clearly indicate tumor progression. The fact that in these points of tumor progression is where plasma cells appear concentrated, added to results of our in vitro experiments, in which cultured tumor cells expressing a specific TAA increased proliferation when the medium was supplemented with the specific anti-TAA IgG [2], strongly suggest that infiltrating plasma cells can participate in a process of tumor promotion.

Moreover, the constant coincidence of eosinophils in these points of focal invasive dysplasia reinforces the idea that these are points of localized inflammation, similar as those of progressive tissue remodeling, existing in autoimmune diseases such as rheumatoid arthritis, where plasma cells infiltrate [106] and even proliferate [107] in the active inflammation sites.

The inflammatory role of tumor infiltrating eosinophils is also controversial. Some medical studies in gastric cancers suggested that tumor eosinophilia is sign of better clinical evolution [108]. However, later studies on cervical cancer tinged that assertion, explaining that infiltrating eosinophilia is associated with a good prognosis only if there is not peripheral blood eosinophilia, otherwise tumor infiltrating is a sign of fatal prognosis. Moreover, in these studies it was observed that infiltrating eosinophils were constant in progressive metastases [109]. Lately, studies on carcinomas of larynx [110], vulva [111], and cervix [112], showed that infiltrating eosinophilia is definitively associated with tumor cell invasion.

In our experience, tumor-infiltrating eosinophils appear associated in the triad "invasive tumor cell–eosinophil–plasma cell" at the focal invasive dysplasia points. Where eosinophils are located, frequent signs of degranulation, matrix disorganization, and cell death, involving mainly plasma cells and rarely tumor cells, are regularly observed close by. This reinforces the hypothesis that this set of cells, in a Th2 cytokine environment, in a matrix embedded of TAA and IgG, may very well signify units of tumor invasion and progression. However, the fate of these units cannot be predicted without the precise knowledge of the cytokine microenvironment.

Eosinophil survival and degranulation is importantly influenced by the immune globulin involved in their activation (IgA or IgG), the Fc gamma RII status [113] and the cytokine microenvironment. The presence of IL-5 and TNFα [114] promote degranulation and attract eosinophils, and the IFNg presence, inhibits such degranulation [115]. Moreover, eosinophil degranulation may have either a relaxing (favoring tumor cell invasion) or a cirrhotic (impairing tumor cell migration) effect on the local matrix, depending on the presence or absence of TNFα[116].

The presence of TNF in the microenvironment could be suspected because the presence of plasma cells, which need TNF for proliferation and differentiation [117]. Moreover, TNF could be released by infiltrating Th2 cells, macrophages of the stroma, the same plasma cells [118], as an effect of the inflammatory environment, and possibly by tumor cells, all present in the focal invasive dysplasia. Therefore, the suspicion that those structures represent points of tumor progression becomes more prominent. And based on this, TNF may acquire a more important role.

2.7. Compared with normal mice, TNFR55–/– mice show reduced liver metastasis efficiency

With an immune and inflammation-altered mouse model of liver metastasis, we explored the tumor cell survival and growth in a "soil" partially insensitive to the TNF and LTα signaling. For this model, tumor cell were injected in p55 TNF-alpha receptor (TNFR1) and p75 TNF-alpha receptor (TNFR2) knockout C57BL mice (TNFRp55–/– and TNFRp75–/–, respectively). Results were compared with experiments executed in normal C57BL mice (CTRL).

The altered host tissues should have a distorted response to both cytokines, either released by tumor cells or by local or infiltrating host cells. Mice were injected with 105 B16F10 melanoma cells into the spleen and sacrificed 7 days after. Most, but not all mice showed visible macroscopic metastasis on the liver surface. Under the microscope, 36% of the livers of the wild type mice (CTRL) did not show metastases, only small granulomata without manifest living tumor cells. The other CTRL livers showed one or more, but not over five, manifest metastases per square centimeter of liver section (Figure 26.13A).

We evaluated the incidence of metastases in TNF-Rp55–/– and CTRL mice as indices of tumor rejection/promotion. Six days after tumor cell injection, livers from TNFR55–/– mice and CTRL mice showed very similar results under microscopy: about 40% of the liver sections did not show more than granulomata and the rest showed 1–3 overt metastases per cm^2 of liver section (Figure 26.132A). Statistical test of differences (F test to compare variances, $N = 35$ per group) in terms of number of metastases per square centimeter, showed a sig-

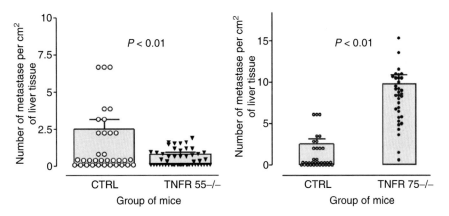

Figure 26.13. Liver metastasis frequencies in TNF-receptors-deficient mice, after intrasplenic tumor cell delivery. Six normal mice (WT) and six genetically target-mutated TNFRp55–/– mice (lacking TNFR1), and six TNFRp75–/– mice (lacking TNFR2) received identical dose of tumor cells per intrasplenic injection. Livers were removed after identical periods of incubation, frozen and from each liver six cryosections were obtained and stained. Number of metastases was counted in a total of 36 sections per group. (A) TNFRp55–/– mice developed less metastasis than controls ($p < 0.01$). (B) TNFRp75–/– mice developed more metastasis than controls ($p < 0.01$).

nificant ($P < 0.0001$) contrast: TNFR55–/– mice developed less metastatic foci than WT mice. These observations established that tumor cell survival and growth in the host is decreased in the absence of TNFR1, therefore, by extrapolation, the tumor rejection is better when the TNF ligation to the p55 receptor is blocked in the host, a concept shared by others [119].

Several explanations could be applied to these results: TNF-Rp55–/– mice have been shown to lack networks of mature FDCs and they do not form germinal centers [120]. In consequence, it has been shown that with no replicating Ags, IgG titers are inefficiently induced, and not maintained in those animals. Moreover, viral infection experiments demonstrated that TNF-Rp55–/– mice immunized with Ags, without adjuvant, induced only IgM but no IgG Abs; only Ag emulsified in CFA or IFA induced IgM and IgG responses that were short-lived and of moderate titer. This would explain the reduction of tumor progression by the decreased of antitumor IgG maturation, as discussed above. Moreover, experiments with immunization of TNF-Rp55–/– mice also showed that infection with live viruses induced excellent neutralizing IgM and IgG responses in these mice [122]. These findings suggested that increased Ag dose, and time of Ag availability, could substitute for FDC-stored Ab-complexed Ag in the induction of efficient IgG responses in TNF-Rp55–/– mice without germinal centers. This could explain the partial tumor reduction (only 50%) observed in mice devoid of classical germinal centers, which could still have some antitumor IgG, compared with B cell KO mice, which do not have any.

Other explanations could be impairment of other inflammatory actions of TNF, driven through the TNFR1, associated with the previously described structure of focal invasive dysplasia. For example, distortion of focal inflammation [121], cellular damage [122,123], involving tumor cells, endothelial cells, and others, can explain changes in focal tumor progression. Additionally, failure of granulocyte recruitment [124,125], including eosinophils [126,127], to the inflammatory foci [128], plasma cell maturation, and Th2 environment promotion, could act reducing tissue damage and consequent enhanced remodeling or healing stimuli [129].

2.8. Compared with normal mice, TNFR75–/– mice show increased liver metastasis efficiency

An additional group of TNF-Rp75–/– mice were injected with 105 B16F10 melanoma cells as well, into the spleen, and sacrificed 7 days after. All mice of this group showed visible macroscopic metastasis on the liver surface. Microscopic explorations showed that 100% of the livers of these mice had overt metastases, and the sinusoid paths were dilated showing a loose image of the liver structure with frequent accumulations of nucleated cells inside the terminal portal venules, which were never observed in the WT livers.

We compared the incidence of metastases in TNF-Rp75–/– versus WT mice as indices of tumor rejection/promotion. Microscopic examination of the liver sections, 6 days after tumor cell injection, showed that 100% of the livers from TNFR75–/– mice had more than one obvious metastasis per square centimeters of liver section (Figure 26.13B). Statistical test of differences (F test to compare variances, $N = 35$ per group) in terms of number of metastases per square centimeters, showed a significant ($P < 0.0001$) contrast: TNF-Rp75–/– mice developed many more metastatic foci than WT mice. These observations established that tumor cell survival and growth in the host is increased in the absence of TNFR2. Therefore, by extrapolation, the tumor promotion is better when the TNF ligation to the p75 receptor is blocked in the host.

The discussion of these results should emphasize that both TNF receptors are not redundant in the case of TNF influence on tumor progression. Signals conducted through each one will have opposite effects (Figure 26.13 A and B).

In other autoimmune disorders, such as lupus erythematosus, TNFR2 polymorphisms are associated with susceptibility to develop this disease [130] showing that different TNF sensitivity of this receptor influences the severity of the autoimmune responses. Here we observed that absolute lack of signal through this receptor increases dramatically the severity of the metastatic outcome.

As it has been shown, TNF signals through TNFR1 receptor have a role in cytotoxicity, whereas through TNFR2, regulates death responses or proliferation. In these experiments of tumor metastasis on TNF-Rp75–/–, both deficiencies, in cell death and in cell proliferation control, may explain the exaggerated rate of metastases. Moreover, whereas TNFR1 receptor activates proinflammatory transcription nuclear factor-kappaB (NF-kappaB) [131], TNFR2-activating role of this factor is poor [132]. Therefore, the level of inflammation was maintained through the TNFR1, causing tumor promotion. Even more, it has been shown that in the absence of TNFR2 the antimycobacterial response and granulomas formation, which are accumulation of chronic inflammatory cells and products, are stronger [133].

Therefore, these results indicate that in the presence of both receptors, TNF signals delivered through the TNFR1 are conductive to inflammatory specifics that promote cancer cell invasion, but TNF signal delivered through TNFR2 somehow quenches the TNFR1 pathway. Thus, if in normal circumstances the interplay of both receptors yields a certain level of tumor promotion, in cases of TNFR2 failure that yield will be higher, because only TNFR1 will act without constrains. In the case of TNFR1 failure, TNFR2 does not operate, and the level of tumor promotion will be lower because the lack of effective TNF stimuli (Figure 26.14).

3. CONCLUSION

Here we collected data indicating that tumor cell invasion and malignant progression could be favored by inflammatory reactions happening in the tumor microenvironment, and that secondary B cell immune response against sTTA, released by tumor or dysplastic cells, could be one of the triggers, or drivers, of that tumor inflammatory process. In favor of this last hypothesis are: (1) Macrophages infiltrating the tumor microenvironment are frequently loaded with TAA and IgG, signs compatible with attraction by, and removal of, immune complexes formed with both components. (2) Tumor connective matrix is embedded of both TAA and IgG, data compatible with extracellular release of TAA, infiltration with IgG and consequent immune complex formation, if there is affinity between both components. (3) Tumor regional LNs present tumor TAA massively on FDCs, and show B cell hyperplasia, data compatible with a strong B cell secondary response against TAA. (4) the genetic absence of germinal centers in tumor-bearing animal, and consequent reduction of secondary immune response, concurs with slower tumor growth and 50% reduction of tumor recurrence. This is compatible with the concept that IgG generated against TAA can fuel tumor inflammation and promote tumor growth and recurrence. (5) The genetic absence of BCR, and consequent complete absence of IgG and secondary B cell response, concurs with higher resistance of tumor development, and dramatic 80–90% reduction of tumor recurrence. (6) B cell transplantation from a tumor-bearing animal to an isogenic animal, challenged with the same tumor, accelerates tumor growth and increases metastasis yield. (7) Ig mRNA obtained from tumor infiltrating lymphoplasmacytic cells has a particular signature of VH sequences. Some VH sequences are minimally found in peripheral lymphoid tissues. This is compatible with a local, intratumor, and

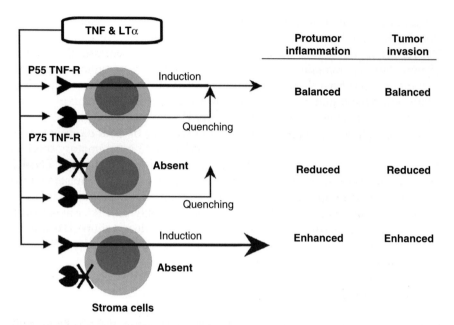

Figure 26.14. Diagram of hypothetical explanation of the differences in metastasis development associated with the malfunction of either TNFR1 or TNFR2. Signal transmitted by TNFR1 (p55-TNF-R) induces metastasis-promoting inflammation. Signal transmitted through TNFR2 (p75-TNF-R) quenches the TNFR1 induction, resulting in a balanced metastasis growth. TNF reception through both receptors produces a balanced development of metastasis. If TNFR1 is not functional the induction of metastasis-promoting inflammation is absent, and metastasis development reduced. If TNFR2 is not functional, the quenching effect is missing, and the metastasis-promoting inflammation is stronger inducing exacerbated metastasis.

specific B cell response amplification to tumor antigens. (8) The expression in hosts, of intratumor prominent VH sequences (imposed by in vivo transfection), prior to tumor challenge, make it more resistant to tumor growth and tumor recurrence. This is compatible with the development of anti-VH antibodies in the host, which after tumor challenge could control the anti-TAAs immune response through the idiotype–anti-idiotype network. (9) Micro spots of tumor progression, named here as "focal invasive dysplasia", show as constants: lymphocyte, plasma cell and eosinophil infiltration, disrupted basal limit, tumor cell invasion, fibroblastic activity, and TAA plus IgG embedding of the matrix. Signs all compatible with a local inflammation under a TNF influence, and a Th2 cytokine environment. (10) Genetic lack of p55 TNF-R renders mice with increased resistance to tumor metastasis. Compatible with a role of TNF in the inflammation events that promote tumor progression, and points out that signals that promote tumor progression are delivered through the p55 TNF-R. (11) Mice with genetic lack of p75 TNF-R are much more susceptible to develop tumor metastases, which suggest the interpretation that TNF signals delivered through p75 TNF-R quench the protumor inflammatory pathway initiated through the p55 TNF-R.

All taken together, these features indicate the possible existence of antitumor autoimmune pathways that could either contain or promote tumor progression. The imbalance in one or other way will be reflected in cancer control or cancer development (Figure 26.14).

4. CONCLUDING COMMENT

Multiple Studies from different sources, including ours, are gathering data about the neoplastic physiopathology, which are conducive to a new vision of it. On one side, inflammation's major responsibilities in the cancer process are gaining acceptation and strength. On the other side, many cancer associated immune disturbances, including the cancer-promoting cascade triggered by immune autoantibodies against TAA, are strongly sustaining the existence of an autoimmune component subjacent in cancer pathology.

Although cancer was considered an inflammatory pathology at the end of the 19th century by R. Virchow, and S. Paget put the burden of cancer cells and stroma at the same level, the 20th century was characterized by a monochromatic vision of cancer as a cell proliferation disorder, and by a complete disregard for those old theories. With no doubt, the reasonable obsession for curing a lethal disease contributed to this focused idea. Bacterial infections were mortal even at the beginning of the 20th century; however, a dramatic turn in medicine occurred with the discovery of the antibiotics. The basic action of antibiotics was the arrest of the infecting bacterial proliferation, allowing the immune system and the tissue remodeling mechanisms to finish the healing process. The possibility of projecting this brilliant therapeutic success in the cancer scenario was encouraging. If cancer cells behave like infecting bacteria, and drugs would be able to inhibit exclusively tumor cell proliferation, an effective anticancer therapy would be achievable.

Many tools were tested against cancer cell proliferation, from radiation to natural venoms, such as colchicines and others, and failed. Each failure encouraged the search and the analysis of factors controlling cell proliferation, with the aim of finding inhibitors. In that endeavor the genetic background of cancer was progressively discovered and, since then, cancer has been considered a genetic disease with a concomitant cell proliferation disorder. The last 50 years astonishing technological progresses allowed parallel advances in genetics, proteomics, cell biology, and drug discovery, but the essence of the tumor pathology, still is uncontrollable.

Medicine has invested more than 100 years of powerful research in the pursuit of this cancer therapeutic strategy, founded on cell proliferation control, and relative little success has been achieved. Probably that elusive behavior of cancer is whispering that this monochromatic view is insufficient to foresee the full picture of an effective anticancer treatment.

All recognized autoimmune diseases share basic components: bizarre immune disorders and specific tissue degradation. And so does cancer. Moreover, following Rose and Bona criteria [134], from the direct, indirect, and circumstantial evidences required to qualify as an autoimmune disease, cancer adds all three.

Without disputing the genetic background of cancer, data are teaching us that malignant traits of cancer progression have anchors in the depth of autoimmune disorders, and should probably be treated like them.

Since Paul Ehrlich in 1901 hypothesized the possibility of auto-aggressive diseases under the theory of Horror Autotoxicus, 3 years passed until the discovery by Donath and Landsteiner of an autoantibody causing the paroxysmal cold hemoglobinuria. And many years passed until other autoimmune diseases were considered so. Today, from the orthodox point of view, the idea of considering cancer from an autoimmune angle could cause "horror", but not to do this, it could be a bigoted walk through the dark.

REFERENCES

1. Barbera-Guillem E, May KF Jr, Nyhus JK, Nelson MB. Promotion of tumor invasion by cooperation of granulocytes and macrophages activated by anti-tumor antibodies. *Neoplasia* 1999, 1(5): 453–460.
2. Nelson MB, Nyhus JK, Oravecz-Wilson KI, Barbera-Guillem E. Tumor cells express FcgammaRI which contributes to tumor cell growth and a metastatic phenotype. *Neoplasia* 2001, 3(2): 115–124.
3. Wen YJ, Mancino A, Pashov A, Whitehead T, Stanley J, Kieber-Emmons T. Antigen binding of human IgG Fabs mediate ERK-associated proliferation of human breast cancer cells. *DNA Cell Biol* 2005, 24: 73–84.
4. Barbera-Guillem E, Nyhus JK, Wolford CC, Friece CR, Sampsel JW. Vascular endothelial growth factor secretion by tumor-infiltrating macrophages essentially supports tumor angiogenesis, and IgG immune complexes potentiate the process. *Cancer Res* 2002, 62(23): 7042–7049.
5. Barbera-Guillem E, Sampsel JW. Immune-promoted tumor cell invasion and metastasis. New considerations in cancer therapy. *Adv Exp Med Biol* 2003, 532: 153–173.
6. Fidler IJ, Gruys E, Cifone MA, Barnes Z, Bucana C. Demonstration of multiple phenotypic diversity in a murine melanoma of recent origin. *J Natl Cancer Inst* 1981, 67(4): 947–956.
7. Fidler IJ. Selection of successive tumour lines for metastasis. *Nat New Biol* 1973, 242(118): 148–149.
8. Nicolson GL, Brunson KW, Fidler IJ. Specificity of arrest, survival, and growth of selected metastatic variant cell lines. *Cancer Res* 1978, 38(11 Pt 2): 4105–4111.
9. Fidler IJ, Nicolson GL. Organ selectivity for implantation survival and growth of B16 melanoma variant tumor lines. *J Natl Cancer Inst* 1976, 57(5): 1199–1202.
10. Paget S. The distribution of secondary growths in cancer of the breast. *Lancet* 1889, 1: 571–573.
11. Fidler IJ, Yano S, Zhang RD, Fujimaki T, Bucana CD. The seed and soil hypothesis: vascularisation and brain metastases. *Lancet Oncol* 2002, 3(1): 53–57.
12. Pantel K, Brakenhoff RH. Dissecting the metastatic cascade. *Nat Rev Cancer* 2004, 4(6): 448–456.
13. Kuo TH, Kubota T, Watanabe M, Furukawa T, Teramoto T, Ishibiki K, Kitajima M, Moossa AR, Penman S, Hoffman RM. Liver colonization competence governs colon cancer metastasis. *Proc Natl Acad Sci USA* 1995, 92(26): 12085–12089.
14. Gassmann P, Enns A, Haier J. Role of tumor cell adhesion and migration in organ-specific metastasis formation. *Onkologie* 2004, 27(6): 577–582.

15. Nubel T, Dippold W, Kleinert H, Kaina B, Fritz G. Lovastatin inhibits Rho-regulated expression of E-selectin by TNFalpha and attenuates tumor cell adhesion. *FASEB J* 2004, 18(1): 140–142. Epub 2003 Nov 20.
16. Langley RR, Carlisle R, Ma L, Specian RD, Gerritsen ME, Granger DN. Endothelial expression of vascular cell adhesion molecule-1 correlates with metastatic pattern in spontaneous melanoma. *Microcirculation* 2001, 8(5): 335–345.
17. Sheski FD, Natarajan V, Pottratz ST. Tumor necrosis factor-alpha stimulates attachment of small cell lung carcinoma to endothelial cells. *J Lab Clin Med* 1999, 133(3): 265–273.
18. Weiss L. Random and nonrandom processes in metastasis, and metastatic inefficiency. *Invasion Metastasis* 1983, 3(4): 193–207.
19. Cameron MD, Schmidt EE, Kerkvliet N, Nadkarni KV, Morris VL, Groom AC, Chambers AF, MacDonald IC. Temporal progression of metastasis in lung: cell survival, dormancy, and location dependence of metastatic inefficiency. *Cancer Res* 2000, 60(9): 2541–2546.
20. Mandelboim O, Vadai E, Fridkin M, Katz-Hillel A, Feldman M, Berke G, Eisenbach L. Regression of established murine carcinoma metastases following vaccination with tumour-associated antigen peptides. *Nat Med* 1995, 1(11): 1179–1183.
21. Ochsenbein AF, Klenerman P, Karrer U, Ludewig B, Pericin M, Hengartner H, Zinkernagel RM. Immune surveillance against a solid tumor fails because of immunological ignorance. *Proc Natl Acad Sci USA* 1999, 96(5): 2233–2238.
22. Punt CJ, Barbuto JA, Zhang H, Grimes WJ, Hatch KD, Hersh EM. Anti-tumor antibody produced by human tumor-infiltrating and peripheral blood B lymphocytes. *Cancer Immunol Immunother* 1994, 38(4): 225–232.
23. Ichiki Y, Takenoyama M, Mizukami M, So T, Sugaya M, Yasuda M, So T, Hanagiri T, Sugio K, Yasumoto K. Simultaneous cellular and humoral immune response against mutated p53 in a patient with lung cancer. *J Immunol* 2004, 172(8): 4844–4850.
24. May KF Jr, Chen L, Zheng P, Liu Y. Anti-4-1BB monoclonal antibody enhances rejection of large tumor burden by promoting survival but not clonal expansion of tumor-specific CD8+ T cells. *Cancer Res* 2002, 62(12): 3459–3465.
25. Li Y, McGowan P, Hellstrom I, Hellstrom KE, Chen L. Costimulation of tumor-reactive CD4+ and CD8+ T lymphocytes by B7, a natural ligand for CD28, can be used to treat established mouse melanoma. *J Immunol* 1994, 153(1): 421–428.
26. Murasko DM, Prehn RT. Stimulatory effect of immunization on tumor induction by Moloney murine sarcoma virus. *J Natl Cancer Inst* 1978, 61(5): 1323–1327.
27. Hewitt HB, Blake ER, Walder AS. A critique of the evidence for active host defence against cancer, based on personal studies of 27 murine tumours of spontaneous origin. *Br J Cancer* 1976, 33(3): 241–259.
28. Wang W, Bergh A, Damber JE. Cyclooxygenase-2 expression correlates with local chronic inflammation and tumor neovascularization in human prostate cancer. *Clin Cancer Res* 2005, 11(9): 3250–3256.
29. Smith GR, Missailidis S. Cancer, inflammation and the AT1 and AT2 receptors. *J Inflamm (Lond)* 2004, 1(1): 3.
30. Nyhus JK, Wolford CC, Friece CR, Nelson MB, Sampsel JW, Barbera-Guillem E. IgG-recognizing shed tumor-associated antigens can promote tumor invasion and metastasis. *Cancer Immunol Immunother* 2001, 50(7): 361–372.
31. Wu CY, Wang CJ, Tseng CC, Chen HP, Wu MS, Lin JT, Inoue H, Chen GH. Helicobacter pylori promote gastric cancer cells invasion through a NF-kappaB and COX-2-mediated pathway. *World J Gastroenterol* 2005, 11(21): 3197–3203.
32. de Visser KE, Korets LV, Coussens LM. De novo carcinogenesis promoted by chronic inflammation is B lymphocyte dependent. *Cancer Cell* 2005, 7(5): 411–423.
33. Houghton J, Stoicov C, Nomura S, Rogers AB, Carlson J, Li H, Cai X, Fox JG, Goldenring JR, Wang TC. Gastric cancer originating from bone marrow-derived cells. *Science* 2004, 306(5701): 1568–1571.
34. Thompson CA. COX-2 inhibitors still eyed for cancer prevention. *Am J Health Syst Pharm* 2005, 62(9): 890–894.

35. Tsutsui S, Yasuda K, Suzuki K, Tahara K, Higashi H, Era S. Macrophage infiltration and its prognostic implications in breast cancer: the relationship with VEGF expression and microvessel density. *Oncol Rep* 2005, 14(2): 425–431.
36. Balaz P, Friess H, Kondo Y, Zhu Z, Zimmermann A, Buchler MW. Human macrophage metalloelastase worsens the prognosis of pancreatic cancer. *Ann Surg* 2002, 235(4): 519–527.
37. Campbell AS, Albo D, Kimsey TF, White SL, Wang TN. Macrophage inflammatory protein-3alpha promotes pancreatic cancer cell invasion. *J Surg Res* 2005, 123(1): 96–101.
38. Kimsey TF, Campbell AS, Albo D, Wilson M, Wang TN. Co-localization of macrophage inflammatory protein-3alpha (Mip-3alpha) and its receptor, CCR6, promotes pancreatic cancer cell invasion. *Cancer J* 2004, 10(6): 374–380.
39. Lewis C, Murdoch C. Macrophage responses to hypoxia: implications for tumor progression and anti-cancer therapies. *Am J Pathol* 2005, 167(3): 627–635.
40. Sickert D, Aust DE, Langer S, Haupt I, Baretton GB, Dieter P. Characterization of macrophage subpopulations in colon cancer using tissue microarrays. *Histopathology* 2005, 46(5): 515–521.
41. Inaba T, Sano H, Kawahito Y, Hla T, Akita K, Toda M, Yamashina I, Inoue M, Nakada H. Induction of cyclooxygenase-2 in monocyte/macrophage by mucins secreted from colon cancer cells. *Proc Natl Acad Sci USA* 2003, 100(5): 2736–2741.
42. Johnson VG, Schlom J, Paterson AJ, Bennett J, Magnani JL, Colcher D. Analysis of a human tumor-associated glycoprotein (TAG-72) identified by monoclonal antibody B72.3. *Cancer Res* 1986, 46(2): 850–857.
43. Carlos CA, Dong HF, Howard OM, Oppenheim JJ, Hanisch FG, Finn OJ. Human tumor antigen MUC1 is chemotactic for immature dendritic cells and elicits maturation but does not promote Th1 type immunity. *J Immunol* 2005, 175(3): 1628–1635.
44. Akis N, Andl C, Zhou D. A mixed hemadsorption assay for detection of cell surface binding anti-tumor antibodies in human sera. *J Immunol Methods* 2002, 261(1–2): 119–127.
45. Luborsky JL, Barua A, Shatavi SV, Kebede T, Abramowicz J, Rotmensch J. Anti-tumor antibodies in ovarian cancer. *Am J Reprod Immunol* 2005, 54(2): 55–62.
46. Liu Y, Masuda E, Blank MC, Kirou KA, Gao X, Park MS, Pricop L. Cytokine-mediated regulation of activating and inhibitory Fc gamma receptors in human monocytes. *J Leukoc Biol* 2005, 77(5): 767–776.
47. Webster WB, Harwood SJ, Carroll RG, Morrissey MA. Pharmacokinetics of indium-111-labeled B72.3 monoclonal antibody in colorectal cancer patients. *J Nucl Med* 1992, 33(4): 498–504.
48. Esteban JM, Colcher D, Sugarbaker P, Carrasquillo JA, Bryant G, Thor A, Reynolds JC, Larson SM, Schlom J. Quantitative and qualitative aspects of radiolocalization in colon cancer patients of intravenously administered MAb B72.3. *Int J Cancer* 1987, 39(1): 50–59.
49. Hitchcock CL, Arnold MW, Young DC, Schneebaum S, Martin EW Jr. TAG-72 expression in lymph nodes and RIGS. *Dis Colon Rectum* 1996, 39(4): 473–475.
50. Loy TS, Haege DD. B72.3 immunoreactivity in benign abdominal lymph nodes associated with gastrointestinal disease. *Dis Colon Rectum* 1995, 38(9): 983–987.
51. Cornelius EA, West AB. False tumor-positive lymph nodes in radioimmunodiagnosis and radioimmunoguided surgery: etiologic mechanisms. *J Surg Oncol* 1996, 63(1): 23–35.
52. Stephens AD, Punja U, Sugarbaker PH. False-positive lymph nodes by radioimmunoguided surgery: report of a patient and analysis of the problem. *J Nucl Med* 1993, 34(5): 804–808.
53. Nieroda CA, Mojzisik C, Sardi A, Ferrara P, Hinkle G, Thurston MO, Martin EW Jr. The impact of radioimmunoguided surgery (RIGS) on surgical decision-making in colorectal cancer. *Dis Colon Rectum* 1989, 32(11): 927–932.
54. Barbera-Guillem E, Nelson MB, Barr B, Nyhus JK, May KF Jr, Feng L, Sampsel JW. B lymphocyte pathology in human colorectal cancer. Experimental and clinical therapeutic effects of partial B cell depletion. *Cancer Immunol Immunother* 2000, 48(10): 541–549.

55. Fu YX, Huang G, Matsumoto M, Molina H, Chaplin DD. Independent signals regulate development of primary and secondary follicle structure in spleen and mesenteric lymph node. *Proc Natl Acad Sci USA* 1997, 94(11): 5739–5743.
56. Fu YX, Molina H, Matsumoto M, Huang G, Min J, Chaplin DD. Lymphotoxin-alpha (LTalpha) supports development of splenic follicular structure that is required for IgG responses. *J Exp Med* 1997, 185(12): 2111–2120.
57. Wang Y, Huang G, Wang J, Molina H, Chaplin DD, Fu YX. Antigen persistence is required for somatic mutation and affinity maturation of immunoglobulin. *Eur J Immunol* 2000, 30(8): 2226–2234.
58. Davis IA, Rouse BT. Immune responsiveness of lymphotoxin-alpha-deficient mice: two reconstitution models. *Cell Immunol* 1998, 189(2): 116–124.
59. Matsumoto M, Lo SF, Carruthers CJ, Min J, Mariathasan S, Huang G, Plas DR, Martin SM, Geha RS, Nahm MH, Chaplin DD. Affinity maturation without germinal centres in lymphotoxin-alpha-deficient mice. *Nature* 1996, 382(6590): 462–466.
60. Ruddle NH, Homer R. The role of lymphotoxin in inflammation. *Prog Allergy* 1988, 40: 162–182.
61. Tsoukas CD, Rosenau W, Baxter JD. Cellular receptors for lymphotoxin: correlation of binding and cytotoxicity in sensitive and resistant target cells. *J Immunol* 1976, 116(1): 184–187.
62. Kratz A, Campos-Neto A, Hanson MS, Ruddle NH. Chronic inflammation caused by lymphotoxin is lymphoid neogenesis. *J Exp Med* 1996, 183(4): 1461–1472.
63. Donjerkovi CD, Scott DV. Activation-induced cell death in B lymphocytes. *Cell Research* 2000, 10: 179–192.
64. Kitamura D, Roes J, Kuhn R, Rajewsky K. A B cell-deficient mouse by targeted disruption of the membrane exon of the immunoglobulin mu chain gene. *Nature* 1991, 350(6317): 423–426.
65. Ugrinovic S, Menager N, Goh N, Mastroeni P. Characterization and development of T-Cell immune responses in B-cell-deficient (Igh-6(–/–)) mice with Salmonella enterica serovar Typhimurium infection. *Infect Immun* 2003, 71(12): 6808–6819.
66. Dunn GP, Old LJ, Schreiber RD. The immunobiology of cancer immunosurveillance and immunoediting. *Immunity* 2004, 21(2): 137–148.
67. Monach PA, Schreiber H, Rowley DA. CD4+ and B lymphocytes in transplantation immunity. II. Augmented rejection of tumor allografts by mice lacking B cells. *Transplantation* 1993, 55(6): 1356–1361.
68. de Visser KE, Korets LV, Coussens LM. De novo carcinogenesis promoted by chronic inflammation is B lymphocyte dependent. *Cancer Cell* 2005, 7(5): 411–423.
69. Qin Z, Richter G, Schuler T, Ibe S, Cao X, Blankenstein T. B cells inhibit induction of T cell-dependent tumor immunity. *Nat Med* 1998, 4(5): 627–630.
70. de Visser KE, Coussens LM. The interplay between innate and adaptive immunity regulates cancer development. *Cancer Immunol Immunother* 2005, 54(11): 1143–1152.
71. Koeppen HK, Singh S, Stauss HJ, Park BH, Rowley DA, Schreiber H. CD4-positive and B lymphocytes in transplantation immunity. I. Promotion of tumor allograft rejection through elimination of CD4-positive lymphocytes. *Transplantation* 1993, 55(6): 1349–1355.
72. Piccirillo CA, Shevach EM. Cutting edge: control of CD8+ T cell activation by CD4+CD25+ immunoregulatory cells. *J Immunol* 2001, 167(3): 1137–1140.
73. Shevach EM, Piccirillo CA, Thornton AM, McHugh RS. Control of T cell activation by CD4+CD25+ suppressor T cells. *Novartis Found Symp* 2003, 252: 24–36; discussion 36–44, 106–114.
74. Curotto de Lafaille MA, Lafaille JJ. CD4(+) regulatory T cells in autoimmunity and allergy. *Curr Opin Immunol* 2002, 14(6): 771–778.
75. Chen ML, Pittet MJ, Gorelik L, Flavell RA, Weissleder R, von Boehmer H, Khazaie K. Regulatory T cells suppress tumor-specific CD8 T cell cytotoxicity through TGF-beta signals in vivo. *Proc Natl Acad Sci USA* 2005, 102(2): 419–424. Epub 2004 Dec 27.

76. Renard N, Duvert V, Blanchard D, Banchereau J, Saeland S. Activated CD4+ T cells induce CD40-dependent proliferation of human B cell precursors. *J Immunol* 1994, 152(4): 1693–1701.
77. Tolaymat N, Weber SP, Cowdery JS. Chronic in vivo depletion of CD4 T cells begun in utero inhibits gut B cell differentiation. *Clin Immunol Immunopathol* 1990, 56(1): 97–107.
78. Koeppen HK, Singh S, Stauss HJ, Park BH, Rowley DA, Schreiber H. CD4-positive and B lymphocytes in transplantation immunity. I. Promotion of tumor allograft rejection through elimination of CD4-positive lymphocytes. *Transplantation* 1993, 55(6): 1349–1355.
79. Boon T, Old LJ. Cancer tumor antigens. *Curr Opin Immunol* 1997, 9(5): 681–683.
80. Cote RJ, Morrissey DM, Oettgen HF, Old LJ. Analysis of human monoclonal antibodies derived from lymphocytes of patients with cancer. *Fed Proc* 1984, 43(9): 2465–2469.
81. Cote RJ, Morrissey DM, Houghton AN, Beattie EJ Jr, Oettgen HF, Old LJ. Generation of human monoclonal antibodies reactive with cellular antigens. *Proc Natl Acad Sci USA* 1983, 80(7): 2026–2030.
82. Fernandez-Madrid F, VandeVord PJ, Yang X, Karvonen RL, Simpson PM, Kraut MJ, Granda JL, Tomkiel JE. Antinuclear antibodies as potential markers of lung cancer. *Clin Cancer Res* 1999, 5(6): 1393–1400.
83. Hansen MH, Nielsen H, Ditzel HJ. The tumor-infiltrating B cell response in medullary breast cancer is oligoclonal and directed against the autoantigen actin exposed on the surface of apoptotic cancer cells. *Proc Natl Acad Sci USA* 2001, 98(22): 12659–12664.
84. Hansen MH, Nielsen HV, Ditzel HJ. Translocation of an intracellular antigen to the surface of medullary breast cancer cells early in apoptosis allows for an antigen-driven antibody response elicited by tumor-nfiltrating B cells. *J Immunol* 2002, 169(5): 2701–2711.
85. Bradley SV, Oravecz-Wilson KI, Bougeard G, Mizukami I, Li L, Munaco AJ, Sreekumar A, Corradetti MN, Chinnaiyan AM, Sanda MG, Ross TS. Serum antibodies to huntingtin interacting protein-1: a new blood test for prostate cancer. *Cancer Res* 2005, 65(10): 4126–4133.
86. Mills IG, Gaughan L, Robson C, Ross T, McCracken S, Kelly J, Neal DE. Huntingtin interacting protein 1 modulates the transcriptional activity of nuclear hormone receptors. *J Cell Biol* 2005, 170(2): 191–200.
87. Traub UC, DeJager RL, Primus FJ, Losman M, Goldenberg DM. Antiidiotype antibodies in cancer patients receiving monoclonal antibody to carcinoembryonic antigen. *Cancer Res* 1988, 48(14): 4002–4006.
88. Rossi F, Dietrich G, Kazatchkine MD. Anti-idiotypes against autoantibodies in normal immunoglobulins: evidence for network regulation of human autoimmune responses. *Immunol Rev* 1989, 110: 135–149.
89. Sneath PH, Sokal RR. *Numerical taxonomy — the principles and practice of numerical classification*. WH. Freeman, San Francisco, 1973.
90. Nyhus JK, Wolford C, Feng L, Barbera-Guillem E. Direct in vivo transfection of antisense Fas-ligand reduces tumor growth and invasion. *Gene Ther* 2001, 8(3): 209–214.
91. Bins AD, Jorritsma A, Wolkers MC, Hung CF, Wu TC, Schumacher TN, Haanen JB. A rapid and potent DNA vaccination strategy defined by in vivo monitoring of antigen expression. *Nat Med* 2005, 11(8): 899–904.
92. Boon T, Old LJ. Cancer tumor antigens. *Curr Opin Immunol* 1997, 9(5): 681–683.
93. Imahayashi S, Ichiyoshi Y, Yoshino I, Eifuku R, Takenoyama M, Yasumoto K. Tumor-infiltrating B-cell-derived IgG recognizes tumor components in human lung cancer. *Cancer Invest* 2000, 18(6): 530–536.
94. Parkes H, Collis P, Baildam A, Ralphs D, Lyons B, Howell A, Craig R. In situ hybridisation and S1 mapping show that the presence of infiltrating plasma cells is associated with poor prognosis in breast cancer. *Br J Cancer* 1988, 58(6): 715–722.
95. Cote RJ, Morrissey DM, Houghton AN, Beattie EJ Jr, Oettgen HF, Old LJ. Generation of human monoclonal antibodies reactive with cellular antigens. *Proc Natl Acad Sci USA* 1983, 80(7): 2026–2030.

96. Hansen MH, Nielsen H, Ditzel HJ. The tumor-infiltrating B cell response in medullary breast cancer is oligoclonal and directed against the autoantigen actin exposed on the surface of apoptotic cancer cells. *Proc Natl Acad Sci USA* 2001, 98(22): 12659–12664.
97. Goronzy JJ, Gold KN, Weyand CM. T-Cell Derived Lymphokines as Regulators of Chronic Inflammation: Potential Targets for Immunomodulation? *Am J Ther* 1996, 3(2): 109–114.
98. Fadare O, Hileeto D, Gruddin YL, Mariappan MR. Sclerosing mucoepidermoid carcinoma of the parotid gland. *Arch Pathol Lab Med* 2004, 128(9): 1046–1049.
99. Khanafshar E, Phillipson J, Schammel DP, Minobe L, Cymerman J, Weidner N. Inflammatory myofibroblastic tumor of the breast. *Ann Diagn Pathol* 2005, 9(3): 123–129.
100. Maggio E, van den Berg A, Diepstra A, Kluiver J, Visser L, Poppema S. Chemokines, cytokines and their receptors in Hodgkin's lymphoma cell lines and tissues. *Ann Oncol* 2002, 13 (Suppl.) 1: 52–56.
101. Chi DS, Harris NS. Relationship of surface immunoglobulin-bearing cells, plasma cells, and tumor development in anaplastic carcinoma-bearing A/J mice. *Cancer Res* 1977, 37(4): 1119–1124.
102. Hansen MH, Nielsen H, Ditzel HJ. The tumor-infiltrating B cell response in medullary breast cancer is oligoclonal and directed against the autoantigen actin exposed on the surface of apoptotic cancer cells. *Proc Natl Acad Sci USA* 2001, 98(22): 12659–12664. Epub 2001 Oct 16.
103. Ito T, Saga S, Nagayoshi S, Imai M, Aoyama A, Yokoi T, Hoshino M. Class distribution of immunoglobulin-containing plasma cells in the stroma of medullary carcinoma of breast. *Breast Cancer Res Treat* 1986, 7(2): 97–103.
104. Hussain MA, Tyagi SP, Tyagi N, Khan MH. Stromal cellular response in breast tumours and allied lesions. *J Indian Med Assoc* 1992, 90(5): 119–121.
105. Hermsen BB, von Mensdorff-Pouilly S, Fabry HF, Winters HA, Kenemans P, Verheijen RH, van Diest PJ. Lobulitis is a frequent finding in prophylactically removed breast tissue from women at hereditary high risk of breast cancer. *J Pathol* 2005, 206(2): 220–223.
106. Tsubaki T, Takegawa S, Hanamoto H, Arita N, Kamogawa J, Yamamoto H, Takubo N, Nakata S, Yamada K, Yamamoto S, Yoshie O, Nose M. Accumulation of plasma cells expressing CXCR3 in the synovial sublining regions of early rheumatoid arthritis in association with production of Mig/CXCL9 by synovial fibroblasts. *Clin Exp Immunol* 2005, 141(2): 363–371.
107. Perry ME, Mustafa Y, Wood SK, Cawley MI, Dumonde DC, Brown KA. Binucleated and multinucleated forms of plasma cells in synovia from patients with rheumatoid arthritis. *Rheumatol Int* 1997, 17(4): 169–174.
108. Iwasaki K, Torisu M, Fujimura T. Malignant tumor and eosinophils. I. Prognostic significance in gastric cancer. *Cancer* 1986, 58(6): 1321–1327.
109. Lowe D, Jorizzo J, Hutt MS. Tumour-associated eosinophilia: a review. *J Clin Pathol* 1981, 34(12): 1343–1348.
110. Said M, Wiseman S, Yang J, Alrawi S, Douglas W, Cheney R, Hicks W, Rigual N, Loree T, Spiegel G, Tan D. Tissue eosinophilia: a morphologic marker for assessing stromal invasion in laryngeal squamous neoplasms. *BMC Clin Pathol* 2005, 5(1): 1.
111. Spiegel GW: Eosinophils as a marker for invasion in vulvar squamous neoplastic lesions. *Int J Gynecol Pathol* 2002, 21: 108–116.
112. Spiegel GW, Ashraf M, Brooks JJS: Eosinophils as a marker for invasion in cervical squamous neoplastic lesions. *Int J Gynecol Pathol* 2002, 21: 117–124.
113. Kim JT, Schimming AW, Kita H. Ligation of Fc gamma RII (CD32) pivotally regulates survival of human eosinophils. *J Immunol* 1999, 162(7): 4253–4259.
114. Horie S, Gleich GJ, Kita H. Cytokines directly induce degranulation and superoxide production from human eosinophils. *J Allergy Clin Immunol* 1996, 98(2): 371–381.
115. Fujisawa T, Abu-Ghazaleh R, Kita H, Sanderson CJ, Gleich GJ. Regulatory effect of cytokines on eosinophil degranulation. *J Immunol* 1990, 144(2): 642–646.
116. Temkin V, Aingorn H, Puxeddu I, Goldshmidt O, Zcharia E, Gleich GJ, Vlodavsky I, Levi-Schaffer F. Eosinophil major basic protein: first identified natural heparanase-inhibiting protein. *J Allergy Clin Immunol* 2004, 113(4): 703–709.

117. Lebman DA, Edmiston JS. The role of TGF-beta in growth, differentiation, and maturation of B lymphocytes. *Microbes Infect* 1999, 1(15): 1297–1304.
118. Di Girolamo N, Visvanathan K, Lloyd A, Wakefield D. Expression of TNF-alpha by human plasma cells in chronic inflammation. *J Leukoc Biol* 1997, 61(6): 667–678.
119. Kitakata H, Nemoto-Sasaki Y, Takahashi Y, Kondo T, Mai M, Mukaida N. Essential roles of tumor necrosis factor receptor p55 in liver metastasis of intrasplenic administration of colon 26 cells. *Cancer Res* 2002, 62(22): 6682–6687.
120. Karrer U, Lopez-Macias C, Oxenius A, Odermatt B, Bachmann MF, Kalinke U, Bluethmann H, Hengartner H, Zinkernagel. Antiviral B cell memory in the absence of mature follicular dendritic cell networks and classical germinal centers in TNFR1–/– mice. *J Immunol* 2000, 164(2): 768–778.
121. Le CH, Nicolson AG, Morales A, Sewell KL. Suppression of collagen-induced arthritis through adenovirus-mediated transfer of a modified tumor necrosis factor alpha receptor gene. *Arthritis Rheum* 1997, 40(9): 1662–1669.
122. Pryhuber GS, O'Brien DP, Baggs R, Phipps R, Huyck H, Sanz I, Nahm MH. Ablation of tumor necrosis factor receptor type I (p55) alters oxygen-induced lung injury. *Am J Physiol Lung Cell Mol Physiol* 2000, 278(5): L1082–1090.
123. Stoelcker B, Ruhland B, Hehlgans T, Bluethmann H, Luther T, Mannel DN. Tumor necrosis factor induces tumor necrosis via tumor necrosis factor receptor type 1-expressing endothelial cells of the tumor vasculature. *Am J Pathol* 2000, 156(4): 1171–1176.
124. Alcamo E, Mizgerd JP, Horwitz BH, Bronson R, Beg AA, Scott M, Doerschuk CM, Hynes RO, Baltimore D. Targeted mutation of TNF receptor I rescues the RelA-deficient mouse and reveals a critical role for NF-kappa B in leukocyte recruitment. *J Immunol* 2001, 167(3): 1592–1600.
125. Yan HC, Delisser HM, Pilewski JM, Barone KM, Szklut PJ, Chang XJ, Ahern TJ, Langer-Safer P, Albelda SM. Leukocyte recruitment into human skin transplanted onto severe combined immunodeficient mice induced by TNF-alpha is dependent on E-selectin. *J Immunol* 1994, 152(6): 3053–3063.
126. Lukacs NW, Strieter RM, Chensue SW, Widmer M, Kunkel SL. TNF-alpha mediates recruitment of neutrophils and eosinophils during airway inflammation. *J Immunol* 1995, 154(10): 5411–5417.
127. Lampinen M, Carlson M, Sangfelt P, Taha Y, Thorn M, Loof L, Raab Y, Venge P. IL-5 and TNF-alpha participate in recruitment of eosinophils to intestinal mucosa in ulcerative colitis. *Dig Dis Sci* 2001, 46(9): 2004–2009.
128. Roach DR, Bean AG, Demangel C, France MP, Briscoe H, Britton WJ. TNF regulates chemokine induction essential for cell recruitment, granuloma formation, and clearance of mycobacterial infection. *J Immunol* 2002, 168(9): 4620–4627.
129. Mori R, Kondo T, Ohshima T, Ishida Y, Mukaida N. Accelerated wound healing in tumor necrosis factor receptor p55-deficient mice with reduced leukocyte infiltration. *FASEB J* 2002, 16: 963–974.
130. Komata T, Tsuchiya N, Matsushita M, Hagiwara K, Tokunaga K. Association of tumor necrosis factor receptor 2 (TNFR2) polymorphism with susceptibility to systemic lupus erythematosus. *Tissue Antigens* 1999, 53(6): 527–533.
131. Hsu H, Xiong J, Goeddel DV. The TNF receptor 1-associated protein TRADD signals cell death and NF-kappa B activation. *Cell* 1995, 81(4): 495–504.
132. McFarlane SM, Pashmi G, Connell MC, Littlejohn AF, Tucker SJ, Vandenabeele P, MacEwan DJ. Differential activation of nuclear factor-kappa B by tumour necrosis factor receptor subtypes. TNFR1 predominates whereas TNFR2 activates transcription poorly. *FEBS Lett* 2002, 515(1–3): 119–126.
133. Jacobs M, Brown N, Allie N, Chetty K, Ryffel B. Tumor necrosis factor receptor 2 plays a minor role for mycobacterial immunity. *Pathobiology* 2000, 68: 68–75.
134. Rose NR, Bona C. Defining criteria for autoimmune diseases (Witebsky's postulates revisited). *Immunol Today* 1993, 14(9): 426–430.

CHAPTER 27

SYNERGISTIC MOLECULAR MECHANISMS IN HORMONE-SENSITIVE BREAST CANCER

ANTONIO LLOMBART CUSSAC

Head Medical Oncology Service, Hospital Universitario Arnau de Vilanova, Lleida, Spain

Breast cancer is the most common cancer among women and the second leading cause of death in developed countries. However, a decrease in overall mortality rates has been shown over the last decade [1]. With various factors contributing to this gain, the generalization of systemic therapies both in the early and advanced stages is certainly impelling this progress. For early stage estrogen receptor (ER)-positive breast cancer allocation to about 5 years of adjuvant tamoxifen reduces the annual breast cancer death rate by 31%; with an estimated 15-year survival benefit of 9.2% ($2p < 0.00001$) [2]. The last generation of aromatase inhibitors (anastrozole, letrozole, and exemestane) has shown to be more effective in advanced disease; adding benefits in early stage when replacing or switching tamoxifen [3]. Even though the beneficial effect of endocrine therapies is restricted to patients with ER and/or progesterone receptor (PR)-positive tumors, a large fraction of hormone receptor-positive tumors do not respond as expected to the treatment [4].

Resistance to endocrine therapy is a huge clinical problem and researchers have therefore actively pursued mechanistic studies of endocrine resistance and tested predictive markers other than steroid receptors. The mechanisms underlying resistance in breast cancer remain relatively poorly defined, but their elucidation has potential to reveal new therapeutic targets to treat or even prevent resistance. Over the last 20 years, a large number of cell biologic factors have been reported that identify those patients who will benefit from endocrine therapy or fail to respond [5]. Few of these, however, seemed valuable and useful in daily clinical practice. Nevertheless, research into the mechanisms of endocrine resistance in breast cancer has revealed that various growth factor pathways and oncogenes involved in the signal transduction cascade become activated and utilized by breast cancer cells to bypass normal endocrine responsiveness [6].

1. ER IN BREAST CANCER

ER acts in the nucleus as a transcriptional regulator of specific genes. The protein has a ligand-binding domain, several transcription activation domains, and a DNA-binding domain that interacts with specific regions in the promoter of target genes, including sites known as estrogen-responsive elements (ERE) [7]. The binding of estrogen to ER facilitates binding of the receptor complex to promoter regions of target genes. The recruitment of coactivators such as AIB1 (SRC3) and other proteins with acetyltransferase activity helps to unwind the chromatin allowing transcription to occur. By contrast, the ER conformation induced by the binding of tamoxifen favors the recruitment of corepressors and deacetylases that inhibit transcriptional activity. In total, estrogen regulates the expression of many genes important for normal cell physiology and growth of some breast tumors. PR, PS2, the heat shock proteins, TGF-α, IGF-II, IRS1, and vascular endothelial growth factor (VEGF), in addition to IGF-IR and cyclin D1, represent just a few of the many genes regulated by estrogen in target cells. Interestingly, ER can also decrease expression of many genes as well [8].

However, ER functions may also occur outside the nucleus, as a small pool of ER is located in the cytoplasm or bound to the plasma membrane [9]. This membrane-bound ER can directly interact with and/or activate IGF-IR, PI3K, Src, EGFR, and HER2 [10]. Src in turn activates matrix metalloproteinase 2 who cleaves heparin-binding EGF from the cell surface [11]. Then, EGF can interact by autocrine or paracrine mechanisms with adjacent EGF receptors to initiate growth factor signaling. Bidirectional cross talk between ER and growth factor pathways results in a positive feedback cycle of cell survival and cell proliferative stimuli. Clinically, it may be crucial to block this cross talk by inhibiting both signaling networks to achieve optimal therapeutic activity.

2. GROWTH FACTOR SIGNALLING AND ENDOCRINE RESISTANCE IN BREAST CANCER

There is no doubt that cross talk between ER and the EGFR signaling pathways is one, if not the major, mechanisms for resistance to endocrine therapy in breast cancer [12]. cErbB2 (HER2) is a member of this EGFR family of transmembrane tyrosine kinases. The family also includes HER3 and a poorly understood HER4. Both HER2 and HER3 need to heterodimerize with other members of the family to generate the kinase cascade. Growth factors such as epidermal growth factor (EGF), transforming growth factor-α, and amphiregulin bind to the external domain of epidermal growth factor receptor (EGFR), which then induces homo- or heterodimerization with another receptor in the family and activate the tyrosine kinase [13]. Heregulin and other ligands, bind to the external domain of HER3. This also initiates heterodimerization and then activation of Akt, Erk1/2 mitogen-activated protein kinase (MAPK) or other intermediates. Because HER2 does not have a ligand, it may be relatively inactive unless the cell also expresses EGFR or HER3 [14].

Preclinical models suggest that both de novo and acquired resistance to endocrine therapy is often associated with increased signaling from the EGFR/HER2 pathway [15] and a switch from nuclear to membrane-initiated steroid signaling (MISS) ER activity. Importantly, although human breast cancer xenografts that overexpress HER2 are stimulated by tamoxifen as a mechanism of de novo resistance, these tumors are still completely inhibited by estrogen withdrawal. The increased MISS ER activity and the resultant increase in tamoxifen agonist activity in these HER2-overexpressing cells is entirely dependent on ER cross talk with the growth factor signaling pathway because both phenomena were completely reversed in the presence of specific EGFR/HER2 inhibitors [16]. Increased HER2 activity results in more ER outside the nucleus, and classical ER-dependent transcription is inhibited by a mechanism that is HER2-dependent and involves ER sequestration in the cytoplasm through the ER corepressor MTA1s [17]. High levels of membrane and cytoplasmic ER have also been found in breast cancer cells amplified for the *HER2* gene 115. Whether this change in ER localization in the cell also occurs in human tumors requires more study.

Activation of the EGFR/HER2 signaling pathway initiates a kinase-signaling cascade that has a variety of effects on the tumor cells, including inhibition of apoptosis, stimulation of cell proliferation, enhanced invasion and cell motility, and induction of angiogenesis stimuli. Cell survival and proliferation is mediated mostly through the PI3K/Akt and the Erk1-2/MAPK pathways. These kinases are also responsiveness for the activation of ER coregulators such as AIB1 and NCoR [7,8]. This phosphorylation augments the transcriptional activation potential of ER and enhances its effects on cell proliferation and survival. In tumors expressing both ER and abundant HER2, these two pathways may contribute to hormonal therapy resistance.

Various growth factor pathways and oncogenes involved in the signal transduction cascade are utilized by breast cancer cells to bypass normal endocrine responsiveness [6]. The monoclonal antibody trastuzumab (Herceptin), which targets the extracellular domain of HER2, was able to increase the endocrine sensitivity of ER[+]/HER2[+] breast cancer cell lines. However, HER2 signaling involves the activation of various downstream intracellular kinases. As such, these intracellular pathways represent attractive targets for pharmacologic intervention with small molecule signal transduction inhibitors (STIs) that target aberrantly or excessively expressed oncogene products. Many such drugs are in active development for breast cancer, including type 1 growth factor tyrosine kinase inhibitors (TKI), farnesyltransferase inhibitors (FTIs), MEK inhibitors, and mTOR antagonists.

3. PRECLINICAL EFFICACY OF INTRACELLULAR STIS IN BREAST CANCER

Enhanced expression of type I growth factor receptors such as EGFR and HER2, together with subsequent downstream activation of signaling pathways regulated by the MAPK/ERK, has been found in breast cancer cells that become resistant over time to endocrine therapy with tamoxifen. Treatment with various STIs has been used in preclinical models in an attempt to overcome this resistance by blocking upregulated signaling pathways [18]. For example, in MCF-7 cells that developed resistance to tamoxifen, both gefitinib, which targets the internal tyrosine kinase domain of EGFR, and trastuzumab, which blocks the external domain of HER2, were effective at reducing downstream ERK1/2 MAPK signaling and inhibiting cell growth [19]. EGFR and HER2 heterodimerize in the resistant cells, such that targeting either one of the receptors can be an effective therapy. Of note, hormone-sensitive cells (in which neither receptors are overexpressed) were unaffected by either gefitinib or trastuzumab therapy. Similar data has been reported by other groups in tamoxifen-resistant, HER2-transfected MCF-7 cells with Trastuzumab [20]. Likewise, in cells resistant to LTED, both growth and ER-mediated gene transcription was abrogated by a number of different intracellular approaches to interrupt signaling, including the TKI gefitinib, the MEK inhibitor UO126, and the ER downregulator fulvestrant, which degrades residual ER [21]. Several groups have shown that these different STIs may also inhibit the growth of breast cancer xenograft tumors in vivo [22].

Other intracellular signaling and cell survival pathways are also activated in hormone-resistant breast cancer, particularly the phosphatidylinositol 3-kinase (PI3K)/Akt pathway. Akt (or PKB) is a serine/threonine kinase that promotes cell survival and is activated in response to many different growth factors, including insulin, insulin-like growth factor 1 (IGF-1), basic fibroblast growth factor (bFGF), EGF, heregulin, and VEGF. Once activated, Akt exerts anti-apoptotic effects through phosphorylation of substrates that directly regulate the apoptotic machinery (i.e., Bad and caspase 9). In addition, the mammalian target of rapamycin (mTOR) is a downstream effector of the PI3K/Akt signaling pathway that activates p70S6 kinase and 4E-binding protein-1, which, in turn, regulate transition through the G1/S phase of the cell cycle. Approaches to targeting these cell survival pathways have included either specific PI3K inhibitors, such as LY294002, or rapamycin analogues, such as temsirolimus (CCI-779) or everolimus (RAD-001), that target mTOR. Breast cancer cell lines with activated Akt (as by loss of the regulatory PTEN tumor suppressor gene) are especially sensitive to mTOR antagonism [23].

4. STIS IN COMBINATION WITH ENDOCRINE THERAPY

Several preclinical reports have implied that, in hormone-sensitive, ER-positive breast cancer, STIs as monotherapy may have only a minimal effect on tumor growth, especially if cells lack the activation and dependence of the various signal transduction pathways described above. Emerging evidence suggests that adaptive changes occur during prolonged endocrine therapy, particularly upregulation of growth factor signaling. Thus, strategies to combine endocrine with STI therapies have been used as a means to prevent development of resistance and improve therapeutic efficacy. In vitro, combined tamoxifen and gefitinib provided nearly complete inhibition of phosphorylated ERK1/2 MAPK and Akt, together with greater G0/G1 cell-cyle arrest and suppression of the cell-survival protein bcl-2 than that observed with just tamoxifen [24]. In particular, combined therapy prevented the acquired expression of EGFR/MAPK signaling and the subsequent resistance that occurred after 5 weeks in tamoxifen-alone-treated cells.

For established hormone-resistant, HER2-positive breast cancer, the strategy of combined STIs and endocrine therapy may also be more effective than using STIs alone. In vivo, gefitinib and tamoxifen provided maximal growth inhibition and significantly delayed the growth of HER2-positive MCF-7 xenografts compared with gefitinib alone [25]. Moreover, similar effects were seen with gefitinib combined with estrogen deprivation, which provided greater inhibition of growth and substantially delayed acquired resistance compared with estrogen deprivation alone [26]. A synergistic effect has also been reported for Trastuzumab combined with tamoxifen in ER-positive, HER2-positive BT-474 breast cancer cells, with enhanced accumulation of cells in G0/G1 and reduction in S phase of the cell cycle compared with either therapy alone [27]. Of interest, there was no evidence for any induction of apoptosis. Recently, the dual EGFR/HER2 inhibitor lapatinib has been shown to cooperate with tamoxifen to provide more rapid and profound cell-cycle arrest than either therapy alone in hormone-resistant cells [28]. The two drugs together caused a greater reduction in cyclin D1, together with a greater increase in the kinase inhibitor p27 and cyclin E-cdk2 inhibition, in various tamoxifen-resistant breast cancer cell lines. Lapatinib was able to restore tamoxifen sensitivity in EGFR- or HER2-expressing cells, while, in vivo, combined laptaninib and tamoxifen caused maximal regression of HER2-overexpressing, tamoxifen-resistant MCF-7 xenografts [28].

Other STIs that have only a minimal effect on hormone-sensitive breast cancer may also be more effective when combined with endocrine therapy. The FTI tipifarnib inhibits the growth of a number of human breast cancer cells lines in vitro, most of which contain normal wild-type *ras* genes [29]. However, in vivo, tipifarnib produced only a modest cytostatic effect on hormone-sensitive MCF-7 xenograft growth, with evidence of induction of apoptosis and enhanced expression of the cell-cycle inhibitory protein p21 [30]. In contrast, when tipifarnib was combined with tamoxifen or estrogen deprivation therapy, combined treatment induced significantly greater tumor regression than either endocrine therapy alone [31]. Three other groups have since reported a similar interaction for FTIs with tamoxifen or aromatase inhibitors, and have suggested either a synergistic [32] or an additive antitumor effect [33]. One recent study implied an additive effect on G0/G1 cell-cycle arrest, and that the FTI-277, when combined with tamoxifen, maintained higher levels of the cdk inhibitor p21waf/cip1, resulting in an additive effect on inactivation of cyclin E/Cdk2 complexes and decreased phosphorylation of pRb [34].

More recently, a similar rationale has emerged to support the combination of mTOR antagonists with either tamoxifen or an aromatase inhibitor in preclinical models of ER-positive, hormone-sensitive, and resistant breast cancer [35]. The estrogen-dependent growth of both wild-type MCF-7 and aromatase-expressing (MCF-7/Aro) breast cancer cells was inhibited in adose-dependent manner by the mTOR antagonist everolimus (RAD-001), suggesting that mTOR signaling is required for the estrogen-dependent proliferation of these cells. In subsequent experiments with the MCF-7/Aro cells, the combination of letrozole and everolimus produced maximal growth inhibition with clear evidence for additive/synergistic effects. Evidence has emerged that activation of Akt/PKB and the downstream mTOR pathway can cause resistance to tamoxifen [36]. MCF-7 cells expressing a constitutively active Akt were able to proliferate under reduced estrogen conditions and were resistant to the growth inhibitory effects of tamoxifen, both in vitro and in vivo, in xenograft models [35]. However, cotreatment with the mTOR inhibitor temsirolimus (CCI-779) inhibited mTOR activity and restored sensitivity to tamoxifen, primarily through induction of apoptosis, thus suggesting that Akt-induced tamoxifen resistance may in part be mediated by signaling through the mTOR pathway.

5. CLINICAL TRIALS WITH INTRACELLULAR STIS IN BREAST CANCER

All these experimental data indicate that a variety of intracellular signaling pathways may be activated or overexpressed in endocrine-resistant breast cancer, and suggest that targeting such pathways may be an effective therapy. Trastuzumab (Herceptin), the monoclonal antibody to the external domain of HER2, provides significant activity even as single agent. Thus, it was hoped that various small molecule STIs may also prove effective anticancer strategies in this setting.

Early phase II clinical studies were initiated with EGFR/HER2 TKIs, FTIs, and more recently mTOR antagonists, both as monotherapy or more recently in combination with endocrine therapy and cytotoxic agents.

6. TRIALS WITH EGFR TYROSINE KINASE INHIBITORS

The most extensively studied TKI in breast cancer is the EGFR TKI gefitinib (Iressa), an orally active, low molecular weight, synthetic anilinoquinazoline and a potent selective inhibitor of EGFR-TK. In various breast cancer cell lines that express EGFR and/or HER2, gefitinib given as a single agent induced a dose-dependent antiproliferative effect which delayed growth [37]. Experiments have demonstrated that gefitinib may inhibit the growth of acquired endocrine-resistant MCF-7 breast cancer cells in vitro [19,21,38] and in vivo [16].

There have been three phase II monotherapy studies of gefitinib in patients with advanced breast cancer [39–41]. Overall, the data are relatively disappointing with low clinical response rates and short times to disease progression. The only trial to report a significant number of responses included patients with ER-positive, tamoxifen-resistant breast cancer [41], the setting in which preclinical models had shown the best evidence of activity for gefitinib. Pharmacodynamic studies have been performed in one of these trials, confirming that EGFR tyrosine kinase signaling is inhibited in both skin and tumor biopsies by doses of gefitinib delivered orally [40]. However, there was discordance in the effect of gefitinib on downstream intracellular signaling in treated tumor biopsies, with lack of inhibition of Ki-67 (a marker of cell proliferation) in tumor, but not in matched skin biopsies. This suggested that activation of other intracellular pathways downstream of EGFR (especially in breast cancer as opposed to normal skin cells) might determine the clinical response to gefitinib. More research is required to establish tumor phenotypes in responding versus nonresponding patients [42]. Fewer clinical data exist regarding the other EGFR TKI in breast cancer, although a phase II trial of the selective EGFR TKI erlotinib (OSI-774) in breast cancer was relatively disappointing [43, 44].

Based on the preclinical evidence for added benefit outlined above, a number of small phase I/II trials have been initiated with TKIs in combination with tamoxifen, fulvestrant or an aromatase inhibitor. Some of these trials are in the second-line setting, with one trial enrolling patients whose tumor was progressing on tamoxifen, and then adding lapatinib to tamoxifen to see whether clinical responses could be observed and resistance reversed. Another trial will compare gefitinib alone to the combination of gefitinib plus tamoxifen after progression on tamoxifen. However, the ultimate clinical test for the hypothesis that TKIs enhance the efficacy of endocrine therapy is the randomized, controlled clinical trial of endocrine therapy alone versus combined endocrine/TKI therapy. Many of these trials are in the first-line setting, where early clinical (and indeed experimental) data have shown that gefitinib alone may have limited activity. Therefore, the primary endpoint for these trials is to investigate whether time to disease progression (TTP) can be significantly prolonged by the addition of a TKI to endocrine therapy, thus delaying the emergence of resistance as demonstrated in various preclinical models described above.

7. TRIALS WITH DUAL EGFR/HER2 TYRSOINE KINASE INHIBITORS

Cooperative activation of different type I growth factor receptors (EGFR, HER2, HER-3, or HER-4) may limit the efficacy of targeting just one single receptor. Lapatinib (GW572016) is a potent dual inhibitor of both EGFR and HER2 and inhibits autophosphorylation of the receptors [44]. In the phase I study, diarrhea and skin rash were the main toxicities [45], and clinical activity was reported in trastuzumab-resistant breast cancer patients [46]. Several biologic and pharmacologic reasons may account for the efficacy of a small molecule inhibitor of HER2 in patients resistant to monoclonal therapy with trastuzumab. A phase II trial of lapatinib has been completed in heavily pretreated patients with advanced breast cancer that progressed on prior Trastuzumab containing regimens. A recent interim analysis in the first 41 patients confirmed clinical activity for Lapatinib in breast cancer, with partial responses in 7% of patients and/or stable disease in 24% of patients after 16 weeks of therapy [47].

The pivotal trials for lapatinib will be the large, randomized, phase III trial of letrozole with or without lapatinib in ER-positive advanced metasatic breast cancer. This study in over 760 patients is powered to detect a 30% improvement in median time to disease progression from 10 to 13 months (hazard ratio of 0.769).

8. TRIALS WITH FTIS

Monotherapy activity has been reported for the FTI tipifarnib (Zarnestra) in advanced breast cancer. A total of 76 patients were treated with tipifarnib, either as a continuous dose of 300 or 400 mg b.i.d. ($n = 41$) or an intermittent dose of 300 mg b.i.d. for 21 days followed by 7 days off-therapy ($n = 35$) [31]. In the continuous treatment arm, there were four partial responses (10%) lasting 4–12 months and six patients with stable disease (15%) for at least 6 months. In the intermittent treatment arm, there were five partial responses (14%) and three patients with stable disease (9%). Objective responses were seen in both soft tissue and viscereal sites of disease. The main toxicities were neutropenia, thrombocytopenia, neurotoxicity, and fatigue. These results were independent of ras status, estrogen/PR, or HER2/EGFR receptor status, and 40% of patients had received only prior adjuvant and/or metastatic endocrine therapy at entry into the trial. Thus, while FTIs may have modest antitumor activity as a single-agent therapy, stabilization of disease in those resistant to endocrine therapy may represent a meaningful response.

Based on the encouraging preclinical data discussed above from several groups for additive or synergistsic effects for combining FTIs with endocrine therapy, several nonrandomized and randomized phase II studies of FTIs in combination with either tamoxifen, fulvestrant, or aromatase inhibitors have been undertaken. It is unlikely that any overlapping toxicities will be seen for combinations of FTIs with endocrine therapy, but at least two trials are ensuring that no pharmacokinetic interactions exist whereby tamoxifen- or aromatase inhibitor-induced hepatic enzymes could enhance clearance of FTIs and lower serum concentrations. For example, pharmacokinetic and pharmacodynamic endpoints have been assessed by a sequential design in 11 patients treated initially with the FTI tipifarnib (either 200 or 300 mg b.i.d. for 21/28 days), and after 1 week tamoxifen was added [48]. There was no significant change in the pharmacokinetic profile for tipifarnib; moreover, the pharmacodynamic end point (inhibition of farnesyltransferase in peripheral blood mononuclear cells) was enhanced from 30% enzyme suppression to 41% by the combination. In addition, such early phase II trials may help to determine the optimal schedule for the combination. There are no published efficacy data from any of the randomized endocrine/FTI trials to date.

9. TRIALS WITH mTOR ANTAGONISTS

Two mTOR antagonist agents are now in phase III study trials, RAD-001 and CCI-779. Both have provided activity as single agent. Temsirolimus (CCI-779) given by weekly intravenous administration to heavily pretreated metastatic breast cancer whose disease relapsed or was refractory to previous anthracyclines and/or taxanes [49]. Based on 94 evaluable patients, partial responses were seen in 11% patients with stable disease in an additional 33% of patients. The main toxicities included alteration in transaminases, mucositis, rash, and mild nausea.

Again based on encouraging preclinical data for possible synergistic effects when mTOR antagonists were combined with endocrine therapy [35], clinical trials looking at combined therapy have been initiated. Preliminary data from a small, three-arm randomized study has compared two different schedules of oral temsirolimus (10 mg continuous or 30 mg for 5–14 days) combined with letrozole. Initial higher doses of temsirolimus were used but were poorly tolerated (grade 2/3 stomatitis) when combined with long-term endocrine therapy [50]. Efficacy results have suggested similar response rates for the combination compared with letrozole alone. However, clinical end points using objective tumor response rates in small, nonrandomized phase II studies must be viewed cautiously, and the primary role of such studies should be to provide safety and supportive biologic data for the combination in advance of definitive randomized trials.

Two large-scale phase III clinical trial with orally active formulations of temsirolimus and everolimus in combination with letrozole have been started. The first targets are metastatic patients; the second is designed as an exploratory study in the neoadjuvant setting.

10. CONCLUSION

There is much enthusiasm surrounding these novel STIs and their potential role in the treatment of breast cancer. However, considerable thought is needed in order to maximize their potential. Central to their development will be a clear understanding of the molecular biology of these pathways, in particular the differences between hormone-sensitive and hormone-resistant disease. Preclinical models both in vitro and in vivo are important to clarify the benefit and utility of combined endocrine/STI therapy. For clinical trials, appropriate patient selection will be important, and parallel biologic studies are now a requirement for development of these drugs. Randomized studies remain central to determine any added benefit of combined endocrine/STI therapy, and must be appropriately powered for relevant endpoints in order to encourage further clinical development. Presurgical studies may be an additional means of gaining biologic and clinical information about such combined therapies. Many trials have now started, and we eagerly await the results to see whether intracellular STIs can provide a significant therapeutic benefit.

REFERENCES

1. Jemal A, Tiwari RC, Murray T, et al. Cancer statistics, *Cancer J Clin* 54: 30–40, 2004.
2. Early Breast Cancer Trialists' Collaborative Group: Effects of chemotherapy and hormonal therapy for early breast cancer on recurrence and 15-year survival: an overview of the randomised trials. *Lancet* 365: 1687–1717, 2005.
3. Strasser-Weippl K, Goss PE. Advances in adjuvant hormonal therapy for postmenopausal women. *J Clin Oncol* 23: 1751–9, 2005.
4. Ali S, Coombes RC. Endocrine-responsive breast cancer and strategies for combating resistance. *Nat Rev Cancer* 2: 101–112, 2002.
5. Klijn JG, Berns EM, Foekens JA. Other endocrine and biological agents in the treatment of advanced breast cancer, in: Miller WR, Ingle JN (ed) *Endocrine Therapy in Breast Cancer*. Marcel Dekker, New York, 2002.
6. Nicholson RI, McClelland RA, Robertson JRF, and Gee JMW. Involvement of steroid hormone and growth factor cross-talk in endocrine response in breast cancer. *Endocr Relat Cancer* 6: 373–387, 1999.
7. Nemere I, Pietras RJ, Blackmore PF. Membrane receptors for steroid hormones: Signal transduction and physiological significance. *J Cell Biochem* 88: 438–445, 2003.
8. Cui X, Schiff R, Arpino G, Osborne CK, Lee AV. Biology of progesterone receptor loss in breast cancer and its implications for endocrine therapy. *J Clin Oncol* 23: 7721–7735, 2005.
9. Levin ER. Cellular functions of the plasma membrane estrogen receptor. *Trends Endocrinol Metab* 10: 374–377, 1999.
10. Razandi M, Alton G, Pedram A, et al. Identification of a structural determinant necessary for the localization and function of estrogen receptor alpha at the plasma membrane. *Mol Cell Biol* 23: 1633–1646, 2003.
11. Biscardi JS, Ishizawar RC, Silva CM, Parsons SJ. Tyrosine kinase signalling in breast cancer: epidermal growth factor receptor and c-Src interactions in breast cancer. *Breast Cancer Res* 2: 203–210, 2000.
12. Kurokawa H, Arteaga CL. ErbB (HER) receptors can abrogate antiestrogen action in human breast cancer by multiple signaling mechanisms. *Clin Cancer Res* 9: 511S–515S, 2003.
13. Menashi S, Serova M, Ma L, et al. Regulation of extracellular matrix metalloproteinase inducer and matrix metalloproteinase expression by amphiregulin in transformed human breast epithelial cells. *Cancer Res* 63: 7575–7580, 2003.
14. Yuste L, Montero JC, Esparis-Ogando A, et al. Activation of ErbB2 by overexpression or by transmembrane neuregulin results in differential signaling and sensitivity to herceptin. *Cancer Res* 65: 6801–6810, 2005.
15. Kurokawa H, Lenferink AE, Simpson JF, et al. Inhibition of HER2/neu (erbB-2) and mitogen-activated protein kinases enhances tamoxifen action against HER2-overexpressing, tamoxifen-resistant breast cancer cells. *Cancer Res* 60: 5887–5894, 2000.
16. Shou J, Massarweh S, Osborne CK, et al. Mechanisms of tamoxifen resistance: Increased estrogen receptor-HER2/neu cross-talk in ER/ HER2-positive breast cancer. *J Natl Cancer Inst* 96: 926–935, 2004.

17. Kumar R, Wang RA, Mazumdar A, et al. A naturally occurring MTA1 variant sequesters oestrogen receptor-alpha in the cytoplasm. *Nature* 418: 654–657, 2002.
18. Nicholson RI, Hutcheson IR, Knowlden JM, et al. Nonendocrine Pathways and Endocrine Resistance: Observations with Antiestrogens and Signal Transduction Inhibitors in Combination. *Clin Cancer Res* 10: 346–354, 2004.
19. Knowlden JM, Hutcheson IR, Jones HE, et al. Elevated levels of epidermal growth factor receptor/c-erbB2 heterodimers mediate an autocrine growth regulatory pathway in tamoxifen-resistant MCF-7 cells. *Endocrinology* 144: 1032–1044, 2003.
20. Witters L, Engle L, and Lipton A. Restoration of estrogen responsiveness by blocking the HER-2/neu pathway. *Oncol Rep* 9: 1163–1166, 2002.
21. Martin LA, Farmer I, Johnston SR, et al. Enhanced estrogen receptor (ER) alpha, ERBB2, and MAPK signal transduction pathways operate during the adaptation of MCF-7 cells to long term estrogen deprivation. *J Biol Chem* 278: 30458–30468, 2003.
22. Kurokawa H, Lenferink AEG, Simpson JF, et al. Inhibition of HER2/neu (erbB-2) and Mitogen-activated Protein Kinases Enhances Tamoxifen Action against HER2-overexpressing, Tamoxifen-resistant Breast Cancer Cells. *Cancer Res* 60: 5887–5894, 2000.
23. Yu K, Toral-Barza L, Discafani C, et al. mTOR, a novel target in breast cancer: the effect of CCI-779, an mTOR inhibitor, in preclinical models of breast cancer. *Endocr Relat Cancer* 8: 249–258, 2001.
24. Gee JM, Harper ME, Hutcheson IR, et al. The antiepidermal growth factor receptor agent gefitinib (ZD1839/Iressa) improves antihormone response and prevents development of resistance in breast cancer in vitro. *Endocrinology* 144: 5105–5117, 2003.
25. Shou J, Massarweh S, Osborne CK, et al. Mechanisms of tamoxifen resistance: increased estrogen receptor-HER2/neu crosstalk in ER/HER2-positive breast cancer. *J Nat Cancer Inst* 96: 926–935, 2004.
26. Massarweh S, Shou J, Mohsin SK, et al. Inhibition of epidermal growth factor/HER2 receptor signaling using ZD1839 (Iressa) restores tamoxifen sensitivity and delays resistance to estrogen deprivation in HER2-overexpressing breast tumors. *Proceedings of the American Society of Clinical Oncology* 21 33a (A130), 2002.
27. Argiris A, Wang CX, Whalen SG, and DiGiovanna MP. Synergistic interactions between tamoxifen and trastuzumab (Herceptin). *Clin Cancer Res* 10: 1409–1420, 2004.
28. Chu I, Blackwell K, Chen S, and Slingerland J. The dual ErbB1/ErbB2 inhibitor lapatinib (GW572016) cooperates with tamoxifen to inhibit both cell proliferation and estrogen-dependent gene expression in antiestrogen resistant breast cancer. *Cancer Res* 65: 18–25, 2005.
29. End DW, Smets G, Todd AV, et al. Characterization of the antitumor effects of the selective farnesyl protein transferase inhibitor R115777 in vivo and in vitro. *Cancer Res* 61: 131–137, 2001.
30. Kelland LR, Smith V, Valenti M, et al. Preclinical antitumor activity and pharmacodynamic studies with the farnesyl protein transferase inhibitor R115777 in human breast cancer. *Clin Cancer Res* 7: 3544–3550, 2001.
31. Johnston SR, Hickish T, Ellis P, et al. Phase II study of the efficacy and tolerability of two dosing regimens of the farnesyl transferase inhibitor, R115777, in advanced breast cancer. *J Clin Oncol* 21: 2492–2499, 2003.
32. Ellis CA, Vos MD, Wickline M, et al. Tamoxifen and the farnesyl transferase inhibitor FTI-277 synergize to inhibit growth in estrogen receptor-positive breast tumor cell lines. *Breast Cancer Res Treat* 78: 59–67, 2003.
33. Long BJ, Liu G, Marrinan CH, et al. Combining the farnesyl transferase inhibitor (FTI) lonafarnib (SCH66336) with antiestrogens and aromatase inhibitors results in enhanced growth inhibition of hormonedependent human breast cancer cells and tumor xenografts. *ASCO Proceedings* 45: A3868, 2004.
34. Doisneau-Sixou SF, Cestac P, Faye JC, et al. Additive effects of tamoxifen and the farnesyl transferase inhibitor FTI-277 on inhibition of MCF-7 breast cancer cell-cycle progression. *Int J Cancer* 106: 789–798, 2003.

35. DeGraffenried LA, Friedrichs WE, Russel, et al. Inhibition of mTOR activity restores tamoxifen response in breast cancer cells with aberrant Akt activity. *Clin Cancer Res* 10: 8059–8067, 2004.
36. Clark AS, West K, Streicher S, and Dennis PA. Constitutive and inducible Akt activity promotes resistance to chemotherapy, trastuzumab, or tamoxifen in breast cancer cells. *Mol Cancer Ther* 1: 707–717, 2002.
37. Moulder SL, Yakes FM, Muthuswamy SK, et al. Epidermal growth factor receptor (HER1) tyrosine kinase inhibitor ZD1839 (Iressa) inhibits HER2/neu (erbB2)-overexpressing breast cancer cells in vitro and in vivo. *Cancer Res* 61: 8887–8895, 2001.
38. Hutcheson IR, Knowlden JM, Madden TA, et al. Oestrogen receptor-mediated modulation of the EGFR/MAPK pathway in tamoxifen-resistant MCF-7 cells. *Breast Cancer Res Treat* 81: 81–93, 2003.
39. Albain K, Elledge R, Gradishar WJ, et al. Open-label phase II multicenter trial of ZD1839 (Iressa) in patients with advanced breast cancer. *Breast Cancer Res Treat* 76 A20, 2002.
40. Baselga J, Albanelli J, Ruiz A, et al. Phase II and tumor pharmacodynamic study of gefitinib in patients with advanced breast cancer. *ASCO Proceedings* 22: 24, 2003.
41. Robertson JFR, Gutteridge E, Cheung KL, et al. Gefitinib (ZD1839) is active in acquired tamoxifenresistant oestrogen receptor positive and ER-negative breast cancer: results from a phase II study. *ASCO Proceedings* 22: 23, 2003.
42. Dancey JE and Freidlen B. Targeting epidermal growth factor receptor – are we missing the mark? *Lancet* 362: 62–64, 2003.
43. Winer E, Cobleigh M, Dickler M, et al. Phase II multicenter study to evaluate the efficacy and safety of Tarceva (erlotinib, OSI-774) in women with previously treated locally advanced or metastatic breast cancer. *Breast Cancer Res Treat* 76: A445, 2002.
44. Rusnak DW, Lackey K, Affleck K, et al. The effects of the novel, reversible epidermal growth factor receptor/ erbB2 tyrosine kinase inhibitor, GW2016, on the growth of human normal and tumor-derived cell lines in-vitro and in-vivo. *Mol Cancer Ther* 1: 85–94, 2001.
45. Xia W, Mullin R, Keith B, et al. Antitumor activity of GW572016; a dual tyrosine kinase inhibitor blocks EGFR activation of EGFR/erbB2 and downstream Erk1/2 and AKT pathways. *Oncogene* 21: 6255–6263, 2002.
46. Spector N, Raefsky E, Hurwitz H, et al. Safety, clinical efficacy, and biological assessments from EGF10004; a randomised phase IB study of GW572016 for patients with metastatic carcinomas overexpressing EGFR or erbB2. *ASCO Proceedings* 22: 193 (A772), 2003.
47. Blackwell KL, Kaplan EH, Franco SX, et al. A phase II, open-label, multicenter study of GW572016 in patients with trastuzumab-refractory metastatic breast cancer. *ASCO Proceedings* 22: 3006, 2004.
48. Lebowitz PF, Eng-Wong J, Balis F, et al. A phase I trial of tipifarnib, a farnesyltransferase inhibitor, and tamoxifen in hormonereceptor positive metastatic breast cancer. *ASCO Proceedings* 22: 644, 2004.
49. Chan S, Scheulen ME, Johnston SRD, et al. A phase II study of two dose levels of CCI-779 in locally advanced or metastic breast cancer failing prior anthracyclines and/or taxanes regimens. *ASCO Proceedings* 22: 774, 2003.
50. Baselga J, Fumoleau P, Gil M, et al. Phase II, 3-arm study of CCI-779 in combination with letrozole in postmenopausal women with locally advanced or metastatic breast cancer: preliminary results. *ASCO Proceedings* 22: 544, 2004.

INDEX

Actinomycin D (ACT-D), 16
Acute lymphoblastic leukemia (ALL), 122-125
   and genetic abnormalities, 123
Adenocarcinoma
   esophageal adenocarcinomas (Barret's adenocarcinoma), 117
   of the upper gastrointestinal tract, 115, 116, 119
AEE788, 337
AF4/MLL fusion gene, 123
AG1478, 331
Akt, 14, 310, 311, 314, 315, 320, 322, 329-335, 337-380
   PI3K/Akt pathway, 377, 378
ALK-TMP3 fusion gene, 99, 101, 124
   inflammatory myofibroblastic tumor, 99, 101
   large cell anaplastic lymphoma, 99, 101
ALK-TMP4 fusion gene, 136
Alveolar rhabdomyosarcoma and specific translocation, 54, 100, 126, 127, 129
Alveolar soft part sarcoma, 55, 100, 126
Amphiregulin, 330, 376
Anaplastic large cell lymphoma (ALCL), 124, 125
Anaplastic lymphoma kinase (ALK) fusion genes, 124
Anastrozole, 375
Angiogenesis, 25, 180, 182, 196, 223, 251-257, 268, 279, 280, 330, 342-344, 348, 377
   and tumor development, 253

Angiosarcoma, 100, 108, 109, 126
   and HHV8, 108, 109
Anti-CD99 0662 Mab, 18, 19
Anti-tumor associated antigens and cancer therapy, 341-368
AP23573 mTOR inhibitor, 309, 310, 315, 320-322
APC gene, 171
Aromatase inhibitors, 336, 375, 379-382
   Anastrozole, 375
   Exemestane, 375
   Letrozol, 318, 336, 375, 380, 382, 383
Aryl Hydrocarbon receptor nuclear translocator (ARNT), 181, 182
ARNT-2, 181
Askin's tumor, 3
Athymic mice, 7, 13, 15, 17, 19
Atypical Ewing's sarcoma, 3
β-catenin, 26, 60, 121, 131-133
Barret's adenocarcinoma, 117
Bax, 116
Beckwith-Wiedman syndrome (BWS), 131
Betacellulin, 330
Bcl-2, 116, 379
BCL9-2, 131
BCR/ABL fusion gene, 123, 169
Bevacizumab in metastatic colon cancer, 262-265
Bevacizumab with oxaliplatin, 262, 264, 269
Bio-repositories, 66, 76

Bladder carcinomas and WWOX gene expression, 157
BMS-599626 (ErbB receptor inhibitor), 330, 332, 337
Bone tumor, 13, 14, 54, 96, 121
Brain metabolites observed by 1H-MRS, 290
Brain tumors and metabolism by MRS vs. Histopathology, 286
Brain tumors and MRS, 285-302
Brain tumors and in vivo metabolic diagnosis, 285
Brain tumors classification and grade by MRS, 290, 291, 298
Brain tumors detection and location by MRS, 291
  discrimination between recurrence and radio-necrosis, 295
Breast cancer, 60, 149-153, 179, 180, 185, 186, 199, 206, 239, 241, 269, 279, 314, 315, 318, 330, 332, 334-337, 375-384
  and the hypoxia model, 179
  hormone sensitive, 375-384
  pharmagenomic predictors for preoperative chemotherapy, 233-247
Burkit lymphoma, 124, 125

Caldesmon, 115, 117
Cancer care in Europe, 88
  activities against cancer, 89
Cancer control in Europe, 87-93
Capecitabine, 257, 258, 336
Carboplatin, 257
Carboxylesterase (CE) polymorphysms, 218
Caspase-independent mechanisms, 13, 18
CCI-779 mTOR inhibitor, 309, 310, 314-322
  Phase I clinical trials, 314, 317, 318
  Phase II clinical trials, 316, 318, 319
CCN family of growth regulators, 23-27
  biological functions, 25
  CCN1 (Cry61), 23, 26, 27, 35
  CCN2 (Ctgf), 25-27, 35-37
  CCN3 (NOV), 8, 23, 25-27, 31
    antiproliferative activity of the full length, 31-34
    expression in tumors, 27-30
    proteins, 34-37
    nuclear addressing of the aminotruncated CCN3 proteins, 37, 38
    model for the avctivity of CCN3 proteins, 35, 36
  CCN4 (Wisp 1), 23, 27
  CCN5 (Wisp 2), 23, 24, 26, 27
  CCN6 (Wisp 3), 23
CD99 (see also MIC2), 2, 4, 6, 7, 9, 13, 14, 17-19, 41, 42, 49, 50, 102
  anti-CD99 0662 Mab, 7, 8
  zyxin antisense oligonucleotides, 18
CDK1A$^{p21/Waf1}$, 48
CDK1C$^{p57/Kip2}$, 48
Cell cycle and mouse models, 139
Cell cycle and tumor development, 143
Cell cycle regulation, 49, 139, 141, 144, 196
Cell lines, 7, 8, 17, 32, 36, 41, 43-50, 105, 141, 149, 151-153, 155, 167, 182, 184, 187, 199, 202, 206, 222, 254, 314, 315, 329, 333, 335, 377-380
  3LL (lung carcinoma), 349, 356, 358
  A673 (Ewing), 46, 48
  BT474 (Breast cancer), 334, 335
  G59 (glioblastoma), 31, 32, 36
  HCT116 (colorectal cancer), 199
  HT29 (colorectal cancer), 202
  JEG (choriocarcinoma), 32
  MCF7 (breast cancer), 151-155, 336, 378-381
  MDA-MB231 (breast cancer), 151, 152
  MDA-MB431 (breast cancer), 152
  MDA-MB435 (breast cancer), 151, 152
  NIH3T3 (fibroblasts), 43, 46
  Peo1 (ovarian cancer), 152
  SK-N-MC (Ewing), 48
  STAT-ET-1 (Ewing), 48
  STAT-ET-7.2 (Ewing), 48
  T47D (breast cancer), 152
  TC252 (Ewing), 48
  TC71 (Ewing), 34
  VH64 (Ewing), 48
  WE68 (Ewing), 48

c-erbB-2 (see also HER-2), 376
  Dual/Pan-HER tyrosine kinase
    inhibitors, 333
Cetuximab, 258, 268, 332
c-fos, 322
CHFR gene in NSCLC, 199
Chicken embryo fibroblasts (CEF), 27,
  29, 30, 315
Chondrosarcoma, 18, 33, 53-55, 126, 317
  expression analysis, 59
  myxoid, extraskeletal, 102, 126
  specific translocation, 54
Choriocarcinoma, 31, 32
Chromatin immunoprecipitation (ChIP)
  approach, 45, 47
  in Ewing's sarcomas, 59
CI-1033 (ErbB receptor inhibitor), 330,
  332, 336, 337
Cisplatin, 16, 115-119, 185, 199, 201, 206,
  211, 222
  related genes, 115
CK1 kinase, 131
Clear cell sarcoma (see also EWS-ATF1
  fusion gene), 55, 59, 99, 100, 103,
  104, 126
CMYC, 49, 125, 131
Colon cancer, 171, 222, 251, 345-347,
  359
  and PET-Scan, 306
  and VEGF inhibitors, 251-270, 359
Colorectal cancer and
  pharmacogenomics, 211-225
Competitive ATP inhibitors, 331
  AG1478, 331
  PD153035, 331
CONFIRM-1 clinical trial, 251, 252,
  265-268
CONFIRM-2 clinical trial, 251, 252,
  266-268
Core needle biopsy (CNB) in breast
  cancer for DNA microarrays, 236,
  237, 239, 241
Cry61 (see CCN1)
Ctgf (see CCN2)
CT-Scan, 118, 294, 303
CXC chemokine family, 223
  CXCR1, 223, 224
  CXCR2, 223

Cyclin D, 141, 142, 144, 212, 312, 332,
  335
Cyclin-dependent kinases, 139-141
  inhibitors, 140, 141
Cyclolignan PPP, 16, 17
CYP3A4, 219
CYP3A5, 219

Denaturing high pressure liquid
  chromatography (DHPLC), 118
Dermatofibrosarcoma protuberans and
  giant cell fibroblastoma, 55, 100
Desmoplastic round cell tumor and
  specific translocation, 54, 55, 100,
  126
Diagnosis of childhood malignancies,
  121-134
Dihydropyrimidine dehydrogenase
  (DPD), 115-117, 119, 211-213,
  217, 219, 220
  and polymorphisms, 213
DNA methylation, 118, 196, 197, 201
  and cell cycle, 197
  and circulating DNA, 197
  and DNA repair, 197
  and lung cancer, 195-206
  inhibitors, 118
  Docetaxel and multigene predictor in
    breast cancer, 241
Doxorubicin (DXR), 15, 16, 18, 19, 233,
  241, 245, 257
DNA-polymorphisms, 116
Dual/Pan-HER tyrosine kinase inhibitors,
  229

E-cadherine, 118, 131
  methylation, 118
ELF3, 117
EKB-569, 332
Embryonal rhabdomyosarcoma, 126
Epidermal growth factor (EGF), 37, 330,
  378, 376
Epidermal Growth Factor Receptor
  (EGFR, HER1, ErbB1), 104, 203,
  222-225, 314, 329-337
  Dual/Pan-HER tyrosine kinase
    inhibitors, 329
  Inhibitors and breast cancer, 376-378

Epidermal Growth Factor
Receptor (*Continued*)
Ligands, 330
Mechanisms of Activation and
Signaling Pathways, 330
Polymorphisms, 224
Ras-Raf-MAPK pathway, 330, 331
EORTC, 68, 73, 84, 92
Epiregulin, 330
ErbB receptors, 330-332, 337
Inhibitors, 332, 333, 336
BMS-599626, 332, 337
CI-1033, 332, 336
ERCC1, 115-117, 119, 211, 213, 221
ERCC2, 221
ERCC4, 115, 117, 119
Erlotinib, 332, 381
Estrogen receptor in breast cancer, 375-384
Estrogen-responsive elements (ERE), 376
Etoposide, 7, 16
E3200 clinical trial, 262-264, 269
eTUMOR project and MRS for brain and prostate cancer, 285-300
ETV-NTRK3 fusion gene, 55, 101, 126
infantile fibrosarcoma, 99, 101
secretory carcinoma of the breast, 99, 101
European Commission, 1, 65, 67, 69, 76, 84
European human frozen tumor tissue bank (network), see also TuBaFrost, 65, 67, 69, 70, 75, 76, 84
European Virtual Tumor Tissue Banking, 64-73
Ewing's sarcoma (see also EWS-FLI1), 2-6, 13-19, 32-34, 41-50, 53-55, 58, 59, 99, 101, 103
expression analysis, 59
Ewing's Sarcoma Family of Tumors (ESFT), 1-9, 41-50
Histogenesis, 3
Prognosis, 4
Therapy, 4
Varieties, 2
EWS-ets gene rearrangements, 3, 14, 43
EWS-ATF1, 55, 99, 103

EWS-ERG, 4-6, 14, 126
EWS-FLI1, 2, 4-9, 13-16, 41-50, 54, 99, 126
and CD99, 49, 50
common affected pathways, 49
dependent transformation, 16
functional hierarchy, 43
isolation of direct target genes, 41, 44
tissue specific targets, 42-46
type1, 4
EWS gene (see also EWS-ets gene rearrangements), 42, 46, 48, 101, 103
Excision repai cross-complemengting (ERCC) gene family, 221
ERCC1 polymorphisms, 221
ERCC2 polymorphisms, 221
Exemestane, 375
4E-BP1, 310, 312, 313, 315, 320
14-3-3σ (stratifin), 118, 195, 199, 200, 202, 203, 206
methylation, 118, 195

Familial polyposis coli (FAP), 131, 169, 171
FANCF gene in NSCLC, 201, 206
Farnesyltransferase inhibitors (FTIs), 377, 383
clinical trials, 382, 383
FGF18, 131
Fibrosarcoma, 55, 99, 101, 126
Fine needle aspirations (FNA) in breast cancer for DNA microarrays, 236, 237, 241
FK506-binding protein (FKBP12), 310, 313
FLI1, 41-50
Fluoracil and metabolism (5-FU), 117, 233, 241, 257, 317, 318
and pharmacogenomics, 233
Fluorodeoxyuridylate (5-FdUMP), 213
Folate metabolism, 115
FOLFOX4, 251, 258, 263-266
Fractional allelic loss (FAL), 118
FRA16D, 149, 151
FUS-CHOP fusion gene, 99, 126
Fusion genes in sarcomas, 126

GADD45A, 115
Gastric carcinomas, 115, 116, 118, 120
Gastrointestinal stromal tumors (GIST), 99, 100, 104
  Carney triad, 106
  Imatinib mesylate, 99, 104, 105
  KIT, 99, 104-108
  PDGFRA, 99, 104-106
  NF1, 107
Gefitinib, 314, 320, 332, 337, 378-381
Gemcitabine and NSCLC, 195, 201, 206
Gene expression profiling, 44, 46, 59, 129, 137, 166, 234, 236
  from needle biopsies in breast cancer, 234, 235
  platforms, 236, 238
Germline polymorphisms and clinical significance in colorectal cancer, 213
Glioblastoma cells, 31, 32, 36
Glioblastoma multiforme and MRS, 319-321
Glioma and MRS, 288, 292, 294, 298
Glutathione S-transferase (GSTP1), 116
  and polymorphisms, 213, 222
GSK3 kinase, 131
GTV, 303, 305, 306
GW572016 (Lapatinib), 329, 333-336, 381

Heparin binding EGF, 330, 376
Hepatoblastoma, 121, 130-133
Hepatocellular carcinoma and FRA16D, 150, 151
HER receptor signalling family, 329
  HER1 (ErbB1), 329, 330, 337
  HER2 (see also c-erbB-2), 117, 247, 318, 324, 329-337, 376-382
  HER3 (ErbB3), 329-331, 334, 336, 377
  HER4 (ErbB4), 330, 337, 376
HER2p95, 329-331, 334
  HER2p95-HER3 heterodimers, 329, 334
  lapatinib, 329, 332
  trastuzumab, 329, 332
HEY1 gene, 48
High-grade undifferentiated pleomorphic sarcoma, 126
Histone deacetylase inhibitors, 118

HPP1 methylation, 118
Human carboxylesterase (hCE-2) polymorphisms, 218
Human herpesvirus 8 (HHV8), 100, 108, 109
  Angiosarcoma, 100, 108, 109, 126
  hemangioma, 109
  latent nuclear antigen-1, 109
  Kaposi sarcoma, 100, 108, 109
Hypoxia
  and the breast cancer model, 179, 185
  and the neuroblastoma model, 179, 183
  cellular response, 180
  HIF driven genes, 187
    GLUT-1, 187, 279
    GLUT-3, 187
  induced differentiation, 185
  inducible factors (HIF), 180-188, 253
  on the tumor phenotype, 179-189
  response elements (HRE), 181, 253
Hypoxia inducible factors (HIF), 180-188, 253
  HIF-1a, 181-187, 312, 313-314
  HIF-2a, 181-183
  HIF-3a, 181
  regulation, 182, 183
  and prolyl hydroxylases (PHD), 182, 183

IGFBP genes, 47, 48
IGF1R, 2-9, 13, 376
Ifosfamide (IFO), 16
Imatinib mesylate (see also Gleevec and TKI), 99, 104, 105, 320
  resistant mutations, 99
Immune response to invasiveness and metastasis, 342-344
Infantile fibrosarcoma, 55, 99, 101
  ETV-NTRK3, 55, 99, 101
Inflamation and tumor development, 341, 342, 365
Inflammatory myofibroblastic tumor, 99, 101, 126
  ALK-TMP4, 99, 101
Integrative Tumor Biology, 96, 97
Interferon alpha (IFN), 316
Interleukin 8 polymorphisms, 223

Insuline-like growth factor-1 (IGF), 4, 13, 14, 24, 237, 378
  binding proteins (IGFBP), 16, 17, 24, 27, 47, 48
  IGFBP-3, 26, 37, 46
  signaling system, 26
Insulin-like growth Factor Receptor I (IGF-IR), 4, 13, 16, 17, 376
  antisense oligonucleotide approach, 15
  dominant negative mutants, 7, 15
  neutralizing antibody, 15
  suramin, 15
Intensity Modulated Radiotherapy (IMRT), 283, 303, 306, 308
Irinotecan (CPT-11), 211, 218-221, 225, 258, 261, 262, 265, 268

K-ras gene and lung cancer, 197, 198
Kaposi sarcoma, 100, 108, 109
  HHV8, 100, 108, 109
KIT (see also tyrosine kinase receptors), 99, 100, 104-108, 184
  gastrointestinal schwannoma, 108
  gastrointestinal stromal tumor (see GIST), 105, 106
  mutations, 104, 105, 107
  oncogene, 104
KU80, 115, 119

Lapatinib (GW572016), 329, 332-337, 379, 381, 382
Large cell anaplastic lymphoma, 99, 101
  ALK-TMP4, 99, 101
Leiomyosarcoma, 53, 54, 59, 100, 102, 106, 126
  and genetic analysis, 53, 59
Letrozol, 318, 336, 375, 380, 382, 385
Leucovorin (LV), 214, 221, 257-259, 261-263, 265, 268, 317, 318
Li-Fraumeni syndrome, 131
Liposarcoma and genetic analysis, 55, 59, 126
Loss of heterozygosity (LOH), 115, 118, 149, 150, 216, 217
  breast cancer, 149
  gastric carcinoma, 115
Lung cancer and radiology, 305, 308

Lymphoblastic lymphoma/leukemia, 18, 122, 124
Lymphomas of children and adolescents, 124
Lymphotoxin-alpha deficient mice and tumor recurrence, 349, 350
Lysyl oxidase methylation, 118

MAGIC trial, 116
Magnetic resonance spectroscopy (MRS) in prostate and brain tumors, 285-300
  treatment planning of brain tumors, 294
Malignant fibrous histiocytoma, 53, 54, 100
Malignant mesothelioma, 109
  SV40, 109
Malignant peripheral nerve seath tumor (MPNST), 100, 102, 107, 126
MAP7, see mitogen-activated protein kinase
MAPK, 285, 286, 289, 290
  Pathway, 155, 377
Matrixmetalloprotease (MMP), 17
  MMP-2, 17
  MMP-9, 17
MAV-1, 27, 29
MEK inhibitors, 377, 378
MEK-MAPK, 16, 311
Meningioma, 99, 100, 108, 291, 298
  NF2, 99, 100
Mesenchymal chondrosarcoma, 18
MET gene, 131
Methylation patterns and chemosensitivity in NSCLC, 195-206
Methylation-specific PCR (MSP), 195-206
  in NSCLC, 195-206
Methylenetetrahydrofolate reductase (MTHFR), 115, 117, 213
MGMT methylation, 118
MIC2 gene (see also CD99), 4, 14, 17
  $p30/32^{MIC2}$, 14
Microsatellite analysis (MSI), 115, 118
  gastric carcinoma, 115, 118

Mitogen-activated protein kinase (MAPK), 24, 313, 329-335, 377
MEK/MAPK pathway, 16
MLL-LARG fusion gene, 58
MMP-7, 131
MMP-26, 131
Molecular imaging of cancer, 277-283
  angiogenesis, 280
  apoptosis, 281
  cell proliferation, 279
  estrogen receptor status and PET, 279
  hypoxia, 280
  monitorization of gene therapy, 281
  PET and SPECT, 277, 278
  PET or SPECT tracers, 279
  radionuclide therapy, 282
  somatostatin receptor ligand and SPECT, 279
MRI 283, 285, 286, 289, 291-295, 298, 303-305
  for brain, 285, 286, 291
MRP1, 115, 117
mTOR and cancer therapy, 309-323, 379, 380
  antagonists, 377, 378, 380, 383
  inhibitors and predictors of response, 314
  pathway, 318, 380
Multigene predictors of response to preoperative chemotherapy, 234, 235, 239-242
Myeloblastosis Associated Virus Type 1 (MAV-1), 27, 29
Myxoid liposarcoma (see also FUS-CHOP fusion gene), 99, 126

Nephroblastoma, 4, 24, 27-30
Neoadjuvant chemotherapy, 115-119
Neuregulins, 330
Neuroblastoma, 33, 46, 54, 103, 179-189
  and the hypoxia model, 179
Neurofibromatosis 2 tumor suppressor gene (NF2), 99, 100, 107
  genetic loss, 99, 107, 108
  meningiomas, 99, 107, 108
  mutations, 100, 107, 108
  perinueurioma, 108
  policytic astrocytomas, 107
  schwanommas, 100, 107, 108
  syndrome, 107
Non-small-cell lung cancer (NSCLC), 195-206, 257, 269, 317, 320
NOTCH pathway, 25, 48, 185
NOVH (NOV), 2, 4-6, 8, 9, 23, 24, 27, 29
Nuclear medicine imaging, 277, 279
NVP-AEW541, 16, 17

Organization of European Cancer Institutes (OECI), 73, 87, 90
  acreditation of European cancer centers, 91
  cancer patients and cancer activities, 90
Orotate-phosphoribosyl transferase (OPRT) polymorphisms, 217, 218
Ornithine decarboxylase, 312, 313
Osteosarcoma and genetic studies, 53, 54, 126
Oxaliplatin, 211, 213, 214, 221-225, 258, 264, 265, 269
  and bevacizumab, 262, 264

p14 in sarcomas, 54, 58
p16 in sarcomas, 54, 58
  methylation, 118
p27, 313, 315, 332, 379
p53, 4, 26, 46, 115, 116, 118, 127, 131, 152, 153, 155, 163, 197-199, 203, 206, 313, 315
  in NSCLC, 198
  mutations and gastric cancer, 118, 199
Paclitaxel, 117, 197, 206, 233, 241-246, 257, 336
Paravertebral small-cell tumor, 3
PAX3/FKHR fusion gene, 55, 126, 129
PAX7/FKHR fusion gene, 55, 126, 129
PBX/E2A fusion gene, 123
PD153035, 331
Pediatric hematopoietic malignancies, 122-125
  Acute lymphoblastic leukemia (ALL), 122
  and genetic abnormalities, 123

Pediatric hematopoietic malignancies
(*Continued*)
Anaplastic large cell lymphoma
(ALCL), 124
Burkit lymphoma, 125
Lymphomas of children and
adolescents, 124
Pediatric solid tumors, 125-130
Alveolar rhabdomyosarcomas, 129
Hepatoblastoma, 130
Sarcomas with complex karyotypes,
126
Synovial sarcomas, 127
Pharmacogenomic predictors for
preoperative chemotherapy in
breast cancer, 233-247
Pharmacogenomics and toxicity to
fluoropyrimidines, 213-218
and toxicity to irinotecan (CPT-11),
213-218
and toxicity to oxiliplatin, 221-225
Phosphatydilinositol 3-kinase
(PI3K/Akt), 14, 310, 311, 314, 315,
322, 329-333, 376-378
Inhibitors, 378
Platelet derived growth factor receptor
alpha (PDGFRA), 99, 100,
104-106
gastrointestinal stromal tumor, 104
mutations, 104, 106
oncogene, 104
Platinum, 117, 195, 197, 201, 211, 218,
221, 222
Pleomorphic rhabdomyosarcoma, 126
Positron emission tomography (PET), 84,
119, 277-283
brain location and classification, 295
PET-CT for lung, 305
PET-Scan, 303, 306
Prader-Willi syndrome, 131
Primitive neuroectodermal tumor
(PNeT), 3, 42, 55
Progesterone receptor (PR), 157, 336,
375
Prognostic and therapeutic targets in the
Ewing Family of tumors
(PROTHETS), 1
Objectives, 5-9

Prolyl hydroxylases (PHD), 182, 183
and HIF-stabilization, 182
PHD1, 182
PHD2, 182, 183
PHD3, 182, 183
Prostate cancer, 180, 269, 279, 286-290
and FRA16D, 151
and MRS techniques, 286, 287
and radiology, 287
Prostate specific antigen (PSA), 287, 298
ProteoSys platform, 117
ProteoTope analysis, 117
PTEN, 311, 314, 316, 322, 378

RAD001 mTOR inhibitor, 309, 310, 314,
315, 320-322
Rapamycin, 309, 310, 312-315, 320-322
Ras homolog enriched in brain, 311
Recurrent amplicons seen in sarcomas,
56, 57
Rhabdomyosarcoma, 16, 18, 46, 49, 54,
55, 100, 103, 126, 127, 129, 313
RNA interference-based technologies
Ewing's sarcoma, 43, 48, 49

S6 kinase protein, 310, 312
Sarcomas and genetic alterations, 53-55
Schwannoma, 99, 100, 107, 108
NF2, 99, 100, 107, 108
Secretory carcinoma of the breast, 99, 101
ETV-NTRK3, 99, 101
Simian virus 40 (SV40), 100, 109
malignant mesothelioma, 100, 109
Single photon emission computer
tomography (SPECT), 277-283,
285
gamma-emitting isotopes, 278
Soft tissue tumors, 55, 100, 104
viral sequences, 104
Small molecule signal transduction
inhibitors (STIs), 377, 378
clinical trials, 380
in breast cancer, 377, 378
in combination with endocrine therapy,
378
Specific chromosomal translocations in
sarcomas, 14, 53-55
Specific targeted research project, 3

Synovial sarcoma, 18, 53-55, 90, 99-102, 126-128
  diagnosis, 99, 102, 128
  expression analysis, 59
  fusion gene, 55, 101
SYT-SSX fusion genes, 99, 101, 102, 126, 128
  SYT-SSX1, 99, 101, 102, 126, 128
  SYT-SSX2, 101, 102, 126, 128
  SYT-SSX4, 126

Tamoxifen, 279, 336, 375-383
Targeted therapies, 5-7, 13-15, 212, 309, 314
Taxane, 117, 318, 336, 383
TEL/AML1 fusion gene, 123
Telepathology and virtual microscope, 79-81
Thymidine phosphorylase (TP), 115-117, 119, 211, 213
Thymidylate synthase (TS), 115, 116, 117, 119, 211
Tipifarnib, 379, 382, 383
Transforming growth factor alpha (TGF-a), 330, 335, 376
Transforming growth factor beta (TGF-b), 26, 49, 353
Translocations and fusion genes in sarcomas, 99, 126
Trastuzumab, 15, 314, 329, 333-336, 377-380, 382
  refractory breast cancer, 329, 330, 334
TREE-2 clinical trial, 264
TTF1, 2, 5, 6
TuBaFrost, 65-73, 76, 81, 83, 84
  Central data base, 66, 81, 83
  Code of conduct, 69, 71, 81
  Ethics and law, 66, 71
  Quality ansurance, 66, 68
  Rules for access and use, 68
  Standard operating procedures, 67
  Virtual microscope, 66, 72, 81, 83
Tumor associated antigens and dendritic cells, 346, 347
Tumor associated macrophages (TAM), 344-346
Tumor infiltrating lymphoplasmacytic cells and production of VH-IgGs, 342-344

Tumor infiltrating plasma cells and tumor autosupply of immune globulins, 358
Tumor progression in immune deficient mice, 350
  µMT/µMT mice, 350, 352
  RAG–/– mice model, 350, 354
  SCID-beige model, 350, 352
Tyrosine kinase inhibitors (TKI), 16, 99, 251-265, 329-337, 377, 380
  Bevacizumab (avastin), 251, 252, 255-265, 267-270
  BMS-599626, 330, 332, 337
  CI-1033, 330, 332, 336, 337
  clinical development, 265
  Dual/Pan-Her tyrosine kinase inhibitors, 329-337
  EKB-569, 332
  ErbB Receptor TK Inhibitors in Clinical Development, 332
  Erlotinib, 332, 381
  Gefitinib, 314, 320, 332, 337, 378-381
  Imatinib mesylate, 99, 104, 105, 320
  Lapatinib (GW572016), 329, 332-337, 379, 381, 382
  PTK-787, 265-268
Tyrosine kinase receptors (TKR), 14, 311
  EGFR, 222-225, 314, 329-337, 376-382
  HER-2, 247, 318, 334, 336
  IGF-IR, 2, 4-9, 13-17, 376
  KIT, 99, 100, 104-108, 184
  Mutations, 104
  PDGFR, 337
  PDGFRA, 99, 100, 104-106
  VEGFR, 337

UDP-gluconoryltransferase (UGT1A1) polymorphisms, 219

Vascular endothelial growth factor (VEGF), 17, 180, 182, 187, 212, 225, 251-257, 265-269, 312-314, 342, 376, 378
  family, 253
  inhibitors and colon cancer, 251-269
  receptors, 254
  role in angiogenesis, 254

Vascular endothelial growth factor
(VEGF) (*Continued*)
  target for anti-cancer therapy, 255
  VEGF-A, 17, 253
Vascular endothelial growth factor
  receptor (VEGFR), 337
VEGF inhibitors in clinical use, 256-270
  clinical development of bevacizumab,
    257
  monoclonal antibodies, 256
    A.4.6.1, 256
    Bevacizumab, 257-264
    IMC-1121B, 256
    2C3, 256
  preclinical studies of bevacizumab,
    257
Vincristine (VCR), 15, 16, 206
Virtual microscope, 66, 73, 76-78, 80-83
  commercial systems, 81
  components, 80

Virtual microscopy, 72, 73, 75-84
  in the TuBaFrost network, 81, 83
Virtual tumor banking, 75-85
Von Hippel-Lindau (VHL), 182, 253

Wilms tumor, 27, 29, 33, 103
  and CCN3, 27, 29
Wnt gene pathway, 131-133
WWOX gene and cancer, 149-157
  biological and biochemical functions,
    152
  exposure to environmental carcinogens,
    153
  expression in different tumor types, 156

Xelox, 252, 258, 264
XG gene, 18

ZD6474, 337
Zyxin antisense oligonucleotides, 18